AF813931

Unraveling New Frontiers and Advances in Bioinformatics

Amit Chaudhary · Sushanta K. Sethi ·
Akarsh Verma
Editors

Unraveling New Frontiers and Advances in Bioinformatics

Springer

Editors
Amit Chaudhary
Postdoctoral Scholar, Heersink School
of Medicine, Department
of Cell, Developmental
and Integrative Biology
The University of Alabama at Birmingham
Birmingham, AL, USA

Sushanta K. Sethi
Department of Mechanical, Materials
and Aerospace Engineering
Indian Institute of Technology Dharwad
Dharwad, Karnataka, India

Akarsh Verma
Department of Mechanical Engineering
University of Petroleum and Energy Studies
(UPES)
Dehradun, Uttarakhand, India

ISBN 978-981-97-7122-6 ISBN 978-981-97-7123-3 (eBook)
https://doi.org/10.1007/978-981-97-7123-3

© The Editor(s) (if applicable) and The Author(s), under exclusive license to Springer Nature Singapore Pte Ltd. 2024

This work is subject to copyright. All rights are solely and exclusively licensed by the Publisher, whether the whole or part of the material is concerned, specifically the rights of translation, reprinting, reuse of illustrations, recitation, broadcasting, reproduction on microfilms or in any other physical way, and transmission or information storage and retrieval, electronic adaptation, computer software, or by similar or dissimilar methodology now known or hereafter developed.
The use of general descriptive names, registered names, trademarks, service marks, etc. in this publication does not imply, even in the absence of a specific statement, that such names are exempt from the relevant protective laws and regulations and therefore free for general use.
The publisher, the authors and the editors are safe to assume that the advice and information in this book are believed to be true and accurate at the date of publication. Neither the publisher nor the authors or the editors give a warranty, expressed or implied, with respect to the material contained herein or for any errors or omissions that may have been made. The publisher remains neutral with regard to jurisdictional claims in published maps and institutional affiliations.

This Springer imprint is published by the registered company Springer Nature Singapore Pte Ltd.
The registered company address is: 152 Beach Road, #21-01/04 Gateway East, Singapore 189721, Singapore

If disposing of this product, please recycle the paper.

Contents

Introduction to Bioinformatics: Past, Present and Future

Shikha Kushwah, Ashish Kumar, and Ashutosh Mani

Abstract Biology is a science that explores life. It produces a wide range of data due to its descriptive nature, covering everything from an organism's size and state to its genetic makeup. The increasing use of gene sequencing and biotechnological techniques has led to a remarkable surge in biological data. This has emphasized the importance of effectively managing, processing, and analyzing this data. To address these challenges, scientists developed bioinformatics—an interdisciplinary field that combines biology, computer science, mathematics, statistics, information science, and engineering to analyze complex biological data sets. Bioinformatics leverages computational methods for in silico analysis of biological questions, facilitating deeper insights and discoveries. The integration of bioinformatics, big data technologies, cloud computing, machine learning, and AI is revolutionizing our understanding of biology and medicine, presenting new opportunities to address global health challenges such as infectious diseases and pandemics. This book chapter examines the evolution of bioinformatics as a multidisciplinary field, highlighting key resources, web tools, and its applications across various domains.

Keywords Bioinformatics · Machine learning · Web tools · Software · Applications · Phylogenetic tree · Sequence alignment · Genomics · Medicine · Databases · Human genome project

1 Introduction

The term bioinformatics was introduced in 1970 for the study of biosystems information processes [1]. It is a multidisciplinary field that combines biology, computer science, mathematics, and data science to manage and analyze biological data. It emerged as a branch of computational biology primarily focused on handling the increasing number of biological data traced back to the 1950 and 1970s when

S. Kushwah · A. Kumar · A. Mani (✉)
Department of Biotechnology, Motilal Nehru National Institute of Technology Allahabad, Prayagraj 211004, India
e-mail: amani@mnnit.ac.in

© The Author(s), under exclusive license to Springer Nature Singapore Pte Ltd. 2024
A. Chaudhary et al. (eds.), *Unraveling New Frontiers and Advances in Bioinformatics*,
https://doi.org/10.1007/978-981-97-7123-3_1

researchers began using computers to manage and interpret biological data. After Sir Frederick Sanger discovered the protein sequence of insulin in the early 1950s, the use of computers became crucial in molecular biology. It became very challenging to compare the multiple sequences manually. Margaret Oakley Dayhoff played a pioneering role in this field. She compiled some of the first primary protein sequence databases, which were published as books, and developed early techniques for aligning sequences [2–4].

Early efforts in bioinformatics primarily focused on creating databases to store and organize biological data, particularly DNA and protein sequences, due to the rapid accumulation of genomic data from new sequencing technologies. Scientists needed a way to store and manage these sequences, so the first bioinformatics task was to develop databases for organizing these data. These databases served as central repositories where researchers could submit, retrieve, and analyze sequences. Key milestones in the early development of databases include the establishment of the Protein Data Bank (PDB) in 1971, which serves as a repository for three-dimensional structures of proteins and nucleic acids [5]. This resource has become invaluable for researchers studying molecular structures and functions. Figure 1 illustrates the different types of databases developed since then.

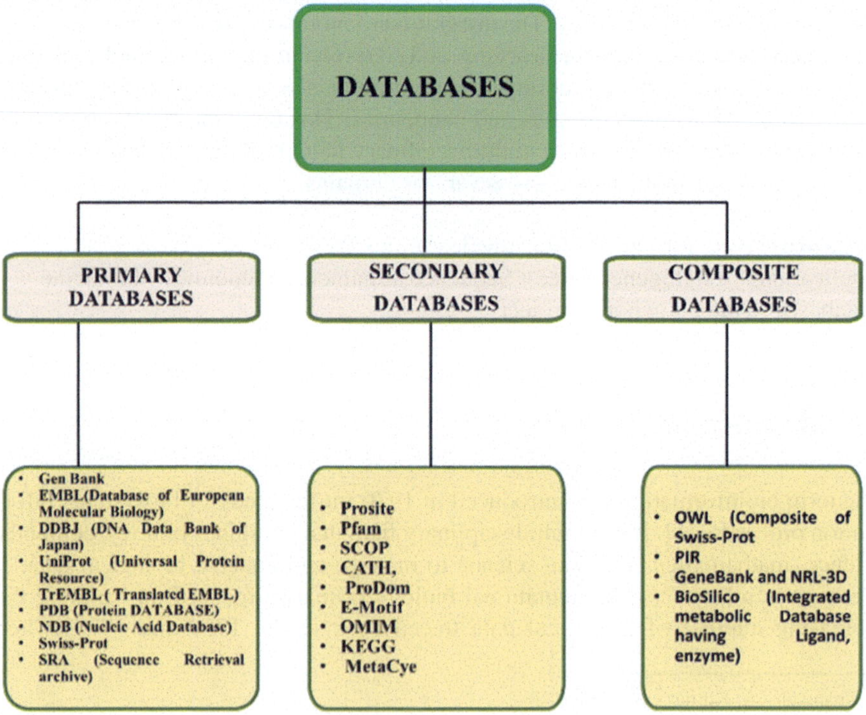

Fig. 1 Describes the different types of databases

Another significant achievement was the development of sequence alignment algorithms such as Needleman-Wunsch and Smith-Waterman in the 1970 and 1980s, which laid the foundation for the comparison of DNA, RNA, and protein sequences to identify similarities and evolutionary relationships [6].

After the 1900s, bioinformatics experienced explosive growth, primarily driven by the Human Genome Project and rapid advances in DNA sequencing technology. The Human Genome Project (HGP), launched in 1990 and completed in 2003, marked a pivotal moment in bioinformatics and genomics [7, 8]. The project's goal was to sequence the entire human genome, an ambitious task that required extensive bioinformatics resources and methods for managing and analyzing vast amounts of genetic data. The successful completion of the HGP provided a wealth of information about the human genome and marked a transformative era in bioinformatics and genomics. Rapid advancements in sequencing technologies continued after the HGP, leading to the development of next-generation sequencing (NGS) methods. These significantly reduced the cost and time required to sequence DNA, allowing researchers to analyze whole genomes and transcriptomes on a larger scale. With access to complete genome sequences, researchers began focusing on functional genomics and transcriptomics, utilizing tools such as microarrays and RNA sequencing (RNA-seq) to study gene expression patterns [9, 10].

The implication of these technologies also produces a sheer volume of data. A single sequencing run can generate terabytes of data. Biological data can exist in various formats, including raw sequences, processed data, and metadata, which requires robust and flexible storage systems for efficient management. Development of big data technologies and cloud computing provides scalable solutions for data storage, processing, and analysis. Machine learning and AI have become integral parts of bioinformatics, offering advanced methods for analyzing complex biological data, such as predicting disease outcomes and identifying drug targets. Global collaboration and data sharing are central to bioinformatics research, with large-scale projects like the Encyclopedia of DNA Elements (ENCODE) and the 1000 Genomes Project involving international teams working together to generate, analyze, and share valuable genomic data.

The exponential growth of biological data also expands Bioinformatics into other omics fields such as proteomics and metabolomics, studying proteins and metabolites respectively. Technologies like mass spectrometry enabled large-scale analysis of proteins and metabolites, providing insights into cellular processes and pathways. This development also saw the emergence of personalized and precision medicine, leveraging genomic data to develop targeted treatments tailored to a patient's unique genetic makeup [11].

2 A Historical Retrospect of Bioinformatics

The history of bioinformatics is interwoven with key discoveries and technological advancements across a range of scientific disciplines, from genetics and microbiology to computer science and data management. But the development of computer technology plays a crucial role in its advancement.

Technological advancement of bioinformatics can be marked in 1969, with the development of the ARPANET, a precursor to the internet, linked computers at Stanford and UCLA, paving the way for global scientific collaboration [12]. In 1970, the Needleman-Wunsch algorithm for sequence comparison was published, providing a method for aligning sequences and identifying similarities. The invention of XML (eXtensible Markup Language) in the late 1970s at IBM facilitated data organization and exchange, critical for the growing field of bioinformatics. In 1972, Paul Berg and his team created the first recombinant DNA molecule, a milestone in genetic engineering. In 1973, the Brookhaven Protein Data Bank (PDB) was established as a repository for three-dimensional protein structures, making crucial data available to researchers worldwide. This database remains an essential resource for studying protein structure and function. The chronological development of bioinformatics continued with the establishment of key databases, tools, and algorithms that would shape the future of the field (Fig. 2). The sequencing of DNA, the advent of the internet, and the creation of global repositories such as GenBank and the European Molecular Biology Laboratory (EMBL) all played crucial roles in advancing bioinformatics. All these dates are arranged chronologically as follows:

1977: Frederick Sanger and his colleagues developed the Sanger sequencing method, also known as chain-termination sequencing, allowing researchers to sequence DNA with high accuracy and paving the way for genomic research [13].

1980: The introduction of the Smith-Waterman algorithm revolutionized sequence alignment by providing a more sensitive method for local sequence comparison than

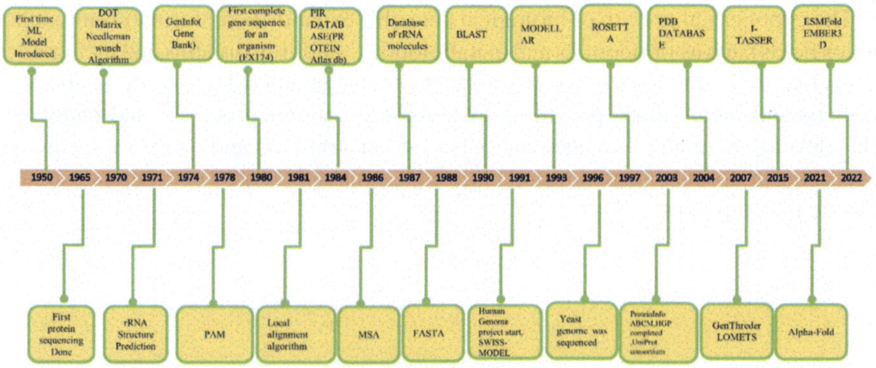

Fig. 2 Highlights the revolutionary developments in the field of bioinformatics

the Needleman-Wunsch algorithm. 1984: FASTP algorithm published Lipman and Pearson [14].

1985: PCR (Polymerase Chain Reaction) technique was discovered by Kary Mullis, while working as a chemist for Cetus Corporation in Emeryville, California [15].

GenBank, a repository of DNA sequences, was established. It quickly became one of the most important databases in bioinformatics, storing and providing access to a vast array of genetic data.

1988: National Centre for Biotechnology Information (NCBI) is established at the National Cancer Institute [16]. The National Research Council started research with the "Human Genome" [17]. The FASTA algorithm for sequence comparison is published by Pearson and Lipman [18].

1990: The Human Genome Project began, aiming to sequence the entire human genome. The project, completed in 2003, spurred advancements in data management and analysis in bioinformatics [8].

1991: Patrick O. Brown invented DNA Microarray Technology [19].

1994: The BLAST (Basic Local Alignment Search Tool) algorithm was introduced by NCBI, providing a fast and efficient way to search biological sequences in databases and revolutionizing sequence comparison and analysis [20].

2000: Ensembl, a joint project between the European Molecular Biology Laboratory (EMBL) and the Wellcome Trust Sanger Institute, was launched. This comprehensive resource provides genome data and annotation for various species [21].

2001: The development of next-generation sequencing (NGS) technologies marked a significant advancement in the field, enabling researchers to sequence large amounts of DNA quickly and cost-effectively [22–24].

2005: Galaxy, an open-source, web-based platform for data analysis in bioinformatics, was introduced [25].

2009: The development of RNA-Seq technology allowed for high-throughput sequencing of RNA, enabling researchers to study gene expression and transcriptomic profiles on a large scale [26].

2012: CRISPR-Cas9 technology emerged as a powerful tool for genome editing, with applications in bioinformatics for modeling genetic variations and exploring potential therapeutic interventions [27, 28].

2013: The development of cloud computing and big data technologies offered scalable solutions for storing, processing, and analyzing vast amounts of biological data, making data more accessible to researchers worldwide [29].

3 The Present: Current Applications and Trends

Bioinformatics has emerged as a multidisciplinary field which consists the intersection of biology, computer science, mathematics, and statistics. By integrating computational and statistical analysis with the biological data, bioinformatics has enabled researchers to analyze large-scale datasets and uncover meaningful patterns and insights [30]. It has experienced rapid growth in recent decades and has become an indispensable part of modern biological research. Bioinformatics provides the tools and methodologies necessary for analyzing large-scale biological data across various domains. These include genomics, proteomics, transcriptomics, metagenomics, and systems biology, among others. Figure 3 describes each and every component of bioinformatics.

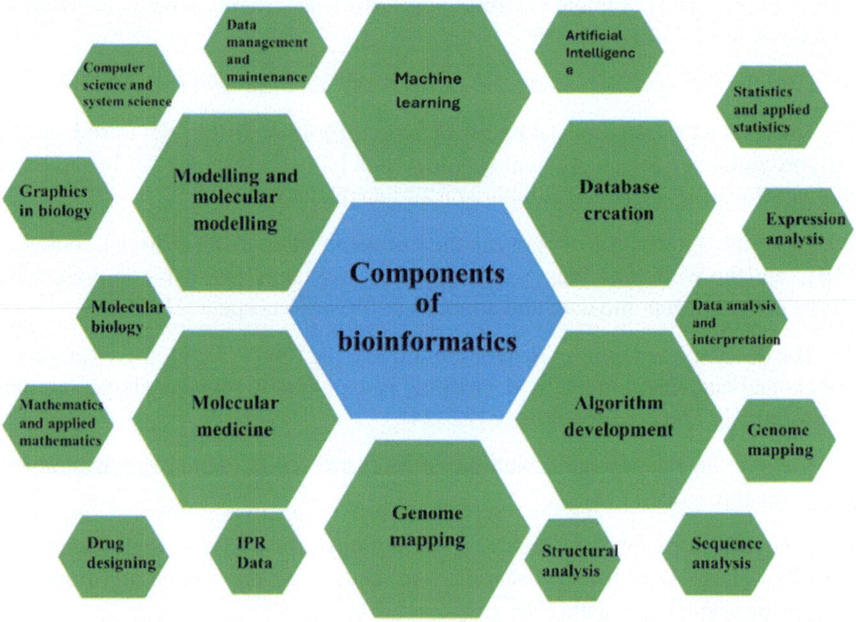

Fig. 3 Illustrates the components of bioinformatics. As an interdisciplinary field, bioinformatics integrates elements from various subjects, each contributing to its overall scope and capabilities

3.1 Modern day Tools and Techniques in Bioinformatics

3.1.1 Sequence Analysis Tools

Sequence analysis involves examining DNA, RNA, or protein sequences to extract meaningful information about their structure, function, and evolutionary relationships. A wide range of bioinformatics tools have been developed for various aspects of sequence analysis, from alignment and searching to annotation and comparison. Table 1 provides an overview of some of the most prominent tools used in sequence analysis.

3.1.2 Structure Prediction Methods

Understanding the 3D structure of biological molecules is essential for elucidating their function and for applications in drug design and protein engineering. Various bioinformatics tools have been developed for the prediction of the three-dimensional (3D) structure of proteins, nucleic acids, and other biological macromolecules based on their primary sequence. These tools are described in Table 2.

3.1.3 Genome Assembly and Annotation Tools

Genome assembly is the process of reconstructing a genome from fragmented DNA sequences obtained through sequencing technologies such as next-generation sequencing (NGS). The goal is to create a representation of the genome, typically in the form of contiguous sequences (contigs) and scaffolds. Different tools are described in Table 3.

4 Applications of Bioinformatics

Bioinformatics plays a significant role across various disciplines. Its applications involve a wide range of scientific disciplines and have led to significant advancements in understanding complex biological systems and diseases. In biotechnology, bioinformatics plays a pivotal role in various sectors, including animal breeding, agriculture, environmental and energy management, fermentation technology, pharmaceuticals, recombinant DNA technology, dairy production, aquaculture, genetically modified organisms (GMOs), and vaccine design.

Table 1 This table provides an overview of various bioinformatics tools used for sequence analysis of DNA, RNA, and protein

Types	Name of the tool	Function	References
Sequence alignment tools	BLAST (Basic Local Alignment Search Tool)	This tool is used for searching sequences against a database to identify regions of similarity	[20]
	Smith-Waterman Algorithm	It performs local sequence alignment, identifying regions of highest similarity between sequences	[6]
	Needleman-Wunsch Algorithm	It is a global sequence alignment algorithm that aligns entire sequences from end to end, making it ideal for comparing sequences of similar lengths	[31]
	ClustalW and Clustal Omega	Perform multiple sequence alignment, to identify conserved regions and construct phylogenetic trees	[32]
Sequence searching and annotation tool	FASTA	Sequence similarity search tool that offers options for both local and global alignment	[18]
	HMMER	Identify remote homologs and characterize protein families. (protein sequence similarity search)	[33]
	TBLASTN	Search protein sequences against a translated nucleotide database	[34]
	BLAT	For aligning sequences against large databases	[35]
Sequence analysis and annotation tool	Geneious	Includes tools for sequence alignment, annotation, primer design, and more	[36]
	MEGA (Molecular Evolutionary Genetics Analysis)	For sequence alignment, phylogenetic tree construction, and molecular evolutionary analysis	[37]
	MAFFT	Tool for rapid multiple sequence alignment	[38]
Gene prediction and annotation tool	Augustus	Gene prediction software for eukaryotes that can identify genes in genomes across various species	[39]
	Ensembl	Includes annotated genome sequences, gene models, and other information related to gene structure and function	[21]

Table 2 Provides an overview of some of the most prominent tools used in structure prediction of Proteins and nucleotides

Types	Name of the tool	Function	References
Homology modelling	MODELLER	Model the 3D structure of a protein from the homologous templates	[40]
	SWISS-MODEL	Automated protein modeling server	[41]
Threading/ fold recognition	I-TASSER	Automated protein modeling server when there are much fewer templates for protein	[42]
	Phyre2	For protein modeling and analysis	[43]
Ab Initio structure prediction	Rosetta	Macromolecular modeling and docking	[44]
	AlphaFold	Uses advanced machine learning for accurate 3D structure prediction of protein	[45]
Secondary structure prediction	PSIPRED	Protein structure prediction web server	[46]
	JPred	Protein secondary structure prediction server. Also predicts solvent accessibility and coiled-coil regions	[47]
Nucleic acid structure prediction	Vienna RNA secondary structure server	Analysis of RNA secondary structure	[48]
	Mfold	Nucleic acid folding and hybridization prediction server	[49]

4.1 *Comparative Genomics*

It supports evolutionary research and phylogenetics by constructing phylogenetic trees using sequence alignment algorithms [37]. The phylogenetic tree construction shows evolutionary relationships and close relatives of species. Bioinformatics also aids in sequencing microbial genomes capable of reducing CO_2 and other greenhouse gases in the atmosphere [60].

4.2 *Personalized Medicine*

Bioinformatics and genomics together contribute to shaping personalized medicine. By analyzing an individual's genome, researchers can tailor treatments and predict responses to drugs. This approach is particularly relevant in treating genetic diseases such as cancers, heart disease, and diabetes [61]. It also supports biomedical research

Table 3 Provides an overview of some of the most prominent tools used in sequencing

Types	Name of the tool	Function	References
Short-read assemblers	SPAdes	It generates single cell genome assemblies	[50]
	SOAPdenovo2	Short-read de novo assembly program	[51]
	Velvet	Quickly build long continuous sequences, de novo assembler	[52]
Long-read assemblers	Canu	Long read assembler	[53]
	Flye	Generates better or comparable assemblies faster	[54]
Hybrid assemblers	Unicycler	Tool for assembling bacterial genomes from a combination of short and long reads	[55]
	MaSuRCA	Well-suited for hybrid assemblies that combine data from both long-read and short-read	[56]
Overlapping	Minimap2	Reference-based sequence mapping	[57]
Repeat identification	RepeatMasker	Identifies and masks repetitive DNA sequences in a given genomic sequence	[58]
	Tandem repeat	Web server to identify tandem repeats	[59]

by studying host–pathogen interactions, which is crucial for understanding and addressing infectious diseases. Bioinformatics has made major contributions to the Human Genome Project (HGP) [7]. It has enabled the storage and analysis of vast genomic data. This would not have been possible without advanced bioinformatics tools.

4.3 Vaccine Development and Computer-Aided Drug Design

Bioinformatics offer tools for identifying vaccine targets from sequenced biological data, aiding in research and the design of immunological vaccines [62]. Bioinformatics challenges the traditional approach of vaccine development and promises to be more effective. Bioinformatics has the ability for the development of community-specific vaccines. VBRC NERVE tool, Mature P software, MHCPred, RANKPEP, and SVMHC software, IEDB, EpiJen, ProPred are some various tools used in development of potent vaccine [63]. Bioinformatics is integral to drug development and Computer-Aided Drug Design (CADD), helping design new drugs for managing diseases. Bioinformatics enables the identification of promising drug targets by analyzing protein structures and genetic data associated with diseases. Bioinformatics tools like AutoDock Vina, PyRx perform the molecular docking and their interactions analysis can be performed by Discovery Studio, LigPlot and PLIP web server. ADMEtlab and Protox 2 can evaluate their safety profiles.

Most recent case of fast drug development can be linked for the vaccine development of coronavirus disease 2019 (COVID-19). Within one year of identifying SARS-CoV-2 as the cause of COVID-19, vaccines such as mRNA and virus-vectored vaccines were successfully created using modern drug development techniques [64, 65].

Additionally, bioinformatics enhances personalized medicine by tailoring drugs to individual patients' genetic profiles. This precision approach optimizes drug efficacy and minimizes adverse reactions, leading to more successful treatment outcomes. Through the use of large datasets, computational simulations, and advanced algorithms, bioinformatics accelerates the discovery and development of novel therapeutic agents and vaccines, ultimately transforming the fields of medicine and public health.

4.4 Metagenomics

Metagenomics allows the study of genetic material from entire microbial communities, providing insights into their diversity, functions, and interactions. Bioinformatics tools such as QIIME and Mothur enable researchers to analyze metagenomic data, profiling the composition and diversity of microbial communities. Beyond community profiling, functional annotation and analysis of metagenomic data provide valuable insights into microbial metabolic pathways and potential biotechnological applications. Tools such as HUMAnN (Human Microbiome Project Unified Metabolic Analysis Network) and MG-RAST (Metagenomic Rapid Annotation using Subsystem Technology) allow researchers to predict and characterize the functions of microbial genes and pathways within a community [66].

HUMAnN, for example, provides a pipeline for characterizing the functional potential of microbial communities based on the presence and abundance of genes and pathways [67]. This information can be used to explore microbial metabolic processes, interactions with the host, and potential responses to environmental changes. MG-RAST, on the other hand, offers rapid annotation and comparison of metagenomic data, helping researchers identify functional elements such as enzymes, pathways, and other molecular markers [68].

The integration of metagenomic sequencing with advanced bioinformatics analysis provides a comprehensive understanding of microbial communities. As sequencing technologies continue to advance, the role of bioinformatics in metagenomics will become increasingly important, enabling the discovery of novel microbial taxa, functions, and interactions that contribute to the health and stability of ecosystems.

5 Future Perspectives

As bioinformatics continues to evolve, the integration of cutting-edge technologies such as single-cell sequencing, CRISPR, and quantum computing is anticipated to further transform the field. These emerging technologies provide powerful tools for advancing our understanding of complex biological systems and developing novel medical interventions.

Single-cell sequencing enables the high-resolution analysis of individual cells within a population. By characterizing the transcriptomes, genomes, and epigenomes of single cells, researchers can uncover cellular diversity and gain insights into the molecular basis of diseases such as cancer and neurological disorders. Following the completion of the Human Genome Project, researchers have initiated a human cell atlas program with the goal of mapping the 37 trillion cells in the human body. Single cell sequencing technologies enable precise analysis of individual cells and are expected to play a pivotal role in advancing this ambitious project [69, 70].

CRISPR (Clustered Regularly Interspaced Short Palindromic Repeats) has emerged as a transformative genome-editing technology. By providing precise and efficient manipulation of genetic sequences, CRISPR has the potential to drive major breakthroughs in gene therapy. In the context of bioinformatics, the technology enables researchers to generate large-scale datasets through high-throughput screening. This large data needs advanced computational tools for data analysis and interpretation. DeepCRISPR, MOFF, PEM-Seq, GUIDE-TagPEAC-Seq, TAPE-Seq, CRISPRDo, EupaDGT, FORECast, CRISPR-GA are some tools currently used in CRISPR. The rapid evolution of CRISPR technology continues to push the boundaries of genetic research, and its integration with bioinformatics holds promise for accelerating discoveries in genetics and beyond [28, 71].

Quantum computing represents another application with significant potential to impact bioinformatics. By harnessing the principles of quantum mechanics, quantum computers can perform complex computations exponentially faster than classical computers. This capability could lead to breakthroughs in bioinformatics tasks such as protein folding, sequence alignment, and large-scale data analysis. Quantum algorithms may enable researchers to tackle previously intractable problems and accelerate drug discovery and other areas of biomedical research [72].

As these technologies continue to mature and integrate with bioinformatics, they are expected to drive transformative changes in the field. These advancements will likely lead to the development of more sophisticated analytical tools and methods, providing researchers with deeper insights into biological processes and disease mechanisms. Ultimately, these innovations hold the promise of enabling new discoveries and medical advancements that can improve human health and well-being.

6 Conclusion

As bioinformatics continues to evolve, it will play a critical role in addressing global health challenges such as infectious diseases and pandemics. The ability to rapidly sequence pathogens and share data across international networks can facilitate timely responses to emerging health threats. Overall, the integration of big data technologies, cloud computing, machine learning, and AI into bioinformatics is revolutionizing our understanding of biology and medicine. It provides the foundation for future discoveries and innovations in healthcare, agriculture, and other related fields. In summary, bioinformatics has evolved from its early days of managing and organizing biological data to its current state as an indispensable part of modern biological research.

Acknowledgements All the authors acknowledge the Department of Biotechnology, MNNIT, Allahabad, for providing the necessary support for conducting the study smoothly.

Conflict of Interest *There is no conflict of interest to declare by the authors.

References

1. Hesper, B., & Hogeweg, P. (1970). Bioinformatica: Een werkconcept. *Kameleon, 1*(6), 28–29.
2. Dayhoff, M. O., Barker, W. C., & Hunt, L. T. (1983). Establishing homologies in protein sequences. In *Methods in enzymology* (pp. 524–545). Academic Press.
3. Dayhoff, M. O. (1969). Atlas of protein sequence and structure. In M. O. Dayhoff (Ed.), *National biomedical research foundation* (Vol. 5).
4. Eck, R. V., & Dayhoff, M. O. (1966). Evolution of the structure of ferredoxin based on living relics of primitive amino acid sequences. *Science, 152*(3720), 363–366. https://doi.org/10.1126/science.152.3720.363
5. Bernstein, F. C., Koetzle, T. F., Williams, G. J., et al. (1977). The Protein Data Bank: A computer-based archival file for macromolecular structures. *Journal of Molecular Biology, 112*(3), 535–542. https://doi.org/10.1016/s0022-2836(77)80200-3
6. Smith, T. F., & Waterman, M. S. (1981). Identification of common molecular subsequences. *Journal of Molecular Biology, 147*(1), 195–197. https://doi.org/10.1016/0022-2836(81)90087-5
7. Hood, L., & Rowen, L. (2013). The human genome project: Big science transforms biology and medicine. *Genome Medicine, 5*(9), 79. https://doi.org/10.1186/gm483
8. Collins, F. S., & Fink, L. (1995). The human genome project. *Alcohol Health and Research World, 19*(3), 190–195.
9. Schuster, S. C. (2008). Next-generation sequencing transforms today's biology. *Nature Methods, 5*(1), 16–18. https://doi.org/10.1038/nmeth1156
10. Shendure, J., Mitra, R. D., Varma, C., & Church, G. M. (2004). Advanced sequencing technologies: Methods and goals. *Nature Reviews Genetics, 5*(5), 335–344. https://doi.org/10.1038/nrg1325
11. Hassan, M., Awan, F. M., Naz, A,. et al. (2022). Innovations in genomics and big data analytics for personalized medicine and health care: A review. *International journal of molecular Sciences, 23*(9). https://doi.org/10.3390/ijms23094645.

12. Roberts, L. (1988). The arpanet and computer networks. In *A history of personal workstations* (pp. 141–172). Association for Computing Machinery.
13. Sanger, F., Nicklen, S., & Coulson, A. R. (1977). DNA sequencing with chain-terminating inhibitors. *Proceedings of the National Academy of Sciences USA, 74*(12), 5463–5467. https://doi.org/10.1073/pnas.74.12.5463
14. Lipman, D. J., & Pearson, W. R. (1985). Rapid and sensitive protein similarity searches. *Science, 227*(4693), 1435–1441. https://doi.org/10.1126/science.2983426
15. Kleppe, K., Ohtsuka, E., Kleppe, R., Molineux, I., & Khorana, H. G. (1971). Studies on polynucleotides. XCVI. Repair replications of short synthetic DNA's as catalyzed by DNA polymerases. *Journal of Molecular Biology, 56*(2), 341–361. https://doi.org/10.1016/0022-2836(71)90469-4.
16. Smith, K. A. (2008). Laws, leaders, and legends of the modern National Library of Medicine. *Journal of the Medical Library Association, 96*(2), 121–133. https://doi.org/10.3163/1536-5050.96.2.121
17. National Research Council Committee on M, Sequencing the Human G. *Mapping and sequencing the human genome.* National Academies Press (US). Copyright © 1988 by the National Academy of Sciences.
18. Pearson, W. R., & Lipman, D. J. (1988). Improved tools for biological sequence comparison. *Proceedings of the National Academy of Sciences USA, 85*(8), 2444–2448. https://doi.org/10.1073/pnas.85.8.2444
19. Schena, M., Shalon, D., Davis, R. W., & Brown, P. O. (1995). Quantitative monitoring of gene expression patterns with a complementary DNA microarray. *Science, 270*(5235), 467–470. https://doi.org/10.1126/science.270.5235.467
20. Altschul, S. F., Gish, W., Miller, W., Myers, E. W., & Lipman, D. J. (1990). Basic local alignment search tool. *Journal of Molecular Biology, 215*(3), 403–410. https://doi.org/10.1016/s0022-2836(05)80360-2
21. Hubbard, T., Barker, D., Birney, E., et al. (2002). The ensembl genome database project. *Nucleic Acids Research, 30*(1), 38–41. https://doi.org/10.1093/nar/30.1.38
22. Venter, J. C., Adams, M. D., Myers, E. W., et al. (2001). The sequence of the human genome. *Science, 291*(5507), 1304–1351. https://doi.org/10.1126/science.1058040
23. Metzker, M. L. (2010). Sequencing technologies-the next generation. *Nature Reviews Genetics, 11*(1), 31–46. https://doi.org/10.1038/nrg2626
24. Gupta, N., & Verma, V. K. (2019). Next-generation sequencing and its application: Empowering in public health beyond reality. *Microbial Technology for the Welfare of Society, 17*, 313–341. https://doi.org/10.1007/978-981-13-8844-6_15. (eCollection 2019)
25. Giardine, B., Riemer, C., Hardison, R. C., et al. (2005). Galaxy: A platform for interactive large-scale genome analysis. *Genome Research, 15*(10), 1451–1455. https://doi.org/10.1101/gr.4086505
26. Wang, Z., Gerstein, M., & Snyder, M. (2009). RNA-Seq: A revolutionary tool for transcriptomics. *Nature Reviews Genetics, 10*(1), 57–63. https://doi.org/10.1038/nrg2484
27. Jinek, M., Chylinski, K., Fonfara, I., Hauer, M., Doudna, J. A., & Charpentier, E. (2012). A programmable dual-RNA-guided DNA endonuclease in adaptive bacterial immunity. *Science, 337*(6096), 816–821. https://doi.org/10.1126/science.1225829
28. Gostimskaya, I. (2022). CRISPR-Cas9: A history of its discovery and ethical considerations of its use in genome editing. *Biochemistry (Moscow), 87*(8), 777–788. https://doi.org/10.1134/s0006297922080090
29. Kumar, D. A., & Mandal S. (2013). *Development of cloud computing in integrated library management and retrieval system.*
30. Bayat, A. (2002). Science, medicine, and the future: Bioinformatics. *BMJ, 324*(7344), 1018–1022. https://doi.org/10.1136/bmj.324.7344.1018
31. Needleman, S. B., & Wunsch, C. D. (1970). A general method applicable to the search for similarities in the amino acid sequence of two proteins. *Journal of Molecular Biology, 48*(3), 443–453.

32. Sievers, F., Wilm, A., Dineen, D., et al. (2011). Fast, scalable generation of high-quality protein multiple sequence alignments using Clustal Omega. *Molecular Systems Biology, 7*, 539. https://doi.org/10.1038/msb.2011.75
33. Finn, R. D., Clements, J., Eddy, S. R. (2011). HMMER web server: Interactive sequence similarity searching. *Nucleic Acids Research, 39*(Web Server issue), W29–37. https://doi.org/10.1093/nar/gkr367.
34. Gertz, E. M., Yu, Y.-K., Agarwala, R., Schäffer, A. A., & Altschul, S. F. (2006). Composition-based statistics and translated nucleotide searches: Improving the TBLASTN module of BLAST. *BMC Biology, 4*(1), 41. https://doi.org/10.1186/1741-7007-4-41
35. Kent, W. J. (2002). BLAT–the BLAST-like alignment tool. *Genome Research, 12*(4), 656–664. https://doi.org/10.1101/gr.229202
36. Kearse, M., Moir, R., Wilson, A., et al. (2012). Geneious basic: An integrated and extendable desktop software platform for the organization and analysis of sequence data. *Bioinformatics, 28*(12), 1647–1649. https://doi.org/10.1093/bioinformatics/bts199
37. Kumar, S., Nei, M., Dudley, J., & Tamura, K. (2008). MEGA: A biologist-centric software for evolutionary analysis of DNA and protein sequences. *Briefings in Bioinformatics, 9*(4), 299–306. https://doi.org/10.1093/bib/bbn017
38. Katoh, K., Misawa, K., Kuma, K., & Miyata, T. (2002). MAFFT: A novel method for rapid multiple sequence alignment based on fast Fourier transform. *Nucleic Acids Research, 30*(14), 3059–3066. https://doi.org/10.1093/nar/gkf436
39. Stanke, M., & Morgenstern, B. (2005). AUGUSTUS: A web server for gene prediction in eukaryotes that allows user-defined constraints. *Nucleic Acids Research, 33*(Web Server issue), W465–467. https://doi.org/10.1093/nar/gki458.
40. Sali, A., & Blundell, T. L. (1993). Comparative protein modelling by satisfaction of spatial restraints. *Journal of Molecular Biology, 234*(3), 779–815. https://doi.org/10.1006/jmbi.1993.1626
41. Schwede, T., Kopp, J., Guex, N., & Peitsch, M. C. (2003). SWISS-MODEL: An automated protein homology-modeling server. *Nucleic Acids Research, 31*(13), 3381–3385. https://doi.org/10.1093/nar/gkg520
42. Zhang, Y. (2008). I-TASSER server for protein 3D structure prediction. *BMC Bioinformatics, 9*(1), 40. https://doi.org/10.1186/1471-2105-9-40
43. Kelley, L. A., Mezulis, S., Yates, C. M., Wass, M. N., & Sternberg, M. J. E. (2015). The Phyre2 web portal for protein modeling, prediction and analysis. *Nature Protocols, 10*(6), 845–858. https://doi.org/10.1038/nprot.2015.053
44. Leaver-Fay, A., Tyka, M., Lewis, S. M., et al. (2011). ROSETTA3: An object-oriented software suite for the simulation and design of macromolecules. *Methods in Enzymology, 487*, 545–574. https://doi.org/10.1016/b978-0-12-381270-4.00019-6
45. Jumper, J., Evans, R., Pritzel, A., et al. (2021). Highly accurate protein structure prediction with AlphaFold. *Nature, 596*(7873), 583–589. https://doi.org/10.1038/s41586-021-03819-2
46. McGuffin, L. J., Bryson, K., & Jones, D. T. (2000). The PSIPRED protein structure prediction server. *Bioinformatics, 16*(4), 404–405. https://doi.org/10.1093/bioinformatics/16.4.404
47. Drozdetskiy, A., Cole, C., Procter, J., & Barton, G. J. (2015). JPred4: A protein secondary structure prediction server. *Nucleic Acids Research, 43*(W1), W389–W394. https://doi.org/10.1093/nar/gkv332
48. Hofacker, I. L. (2003). Vienna RNA secondary structure server. *Nucleic Acids Research, 31*(13), 3429–3431. https://doi.org/10.1093/nar/gkg599
49. Zuker, M. (2003). Mfold web server for nucleic acid folding and hybridization prediction. *Nucleic Acids Research, 31*(13), 3406–3415. https://doi.org/10.1093/nar/gkg595
50. Bankevich, A., Nurk, S., Antipov, D., et al. (2012). SPAdes: A new genome assembly algorithm and its applications to single-cell sequencing. *Journal of Computational Biology, 19*(5), 455–477. https://doi.org/10.1089/cmb.2012.0021
51. Luo, R., Liu, B., Xie, Y., et al. (2012). SOAPdenovo2: An empirically improved memory-efficient short-read de novo assembler. *Gigascience, 1*(1), 18. https://doi.org/10.1186/2047-217x-1-18

52. Zerbino, D. R., & Birney, E. (2008). Velvet: Algorithms for de novo short read assembly using de Bruijn graphs. *Genome Research, 18*(5), 821–829. https://doi.org/10.1101/gr.074492.107
53. Koren, S., Walenz, B. P., Berlin, K., Miller, J. R., Bergman, N. H., & Phillippy, A. M. (2017). Canu: Scalable and accurate long-read assembly via adaptive k-mer weighting and repeat separation. *Genome Research, 27*(5), 722–736. https://doi.org/10.1101/gr.215087.116
54. Kolmogorov, M., Yuan, J., Lin, Y., & Pevzner, P. A. (2019). Assembly of long, error-prone reads using repeat graphs. *Nature Biotechnology, 37*(5), 540–546. https://doi.org/10.1038/s41587-019-0072-8
55. Wick, R. R., Judd, L. M., Gorrie, C. L., & Holt, K. E. (2017). Unicycler: Resolving bacterial genome assemblies from short and long sequencing reads. *PLoS Computational Biology, 13*(6), e1005595. https://doi.org/10.1371/journal.pcbi.1005595
56. Zimin, A. V., Marçais, G., Puiu, D., Roberts, M., Salzberg, S. L., & Yorke, J. A. (2013). The MaSuRCA genome assembler. *Bioinformatics, 29*(21), 2669–2677. https://doi.org/10.1093/bioinformatics/btt476
57. Liyanage, K., Samarakoon, H., Parameswaran, S., & Gamaarachchi, H. (2023). Efficient end-to-end long-read sequence mapping using minimap2-fpga integrated with hardware accelerated chaining. *Scientific Reports, 13*(1), 20174. https://doi.org/10.1038/s41598-023-47354-8
58. Bailly-Bechet, M., Haudry, A., & Lerat, E. (2014). "One code to find them all": A perl tool to conveniently parse RepeatMasker output files. *Mobile DNA, 5*(1), 13. https://doi.org/10.1186/1759-8753-5-13
59. Benson, G. (1999). Tandem repeats finder: A program to analyze DNA sequences. *Nucleic Acids Research, 27*(2), 573–580. https://doi.org/10.1093/nar/27.2.573
60. Bansal, A. K. (2005). Bioinformatics in microbial biotechnology–a mini review. *Microbial Cell Factories, 4*, 19. https://doi.org/10.1186/1475-2859-4-19
61. Al-Dewik, N. I., & Qoronfleh, M. W. (2019). Genomics and precision medicine: Molecular diagnostics innovations shaping the future of healthcare in Qatar. *Advances in Public Health, 2019*, 3807032. https://doi.org/10.1155/2019/3807032
62. Nandy, A., & Basak, S. C. (2019). Bioinformatics in design of antiviral vaccines. *Encyclopedia of Biomedical Engineering,* 280–290. https://doi.org/10.1016/B978-0-12-801238-3.10878-5. (Epub 2018 Sep 13)
63. Sunita, S. A., Singh, Y., & Shukla, P. (2020). Computational tools for modern vaccine development. *Human Vaccines & Immunotherapeutics, 16*(3), 723–735. https://doi.org/10.1080/21645515.2019.1670035
64. Kim, S. (2022). COVID-19 drug development. *Journal of Microbiology and Biotechnology, 32*(1), 1–5. https://doi.org/10.4014/jmb.2110.10029
65. Zhang, Y., Luo, M., Wu, P., Wu, S., Lee, T. Y., & Bai, C. (2022). Application of computational biology and artificial intelligence in drug design. *International Journal of Molecular Sciences, 23*(21). https://doi.org/10.3390/ijms232113568.
66. Kuczynski, J., Lauber, C. L., Walters, W. A., et al. (2011). Experimental and analytical tools for studying the human microbiome. *Nature Reviews Genetics, 13*(1), 47–58. https://doi.org/10.1038/nrg3129
67. Franzosa, E. A., McIver, L. J., Rahnavard, G., et al. (2018). Species-level functional profiling of metagenomes and metatranscriptomes. *Nature Methods, 15*(11), 962–968. https://doi.org/10.1038/s41592-018-0176-y
68. Meyer, F., Paarmann, D., D'Souza, M., et al. (2008). The metagenomics RAST server–a public resource for the automatic phylogenetic and functional analysis of metagenomes. *BMC Bioinformatics, 9*(1), 386. https://doi.org/10.1186/1471-2105-9-386
69. Tang, X., Huang, Y., Lei, J., Luo, H., & Zhu, X. (2019). The single-cell sequencing: New developments and medical applications. *Cell & Bioscience, 9*(1), 53. https://doi.org/10.1186/s13578-019-0314-y
70. Roy, A. L., & Conroy, R. S. (2018). Toward mapping the human body at a cellular resolution. *Molecular Biology of the Cell, 29*(15), 1779–1785. https://doi.org/10.1091/mbc.E18-04-0260

71. Naeem, M., & Alkhnbashi, O. S. (2023). Current bioinformatics tools to optimize CRISPR/Cas9 experiments to reduce off-target effects. *International Journal of Molecular Sciences, 24*(7). https://doi.org/10.3390/ijms24076261.
72. Woolnough, A. P., Hollenberg, L. C. L., Cassey, P., & Prowse, T. A. A. (2023). Quantum computing: A new paradigm for ecology. *Trends in Ecology & Evolution, 38*(8), 727–735. https://doi.org/10.1016/j.tree.2023.04.001

Computational Insights into the Microbial Universe

Mansi Patel, Ajay Bhatia, Nityendra Shukla, Sudhir Kumar Shukla, and Jitendra Narayan

Abstract In the field of microbiology, metagenomics has become a powerful technique that provides insights on a variety of ecosystems, including soil, wastewater, and the microbiome of human gut and skin. The capacity of metagenomics to examine high-complexity samples, where conventional approaches might not be as successful, is one of its main advantages. Metagenomics enables the direct sequencing of DNA from environmental samples, making it possible to identify a wide range of microbes, including ones that are challenging to cultivate in a lab. However, metagenomics has challenges such as accurately characterizing the microbial communities present in the samples, quantification of microorganisms and differentiation between closely related species can be challenging due to their immense diversity. Furthermore, it is still a difficult challenge to comprehend the functional roles of these microbial communities and their possible applications. To empower readers to perform metagenomics analysis on their own while tackling the challenges with precautions, this chapter is devoted to giving details of the metagenomics analysis workflow and tools. The chapter provides a clear and accessible explanation of every stage of the metagenomics analysis, from sequencing techniques to data analysis and interpretation, suitable for readers with varying levels of knowledge. This chapter promotes a greater understanding of metagenomic concepts and approaches by clarifying each step and the related resources. It urges readers to confidently set out on their metagenomics analysis.

Keywords Metagenomics · Shotgun metagenomics · 16 s Metagenomics · NGS · Microbiology

M. Patel · A. Bhatia · N. Shukla · J. Narayan (✉)
CSIR-Institute of Genomics and Integrative Biology (CSIR-IGIB), Mall Road, New Delhi 110007, India
e-mail: jnarayan@igib.res.in

S. K. Shukla
Department of Zoology, Madan Mohan Malviya P.G. College, Bhatpar Rani, Deoria 274702, Uttar Pradesh, India

© The Author(s), under exclusive license to Springer Nature Singapore Pte Ltd. 2024

A. Chaudhary et al. (eds.), *Unraveling New Frontiers and Advances in Bioinformatics*,
https://doi.org/10.1007/978-981-97-7123-3_2

1 Introduction

Microbes, the tiny but essential inhabitants of our planet, are crucial for shaping life. They were the first life form to appear on the Earth. The term "microbe," derived from the Greek word "meta" (meaning "transcendent"), encapsulates these organisms, measuring less than 0.2 mm [1]. While we often use terms like microorganism, bacteria, germ, or bug interchangeably, it's important to recognize that microorganisms come in diverse forms. Microbiologists classify these organisms into various, extremely diverse taxonomic groups [2]. Each group contributes uniquely to Earth's balance, participating in vital processes that sustain life. From nutrient cycling to disease regulation, these unseen architects silently orchestrate existence, reminding us that even the tiniest beings wield immense power in our interconnected biosphere. Only a very small group of microbes, about 1400 species, are pathogenic and responsible for mass human morbidity and mortality.

Metagenomics is the study of the microbial population's genetic material through direct access to their genetic information, utilising various sequencing technologies to sequence entire microbial communities. The communities of microorganisms, such as those found on human skin, in soil, or in a water sample, are frequently studied using metagenomics. It encompasses a suite of methodologies, spanning from genome sequencing to its bioinformatics analysis, aimed at deciphering and deriving valuable insights from the large amount of genetic data generated. Starting with single-nucleotide sequencing and advancing to massively parallel sequencing methods, there have been substantial advancements in sequencing technology. Owing to advancements in sequencing techniques, bioinformatics methodologies, and computation, the speed of metagenomic investigations have increased largely and led to substantial advances in the field of microbial evolution, diversity and microbial ecology over the past decade (Fig. 1).

2 Metagenomics Sequencing

Sanger sequencing, often known as the "Chain Termination Method," is a first-generation sequencing technique that was first introduced in 1977 [3]. Sanger sequencing is historically groundbreaking in the field of genomics, but it has limitations that make it less suitable for certain applications, including metagenomics analysis. It has a comparatively limited capacity for processing DNA fragments, compared to newer sequencing technologies like next-generation sequencing (NGS). It can only sequence one DNA fragment at a time, which is a significant limitation when dealing with complex microbial communities in metagenomics, where thousands to millions of DNA fragments need to be sequenced parallelly. Sanger

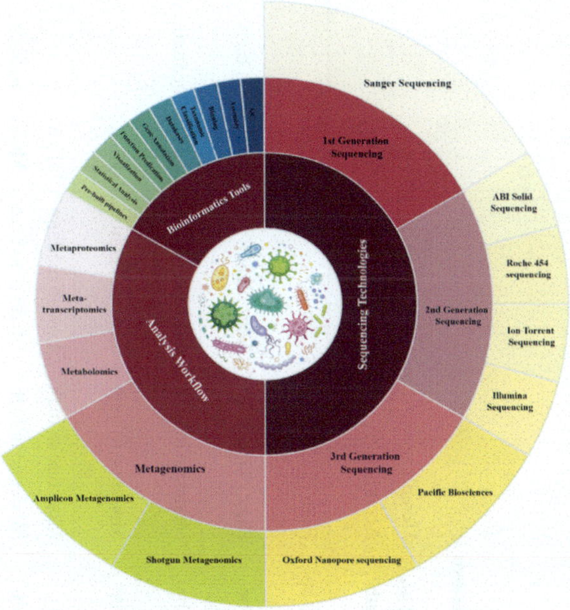

Fig. 1 Overview of content in this chapter

sequencing is comparatively more expensive per base pair compared to NGS technologies, primarily attributed to its labour-intensive workflow and the need for expensive reagents and equipment. NGS techniques offer significant advancements over traditional Sanger sequencing, addressing its limitations.

Shotgun metagenomics and amplicon metagenomics are the two main methods used in metagenomics sequencing and analysis. A summary of each technique is provided below, along with information on how it can be used with various sequencing platforms, such as NGS platforms like Illumina and third-generation platforms like Nanopore offered by Oxford Nanopore Technologies (ONT). Details about alternative sequencing platforms are outlined in Table 1 for reference.

2.1 Metagenomics Sequencing

2.1.1 Shotgun Metagenomics Sequencing

Shotgun metagenomics involves sequencing all fragments of DNA extracted from a microbial community sample, providing a detailed view of the genetic diversity present in the sample. Shotgun sequencing is a method used to sequence entire genomes by breaking DNA into small, random fragments and then sequencing those fragments. In this technique, the DNA sample is fragmented randomly into small pieces. Each of these fragments is then sequenced independently. This approach

Table 1 Comparison of sequencing technologies in terms of chemistry, accuracy, throughput, read length and cost

Sequencing generation	Sequencing technology	Chemistry of sequencer	Available platforms	Read length	Accuracy	Throughput	Cost
First Generation	Sanger Seq [4]	Chain termination	Applied biosystems 3730xl, others	800–1000 bp	High (99.99%+)	Low	High
Second Generation	Illumina Seq [5]	Reversible terminator	Illumina HiSeq, MiSeq, NovaSeq	150–300 bp	High (99.9%+)	Very high	Medium to high
	Ion Torrent Seq [6]	Semiconductor	Ion Torrent Proton, PGM	100–200 bp	Moderate to high	Medium to high	Low to medium
	Roche 454 Seq [7]	Pyrosequencing	Roche 454 GS FLX+, GS junior	400–700 bp	Moderate to high	Low to medium	High
	ABI Solid Seq [8]	Ligation-based	ABI SOLiD	35–75 bp	Moderate to high	Low to medium	High
Third Generation	Oxford Nanopore Seq [9]	Nanopore	Oxford Nanopore MinION, PromethION	Up to 2 Mb	Moderate to high	Medium to high	Medium to high
	Pacific biosciences (PacBio) [10]	Single molecule, real-time	PacBio RS II, sequel, sequel II	10–20 kb	High (99%+)	High	High

* Seq: sequencing

allows for the identification of microbes, functional genes, and metabolic pathways within the community. Shotgun metagenomics is particularly useful for exploring complex microbial ecosystems, such as wastewater, soil, and the human gut microbiome. Identification and study of varied microbial communities has been made possible by shotgun metagenomics, which has revolutionised microbiology. However, there are still challenges with precisely profiling high-complexity samples and comprehending the roles of these communities for possible applications. The presence of different microorganisms in samples taken from the same environment can make it more difficult to identify physiologically and statistically significant differences across small sample sets. Ensuring that analysis methods are capable of detecting even the smallest differences is crucial, especially true for research on the human microbiota, which is influenced by many things like genes, age, diet, and environmental conditions [11]. Limitations of shotgun metagenomic sequencing are: (a) more costly and computationally intensive than amplicon sequencing, (b) may have a higher detection limit for low-abundance taxa and lower taxonomic resolution compared to amplicon sequencing.

2.1.2 Amplicon Metagenomics Sequencing

Amplicon sequencing is a targeted sequencing approach that amplifies specific DNA regions of interest before sequencing. In this technique, specific regions of DNA, known as amplicons, are amplified using Polymerase Chain Reaction (PCR) with primers designed to target those regions. After amplification, the resulting DNA fragments are sequenced to determine the nucleotide sequence of the target region. Amplicon sequencing is often used for targeted analysis of specific genes or genomic regions. It is commonly used in microbial community analysis targeting specific marker genes (e.g., 16S rRNA gene for bacteria and ITS regions for fungi). This approach allows for the taxonomic profiling of microbial communities and can provide insights into community composition and diversity. Amplicon metagenomics is widely used for microbial ecology studies, as it provides a cost-effective way to analyse large sample sets.

Each method has its own strengths and limitations, and the strategy of choice is determined based on the research objectives, sample complexity, and available resources.

2.2 *Sequencing Platforms*

2.2.1 Illumina

The Illumina platform's extremely high outputs (up to 1.5 Tb per run), great precision (error rate of 0.1–1%), and widespread availability have made it the industry standard for shotgun metagenomics sequencing. Primary differences between the Illumina platforms are in overall output and maximal read length. A traditional option for metagenomics using Illumina platform is the Illumina HiSeq 2500. Modern HiSeq 3000 and 4000 systems boost a run's throughput but have a 150 nt read length restriction. Even though the MiSeq instruments can only produce 15 gigabytes in 2 × 300 mode, they are nonetheless beneficial for microbiome investigations involving a single marker gene or a few samples. For shotgun metagenomics, Illumina sequencing offers high throughput sequencing, and cost-effectiveness, making it suitable for large-scale studies and comprehensive analysis of microbial communities. For amplicon metagenomics, Illumina's short-read technology provides high sequencing depth, allowing for accurate taxonomic classification based on amplified marker genes like 16S rRNA or ITS regions. Using Illumina platform shotgun metagenomic sequencing has 2 basic steps: library preparation and sequencing. It begins with random shearing of extracted DNA. Adapters are then ligated to both the 5′ and the 3′ ends of the DNA fragments to enable binding to the sequencer-specific substrate and fragment identification. The resulting library is sequenced using high-throughput NGS. Researchers can combine several samples into a single sequencing run, sequencing thousands of organisms in parallel. This method allows researchers to perform deep subsampling and to identify very low abundance members of the microbial sample that may be missed using other approaches.

2.2.2 Oxford Nanopore

Oxford Nanopore sequencing is increasingly used for both shotgun and amplicon metagenomics, which can be put to use in settings, such as the lab or field. By utilising portable MinION and Flongle devices for nanopore metagenomic sequencing at the sample source, one can streamline the process with a 10-min library prep. For shotgun metagenomics, Nanopore sequencing offers long-read capability, enabling the sequencing of complete genomic fragments. For amplicon metagenomics, Nanopore sequencing can provide long reads spanning entire amplicons, allowing for the resolution of complex microbial communities and improved taxonomic assignment. In metagenomics research, Nanopore sequencing technology offers a number of advantages. It is possible to generate reads of any duration, from ultra-long to short. Long sequencing reads improve genome assembly, allowing for more accurate distinction of closely related microorganisms and more accurate analysis of both known and unknown microbes. With the use of nanopore technology, reads larger than 4 Mb have been produced, enabling the acquisition of whole microbial genomes in single reads or with minimal contigs. Long reads significantly enhance genome assembly and antimicrobial resistance (AMR) gene research by improving the resolution of repetitive sequences and structural variations.

In summary, the choice of sequencing platform for shotgun and amplicon metagenomics depends on factors such as desired sequencing depth and read length, accuracy, and cost considerations. Researchers often select the sequencing platform that best suits their specific research goals and budget constraints (Fig. 2).

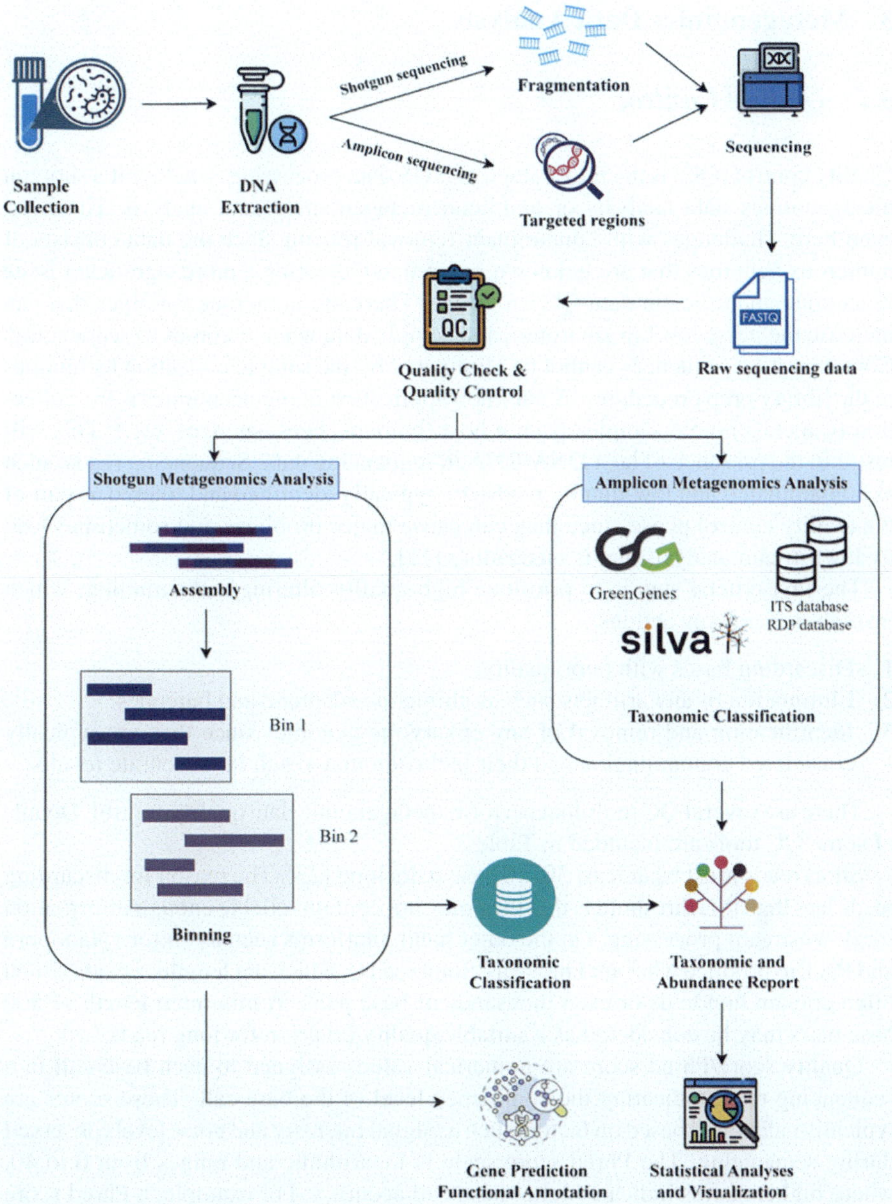

Fig. 2 Metagenomics workflow: The flowchart illustrates the process of metagenomics analysis, starting with sample collection and DNA extraction. It then diverges into shotgun metagenomics involving fragmentation, sequencing, quality check, quality control, assembly, binning, taxonomic classification, gene prediction, functional annotation, and statistical analysis. In amplicon metagenomics, targeted regions are sequenced, followed by quality check, quality control, taxonomic classification using different databases, and then abundance reporting

3 Metagenomics Data Analysis

3.1 Quality Control

Quality control (QC) is the initial stage of NGS data processing, whether it's shotgun metagenomics data analysis or amplicon metagenomics data analysis. However, even here, challenges with contaminant removal remain since the data consists of numerous genomes that are unknown beforehand, creating a more significant issue if accurate and efficient data QC isn't done. There are numerous variables that can increase the noise level in environmental sample data when it comes to sequencing. Contaminating sequences cannot be eliminated by the sample collection techniques or the library prep procedures. A common application of metagenomics is the collection of metagenomic samples from a host (humans, bees, sponges, etc.). This will result in the presence of host DNA/RNA in sequencing data. Sequencing reads, such as contaminated and low-quality reads, are typically identified and filtered as part of the quality control phase since they can cause major problems and sometimes lead to downstream analysis that is misleading [12].

The first crucial step is to prioritize high-quality filtering and trimming, which involves several key actions:

1. Discarding bases with poor quality.
2. Elimination of any artifacts such as chimeras, adapters, and barcodes.
3. Identification and removal of any eukaryotic genomes since they are typically considered contamination and their inclusion may result in inaccurate results.

There are several QC tools that perform metagenomic data quality control. Details of some QC tools are included in Table 2.

Short reads must be at least 30–40 base pairs long [23]. The reason for discarding reads less than this threshold is that they may not contain reliable enough information for downstream processing. On the other hand, platforms such as Oxford Nanopore and PacBio produce what are known as "long reads," which are lengthier reads which often contain hundreds or even thousands of base pairs. A minimum length of 500 base pairs may be considered as a suitable quality criterion for long reads.

Quality score/Phred score are numerical values assigned to each base call in a sequencing read, indicating the confidence level of the base call. These scores are typically calculated based on factors such as signal intensity and noise levels observed during sequencing. The Phred score scale is logarithmic and ranges from 0 to 40, where higher scores indicate higher base call accuracy. For example, a Phred score of Q20 corresponds to an accuracy of 99%, meaning that there is a 1% chance of an incorrect base call at that position in the read. Ideally, a minimum Phred score of Q20 is considered ideal to consider confidence on base calling.

To efficiently eliminate contamination from shotgun metagenomics sequencing data, mapping tools are necessary. Mapping tools are used for aligning sequencing reads to a reference genome or set of reference sequences of hosts like humans. This enables the differentiation of sequences from contaminants and microbiological

Table 2 Tools for sequence quality control (column 1). For each tool, the compatible read type (column 2), features (column 3), and the websites to download these tools (column 4) are listed below

Tool	Platform	Features	Websites
Fastp [13]	Illumina	Rapid processing; filtering and trimming	https://github.com/OpenGene/fastp
FastQC [14]	Illumina	Intuitive data visualization; comprehensive QC metrics	https://www.bioinformatics.babraham.ac.uk/projects/fastqc/
Trimmomatic [15]	Illumina	Customizable trimming; robust quality control	http://www.usadellab.org/cms/index.php?page=trimmomatic
SOAPnuke [16]	Illumina	Memory-efficient preprocessing; versatile filtering	https://github.com/BGI-flexlab/SOAPnuke
SequelTools [17]	PacBio	User-friendly interface; comprehensive QC utilities	https://github.com/ISUgenomics/SequelTools
Proovread [18]	PacBio	Iterative error correction; efficient consensus calling	https://github.com/BioInf-Wuerzburg/proovread
NanoPack [19]	Oxford Nanopore	Comprehensive nanopore data processing; efficient QC	https://github.com/nanopack
MinIONQC [20]	Oxford Nanopore	Scalable QC for large-scale projects; multi-sample support	https://github.com/roblanf/minion_qc
LongQC [21]	Oxford Nanopore and PacBio	Excellent visualization of QC assessment; streamlined workflow	https://github.com/yfukasawa/LongQC
Porechop [22]	Oxford Nanopore	Adaptable adapter trimming; efficient demultiplexing	https://github.com/rrwick/Porechop
NanoPack [19]	Oxford Nanopore	Comprehensive visualization; detailed sequencing metrics	https://github.com/wdecoster/nanopack

species by comparing each read to the known reference genome. Table 3 lists some commonly used alignment tools.

Table 3 Alignment tools (column 1), their compatible read type (column 2), features (column 3), and websites to download these tools (column 4)

Tool	Platform	Features	Websites
BWA [24]	Illumina	Versatile alignment; efficient indexing and mapping	https://github.com/lh3/bwa
Minimap2 [25]	Illumina, PacBio, Oxford Nanopore	Ultra-fast alignment; multi-reference compatibility	https://github.com/lh3/minimap2
Bowtie2 [26]	Illumina	Fast and memory-efficient; sensitive short-read alignment	https://github.com/BenLangmead/bowtie2

3.2 Shotgun Metagenomic Data Analysis

3.2.1 Assembly

The sequencing reads are assembled after quality control. There are numerous assemblers available, but the choice of which to use depends on the user's specific requirements and research objectives. For instance; SETU, a robust pipeline/assembler, is particularly designed for analyzing SARS-CoV-2 data [27]. Some commonly used assembly tools for metagenomic assembly are listed in Table 4.

The traditional metagenome assembly tools for short-read sequencing were developed using overlap-layout consensus (OLC) techniques. The process consists of three primary steps: overlap detection, layout, and consensus. Without using a reference genome, OLC techniques assemble overlapping reads to rebuild the original genome. Finding overlaps between sequencing reads is the first step in OLC. These overlaps show the areas where two reads have similar subsequences in common. The layout stage arranges the reads to their relative places after overlaps have been found. The consensus stage creates a consensus sequence for every overlapped region after layout [38].

Many genome assemblers use a powerful algorithm known as the de Bruijn graph (DBG). Originally suggested for de novo genome assembly, it has evolved into an essential part of several assembly techniques. Sequencing reads are divided into fixed-length subsequences called k-mers, which are typically between 20 and 100 base pairs long. The graph structure is then used to piece together overlapping k-mers to create longer sequences. When compared to other approaches, the De Bruijn graph methodology requires less computer memory. Metagenomic data can be managed easily by de Bruijn graph-based assemblers, which divide the data into k-mers and reassemble genomes from overlapping subsequences [39]. DBG algorithms offer advantages over OLD algorithms in terms of memory efficiency, scalability, and accuracy in genome assembly [38].

Table 4 Assembly tools (column 1) their compatible sequencing read type (column 2), core algorithm (column 3), key features (column 4), and websites to download these tools (column 5)

Tool	Platform	Core algorithms	Features	Website
Omega [28]	Illumina	OLC	Efficient assembly; scaffolding capabilities	http://omega.omicsbio.org
MetaVelvet [29]	Illumina	DBG	Metagenomic assembly; enhanced sensitivity	http://metavelvet.dna.bio.keio.ac.jp
IDBA-UD [30]	Illumina	DBG	Iterative assembly; handling uneven coverage	http://www.cs.hku.hk/~alse/idba_ud
MEGAHIT [31]	Illumina	DBG	Ultra-fast assembly; scalable for large datasets	https://github.com/voutcn/megahit
MetaSPAdes [32]	Illumina	DBG	Metagenomic assembly; built-in error correction	https://github.com/ablab/spades
Ray meta [33]	Illumina	DBG	Parallel assembly; efficient memory usage	http://denovoassembler.sf.net
Nanoscope [34]	Oxford Nanopore	DBG	Nanopore-long read specific; efficient assembly	https://github.com/kuleshov/nanoscope
Canu [35]	PacBio and Oxford Nanopore	OLC	PacBio and Nanopore assembly; error correction	https://github.com/marbl/canu
DBG2OLC [36]	Illumina, PacBio and Oxford Nanopore	DBG and OLC	Hybrid approach, integrating DBG and OLC algorithms	https://github.com/yechengxi/dbg2OLC
Unicycler [37]	Illumina, PacBio and Oxford Nanopore	DBG	Produces circularized contigs	https://github.com/rrwick/Unicycler

(OLC: Overlap Layout Consensus; DBG: De Bruijn Graph)

The tools available to assess the reliability of the contigs and scaffolds generated by the assemblers are displayed in Table 5.

Table 5 Enhance your assembly quality control with a comprehensive suite of tools (column 1) accompanied by convenient links to download them from trusted websites (column 2)

Tool	Websites
MetaQUAST [40]	http://bioinf.spbau.ru/metaquast
REAPR [41]	https://www.sanger.ac.uk/tool/reapr/
VALET [42]	https://github.com/marbl/VALET
DeepMAsED [43]	https://github.com/leylabmpi/DeepMAsED
CheckM [44]	https://github.com/Ecogenomics/CheckM

3.2.2 Binning

With only scaffolds, most of the assembled sequences do not represent entire microbial genomes. Numerous tools for binning have been created to organise the scaffolds into clusters that correspond to an organism's whole genome. The most important stage in metagenomic study is to cluster, or "bin," assembled sequences into different groups that represent microbial genomes. Taxonomy binning and genome binning, also known as taxonomic-dependent and taxonomic-independent binning, are two types of binning methods. By using alignment algorithms like BLAST, Bowtie2, BWA, minimap2, or databases of previously sequenced microbial genomes, taxonomy binning is a supervised technique for comparing metagenomic sequences to a database of genomic sequences. However, the limitation of the taxonomy binning approach is incomplete reference databases, particularly when attempting to understand the functional and metabolic roles of unknown microbes present in the sample. The genome binning methodology is an unsupervised technique that uses machine learning methods to group contigs into distinct genome bins based on the feature patterns of the sequences and the linkage patterns between the sequences, all without the using reference databases [45]. Unsupervised binning uses statistical or coverage based methods to perform efficient and robust contig binning.

While long read technologies like Oxford Nanopore and PacBio can offer advantages in resolving complex genomic regions and aiding in the assembly of complete genomes because of long reads. However, binning may still be necessary. Available tools for binning included in Table 6.

The characteristics that determine the reliability of "bins" include completeness. For further annotation, only the bins of comparatively high completeness are filtered as the Metagenome-Assembled Genomes (MAGs). Bins are commonly classified as high quality, medium-quality or low-quality bins [58]. It is common to use high-quality (completeness more than 90%) and medium-quality (completeness larger than 50%) bins [59]. CheckM is a popular tool for evaluating the quality of bins. Then, just the bins that are reasonably high-quality are selected to be the MAGs for the further processing.

Table 6 Elevate your metagenome binning process with a diverse range of tools (column 1) boasting unique features (column 2), conveniently accessible via download links from reputable websites (column 3)

Tool	Features	Websites
GroopM [46]	Accurate binning with two-stage clustering	https://github.com/minillinim/GroopM
MaxBin2 [47]	Optimized for large datasets; customizable parameters	http://sourceforge.net/projects/maxbin/
CONCOCT [48]	Metagenomic contig binning; efficient memory usage	https://github.com/BinPro/CONCOCT
MetaBAT2 [49]	High-throughput binning; scalable for large datasets	https://bitbucket.org/berkeleylab/metabat
SolidBin [50]	Robust binning; noise reduction in metagenomic datasets	https://github.com/sufforest/SolidBin
BMC3C [51]	Consensus binning approach; improved accuracy	http://mlda.swu.edu.cn/codes.php?name=BMC3C
GraphBin [52]	Scalable and efficient binning; low computational cost	https://github.com/Vini2/GraphBin
METAMVGL [53]	Accurate binning; robust to noise and contamination	https://github.com/ZhangZhenmiao/METAMVGL
VAMB [54]	High-quality binning; optimized for complex datasets	https://github.com/RasmussenLab/vamb
MAGO [55]	Improved binning accuracy; integration of multiple models	http://mago.fe.uni-lj.si/
MetaWRAP [56]	Comprehensive binning pipeline; modular and customizable	https://github.com/bxlab/metaWRAP
DAS tool [57]	Efficient binning of low-abundance taxa; user-friendly	https://github.com/cmks/DAS_Tool

3.2.3 MAG Taxonomic and Abundance Profiling

The aim of taxonomic profiling is to identify the microorganisms present in metagenomic isolates, and quantifying the relative abundance of various microbial taxa within the microbial sample is the objective of abundance profiling. Taxonomic classification in the past relied on BLAST-based alignment of MAGs. More effective techniques for taxonomic classification have been created as a result of reference databases and datasets expanding in size. To determine the abundance of MAGs, several tools are available which have been categorised into 3 groups: (1) protein-based tools, (2) k-mer based tools, (3) marker gene-based tools.

Kraken was the first algorithm developed specifically for metagenomics data, allowing for rapid classification, matching exact k-mers to the lowest common ancestor (LCA) of every taxon [60]. Tools such as Centrifuge, uses a combination of Burrows-Wheeler transform (BWT) and Ferragina-Manzini (FM) index to compress the reference database and efficiently and rapidly classify sequences [61].

Table 7 Tools to strengthen your MAG taxonomy and abundance profiling efforts with a well curated set (column 1) that use several classification methods (column 2) to deliver diverse characteristics (column 3), all easily accessible from reputable sources (column 4)

Tool	Types of method	Features	Websites
Kaiju [63]	Translated protein-based	Fast classification; sensitivity to distant homologs	http://kaiju.binf.ku.dk
Kraken [60]	K-mer based	Scalable; accurate classification	https://ccb.jhu.edu/software/kraken/
Kraken2 [64]	K-mer based	Improved sensitivity; reduced memory usage	https://ccb.jhu.edu/software/kraken2/
Bracken [65]	K-mer based	Abundance estimation; handling low-abundance taxa	https://ccb.jhu.edu/software/bracken/
CLARK [66]	K-mer based	High sensitivity; fast execution	http://clark.cs.ucr.edu/
K-SLAM [67]	K-mer based	Improved accuracy; capability to handle complex datasets	https://github.com/aindj/k-SLAM
MetaPhlAn3 [68]	Marker gene-based	Species-level resolution; profiling of complex samples	https://huttenhower.sph.harvard.edu/metaphlan/
PanPhlAn3 [68]	Marker gene-based	Strain-level resolution; pan-genome analysis	http://segatalab.cibio.unitn.it/tools/panphlan/
IGGsearch [69]	Marker gene-based	Integrated analysis of marker genes; taxonomic profiling	https://github.com/snayfach/IGGsearch

In protein-based tools like Kaiju, once the reads have been converted into amino acid sequences, a database search is performed to find the highest possible number of exact matches. K-mers are DNA sequences that are substrings of length k. Tools based on k-mers use the patterns of k-mers to infer the presence and quantity of MAGs. Then this tool creates k-mer frequency profiles and compares them to reference databases. Tools based on marker genes, normalises the counts of each gene for a given sample and aligns reads to a marker gene database. It is possible to assess each MAG's abundance using such genome-scale data [62]. Table 7 contains tools for MAG taxonomic and abundance profiling.

3.2.4 Gene Prediction Tools

The next important step is to identify the genes. This section covers methods for identifying and predicting genes from MAGs by recognizing putative coding sequences. The most popular methods are model-based gene prediction which rely

Table 8 A comprehensive set of tools to optimize your gene prediction and annotation workflows (column 1), each with a unique methodology (column 2) and a wide range of features (column 3), all available for download from reliable sources

Tool	Type of method	Features	Websites
MetaGeneMark [70]	Model based	Highly accurate gene prediction; suitable for short reads	http://exon.gatech.edu/meta_gmhmmp.cgi
Glimmer-MG [71]	Model based	Improved prediction accuracy; capable of handling complex metagenomes	https://github.com/davek44/Glimmer-MG
FragGeneScan [72]	Model based	Accurate prediction of short genes; suitable for diverse datasets	https://omics.informatics.indiana.edu/FragGeneScan/
Prodigal [73]	Model based	Fast and accurate gene prediction; supports multiple output formats	https://github.com/hyattpd/Prodigal
MetaGene [74]	Model based	Efficient gene prediction; suitable for long reads	http://metagene.nig.ac.jp/metagene/metagene.html
MetaGeneAnnotator [75]	Model based	Comprehensive annotation; integration with other tools	http://metagene.nig.ac.jp/
Meta-MFDL [76]	Machine learning	High accuracy; capable of capturing complex sequence patterns	https://github.com/nwpu903/Meta-MFDL
CNN-MGP [77]	Machine learning	Robust prediction; handles noise and variation in data	https://github.com/rachidelfermi/cnn-mgp
Balrog [78]	Machine learning	Improved prediction accuracy; scalable to large datasets	https://github.com/salzberg-lab/Balrog

on models, mostly Hidden Markov Model (HMMs). Several deep learning algorithms have drawn a lot of interest for the purpose of gene prediction. This category also encompasses a few tools. Resources for annotating genes are listed in Table 8.

3.2.5 Gene Functional Annotation Tools

MAGs as well as the list of genes derived from gene annotation step can be used to analyze the functional characteristics of microbial populations. Table 9 contains resources for gene functional annotation.

After conducting shotgun metagenomics data analysis, one can expect a comprehensive understanding of the microbial communities present in samples. This

Table 9 Gene functional annotation tools (column 1), features (column 2), and websites to download these tools (column 3)

Tool	Features	Websites
eggNOG-mapper [79]	Comprehensive functional annotation; user-friendly interface	http://eggnog-mapper.embl.de
GhostKOALA [80]	Rapid annotation; integration with KEGG pathways	http://www.kegg.jp/blastkoala/
MG-RAST [81]	High-throughput annotation; metagenomic-specific	http://api.metagenomics.anl.gov/api.html
PANNZER2 [82]	Accurate functional predictions; supports multiple species	http://ekhidna2.biocenter.helsinki.fi/sanspanz/
InterProScan [83]	Identification of conserved motifs and domains	http://www.ebi.ac.uk/InterProScan/
GeConT [84]	Context-specific annotation; customizable parameters	http://www.ibt.unam.mx/biocomputo/gecont.html
FunGeCo [85]	Accurate prediction of functional categories	https://web.rniapps.net/fungeco
FlaGs [86]	Functional annotation of viral genomes	https://github.com/GCA-VH-lab/FlaGs

includes insights into taxonomic composition, phylogenetic relationships, functional potential, metabolic pathways, and protein annotations. These results are crucial for studying microbial diversity, ecological roles, metabolic activities, and potential applications in various fields, including environmental science, biotechnology, and human health.

3.3 Marker Gene/Amplicon Based Metagenomics Analysis

The general agreement is that the 16S rRNA gene's sequencing indicates the evolution of eubacteria [87]. The study of microbial diversity in natural environments has come a long way since the advent of SSU rDNA-based molecular methods [88, 89]. Large number of sequences have been deposited in specific databases as a result of the widespread application of sequencing of the 16S rRNA gene in the field of microbial ecology [90–92], strengthening the usage of 16S as the "gold standard" in microbial ecology. While the 16S rRNA gene fragment, containing variable regions, is the preferred target marker gene for bacteria and archaea but for fungi and eukaryotes preferred marker genes are the internal transcribed spacer (ITS) and 18S rRNA gene, respectively.

The overall sequencing output of amplicon sequencing is a set of clusters of nearly identical sequences, referred to as Operational Taxonomic Units (OTUs) [93] or Amplicon Sequence Variants (ASVs) [94] From the analysis of such clusters, information on the community diversity, richness and evenness can be derived [95], while accounting for the degree of divergence between different ecosystems or sample types [96]. However, many taxa often cannot be resolved because the V4–V5 region in bacterial genome doesn't provide enough nucleotide variability to resolve different taxa [97]. The comparison of the 16S rRNA gene sequences allows differentiation between organisms at the genus level across all major phyla of bacteria [98].

The analysis typically begins with cleaning up the sequencing data. Sequencing processes can introduce errors, and there might be sequences originating from the human genome or other contaminants present in the sample. QC tools are employed to filter out these errors and unwanted sequences plus removes low quality reads and adapter sequences. Once the data is cleaned, the next step involves identification of OTUs and aligning those with reference databases.

In OTU identification for bacteria/archaea, it is widely acknowledged that OTUs with a similarity exceeding 97% are typically indicative of the same species. However, alternative dissimilarity thresholds can also be utilized depending on the requirements of subsequent analyses. Various strategies exist for OTU identification:

(1) De novo clustering is employed when amplicons overlap, and no reference sequence collection is accessible. This method groups all reads independently of a reference, albeit it is computationally intensive and may not be ideal for handling extensive datasets.
(2) Closed reference clustering is utilized when amplicons do not overlap, and a reference sequence collection is available. This approach filters out reads that fail to align with a reference sequence.
(3) Open reference clustering is adopted when amplicons overlap, and a reference dataset is accessible. This technique initially attempts to cluster reads against a reference dataset; however, if a read does not match any reference, it is then clustered independently.

The selection of the most suitable clustering method for downstream analysis hinges upon the nature of the study and the specific research objectives [97]. Several databases are commonly used for 16S, 18S, and ITS metagenomics data analysis. These databases offer specialized resources for specific types of organisms or genetic markers, providing researchers with alternative reference points for their analyses. Databases are listed in Table 10.

4 Statistical Analysis

In conducting statistical analyses for metagenomic data, one can use the R programming language, employing various functions available within the Vegan package.

Table 10 Useful databases for 16 s/18 s and ITS regions (column 1), features (column 2) and websites to download these tools (column 3)

Region	Database	Features	Websites
16S/18S	Silva database [99]	Comprehensive database for 16S and 18S rRNA gene sequences	https://www.arb-silva.de/
16S/18S	GreenGenes database [100]	Curated database for 16S rRNA gene sequences	https://greengenes.secondgenome.com/
16S/18S	RDP (Ribosomal database project) [93]	A comprehensive resource for ribosomal RNA data	https://rnacentral.org/expert-database/rdp
16S/18S	ENA (European Nucleotide Archive) [101]	Repository for nucleotide sequence data	https://www.ebi.ac.uk/ena/browser/home
16S/18S	NCBI (National Center for Biotechnology Information)	National Center for Biotechnology Information	https://www.ncbi.nlm.nih.gov/
ITS	UNITE database [102]	Comprehensive database for fungal ITS sequences	https://unite.ut.ee/
ITS	ITSoneDB [103]	Database for fungal Internal Transcribed Spacer (ITS) sequences	https://itsonedb.cloud.ba.infn.it/

- Alpha diversity refers to the diversity of species within a particular habitat or ecosystem. The Shannon diversity index is one of the methods used to quantify alpha diversity. It measures both richness (the total number of species) and evenness (the distribution of individuals among those species) within a community. Here, Richness is defined as the total number of species present, and evenness refers to how evenly distributed the individuals are among those species [104].
- Pairwise Bray–Curtis dissimilarity is a common metric used to assess beta diversity, which measures the variation in species composition between different samples or sites. The Bray–Curtis dissimilarity metric compares communities based on the presence or relative abundance of taxa within those communities. It quantifies how different two communities are in terms of their species composition, taking into account both the presence and abundance of taxa [105].

Utilizing distance matrices derived from this dissimilarity metric, one can conduct principal coordinates analysis (PCoA) to visualize and interpret the patterns present within the data. Biplots can be generated based on the principal components that explain the most significant variation. A biplot is a graphical representation of multivariate data that allows for the simultaneous visualization of both samples and variables. In the context of principal component analysis (PCoA), the principal components capture the main sources of variation in the data. By creating a biplot based on these principal components, one can visually explore the relationships between samples and variables in the dataset.

In summary, statistical analysis provides a robust framework for exploring and interpreting metagenomic data, offering insights into the complex microbial ecosystems under investigation. Through careful consideration of diversity metrics, beta-diversity assessments, and rigorous statistical testing, one can gain a deeper understanding of the intricate relationships within the microbial ecosystem.

5 Pipelines and Tools

Owing to the intricacy and volume of metagenomic data, specific tools and pipelines have been created to speed up and simplify the analysis process. These tools are designed to handle various steps involved in metagenomic analysis, including data preprocessing, taxonomic classification, functional annotation, and visualization. Commonly used pipelines and tools for metagenomics analysis are mentioned in Table 11.

These pipelines and tools are invaluable for researchers in the field of metagenomics as they provide standardized workflows, reduce the computational burden, and facilitate the interpretation of complex biological data. Additionally, these tools are open-source and actively maintained, allowing for community-driven improvements and updates.

6 Visualization of Results

In the realm of metagenomics, visualization of results plays a pivotal role in unraveling the intricate dynamics of microbial communities. Upon generating output from QIIME (Quantitative Insights Into Microbial Ecology), researchers are provided with a taxonomic tree representation in Newick format, easily visualizable using tools like FigTree. Additionally, QIIME produces files in Biom (Biological Observation Matrix) format, capturing OTU (Operational Taxonomic Unit) tables. These Biom files can be seamlessly imported into visualization platforms such as MEGAN or any additional statistical programme that needs data in the matrix format, facilitating comprehensive data exploration. Alpha-diversity analysis, which assesses diversity within individual samples, is supported through various metrics such as Phylogenetic Diversity (PD) and Chao richness estimation. Meanwhile, beta-diversity analysis, essential for understanding diversity across samples, is facilitated by methods like UniFrac and Principal Coordinates Analysis (PCoA). QIIME further enables taxonomic composition and phylogenetic analyses, enriching the analytical toolkit for researchers.

Beyond QIIME, a plethora of tools and software packages exist to further delve into the statistical and visualization intricacies of the data. The Primer-E package, favored by microbial ecologists, offers a suite of multivariate statistical analyses including multidimensional scaling (MDS), analysis of similarities (ANOSIM), and

Table 11 Commonly used pre-built pipelines and tools for metagenomics analysis (column 1), brief description (column 2) and websites to download these tools (column 3)

Tool/software	Description	Websites
Dada2 [106]	Dada2 is a fast and accurate pipeline for processing amplicon-based microbial sequencing data. It includes functions for quality control, error correction, and feature inference from amplicon sequencing data, particularly targeting 16S rRNA gene sequences	https://benjjneb.github.io/dada2/index.html
QIIME2 [107]	QIIME (Quantitative Insights Into Microbial Ecology) is a comprehensive metagenomics analysis toolkit that supports a diverse range of microbial community analysis, including data preprocessing, diversity analysis, taxonomic classification, and functional prediction. It provides a user-friendly interface and integrates various bioinformatics tools and algorithms for robust microbiome analysis	https://qiime2.org/
Mothur [108]	Mothur is a software/package for microbial ecology analysis that enables users to perform a diverse range of tasks, including sequence preprocessing, diversity analysis, taxonomic classification, and statistical analysis of microbial community data. It provides a command-line interface and extensive documentation for microbiome research	https://mothur.org/
MetaPhlAn [109]	MetaPhlAn (Metagenomic Phylogenetic Analysis) is a tool for metagenomic profiling and taxonomic analysis of microbial communities from shotgun metagenomic sequencing data. It uses clade-specific marker genes to rapidly and accurately profile, determine the composition of microbial populations and derive prospective applications	https://github.com/biobakery/MetaPhlAn
USEARCH-UPARSE [110]	USEARCH-UPARSE is a sequence analysis tool designed for microbial community profiling from amplicon sequencing data. It provides functions for quality filtering, dereplication, clustering, chimera removal, and OTU (Operational Taxonomic Unit) picking, allowing users to analyze and interpret microbial community data efficiently	https://drive5.com/uparse/

hypothesis testing. Notably, the R statistical programming language has emerged as a powerhouse for multivariate statistics, with packages like vegan, phyloseq, and Bioconductor offering a plethora of built-in functions and libraries tailored for metagenomic datasets visualization.

7 Metaomics

The techniques available for studying the microbiome have been significantly enhanced by genome-wide sequencing methodologies like metagenomics and metatranscriptomics. Through the genes, transcripts, and eventually proteins and metabolites data, the "meta'omic" approaches enable the investigation of biochemical function and systems-level microbial interactions. Whole-community functional assays, such as metagenomic, metatranscriptomic, metaproteomic, and metabolomics research, enables examination of complex ecosystems involving host organisms, biogeochemical conditions, pathogens, biochemistry, and metabolism, as well as their interconnections. Metatranscriptomics provides insight into the active functional profile of a microbial community by concentrating on the genes that are expressed by the microbial population. By capturing the entire mRNA, An overview of the gene expression in a sample at a specific time and under specific conditions can be obtained from the metatranscriptome [111]. The process of fully identifying and quantifying all of a sample's metabolites—small compounds that an organism releases into its immediate environment—is known as metabolomics [112]. Considering variations in the synthesis of different metabolites are linked to modifications in the functioning of metabolic pathways, metabolomics is a useful method for studying pathways [113]. There is a strong dependence of the microbiome-related metabolomic profile on environmental variables, offering important insights on the microbiome's properties as well as the ways in which the microbial community interacts with the host environment. Therefore, the goal of metabolomics is to enhance our knowledge of the function of the microbiome in the conversion of nutrients and toxins, along with other abiotic elements that could impact the homeostasis of the host ecosystem. Critical biogeochemical cycles are significantly impacted by microbial communities, and research on their metabolome can aid in the creation of prognostic biomarkers for environmental stressors. Nontargeted shotgun mass spectrometry is used in metaproteomics to evaluate the variety and quantity of proteins present in the microbiome. Thousands of proteins can be identified and measured in a single 24-h liquid chromatography–tandem mass spectrometry (LC–MS/MS) analysis from several dominant and subdominant naturally occurring microbial populations. Because it permits the characterization of expressed protein activities within communities and the identification of matches to microbial lineages, which may result in subspecies resolution, the identification of amino acid sequences and relative abundance data for identified peptides is important. It is possible to model microbiomes as heterogeneous networks in order to incorporate information from several -omic sources. Researchers could come up with intriguing new theories, including variables. Heterogeneous networks include the inference of cellular function through the utilisation of links between genetic relationships and protein–protein interactions also. In a study, these similar network types are combined with various co-expression network types to deduce gene dependencies and novel activities, like DNA damage repair [114]. To investigate gene regulation in the bacteria Mycobacterium TB, other

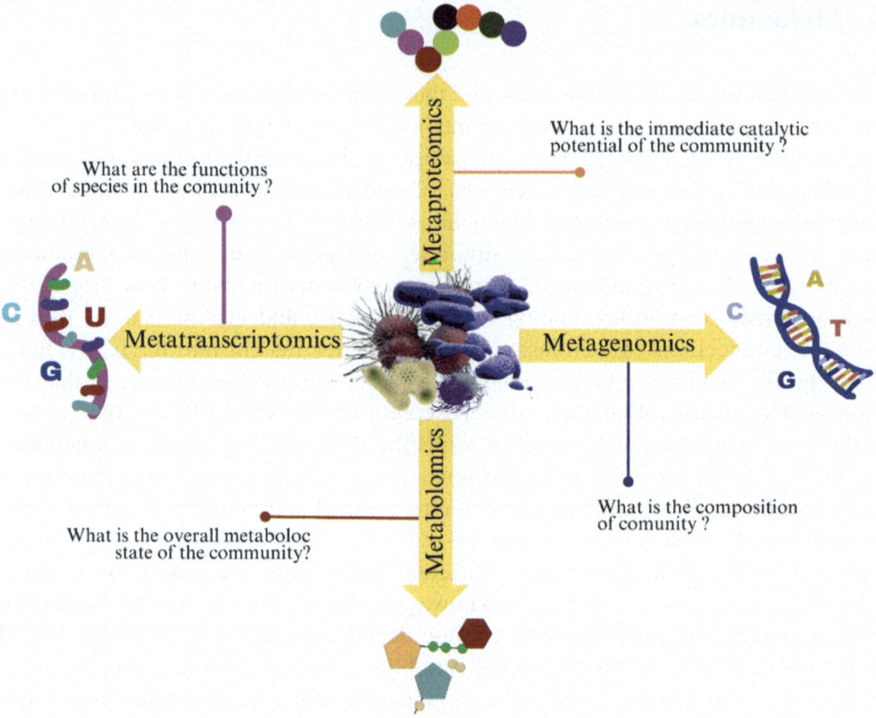

Fig. 3 Meta-omics subfields and the topics they cover

-omics network types were also combined [115] Fig. 3 provides an overview of the various subfields within meta-omics.

8 Challenges and Discussion

Because of fundamental properties of metagenomics data, genome-wide datasets provide several obstacles to researchers. One major obstacle that prevents contextual interpretation of these datasets is the lack of metadata that comes with them. Accurate analysis depends on an understanding of the environmental conditions during sample collection, which is frequently lacking. Plus, powerful computer resources and efficient algorithms are needed to process the enormous volume of sequence data generated by next-generation sequencing techniques. The diversity of species that comprise microbial communities, each with unique genetic characteristics, contributes to their complexity due to genomic heterogeneity. Resolving these challenges is necessary to derive significant insights from the data.

9 Conclusion

When it comes to environmental monitoring and bioremediation, metagenomics is a powerful tool. Metagenomics has completely changed the world of health and illness diagnostics by making it possible to identify viral diseases like SARS-CoV-2 using methods like metagenomic next-generation sequencing (mNGS). This method provides a thorough understanding of the viral genome, guaranteeing prompt and precise diagnosis. Metagenomics is also used in forensic sciences, where it helps identify microbiological residues on objects or at crime scenes. These traces may be very important pieces of evidence that connect suspects to places or occasions. Deciphering the significance of the human microbiome in health and disease requires an understanding of it through metagenomics. The management of gut health, immune response understanding, and personalized therapy can all benefit greatly from the insights gathered from microbiome investigations. Furthermore, metagenomics greatly aids in the discovery of enzymes, especially those having a variety of uses in the food, textile, and biofuel industries. This greatly advances industrial applications. Moreover, metagenomic research in harsh environments, such hot springs and deep-sea vents, reveals novel genes and metabolic pathways with potential for biotechnological uses through bioprospecting. Metagenomics' application even extends to astrobiological research. In short, metagenomics goes beyond the confines of the lab and affects a wide range of sectors, including environmental research, industry, medicine, and space exploration. Its capacity to reveal previously undiscovered microbial worlds keeps inspiring new studies and useful applications, making it a vital instrument in the modern scientist's toolbox.

Conflict of Interest Authors have no conflict of interest.

References

1. National Research Council (U.S.). (2007). Committee on metagenomics: Challenges and functional applications. In *The new science of metagenomics: Revealing the secrets of our microbial planet*. National Academies Press.
2. Pitt, T. L., & Barer, M. R. (2012). 3 Classification, identification and typing of micro-organisms.
3. Quince, C., Walker, A. W., Simpson, J. T., Loman, N. J., & Segata, N. (2017). Shotgun metagenomics, from sampling to analysis. *Nature Biotechnology, 35*(9), 833–844. https://doi.org/10.1038/nbt.3935
4. Sanger, F., Nicklen, S., & Coulson, A. R. (1977). *DNA sequencing with chain-terminating inhibitors (DNA polymerase/nucleotide sequences/bacteriophage 4X174)* (Vol. 74).
5. Illumina. *An introduction to next-generation sequencing technology*. www.illumina.com/technology/next-generation-sequencing.html.
6. ThermoFisher. Ion Torrent next-generation sequencing. Published 2021. Retrieved March 25, 2024, from https://www.thermofisher.com/in/en/home/brands/ion-torrent.html.
7. Luo C, Tsementzi D, Kyrpides N, Read T, Konstantinidis KT (2012) Direct comparisons of Illumina vs. Roche 454 sequencing technologies on the same microbial community

DNA sample. PLoS One. 2012;7(2):e30087. doi: 10.1371/journal.pone.0030087. Epub 2012 Feb 10. Erratum in: PLoS One 7(3). https://doi.org/10.1371/annotation/64ba358f-a483-46c2-b224-eaa5b9a33939. PMID: 22347999; PMCID: PMC3277595.

8. Garrido-Cardenas, J. A., Garcia-Maroto, F., Alvarez-Bermejo, J. A., Manzano-Agugliaro, F. (2017). DNA sequencing sensors: An overview. *Sensors (Switzerland), 17*(3). https://doi.org/10.3390/s17030588.
9. Oxford Nanopore Technologies. Nanopore DNA sequencing. Retrieved March 25, 2024, from https://nanoporetech.com/applications/dna-nanopore-sequencing.
10. Rhoads, A., & Au, K. F. (2015). PacBio sequencing and its applications. *Genomics, Proteomics & Bioinformatics, 13*(5), 278–289. https://doi.org/10.1016/j.gpb.2015.08.002
11. Dixon, W. G., Beukenhorst, A. L., Yimer, B. B., et al. (2019). How the weather affects the pain of citizen scientists using a smartphone app. *NPJ Digital Medicine, 2*(1). https://doi.org/10.1038/s41746-019-0180-3.
12. Zhou, Q., Su, X., & Ning, K. (2014). Assessment of quality control approaches for metagenomic data analysis. *Scientific Reports,* 4. https://doi.org/10.1038/srep06957.
13. Chen, S., Zhou, Y., Chen, Y., & Gu, J. (2018). Fastp: An ultra-fast all-in-one FASTQ preprocessor. In *Bioinformatics* (Vol. 34, pp. i884–i890). Oxford University Press. https://doi.org/10.1093/bioinformatics/bty560.
14. Andrews, S. (2010). FastQC: A quality control tool for high throughput sequence data. Babraham Bioinformatics. Published 2010. Accessed April 18, 2024. http://www.bioinformatics.babraham.ac.uk/projects/fastqc/. (Online)
15. Bolger, A. M., Lohse, M., & Usadel, B. (2014). Trimmomatic: A flexible trimmer for Illumina sequence data. *Bioinformatics, 30*(15), 2114–2120. https://doi.org/10.1093/bioinformatics/btu170
16. Chen, Y., Chen, Y., Shi, C., et al. (2018). SOAPnuke: A MapReduce acceleration-supported software for integrated quality control and preprocessing of high-throughput sequencing data. *Gigascience, 7*(1), 1–6. https://doi.org/10.1093/gigascience/gix120
17. Hufnagel, D. E., Hufford, M. B., Seetharam, A. S. (2020). SequelTools: A suite of tools for working with PacBio Sequel raw sequence data. *BMC Bioinformatics, 21*(1). https://doi.org/10.1186/s12859-020-03751-8.
18. Hackl, T., Hedrich, R., Schultz, J., & Förster, F. (2014). Proovread: Large-scale high-accuracy PacBio correction through iterative short read consensus. *Bioinformatics, 30*(21), 3004–3011. https://doi.org/10.1093/bioinformatics/btu392
19. De Coster, W., D'Hert, S., Schultz, D. T., Cruts, M., & Van Broeckhoven, C. (2018). NanoPack: Visualizing and processing long-read sequencing data. *Bioinformatics, 34*(15), 2666–2669. https://doi.org/10.1093/bioinformatics/bty149
20. Lanfear, R., Schalamun, M., Kainer, D., Wang, W., & Schwessinger, B. (2019). MinIONQC: Fast and simple quality control for MinION sequencing data. *Bioinformatics, 35*(3), 523–525. https://doi.org/10.1093/bioinformatics/bty654
21. Fukasawa, Y., Ermini, L., Wang, H., Carty, K., & Cheung, M. S. (2020). LongQC: A quality control tool for third generation sequencing long read data. *G3: Genes, Genomes, Genetics, 10*(4), 1193–1196. https://doi.org/10.1534/g3.119.400864.
22. Bonenfant, Q., Noe, L., & Touzet, H. (2023). Porechop ABI: Discovering unknown adapters in Oxford Nanopore Technology sequencing reads for downstream trimming. *Bioinformatics Advances, 3*(1). https://doi.org/10.1093/bioadv/vbac085.
23. Li, H., Ruan, J., & Durbin, R. (2008). Mapping short DNA sequencing reads and calling variants using mapping quality scores. *Genome Research, 18*(11), 1851–1858. https://doi.org/10.1101/gr.078212.108
24. Li, H., & Durbin, R. (2009). Fast and accurate short read alignment with Burrows-Wheeler transform. *Bioinformatics, 25*(14), 1754–1760. https://doi.org/10.1093/bioinformatics/btp324
25. Li, H. (2018). Minimap2: Pairwise alignment for nucleotide sequences. *Bioinformatics, 34*(18), 3094–3100. https://doi.org/10.1093/bioinformatics/bty191

26. Langmead, B., & Salzberg, S. L. (2012). Fast gapped-read alignment with Bowtie 2. *Nature Methods, 9*(4), 357–359. https://doi.org/10.1038/nmeth.1923
27. Shukla, N., Srivastava, N., Gupta, R., Srivastava, P., & Narayan, J. (2024). Setu: a pipeline for the robust assembly of SARS-CoV-2 genomes. *Microbiology Resource Announcements, 13*. e0023724. https://doi.org/10.1128/mra.00237-24.
28. Haider, B., Ahn, T. H., Bushnell, B., Chai, J., Copeland, A., & Pan, C. (2014). Omega: An overlap-graph de novo assembler for metagenomics. *Bioinformatics, 30*(19), 2717–2722. https://doi.org/10.1093/bioinformatics/btu395
29. Namiki, T., Hachiya, T., Tanaka, H., & Sakakibara, Y. (2012). MetaVelvet: An extension of Velvet assembler to de novo metagenome assembly from short sequence reads. *Nucleic Acids Research, 40*(20). https://doi.org/10.1093/nar/gks678.
30. Peng, Y., Leung, H. C. M., Yiu, S. M., & Chin, F. Y. L. (2012). IDBA-UD: A de novo assembler for single-cell and metagenomic sequencing data with highly uneven depth. *Bioinformatics, 28*(11), 1420–1428. https://doi.org/10.1093/bioinformatics/bts174
31. Li, D., Liu, C. M., Luo, R., Sadakane, K., & Lam, T. W. (2015). MEGAHIT: An ultra-fast single-node solution for large and complex metagenomics assembly via succinct de Bruijn graph. *Bioinformatics, 31*(10), 1674–1676. https://doi.org/10.1093/bioinformatics/btv033
32. Nurk, S., Meleshko, D., Korobeynikov, A., & Pevzner, P. A. (2017). MetaSPAdes: A new versatile metagenomic assembler. *Genome Research, 27*(5), 824–834. https://doi.org/10.1101/gr.213959.116
33. Boisvert, S., Raymond, F., Godzaridis, É., Laviolette, F., & Corbeil, J. (2012). Ray Meta: Scalable de novo metagenome assembly and profiling. *Genome Biology, 13*(12). https://doi.org/10.1186/gb-2012-13-12-r122.
34. Kuleshov, V., Jiang, C., Zhou, W., Jahanbani, F., Batzoglou, S., & Snyder, M. (2016). Synthetic long-read sequencing reveals intraspecies diversity in the human microbiome. *Nature Biotechnology, 34*(1), 64–69. https://doi.org/10.1038/nbt.3416
35. Koren, S., Walenz, B. P., Berlin, K., Miller, J. R., Bergman, N. H., & Phillippy, A. M. (2017). Canu: Scalable and accurate long-read assembly via adaptive κ-mer weighting and repeat separation. *Genome Research, 27*(5), 722–736. https://doi.org/10.1101/gr.215087.116
36. Ye, C., Hill, C. M., Wu, S., Ruan, J., & Ma, Z. (2016). DBG2OLC: Efficient assembly of large genomes using long erroneous reads of the third generation sequencing technologies. *Scientific reports,* 6. https://doi.org/10.1038/srep31900.
37. Wick, R. R., Judd, L. M., Gorrie, C. L., & Holt, K. E. (2017). Unicycler: Resolving bacterial genome assemblies from short and long sequencing reads. PLoS computational biology, 13(6). https://doi.org/10.1371/journal.pcbi.1005595.
38. Ayling, M., Clark, M. D., & Leggett, R. M. (2020). New approaches for metagenome assembly with short reads. *Briefings in Bioinformatics, 21*(2), 584–594. https://doi.org/10.1093/bib/bbz020
39. Dufault-Thompson, K., & Jiang, X. (2022). Applications of de Bruijn graphs in microbiome research. *iMeta, 1*(1). https://doi.org/10.1002/imt2.4.
40. Mikheenko, A., Saveliev, V., & Gurevich, A. (2016). MetaQUAST: Evaluation of metagenome assemblies. *Bioinformatics, 32*(7), 1088–1090. https://doi.org/10.1093/bioinformatics/btv697
41. Hunt, M., Kikuchi, T., Sanders, M., Newbold, C., Berriman, M., & Otto, T. D. (2013). REAPR: A universal tool for genome assembly evaluation. *Genome Biology, 14*(5). https://doi.org/10.1186/gb-2013-14-5-r47.
42. Olson, N. D., Treangen, T. J., Hill, C. M., et al. (2018). Metagenomic assembly through the lens of validation: Recent advances in assessing and improving the quality of genomes assembled from metagenomes. *Briefings in Bioinformatics, 20*(4), 1140–1150. https://doi.org/10.1093/bib/bbx098
43. Mineeva, O., Rojas-Carulla, M., Ley, R. E., Scholkopf, B., & Youngblut, N. D. (2020). DeepMAsED: Evaluating the quality of metagenomic assemblies. *Bioinformatics, 36*(10), 3011–3017. https://doi.org/10.1093/bioinformatics/btaa124
44. Parks, D. H., Imelfort, M., Skennerton, C. T., Hugenholtz, P., & Tyson, G. W. (2015). CheckM: Assessing the quality of microbial genomes recovered from isolates, single cells, and metagenomes. *Genome Research, 25*(7), 1043–1055. https://doi.org/10.1101/gr.186072.114

45. Yue, Y., Huang, H., Qi, Z., et al. (2020). Evaluating metagenomics tools for genome binning with real metagenomic datasets and CAMI datasets. *BMC Bioinformatics, 21*(1). https://doi.org/10.1186/s12859-020-03667-3.
46. Imelfort, M., Parks, D., Woodcroft, B. J., Dennis, P., Hugenholtz, P., & Tyson, G. W. (2014). GroopM: An automated tool for the recovery of population genomes from related metagenomes. *PeerJ, 2014*(1). https://doi.org/10.7717/peerj.603.
47. Wu, Y. W., Simmons, B. A., & Singer, S. W. (2016). MaxBin 2.0: An automated binning algorithm to recover genomes from multiple metagenomic datasets. *Bioinformatics, 32*(4), 605–607. https://doi.org/10.1093/bioinformatics/btv638.
48. Alneberg, J., Bjarnason, B. S., De Bruijn, I., et al. (2014). Binning metagenomic contigs by coverage and composition. *Nature Methods, 11*(11), 1144–1146. https://doi.org/10.1038/nmeth.3103
49. Kang, D. D., Li, F., Kirton, E., et al. (2019). MetaBAT 2: An adaptive binning algorithm for robust and efficient genome reconstruction from metagenome assemblies. *PeerJ, 2019*(7). https://doi.org/10.7717/peerj.7359.
50. Wang, Z., Wang, Z., Lu, Y. Y., Sun, F., & Zhu, S. (2019). SolidBin: Improving metagenome binning with semi-supervised normalized cut. *Bioinformatics, 35*(21), 4229–4238. https://doi.org/10.1093/bioinformatics/btz253
51. Blasse, C., Saalfeld, S., Etournay, R., Sagner, A., Eaton, S., & Myers, E. W. (2017). PreMosa: Extracting 2D surfaces from 3D microscopy mosaics. *Bioinformatics, 33*(16), 1–7. https://doi.org/10.1093/bioinformatics/xxxxxx
52. Mallawaarachchi, V., Wickramarachchi, A., & Lin, Y. (2020). GraphBin: Refined binning of metagenomic contigs using assembly graphs. *Bioinformatics, 36*(11), 3307–3313. https://doi.org/10.1093/bioinformatics/btaa180
53. Zhang, Z., & Zhang, L. (2021). METAMVGL: A multi-view graph-based metagenomic contig binning algorithm by integrating assembly and paired-end graphs. *BMC Bioinformatics,* 22. https://doi.org/10.1186/s12859-021-04284-4.
54. Nissen, J. N., Johansen, J., Allesøe, R. L., et al. (2021). Improved metagenome binning and assembly using deep variational autoencoders. *Nature Biotechnology, 39*(5), 555–560. https://doi.org/10.1038/s41587-020-00777-4
55. Murovec, B., Deutsch, L., & Stres, B. (2020). Computational framework for high-quality production and large-scale evolutionary analysis of metagenome assembled genomes. *Molecular Biology and Evolution, 37*(2), 593–598. https://doi.org/10.1093/molbev/msz237
56. Uritskiy, G. V., Diruggiero, J., & Taylor, J. (2018). MetaWRAP-A flexible pipeline for genome-resolved metagenomic data analysis 08 Information and Computing Sciences 0803 Computer Software 08 Information and Computing Sciences 0806 Information Systems. *Microbiome, 6*(1). https://doi.org/10.1186/s40168-018-0541-1.
57. Sieber, C. M. K., Probst, A. J., Sharrar, A., et al. (2018). Recovery of genomes from metagenomes via a dereplication, aggregation and scoring strategy. *Nature Microbiology, 3*(7), 836–843. https://doi.org/10.1038/s41564-018-0171-1
58. Bowers, R. M., Kyrpides, N. C., Stepanauskas, R., et al. (2017). Minimum information about a single amplified genome (MISAG) and a metagenome-assembled genome (MIMAG) of bacteria and archaea. *Nature Biotechnology, 35*(8), 725–731. https://doi.org/10.1038/nbt.3893
59. Almeida, A., Mitchell, A. L., Boland, M., et al. (2019). A new genomic blueprint of the human gut microbiota. *Nature, 568*(7753), 499–504. https://doi.org/10.1038/s41586-019-0965-1
60. Wood, D. E., & Salzberg, S. L. (2014). *Kraken: Ultrafast metagenomic sequence classification using exact alignments*. http://ccb.jhu.edu/software/kraken/.
61. Kim, D., Song, L., Breitwieser, F. P., & Salzberg, S. L. (2016). Centrifuge: Rapid and sensitive classification of metagenomic sequences. *Genome Research, 26*(12), 1721–1729. https://doi.org/10.1101/gr.210641.116
62. Yang, C., Chowdhury, D., Zhang, Z., et al. (2021). A review of computational tools for generating metagenome-assembled genomes from metagenomic sequencing data. *Computational and Structural Biotechnology Journal, 19*, 6301–6314. https://doi.org/10.1016/j.csbj.2021.11.028

63. Menzel, P., Ng, K. L., & Krogh, A. (2016) Fast and sensitive taxonomic classification for metagenomics with Kaiju. *Nature Communications,* 7. https://doi.org/10.1038/ncomms 11257.
64. Wood, D. E., Lu, J., & Langmead, B. (2019). Improved metagenomic analysis with Kraken 2. *Genome Biology, 20*(1). https://doi.org/10.1186/s13059-019-1891-0.
65. Lu, J., Breitwieser, F. P., Thielen, P., & Salzberg, S. L. (2017). Bracken: Estimating species abundance in metagenomics data. *PeerJ Computer Science, 2017*(1). https://doi.org/10.7717/peerj-cs.104.
66. Ounit, R., Wanamaker, S., Close, T. J., & Lonardi, S. (2015). CLARK: Fast and accurate classification of metagenomic and genomic sequences using discriminative k-mers. *BMC Genomics, 16*(1). https://doi.org/10.1186/s12864-015-1419-2.
67. Ainsworth, D., Sternberg, M. J. E., Raczy, C., & Butcher, S. A. (2017). K-SLAM: Accurate and ultra-fast taxonomic classification and gene identification for large metagenomic data sets. *Nucleic Acids Research, 45*(4), 1649–1656. https://doi.org/10.1093/nar/gkw1248
68. Beghini, F., McIver, L. J., Blanco-Míguez, A., et al. (2021). Integrating taxonomic, functional, and strain-level profiling of diverse microbial communities with biobakery 3. *Elife,* 10. https://doi.org/10.7554/eLife.65088.
69. Nayfach, S., Shi, Z. J., Seshadri, R., Pollard, K. S., & Kyrpides, N. C. (2019). New insights from uncultivated genomes of the global human gut microbiome. *Nature, 568*(7753), 505–510. https://doi.org/10.1038/s41586-019-1058-x
70. Zhu, W., Lomsadze, A., & Borodovsky, M. (2010). *Ab initio* gene identification in metagenomic sequences. *Nucleic Acids Research, 38*(12). https://doi.org/10.1093/nar/gkq275.
71. Kelley, D. R., Liu, B., Delcher, A. L., Pop, M., & Salzberg S. L. (2012). Gene prediction with Glimmer for metagenomic sequences augmented by classification and clustering. *Nucleic Acids Research, 40*(1). https://doi.org/10.1093/nar/gkr1067.
72. Rho, M., Tang, H., & Ye, Y. (2010). FragGeneScan: Predicting genes in short and error-prone reads. *Nucleic Acids Research, 38*(20). https://doi.org/10.1093/nar/gkq747.
73. Hyatt, D., Chen, G. L., Locascio, P. F., Land, M. L., Larimer, F. W., & Hauser, L. J. (2010). *Prodigal: Prokaryotic gene recognition and translation initiation site identification.* http://www.biomedcentral.com/1471-2105/11/119.
74. Noguchi, H., Park, J., & Takagi, T. (2006). MetaGene: Prokaryotic gene finding from environmental genome shotgun sequences. *Nucleic Acids Research, 34*(19), 5623–5630. https://doi.org/10.1093/nar/gkl723
75. Noguchi, H., Taniguchi, T., & Itoh, T. (2008). Meta gene annotator: Detecting species-specific patterns of ribosomal binding site for precise gene prediction in anonymous prokaryotic and phage genomes. *DNA Research, 15*(6), 387–396. https://doi.org/10.1093/dnares/dsn027
76. Zhang, S. W., Jin, X. Y., Zhang, T. (2017). Gene prediction in metagenomic fragments with deep learning. *BioMed Research International,* 2017. https://doi.org/10.1155/2017/4740354.
77. Al-Ajlan, A., & El Allali, A. (2019). CNN-MGP: Convolutional neural networks for metagenomics gene prediction. *Interdisciplinary Sciences, 11*(4), 628–635. https://doi.org/10.1007/s12539-018-0313-4
78. Sommer, M. J., & Salzberg, S. L. (2021). Balrog: A universal protein model for prokaryotic gene prediction. *PLoS Computational Biology, 17*(2). https://doi.org/10.1371/JOURNAL.PCBI.1008727.
79. Huerta-Cepas, J., Forslund, K., Coelho, L. P., et al. (2017). Fast genome-wide functional annotation through orthology assignment by eggNOG-mapper. *Molecular Biology and Evolution, 34*(8), 2115–2122. https://doi.org/10.1093/molbev/msx148
80. Kanehisa, M., Sato, Y., & Morishima, K. (2016). BlastKOALA and GhostKOALA: KEGG tools for functional characterization of genome and metagenome sequences. *Journal of Molecular Biology, 428*(4), 726–731. https://doi.org/10.1016/j.jmb.2015.11.006
81. Keegan, K. P., Glass, E. M., Meyer, F. (2016). MG-RAST, a metagenomics service for analysis of microbial community structure and function. In *Methods in molecular biology* (Vol. 1399, pp. 207–233). Humana Press Inc. https://doi.org/10.1007/978-1-4939-3369-3_13.

82. Törönen, P., Medlar, A., & Holm, L. (2018). PANNZER2: A rapid functional annotation web server. *Nucleic Acids Research, 46*(W1), W84–W88. https://doi.org/10.1093/nar/gky350
83. Quevillon, E., Silventoinen, V., Pillai, S., et al. (2005). InterProScan: Protein domains identifier. *Nucleic Acids Research, 33*(SUPPL. 2). https://doi.org/10.1093/nar/gki442.
84. Ciria, R., Abreu-Goodger, C., Morett, E., & Merino, E. (2004). GeConT: Gene context analysis. *Bioinformatics, 20*(14), 2307–2308. https://doi.org/10.1093/bioinformatics/bth216
85. Anand, S., Kuntal, B. K., Kuntal, B. K., et al. (2020). FunGeCo: A web-based tool for estimation of functional potential of bacterial genomes and microbiomes using gene context information. *Bioinformatics, 36*(8), 2575–2577. https://doi.org/10.1093/bioinformatics/btz957
86. Saha, C. K., Sanches Pires, R., Brolin, H., Delannoy, M., & Atkinson, G. C. (2020). Brief communication predicting functional associations using Flanking Genes (FlaGs). May 02, 2020. https://doi.org/10.1101/362095
87. Woese, C. R. (1987). *Bacterial evolution* (Vol. 51).
88. Rusch, D. B., Halpern, A. L., Sutton, G., et al. (2007). The sorcerer II global ocean sampling expedition: Northwest Atlantic through eastern tropical Pacific. *PLoS Biology, 5*(3), 0398–0431. https://doi.org/10.1371/journal.pbio.0050077
89. Amann, R. I., Ludwig, W., & Schleifer, K. H. (1995). *Phylogenetic identification and in situ detection of individual microbial cells without cultivation* (Vol. 59).
90. Sogin, M. L., Morrison, H. G., Huber, J. A., et al. (2006). *Microbial diversity in the deep sea and the underexplored* "'*Rare Biosphere.*'". https://doi.org/10.1073/pnas.0605127103.
91. Jones, R. T., Robeson, M. S., Lauber, C. L., Hamady, M., Knight, R., & Fierer, N. (2009). A comprehensive survey of soil acidobacterial diversity using pyrosequencing and clone library analyses. *ISME Journal, 3*(4), 442–453. https://doi.org/10.1038/ismej.2008.127
92. Luna, R. A., Fasciano, L. R., Jones, S. C., Boyanton, B. L., Ton, T. T., & Versalovic, J. (2007). DNA pyrosequencing-based bacterial pathogen identification in a pediatric hospital setting. *Journal of Clinical Microbiology, 45*(9), 2985–2992. https://doi.org/10.1128/JCM.00630-07
93. Cole, J. R., Wang, Q., Fish, J. A., et al. (2014). Ribosomal database project: data and tools for high throughput rRNA analysis. *Nucleic Acids Research, 42*(D1). https://doi.org/10.1093/nar/gkt1244.
94. Callahan, B. J., McMurdie, P. J., & Holmes, S. P. (2017). Exact sequence variants should replace operational taxonomic units in marker-gene data analysis. *ISME Journal, 11*(12), 2639–2643. https://doi.org/10.1038/ismej.2017.119
95. Colwell, R. K. (2014). III.1 Biodiversity: Concepts, patterns, and measurement. In *The princeton guide to ecology* (pp. 257–263). Princeton University Press. https://doi.org/10.1515/9781400833023.257.
96. Sala, C., Vitali, S., Giampieri, E., et al. (2016). Stochastic neutral modelling of the Gut Microbiota's relative species abundance from next generation sequencing data. *BMC Bioinformatics, 17*(2). https://doi.org/10.1186/s12859-015-0858-8.
97. Peterson, D., Bonham, K. S., Rowland, S., et al. (2021). Comparative analysis of 16S rRNA gene and metagenome sequencing in pediatric Gut Microbiomes. *Frontiers in Microbiology,* 12. https://doi.org/10.3389/fmicb.2021.670336.
98. Clarridge, J. E. (2004). Impact of 16S rRNA gene sequence analysis for identification of bacteria on clinical microbiology and infectious diseases. *Clinical Microbiology Reviews, 17*(4), 840–862. https://doi.org/10.1128/CMR.17.4.840-862.2004
99. Quast, C., Pruesse, E., Yilmaz, P., et al. (2013). The SILVA ribosomal RNA gene database project: Improved data processing and web-based tools. *Nucleic Acids Research, 41*(D1). https://doi.org/10.1093/nar/gks1219.
100. DeSantis, T. Z., Hugenholtz, P., Larsen, N., et al. (2006). Greengenes, a chimera-checked 16S rRNA gene database and workbench compatible with ARB. *Applied and Environment Microbiology, 72*(7), 5069–5072. https://doi.org/10.1128/AEM.03006-05
101. Leinonen, R., Akhtar, R., Birney, E., et al. (2011). The European nucleotide archive. *Nucleic Acids Research, 39*(SUPPL. 1). https://doi.org/10.1093/nar/gkq967.

102. Nilsson, R. H., Larsson, K. H., Taylor, A. F. S., et al. (2019). The UNITE database for molecular identification of fungi: Handling dark taxa and parallel taxonomic classifications. *Nucleic Acids Research, 47*(D1), D259–D264. https://doi.org/10.1093/nar/gky1022
103. Santamaria, M., Fosso, B., Licciulli, F., et al. (2018). ITSoneDB: A comprehensive collection of eukaryotic ribosomal RNA Internal Transcribed Spacer 1 (ITS1) sequences. *Nucleic Acids Research, 46*(D1), D127–D132. https://doi.org/10.1093/nar/gkx855
104. Shannon, C. E. (1948). *A mathematical theory of communication.*
105. Bray, J. R., Curtis, J. T., & Roger, J. (1957). *This content downloaded from 147.8.31.43 on Mon.* (Vol. 27).
106. Callahan, B. J., McMurdie, P. J., Rosen, M. J., Han, A. W., Johnson, A. J. A., & Holmes, S. P. (2016). DADA2: High-resolution sample inference from Illumina amplicon data. *Nature Methods, 13*(7), 581–583. https://doi.org/10.1038/nmeth.3869
107. Caporaso, J. G., Kuczynski, J., Stombaugh, J., et al. (2010). QIIME allows analysis of high-throughput community sequencing data. *Nature Methods, 7*(5), 335–336. https://doi.org/10.1038/nmeth.f.303
108. Schloss, P. D., Westcott, S. L., Ryabin, T., et al. (2009). Introducing mothur: Open-source, platform-independent, community-supported software for describing and comparing microbial communities. *Applied and Environment Microbiology, 75*(23), 7537–7541. https://doi.org/10.1128/AEM.01541-09
109. Blanco-Míguez, A., Beghini, F., Cumbo, F., et al. (2023). Extending and improving metagenomic taxonomic profiling with uncharacterized species using MetaPhlAn 4. *Nature Biotechnology, 41*(11), 1633–1644. https://doi.org/10.1038/s41587-023-01688-w
110. Edgar, R. C. (2010). Search and clustering orders of magnitude faster than BLAST. *Bioinformatics, 26*(19), 2460–2461. https://doi.org/10.1093/bioinformatics/btq461
111. Moran, M. A. (2009). Metatranscriptomics: Eavesdropping on complex microbial communities. *Microbe. 4*, 329–335. https://doi.org/10.1128/microbe.4.329.1.
112. Fiehn, O. (2002). Metabolomics-the link between genotypes and phenotypes. *Plant molecular Biology, 48*, 155–71. https://doi.org/10.1023/A:1013713905833.
113. van Dam, J. C. J., Schaap, P. J., Martins dos Santos, V. A. P., & Suárez-Diez, M. (2014). Integration of heterogeneous molecular networks to unravel gene-regulation in Mycobacterium tuberculosis. *BMC System Biology, 8*, 111. https://doi.org/10.1186/s12918-014-0111-5.
114. Amar, D., & Shamir, R. (2014). Constructing module maps for integrated analysis of heterogeneous biological networks. *Nucleic Acids Research, 42*(7), 4208–4219. https://doi.org/10.1093/nar/gku102
115. Mahajna, A., Dinkla, I. J. T, Euverink, G. J. W., Keesman, K. J., & Jayawardhana, B. (2022). Clean and safe drinking water systems via metagenomics data and artificial intelligence: State-of-the-art and future perspective. *Frontiers in Microbiology*, 13. https://doi.org/10.3389/fmicb.2022.832452.

Next-Generation Sequencing and Genomic Data Analysis

Vivek Kumar, Surabhi Gangani, Rohit Shukla, Santosh Kumar Prajapati, Himanshu Shekhar, and Vaishali Shukla

Abstract Next-generation sequencing (NGS) technologies have revolutionized the field of genomics, enabling rapid and cost-effective sequencing of DNA/RNA molecules at unprecedented scales. This chapter presents a comprehensive overview of NGS methodologies and genomic data analysis techniques, highlighting recent advancements and future prospects. We delve into the principles underlying NGS platforms, including Illumina sequencing, Ion Torrent sequencing, and Pacific Biosciences Single Molecule Real-Time (SMRT) sequencing, elucidating their strengths and limitations. In parallel, this abstract explores the challenges and opportunities in genomic data analysis, encompassing a spectrum of computational methods for data preprocessing, alignment, variant calling, and functional annotation. We emphasize the significance of bioinformatics pipelines in handling large-scale genomic datasets, integrating tools for quality control, read mapping, and variant detection. Furthermore, we elucidate the role of machine learning and artificial intelligence in genomic data analysis, elucidating their applications in genotype–phenotype associations, population genetics, and personalized medicine. Through this abstract, we aim to provide a roadmap for researchers and practitioners navigating the complex landscape of NGS and genomic data analysis, fostering a deeper understanding of the underlying technologies and methodologies driving genomic research in the next decade.

V. Kumar (✉)
Molecular Biology Department, Lab MicroCare Diagnostics India, New Delhi 110063, India
e-mail: vivek.kumar2012@vitalum.ac.in

V. Kumar · S. Gangani · R. Shukla · S. K. Prajapati
USF Center for Microbiome Research, Institute of Microbiomes, University of South Florida Morsani College of Medicine, Tampa, FL 33612, USA

Center of Excellence for Aging and Brain Repair, University of South Florida Morsani College of Medicine, Tampa, FL, USA

H. Shekhar
Biotechnology Division, CSIR-IHBT, Palampur 176061, Himachal Pradesh, India

V. Shukla
Department of Molecular Medicine, Sanjay Gandhi Post Graduate Institute of Medical Sciences, Uttar Pradesh, Lucknow 226014, India

© The Author(s), under exclusive license to Springer Nature Singapore Pte Ltd. 2024

A. Chaudhary et al. (eds.), *Unraveling New Frontiers and Advances in Bioinformatics*,
https://doi.org/10.1007/978-981-97-7123-3_3

Keywords Sequencing · Bowtie2 · Variant calling · Epigenomics · scRNA sequencing

1 Introduction

The ability to sequence DNA and RNA quickly and efficiently has transformed genomics research because of next-generation sequencing (NGS) technologies [1]. NGS technologies greatly improve the rate and efficacy of sequencing operations by enabling the simultaneous sequencing of millions to billions of DNA fragments or RNA transcripts, in contrast with conventional Sanger sequencing. Unprecedented progress has been made in several scientific and medical research fields as a result, including metagenomics, cancer genomics, genome sequencing, gene expression analysis, and epigenetics [2–4].

Numerous NGS platforms are available, each with its sequencing chemicals, read lengths, productivity, and uses. These platforms include Oxford Nanopore Technologies, Illumina, Ion Torrent, and Pacific Biosciences (PacBio). For example, Illumina sequencing uses reversible terminator chemistry to produce very accurate short sequencing reads, which makes it appropriate for uses like RNA-seq and whole-genome sequencing [5]. Contrarily, PacBio sequencing uses single-molecule [6], real-time (SMRT) sequencing technology to generate lengthy sequencing reads, which makes it possible to discover structural variations and epigenetic alterations as well as sequence complicated genomic areas [7].

NGS has many broad and varied applicability. NGS has made it easier to identify disease-causing variants in clinical biology, which has aided in the diagnosis and individualized treatment of genetic illnesses. NGS has transformed our knowledge of drug resistance, metastatic disease, and variability in tumors in cancer research, which has sparked the creation of precise cancer treatment techniques and targeted medicines. NGS is applied in agriculture to increase crop quality, disease resistance, and production through marker-assisted selection (MAS) and genome-wide association studies (GWAS).

NGS has the potential to be transformative, but it also has drawbacks, such as complicated data processing, problems with data administration and storage, and problems with quality control. Nevertheless, continuous developments in the fields of bioinformatics, sequencing methods, and computational power promise to solve these issues, hence increasing the applications and capabilities of next-generation sequencing (NGS) in research and beyond.

2 Different Platforms for NGS Sequencing

Considering the unparalleled speed, precision, and affordability of high-quality sequencing of DNA and RNA, next-generation sequencing (NGS) has completely changed the field of genomics research. Numerous platforms with distinct sequencing chemical read lengths, throughput, and uses are currently being developed. Below are a few well-known NGS platforms.

2.1 Illumina Sequencing

Next-generation sequencing (NGS) technologies, such as Illumina sequencing, are extensively utilized and have transformed the field of genomics by offering accurate, efficient, rapid sequencing of DNA and RNA molecules at reasonable prices. Reversible terminator chemistry, which is used by Illumina sequencing platforms, allows for the concurrent simultaneous sequencing of millions to billions of DNA pieces or RNA transcripts [8]. The following are some of the main attributes and elements of Illumina sequencing:

- Reversible terminator chemistry, used in Illumina sequencing, involves incorporating fluorescently tagged nucleotide time into the expanding DNA sequence.
- A reversible terminator group is present in every nucleotide, which stops additional nucleotide incorporation following each cycle of nucleotide addition.
- The integrated nucleotide (A, T, C, or G) is identified by detecting the fluorescence light it emits.

A variety of platforms with different capacities read sizes, and purposes are available from Illumina [9–11].

- HiSeq Platform: High-throughput devices [12] that can produce substantial amounts of sequencing data, which includes HiSeq X, HiSeq 4000, and HiSeq 2500 series
- NovaSeq Series: offers great throughput and cost-effectiveness while being flexible and scalable
- NextSeq Series: Adaptable systems ideal for targeted, RNA-seq, and whole-genome sequencing
- MiSeq Series: Accurate and straightforward benchtop sequencers perfect for smaller-scale tasks and quick delivery times.

2.2 Ion Torrent Sequencing

ThermoFisher Scientific created the NGS technology known as ion torrent sequencing. It works based on semiconductor sequencing, which detects the DNA sequence by measuring the release of hydrogen ions (H+) during nucleotide addition. Ion torrent sequencing depends on detecting pH shifts brought about by hydrogen ions (H+) being released throughout DNA amplification [13]. Whenever dNTPs are integrated into the developing DNA strand, Ion Torrent instantly monitors the ion discharge that takes place, in contrast to other NGS technologies that analyze fluorescence signals. The Ion Torrent-based platforms are Ion GeneStudio S5 Series, Ion GeneStudio S5 Prime System, Ion GeneStudio S5 System, Ion GeneStudio S5 Plus System, and Ion Torrent Genexus System. The usage and application of each platform vary and depend on the individual study design [11, 14–16].

2.3 PacBio Sequencing

The single-molecule real-time (SMRT) sequencing technology used by Pacific Biosciences (PacBio) sequencing is a next-generation sequencing (NGS) [17]. PacBio sequencing is not dependent on optical fluorescence identification or amplification like other sequencing systems. Rather, it tracks the real-time amplification of DNA. During this sequencing, DNA is fragmented, and adaptors are ligated at both ends of fragmented DNA and further ligated on the nanofabricated array. Next, the template immobilized on the array is amplified using fluorescently labeled dNTPs by polymerase enzyme and each added dNTPs are detected in real-time [18–21].

2.4 Oxford Nanopore Sequencing

Oxford Nanopore sequencing operates by guiding DNA or RNA molecules through a tiny pore embedded in a synthetic membrane. As each nucleotide passes through the nanopore, it disrupts an electrical current flowing through the pore characteristically [22]. This disruption is detected and measured, providing a real-time signal that reflects the sequence of bases (adenine, guanine, cytosine, thymine for DNA; uracil replaces thymine in RNA). Special enzymes are used to control the rate at which the nucleic acid strands move through the nanopore, ensuring that the sequencing process is steady and that the electrical signals are clear enough to be interpreted [23]. This method allows for the direct sequencing of strands, and the technology is capable of generating long continuous reads which are advantageous for mapping complex regions of genomes. The entire process is supported by sophisticated software that translates these electrical signals into genetic information, making Oxford Nanopore sequencing a powerful tool for genomic research and a variety of practical applications [24–26].

3 Different Types of NGS Sequencing

Next-Generation Sequencing (NGS) comprises various advanced technologies that have significantly advanced genetic research by facilitating fast DNA and RNA sequencing. Whole Genome Sequencing (WGS) provides a comprehensive analysis by sequencing the entire genome, offering insights into genetic variations for applications ranging from evolutionary studies to personalized medicine [27]. Whole Exome Sequencing (WES focuses specifically on the exons, the coding regions of the genome, making it an efficient method for identifying disease-related genetic mutations due to its focus on regions known to harbor about 85% of disease-related variants [28]. RNA Sequencing (RNA-seq) dives into the transcriptome, revealing RNA expression levels and variants, crucial for understanding gene expression across different conditions and developmental stages [29]. Single-cell sequencing (scRNA-seq) allows for the examination of genomic and transcriptomic data at the individual cell level, providing detailed insights into cell-to-cell variability and aiding in the discovery of new cell types and regulatory mechanisms [30]. Epigenome sequencing techniques, including Bisulfite sequencing and ChIP-Seq, explore modifications that regulate gene expression without altering the DNA sequence, shedding light on how genetic control is exerted at the molecular level [31]. Lastly, ATAC-seq identifies regions of open chromatin across the genome, pinpointing active regulatory elements in various biological states and cell types [32]. Collectively, these NGS methods are pivotal in expanding our understanding of the intricate biological mechanisms that govern life.

3.1 Whole Genome Sequencing Analysis

The rapid expansion of Next-Generation Sequencing (NGS) data has spurred the development and release of various genomic analysis tools in recent times. A plethora of workflows are now accessible for handling high-throughput genome sequencing data [33], spanning different formats such as shell scripts, tool-specific interfaces, and graphical workflow environments [34–36]. A standard computational pipeline for Whole Genome Sequencing typically includes data preparation, alignment and assembly, variant identification, annotation, and analysis. In a wet lab, sample preparation for Whole Genome Sequencing (WGS) analysis involves several critical steps to ensure high-quality DNA extraction, which is essential for obtaining reliable sequencing results. The process starts with the collection of a biological sample, which could be blood, saliva, tissue, or another source containing cells of interest. Once collected, the sample undergoes DNA extraction using methods such as column-based purification or magnetic bead-based techniques, which help to isolate DNA while minimizing contamination and degradation. The extracted DNA must then be quantified and its purity assessed, typically using spectrophotometry or fluorometry, to ensure it meets the specific requirements for sequencing—such as minimal protein contamination and appropriate DNA concentration [37]. Following quantification, the DNA is often fragmented to a specific size range, suitable for the sequencing platform being used, using physical methods like sonication or enzymatic digestion [38]. These DNA fragments are then subjected to library preparation steps, which include the ligation of platform-specific adapters and possibly amplification steps to increase the amount of DNA available for sequencing. The final library is then validated for quality and size distribution before being loaded onto a sequencer, where millions of simultaneous sequencing reactions translate the genetic information into data that can be further analyzed computationally.

Commercial DNA kits for Whole Genome Sequencing (WGS) significantly streamline the processes of DNA extraction, preparation, and sequencing, thus making WGS more accessible for both clinical and research applications. These kits typically include reagents and optimized protocols to ensure high-quality and high-yield DNA extraction from various sample types such as blood, saliva, or tissue. For instance, Illumina offers the TruSeq DNA PCR-Free Library Preparation Kit, which minimizes PCR amplification steps to reduce potential bias and preserve the integrity of the sequence, a crucial factor for clinical and sensitive applications (Illumina, Inc.). Another key player, QIAGEN, provides the QIAamp DNA Mini Kit, designed to efficiently extract DNA from bodily fluids and tissues, making the DNA suitable for direct use in sequencing reactions, including WGS (QIAGEN N.V.). Thermo Fisher Scientific enhances the integration of workflow with the Ion Torrent Genexus Integrated Sequencer, which combines DNA extraction, library preparation, sequencing, and initial data analysis in a single platform, beneficial especially in clinical settings for its rapid turnaround (Thermo Fisher Scientific). Additionally, NEB (New England Biolabs) offers the NEBNext Ultra II DNA Library Prep Kit for Illumina, which is designed to handle a wide range of DNA input quantities and qualities, thereby supporting various research and clinical needs (New England Biolabs). These kits are pivotal in promoting the adoption of WGS, enabling researchers and clinicians to undertake complex genetic analyses and extensive studies with efficiency and reliability. Each kit is enhanced with specific protocols and proprietary technologies to improve various sequencing workflow stages, from DNA extraction to data analysis, thus meeting diverse user requirements.

Performing library preparation, conducting quality control (QC) checks, and sequencing for Whole Genome Sequencing (WGS) involve a series of detailed and critical steps that require precision and accuracy. Library preparation is the first step, where DNA is fragmented, and sequencing adapters are ligated to the ends of these fragments. This process can vary based on the DNA input and the sequencing platform used. For instance, the Illumina TruSeq DNA PCR-Free Library Prep Kit is favored for its ability to handle a wide range of input DNA without the need for PCR amplification, which helps to reduce bias and enhance the integrity of the sequence (Illumina, Inc.). After library preparation, QC checks are essential to confirm that the library is suitable for sequencing. Techniques such as gel electrophoresis, and more frequently, advanced methods like the Agilent 2100 Bioanalyzer or the Qubit fluorometer, are used to assess the size distribution and concentration of the prepared

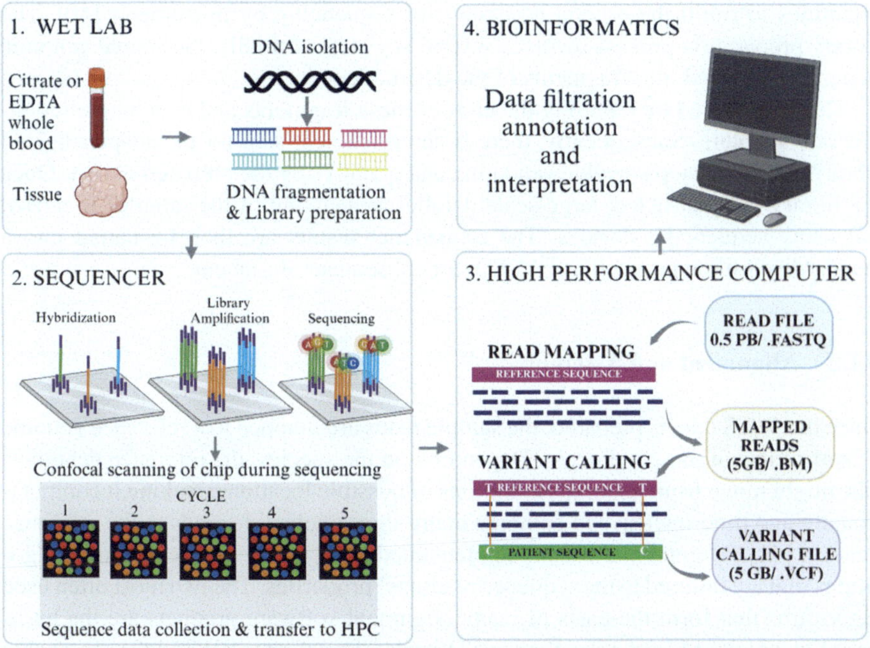

Fig. 1 Schematic representation of the WGS laboratory and bioinformatics flow

libraries, ensuring they meet the necessary criteria for sequencing (Agilent Technologies; Thermo Fisher Scientific). Following successful QC, sequencing can be performed. Modern high-throughput platforms such as the Illumina NovaSeq or Thermo Fisher Ion Torrent systems offer robust scalability and flexibility, making them suitable for various genomic research applications. These platforms facilitate the sequencing process by generating massive amounts of data that can be analyzed to provide comprehensive genomic insights (Illumina, Inc.; Thermo Fisher Scientific). Each of these steps must be meticulously carried out to ensure the generation of high-quality data that can support detailed and accurate genomic analysis. Figure 1 illustrates the laboratory and bioinformatics workflow for whole genome sequencing (WGS).

3.1.1 Data Preparation

The initial phase in the whole genome sequencing (WGS) process involves preparing the data, specifically through sample preparation. Nucleotide samples differ widely in their source and quality, necessitating the use of specialized kits to extract the nucleic acid following purification. To sequence these, the purified genomic material must be organized into a sample library consisting of short fragments [39]. Sample libraries typically need to be of high quality and contain intact sequences in adequate quantities to fulfill the criteria necessary for sequencing by instruments [40]. The library preparation process involves several key steps. Initially, the source genomic material is divided into fragments of the desired lengths.

This is followed by repairing the ends of these fragments and attaching oligonucleotide adapters. Subsequently, there is an enrichment step for the adapter-ligated library. The final steps involve validating and quantifying the prepared library. Once the library is constructed, large-scale parallel sequencing of the samples is carried out using sequencing devices. The sequencing results are then formatted into a standardized file type, such as FASTQ, for subsequent alignment.

3.1.2 Alignment and Assembly

Once the WGS data is prepared, the sample reads are mapped to a reference genome as part of the alignment process. Depending on the species, the reference genome's size might range from hundreds to billions of possible locations, making it computationally and time-intensive to compare many short nucleotide reads to the reference genome [41]. Alignment is usually implemented using one or more customized algorithms that are tailored to the sequence readings' properties. The two most often used algorithms that form the basis of many alignment software products are the hash-based technique [42] and the Burrows-Wheeler Transform (BWT) [43]. To enable scanning in either direction, the hash-based approach creates a hash table using either the genomic reference or the incoming reads. Programs like SHRiMP and mrFAST use the hash table technique [43–45]. Another prevalent technique is the Burrows-Wheeler Transform (BWT), which constructs a suffix array through the BWT to generate transformed sequences that serve as a new reference genome for mapping. The BWT method, known for its speed, outperforms the hash-based method and is implemented in tools like BOWTIE/BOWTIE2, SOAP2, and BWA [46–48] additionally, a wider range of tools are listed in Table 1 alongside these software packages. Whole genome assembly takes place following the sequence alignment stage. Rebuilding the original sequences into lengthy, uninterrupted segments that can be methodically aligned and placed along each chromosome [42] is the goal of this approach. Longer, continuous sequences are mostly assembled via de novo assembly and reference-based assembly.

De novo assembly uses computational techniques for alignment overlapping whole genome sequencing reads to construct known reads into contigs, and then sequences these contigs to structure the genome framework [49]. Reference-based assembly includes mapping each read to a reference genome sequence and constructing a consensus sequence, which is analogous, though not certainly same, to the reference backbone. Numerous tools utilizing these two algorithms have been made available online for assembly purposes [50]. Numerous online tools have been developed for assembly using the two algorithms. De novo assembly can be performed using two types of methods: Overlap graphs [51] and de Bruijn graphs [52].

Table 1 List of tools for aligning the DNA sequence reads to the reference genome

Tools/ software	About	Sources
MAQ	Construct assembly maps from short reads produced by next-generation sequencing (NGS) machines	https://maq.sourceforge.net/
ELAND	Alignment software using hash based techniques for short oligonucleotides	https://www.illumina.com/
Bowtie/ Bowtie2	A short-read aligner that maps short DNA reads to the genome	http://bowtiebio.sourceforge.net/index.shtml
SHRiMP	Sequence alignment tool designed to handle the numerous short reads produced by next-generation sequencing (NGS)	http://compbio.cs.toronto.edu/shrimp/
mrFAST	Seed-and-extend alignment software for mapping short reads using the Illumina platform	https://mrfast.sourceforge.net/
SOAP	A GPU-powered tool to align short reads to a reference sequence	http://soap.genomics.org.cn/index.html
BWA	A software suite designed to map low-divergence sequences versus a large reference genome	https://github.com/lh3/bwa
SEAL	A group of dispersed programs intended for managing, evaluating, and aligning short DNA readings respectively	https://biodoop-seal.sourceforge.net/
BLAST	A set of algorithm designed to align query sequences within a chosen target database	https://blast.ncbi.nlm.nih.gov/Blast.cgi

The overlap graph approach, utilized by software such as Canu [53] and Arachne [54] calculates all the pairwise similarities among reads and displays this data in a chart format. In contrast, the de Bruijn graph method divides reads into k-mers to create network nodes, where 'k' represents the length of these subsequences in bases. Software usually employing this technique includes Velvet [55], SOAPdenovo [56], and ABySS [57]. Compared to de-novo assemblers, reference-based assemblers need less computational power; however, they face greater challenges when encountering unique sequences. Models of reference-based assemblers comprise MAQ, SeqMap, and RMAP [42, 58, 59].

Quality control of sequence reads is critical throughout the entire process, from sampling through alignment and assembly to variant calling and annotation. Various complications arise during the processing of genomic datasets. For instance, misalignment during alignment and assembly can lead to systematic sequencing errors or errors related to the sequencing machine's cycle [60]. These problems may affect the variant calling's accuracy by affecting the quality score or base. For quality control, it is important to combine the aligned reads into the Sequence Alignment/Map or Binary Sequence Alignment/Map format [40]. This step ensures the output is a clean, well-organized, and indexed file ready for further analysis. Additionally, in some WGS workflows, there is a need for advanced quality control measures to recalibrate the base quality scores of sequence reads to account for variations.

3.1.3 Variant Calling

Following the alignment and assembling of short reads, the subsequent critical phase in the WGS pipeline is variant calling. Variant calling determines the presence of a discrepancy between the targeted sequence and the references at a specific location within an entity's genome. This process is carried out subsequently after the genomic sequence alignment and assembly. Further, the results of variant calling are recorded in a uniform structure known as the variant call format. The aligned sequence that diverges from the genomic reference indicates variants that may be associated with certain diseases or could simply be benign natural mutations with no functional impact on the cell [61]. Table 2 displays various types of genetic variations, including SNPs/variants, indels, larger structural variants, and CNVs, based on their characteristics. A summary of these four types of variant calling is provided, along with a list of the tools used for executing variant calling. SNP calling aims to delineate the genetic links between people [62] by identifying places where polymorphisms exist or at least one base differs from the reference genome. High-quality sequencing datasets, broad coverage, and an all-encompassing bioinformatics approach are necessary for accurate SNP naming and statistically significant SNP detection. Similar to this, another often used kind of variation calling is indel calling, which entails the insertion or deletion of nucleotides in an organism's genome [63]. Accurate indel calling presents more challenges than SNP calling due to indels occurring at a rate nearly eight times lower than SNPs. Various methods are used for both SNP and indel calling, with the Bayesian probabilistic method being the most common in software packages. This approach calculates the probability scores for each genotype and selects the genotype with the utmost probability at a given position [64]. The tools mentioned are GATK, SOAPsnp, SAMtools, and Isaac Variant Caller [43, 46, 65, 66]. Furthermore, the pattern growth method is the alternative prevalent process for indel calling, which involves analyzing pieces from paired-end reads in comparison to the reference genome, as seen in tools like Dindel [67].

Additionally, the process of calling SVs/CNVs is akin to calling SNPs and indels. Structural variants can affect large portions of the genome, potentially disrupting genes and regulatory elements linked to genetic disorders and utilized as biomarkers in the clinical setting. To identify large-scale structural variants, which introduce variations in the sequence through deletions, insertions, and rearrangements of genomic material, the method relies on analyzing discordant fragments based on paired-end reads. By mapping each end of the paired-end read and interpreting discordant fragments into potential breakpoints, the type of structural variant can be inferred [68]. For instance, chromosomal translocations represent a type of large-scale structural variant that involves the rearrangement of chromosomes. These can unsettle DNA and lead to the creation of new fusion products. Numerous characteristic translocations have been recognized as mutations that contribute to the onset of diseases. The

Table 2 List of Variant calling packages aimed for uncovering of CNVs, indels, SVs, and SNVs

Type of calling	Software	Description	Resources
SNP/ Indel	GATK	A set of command-line tools for variant discovery analysis	https://gatk.broadinstitute.org/hc/en-us
	SAMtools	A collection of software tools for handling high-throughput sequencing data	https://www.htslib.org/
	Issac Variant Caller	Identifies single nucleotide variants (SNVs) and indels from the aligned sequencing reads of a diploid sample	https://github.com/
	SOAPsnp	Software tool for identifying somatic mutations in SNVs through resequencing	http://soap.genomics.org.cn/soapsnp.html
	Dindel	A Bayesian approach for identifying indels from short-read sequence data in individuals	https://github.com/genome/dindel-tgi
SV/ CNV	Breakdancer	Identify indels from 10 base pairs up to 1 megabase using a unified traditional method	https://gmt.genome.wustl.edu/packages/breakdancer/
	iSVP	A pipeline that utilizes various tools to detect structural variants	https://nagasakilab.csml.org/en/isvp
	Pindel	Identifies breakpoints of large deletions, medium-sized insertions, and inversions	https://github.com/genome/pindel
	SvABA	Identifies structural variants from short-read sequencing data through genome-wide local assembly	https://github.com/walaj/svaba
	SVMerge	A pipeline identifies structural variants by aggregating results from existing SV calling tools	https://svmerge.sourceforge.net/

techniques used to identify relevant variants primarily include read pair, split read, read depth, and de novo assembly methods [44, 69] Nonetheless, there remains a shortage of bioinformatic frameworks for SV/CNV calling. To date, barely a limited number of tools are accessible, including Breakdancer, iSVP, Pindel, SvABA, and SVMerge [70–73]. A list of tools available for variant calling is depicted in Table 2.

3.1.4 Annotation and Analysis

After variant calling is finished, the subsequent essential stage is to annotate the target genome in comparison with the reference genome to identify possible variances. Genome annotation involves recognizing segments of the genome that represent well-known and likely open reading frames, plus aligning these sections with gene sequences found in current databases [41]. This process requires the addition of biological information to the annotations. The range of biological data spans from gene models to gene functions, comprising resources like the KEGG [74] and Gene Ontology [75]. A typical genome annotation effort involves extensive work, as evidenced in detailed reports on human genome sequences that provide in-depth genome data.

Each unit of genome annotation catalogs a discrete gene followed by a corresponding protein product. Additionally, basic annotations of gene function, such as domain boundaries and functionally characterized homologs, might be included in the file, although these details often require more thorough investigation. The annotation process conceptually consists of two parts: the computation and the annotation part. The computation segment uses datasets from genomes or certain transcriptomes, which are utilized concurrently to generate preliminary gene and transcriptomics data. The annotation segment then compiles this data into a comprehensive gene annotation using a set of rules within an annotation pipeline. The success of genome annotation largely depends on the quality of the genome assembly, which is crucial for achieving satisfactory results with near-complete genomes that have minimal gaps.

Examining the effects of chromosomal variations on wild-type gene function and, ultimately, on an individual's health is the goal of genome analysis. With the progress of whole genome sequencing (WGS), an increasing number of genetic variations are being found. It can be difficult to identify disease-related variations and to pinpoint their precise kinds, positions, functions, and possible effects because of things like complicated genomic areas, inaccurate variant calling, and problems recognizing the locus phase. A critical component of genomic annotation and analysis is comprehending the relationship between genomic variants and their associations with illnesses, as this understanding can offer distinct insights into the impact of variations on individuals [76].

Different types of programmed processes and useful tools are available for the annotation and analysis of discovered variations. While most tools can handle SNPs and indels, specialized tools for annotating structural variants (SVs) are fewer, such as ANNOVAR [76] and VAT [77]. These tools offer distinct functions, including reference database annotations, evolutionary information, predictions of the effects of identified variants, allele frequency, and chromosomal positions about nearby sites of interest such as genes, segmental duplications, splice sites, and regulatory regions [78]. All these analytical outcomes from individual variants can be utilized in disease diagnosis. Common tools used for performing annotations and analysis are listed in Table 3.

The above-mentioned frameworks link the actions and tools available at each stage of the process in a sequential manner to establish a comprehensive computational workflow for waveguide separation. Many systems have been created to manage process integration, optimization, and genomic sequencing analysis. Illumina is one of the most widely used WGS technologies; it uses bridge PCR for sequencing-by-synthesis, fluorescence imaging, and DNA colony amplification [78]. Illumina provides a variety of instruments with differing features and prices, including as the HiSeq X Ten, MiSeq, HiSeq, and NextSeq.

Competing with Illumina is the Ion Torrent platform, analogous to cyclic array where DNA colonies on beads are amplified inside wells on a patterned surface. This technique employs a unique, non-optical method of detection by assessing the

Table 3 The genome annotation softwares and their brief details with the accessible links

Software	Description	Resources
ANNOVAR	Use current data to annotate genetic variants identified across various genomes	https://annovar.openbioinformatics.org/en/latest/
VAT	A computational system designed to functionally annotate variants in individual genomes	https://www.illumina.com/
VARIANT	A short-read aligner that processes more than 25 million 35-base pair reads each hour, mapping short DNA sequences (reads) to the human genome	http://variant.bioinfo.cipf.es/
VAAST	Sequence alignment tool designed to handle the numerous short reads produced by next-generation sequencing (NGS)	https://www.yandell-lab.org/software/vaast.html
snpEFF	Seed-and-extend alignment software for mapping short reads using the Illumina platform	https://pcingola.github.io/SnpEff/
VariBench	A GPU-powered software tool for aligning short reads to a reference sequence	http://structure.bmc.lu.se/VariBench/
PolyPhen/ PolyPhen-2	A software package designed to map low-divergence sequences against a large reference genome	http://genetics.bwh.harvard.edu/pph2/
SeattleSeq	A collection of distributed applications designed for aligning short DNA reads and managing and analyzing short-read alignments	http://snp.gs.washington.edu/SeattleSeqAnnotation150/

discharge of a hydrogen ion [14]. Another significant player is Pacific Biosciences, which provides comprehensive WGS solutions and innovative applications for gene discovery. Although they are more costly, however, the longer reads and negligible GC bias offered by Pacific Biosciences enables study of broader percentage of human genetic variation, providing a key benefit above short-read sequencing methodology [79].

Additionally, the Oxford Nanopore platform stands out by requiring low amounts of initial nucleotides and delivering long-reads without requiring manipulation [24]. Furthermore, numerous other entities like Galaxy, LONI, GATK, and Taverna are actively exploring innovative and advanced WGS technologies [65, 80–82]. Given these developments, the prospects for WGS platform development appear very promising.

4 Applications

The use of whole genome sequencing (WGS) spans a wide array of fields, such as agriculture, food, public health, and even space, for various purposes and settings. However, this article focuses solely on its clinical uses, which are directly associated with disease and health. The clinical use of WGS helps clarify the genomic factors underlying human hereditary traits. Indeed, numerous genome projects have been launched that focus on employing WGS to diagnose individuals with diseases, including cancer and more. This overview will further enhance our knowledge of pathophysiology through WGS and underscore its significance in identifying genes and variants responsible for diseases and enhancing public health.

Present clinical applications primarily involve Mendelian diseases, complex diseases, and cancer [83]. The detection of Mendelian disease is frequently used using WGS, which has proven effective in identifying causal mutations [84, 85]. The largest compendium of Mendelian diseases includes 7,000 unique diseases cataloged according to the OMIM database [86].

Unidentified genetic components are thought to be responsible for about half of Mendelian illnesses. The integration of gene information, variation frequency, function prediction, and evolutionary conservation are all included in the sequencing process' identification of Mendelian illnesses. This strategy is predicated on the idea that uncommon mutations that change the way proteins function in homologous genes and are associated with Mendelian disorders are more likely to exist. Commonly or complex clinical manifestations represent additional clinical application areas that are influenced by multiple genetic variants. Genome-wide association studies (GWAS) investigate complex diseases by examining variantion that modestly provide to disease susceptibility. While GWAS typically only genotype common variants, the WGS technique enables the annotation of all variants, both rare and common, and ensures the sequencing of actual functional genes. Cancer diagnosis considered a significant type of complex disease, has seen a heightened need for diagnosis at the present time. Considerable focus has been placed on using WGS to identify different cancer types, including breast, gastric, and prostate cancer [87–89].

Even while certain cancer detection techniques have shown remarkable promise, they also highlight the shortcomings of whole exome sequencing, which ignores non-coding areas in favor of protein-coding regions, making a more thorough and in-depth analysis of cancer impossible. WGS is tackling the problem of finding mutations in non-coding areas often observed across different tumors thanks to developments in sequencing technology, opening the door for potential applications in customized therapy in the future [90].

We demonstrate the clinical workflow of WGS for both healthy persons and those with illness using data from the first randomized clinical study to integrate WGS into clinical medicine, the MedSeq Project [91]. Figure 2 depicts the overall layout of the clinical WGS process, which is composed of five main components: hospitals, patients, laboratory tests, reports and feedback, and solutions.

5 Epigenomics

Epigenomics is a different kind of whole genome sequencing. The entire collection of epigenetic alterations on a cell's genetic material, or the epigenome, is studied in a field called epigenomics. Chemical alterations to DNA and histone proteins,

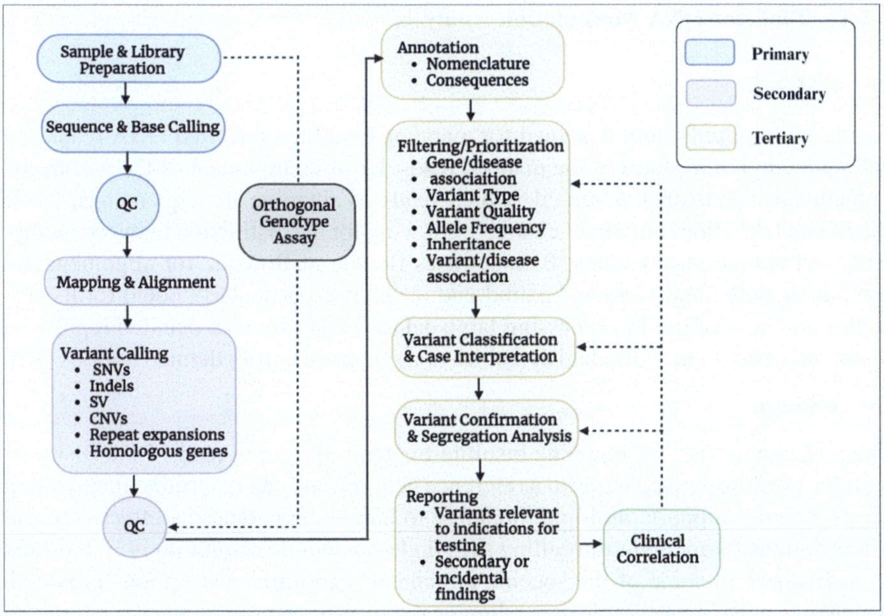

Fig. 2 Whole genome sequencing workflow

known as epigenetic modifications, impact gene expression without changing the DNA sequence. These alterations are essential for controlling gene activity, impacting biological functions, and causing certain disease states. In order to map and comprehend patterns of these modifications throughout the genome in many cells and situations, the subject of epigenomics combines genetics, molecular biology, and bioinformatics.

Histone alterations, chromatin remodeling, and DNA methylation are the three primary epigenetic changes.

5.1 DNA Methylation

It adds the methyl group to the DNA molecule which typically occur at cytosine bases adjacent to guanine bases (CpG sites). This modification can affect gene expression and is involved in various biological processes and diseases such as cancer, developmental disorders, neurological disorders and aging also. The study of DNA methylation patterns, therefore, is crucial for understanding these processes and developing potential therapeutic strategies.

5.1.1 Tools for DNA Methylation Analysis

(a) **Bismark:**

Bismark is a popular tool designed for mapping bisulfite-converted DNA reads and calling methylation states of the mapped reads. Bisulfite treatment of DNA converts unmethylated cytosines to uracil (which reads as thymine in sequencing), while methylated cytosines remain unchanged, allowing for the distinction between methylated and unmethylated states. Bismark uses Bowtie or Bowtie2 for alignment and can handle both single-end and paired-end reads. It is particularly noted for its efficiency and ease of use in processing large datasets and provides detailed reports on methylation levels at individual cytosines or aggregated across defined regions [92].

(b) **BSmap:**

Smap is another tool for mapping bisulfite-treated sequencing reads. Like Bismark, it aligns bisulfite-treated reads to a reference genome and can determine methylation levels. BSmap supports multiple threading to handle large datasets effectively and offers options for methylation calling at a single-nucleotide resolution [93]. It differs from Bismark in some of the specific alignment algorithms and options provided, which may make it preferable depending on the specific requirements of the research project.

(c) **MethylKit:**

It is an R package designed for the analysis of methylation data derived from bisulfite sequencing. Its primary function is to facilitate the statistical analysis of methylation profiles across multiple samples, allowing researchers to identify changes in methylation across different conditions or treatments. It can handle large datasets, perform normalization, and provide tools for visualizing methylation patterns. MethylKit is particularly useful for conducting differential methylation analysis, where the goal is to compare methylation patterns between groups to identify biologically significant changes [94].

(d) **DMRcate:**

DMRcate is another R package that is used to identify differentially methylated regions (DMRs) across different biological conditions or treatments. The software implements a statistical approach that combines smoothing and multiple testing correction to accurately define regions with significant changes in methylation. DMRcate is especially useful when working with data that spans large genomic regions and can be integrated with other data types, such as gene expression data, to enhance biological interpretation [95].

(e) **DSS (Dispersion Shrinkage for Sequencing Data):**

DSS is an R package designed for differential methylation analysis using a beta-binomial generalized linear model. This approach accounts for biological variability and technical noise by estimating and shrinking dispersion parameters, making it robust against data overdispersion, a common issue in counting data from sequencing assays. DSS can analyze both whole-genome bisulfite sequencing and reduced representation bisulfite sequencing data, making it versatile for different types of methylation studies [96].

5.1.2 ChIP-Seq (Chromatin Immunoprecipitation Followed by Sequencing)

ChIP-seq is a vital technique for studying protein-DNA interactions, widely used to explore the binding sites of transcription factors and the presence of histone modifications across the genome. Analyzing the vast amounts of data generated by ChIP-seq experiments requires robust computational tools, each designed to handle specific aspects of ChIP-seq data analysis such as peak calling, bias correction, motif analysis, and annotation.

MACS (Model-based Analysis of ChIP-Seq) is one of the most frequently utilized tools for identifying enriched regions, or "peaks," where protein-DNA interactions are significantly higher than background [97]. MACS calculates the distance between the forward and reverse strand tags to predict the enrichment location more precisely and uses a dynamic Poisson distribution to assess the significance of enrichment. This allows MACS to effectively differentiate signal from noise, making it suitable for both broad and narrow peak detection [98].

SICER, on the other hand, is particularly adept at handling datasets where enriched regions span broader domains, typical of certain histone modifications [99]. Unlike tools that focus solely on peak intensity, SICER groups nearby enriched regions to identify broader areas of chromatin modification. It also incorporates strategies to correct for local biases in the genome, which is crucial for accurate representation of histone modification landscapes [99].

HOMER (Hypergeometric Optimization of Motif EnRichment) extends beyond peak calling and is particularly valuable for its comprehensive analysis capabilities. After identifying enriched regions, HOMER analyzes the underlying DNA sequence to discover motifs—short, recurring sequence patterns that are potentially responsible for the recruitment of specific proteins. This dual functionality makes HOMER a versatile tool for both identifying regulatory regions and understanding the molecular mechanisms governing transcription factor binding [100].

Finally, ChIPseeker is an R package designed to streamline the annotation and visualization of ChIP-seq data. It links identified peaks to nearby genes and other genomic features, providing a clearer picture of the potential biological significance of each interaction. ChIPseeker can generate various plots to visualize the distribution of peaks across genomic annotations, facilitating easy interpretation and comparison of results from different samples or experimental conditions [101].

Together, these tools form a comprehensive suite for ChIP-seq analysis, enabling researchers to not only pinpoint where proteins interact with the DNA but also to interpret the biological implications of these interactions. Whether investigating transcription factor dynamics, chromatin structure, or regulatory networks, these tools provide critical insights into the complex regulation of gene expression.

5.1.3 Chromatin Accessibility Analysis

Chromatin accessibility analysis using ATAC-seq (Assay for Transposase-Accessible Chromatin with high throughput sequencing) provides crucial insights into regulatory regions of the genome by identifying open chromatin sites where transcription factors and other regulatory proteins can bind. The analysis of ATAC-seq data involves a combination of bioinformatic tools each designed to address specific steps in the pipeline, from quality control to motif analysis.

The ATAC-seq Analysis Pipeline, which integrates tools like ATAC-seqQC, MACS2, and deepTools, offers a comprehensive approach to processing and analyzing ATAC-seq data [102]. ATAC-seqQC is particularly useful for initial quality control, providing metrics to assess the success of library preparation and sequencing and the distribution of fragment sizes, which are indicative of nucleosome positioning and chromatin structure [97]. MACS2 is then employed to identify peaks, which represent regions of increased accessibility. This tool is adept at handling narrow peaks typical of transcription factor binding sites, a common focus in ATAC-seq studies. deepTools complements these by offering advanced capabilities for visualizing and analyzing sequencing data, helping researchers to create intuitive visual representations of chromatin accessibility across different conditions or genetic backgrounds [103].

DiffBind is another crucial tool used to compare ATAC-seq data across different conditions or time points [104]. It aggregates data from multiple samples and identifies differentially accessible chromatin regions, providing insights into how chromatin accessibility changes in response to different biological stimuli or in different disease states. This tool is invaluable for studies aimed at understanding dynamic changes in transcriptional regulation [105–107].

HMMRATAC is a specialized peak caller designed specifically for ATAC-seq data, using a hidden Markov model to differentiate peaks corresponding to open chromatin from background noise [108]. This approach is particularly effective in capturing the subtle nuances of chromatin accessibility, allowing for a more refined analysis of open chromatin sites and nucleosome positioning [109].

Lastly, CentriMo, part of the comprehensive MEME suite, is focused on motif discovery within DNA sequences identified as peaks in ATAC-seq studies [110]. By analyzing these sequences, CentriMo can identify motifs that are centrally enriched within the data set, suggesting potential binding sites for transcription factors that may regulate gene expression at these accessible regions [111].

Together, these tools form a robust framework for ATAC-seq data analysis, from initial quality assessment through to functional interpretation of the data. Each tool addresses a specific aspect of the analysis pipeline, allowing researchers to derive detailed and biologically meaningful insights into chromatin dynamics and transcriptional regulation.

5.1.4 Integrative Analysis and Visualization

Integrative analysis and visualization are key to extracting meaningful insights from complex epigenomic datasets. Such analysis often involves combining data from multiple sources or experiments to understand the regulatory mechanisms underlying gene expression [112]. Several tools have been developed to aid researchers in these tasks, each offering unique functionalities tailored to the specific needs of epigenomic research [113].

EpiExplorer is an innovative tool designed to facilitate the interactive visualization and exploration of large epigenomic datasets. This web-based platform allows users to quickly compare and analyze genomic regions based on various epigenetic marks, DNA methylation, genetic variations, and gene annotations. One of the standout features of EpiExplorer is its ability to perform real-time analysis on user-defined subsets of data, enabling researchers to dynamically explore relationships between epigenetic modifications and other genomic features. This can be particularly useful in hypothesis testing or generating new research questions based on observed epigenomic patterns [114].

The Epigenome Roadmap Browser is another pivotal tool that serves as a comprehensive resource for exploring epigenomic data across a diverse array of cell types and tissues. Developed as part of the NIH Roadmap Epigenomics Mapping Consortium, this browser provides access to a vast repository of human epigenomic data, including DNA methylation, histone modification, and chromatin accessibility across different developmental stages and conditions. The visualization capabilities of the Roadmap Browser allow researchers to easily compare epigenomic landscapes across cell types and identify cell-specific or universal epigenetic modifications, which are crucial for understanding cellular differentiation and organismal development [115].

DeepTools represents a suite of software tools designed for the efficient processing, normalization, and visualization of deep-sequencing data. It includes a variety of command-line programs that can generate highly informative visualizations, such as heatmaps of signal intensity around genomic features, coverage plots, and principal component analysis (PCA) plots. These tools are particularly valuable for visualizing complex datasets, such as those comparing several histone modifications or integrating multiple epigenomic data types. deepTools can handle data from a variety of sequencing experiments, including ChIP-seq, RNA-seq, and ATAC-seq, making it highly versatile for studies that involve multiple layers of genomic and epigenomic regulation [103].

Together, these tools empower researchers to perform sophisticated integrative analyses and visualizations of epigenomic data. By leveraging EpiExplorer, the Epigenome Roadmap Browser, and deepTools, researchers can dissect the functional roles of epigenetic modifications in gene regulation, understand how epigenetic [55] landscapes change in different biological contexts, and ultimately elucidate the complex interactions that define the genomic basis of health and disease. These tools also facilitate the sharing and dissemination of epigenomic data, promoting collaborative research efforts and the advancement of our understanding of epigenomic regulation across the global scientific community.

5.1.5 General Epigenomics Analysis

Epigenomic analysis often requires the integration and manipulation of complex datasets that describe DNA methylation, histone modifications, chromatin accessibility, and more. A combination of generalized bioinformatics tools and specialized software packages tailored for epigenetic data can be extremely effective. Two foundational resources in this domain are bedtools and the R/Bioconductor ecosystem, each offering unique utilities for the processing and analysis of genomic and epigenetic data. bedtools is a powerful command-line tool used extensively in bioinformatics for the manipulation of genomic intervals. While not specifically designed for epigenetics, its functionality is critical for the typical workflows associated with epigenomic data analysis. bedtools allows users to perform a wide range of operations on genomic data such as: Intersect This function enables researchers to find overlapping regions between different datasets, such as overlaps between regions of DNA methylation and histone modifications. This is particularly useful in epigenomics to study co-occurring epigenetic marks that might regulate gene expression synergistically. Merge: Combining adjacent or overlapping intervals is essential in epigenomics, especially when aggregating broad histone marks or consolidating regions of open chromatin identified in ATAC-seq or DNase-seq. Subtract: Removing specific intervals from a dataset, such as excluding regions of DNA methylation from analysis of open chromatin, helps refine analysis inputs and focus on relevant genomic contexts. Sort and Shuffle: These tools help in organizing genomic data into usable formats and generating randomized data for control analyses, respectively. bedtools is highly valued for its efficiency and versatility, making it an indispensable tool in genomic research, including epigenomic studies where data from various assays need to be integrated and compared [116].

The R/Bioconductor project provides a comprehensive suite of statistical and graphical tools for the analysis of genomic data, including specialized packages for epigenomics. Bioconductor is particularly well-suited for handling high-throughput sequencing data and offers several advantages: GenomicRanges: This package is a cornerstone of the Bioconductor project, enabling the handling and manipulation of genomic ranges and annotations efficiently. It allows researchers to store and operate on genomic intervals and associated data, which is essential for organizing large epigenomic datasets. methylKit: A package specifically designed for the analysis of DNA methylation data obtained from high-throughput sequencing methods. It offers functionality for data preprocessing, normalization, visualization, and statistical analysis to identify differentially methylated regions. ChIPseeker: For annotating ChIP-seq data, this package links peaks to genes and other genomic features, and provides visualization tools that facilitate the understanding of the biological context of the interactions. ComplexHeatmap: A tool for visually representing complex genomic data as heatmaps, which is particularly useful for displaying patterns of epigenetic marks across different samples or conditions. Bioconductor's seamless integration with R, a powerful statistical programming environment, makes it an excellent choice for the complex data analyses often required in epigenetics. Researchers can leverage its capabilities to perform detailed statistical analyses, integrate various data types, and produce high-quality visualizations [117].

In summary, tools like bedtools and the Bioconductor packages in R provide essential capabilities for the manipulation, analysis, and visualization of epigenomic data. These tools help bridge the gap between raw sequencing data and biological interpretation, enabling researchers to uncover the intricate mechanisms of epigenetic regulation.

6 Conclusion

The transformative impact of Next-Generation Sequencing (NGS) on genomics research has been profound, enabling rapid and efficient sequencing of DNA and RNA across diverse fields such as cancer genomics, metagenomics, and epigenetics. NGS platforms, including Illumina, Ion Torrent, PacBio, and Oxford Nanopore, each offer unique sequencing capabilities suited to different scientific needs. Illumina platforms are renowned for their accurate short reads and wide application in RNA-seq and whole-genome sequencing. In contrast, PacBio and Oxford Nanopore excel in generating longer reads which are crucial for studying complex genomic regions and epigenetic modifications.

The application of NGS extends beyond basic research into clinical settings where it aids in disease diagnosis and the development of personalized treatments, particularly in oncology and genetic disorders. In agriculture, NGS facilitates the enhancement of crop quality and disease resistance through genomic studies.

Despite its many advantages, NGS faces challenges such as complex data management, quality control, and the need for advanced computational resources. However, continuous advancements in bioinformatics and sequencing technologies are progressively overcoming these hurdles, broadening the utility and accessibility of NGS.

Practical application of NGS in the lab involves several key procedures, starting from sample preparation and DNA extraction to library preparation and sequencing. The use of commercial DNA kits has streamlined these processes, ensuring high-quality DNA extraction and preparation for sequencing. Following library preparation, rigorous quality control checks are essential to validate the suitability of libraries for sequencing. Platforms like Illumina's NovaSeq and Thermo Fisher's Ion Torrent facilitate high-throughput sequencing, providing detailed genomic insights.

Furthermore, the data generated through NGS requires extensive analysis, typically involving alignment to reference genomes, variant calling, and annotation to determine the biological implications of genetic variations. These computational processes are supported by sophisticated software tools designed for specific types of data analysis, such as variant detection and epigenomic profiling.

Overall, NGS represents a cornerstone technology in genomics, driving innovations in scientific research and clinical diagnostics. As the technology evolves, it promises to deepen our understanding of complex biological systems and enhance the precision of genetic research and medical interventions.

Conflict of Interest None of authors have competing interest.

References

1. Metzker, M. L. (2010). Sequencing technologies-the next generation. *Nature Reviews Genetics, 11*(1), 31–46. https://doi.org/10.1038/nrg2626
2. Mardis, E. R. (2008). Next-generation DNA sequencing methods. *Annual Review of Genomics and Human Genetics, 9*, 387–402. https://doi.org/10.1146/annurev.genom.9.081307.164359
3. Goodwin, S., McPherson, J. D., & McCombie, W. R. (2016). Coming of age: Ten years of next-generation sequencing technologies. *Nature Reviews Genetics, 17*(6), 333–351. https://doi.org/10.1038/nrg.2016.49
4. Koboldt, D. C., Steinberg, K. M., Larson, D. E., Wilson, R. K., & Mardis, E. R. (2013). The next-generation sequencing revolution and its impact on genomics. *Cell, 155*(1), 27–38. https://doi.org/10.1016/j.cell.2013.09.006
5. Mardis, E. (2013). Next-generation sequencing platforms. *Annual Review of Analytical Chemistry (Palo Alto, California)*, 6. https://doi.org/10.1146/annurev-anchem-062012-092628.
6. Rhoads, A., & Au, K. F. (2015). PacBio sequencing and its applications. *Genomics, Proteomics & Bioinformatics, 13*(5), 278–289. https://doi.org/10.1016/j.gpb.2015.08.002
7. Rhoads, A., & Au, K. F. (2015). PacBio Sequencing and its Applications. *Genomics, Proteomics & Bioinformatics., 13*(5), 278–289. https://doi.org/10.1016/j.gpb.2015.08.002
8. Bentley, D. R., Balasubramanian, S., Swerdlow, H. P., et al. (2008). Accurate whole human genome sequencing using reversible terminator chemistry. *Nature, 456*(7218), 53–59. https://doi.org/10.1038/nature07517
9. van Dijk, E. L., Auger, H., Jaszczyszyn, Y., & Thermes, C. (2014). Ten years of next-generation sequencing technology. *Trends in Genetics, 30*(9), 418–426. https://doi.org/10.1016/j.tig.2014.07.001
10. Mardis, E. R. (2013). Next-generation sequencing platforms. *Annual Review of Analytical Chemistry (Palo Alto, California), 6*, 287–303. https://doi.org/10.1146/annurev-anchem-062012-092628
11. Quail, M. A., Smith, M., Coupland, P., et al. (2012). A tale of three next generation sequencing platforms: comparison of Ion Torrent, Pacific Biosciences and Illumina MiSeq sequencers. *BMC Genomics, 13*, 341. https://doi.org/10.1186/1471-2164-13-341.
12. Reuter, J. A., Spacek, D. V., & Snyder, M. P. (2015). High-throughput sequencing technologies. *Molecular Cell, 58*(4), 586–597. https://doi.org/10.1016/j.molcel.2015.05.004
13. Rothberg, J. M., Hinz, W., Rearick, T. M., et al. (2011). An integrated semiconductor device enabling non-optical genome sequencing. *Nature, 475*(7356), 348–352. https://doi.org/10.1038/nature10242
14. Merriman, B., & Rothberg, J. M. (2012). Progress in ion torrent semiconductor chip based sequencing. *Electrophoresis, 33*(23), 3397–3417. https://doi.org/10.1002/elps.201200424
15. Glenn, T. C. (2011). Field guide to next-generation DNA sequencers. *Molecular Ecology Resources, 11*(5), 759–769. https://doi.org/10.1111/j.1755-0998.2011.03024.x

16. Voelkerding, K. V., Dames, S. A., & Durtschi, J. D. (2009). Next-generation sequencing: From basic research to diagnostics. *Clinical Chemistry, 55*(4), 641–658. https://doi.org/10.1373/clinchem.2008.112789
17. Eid, J., Fehr, A., Gray, J., et al. (2009). Real-time DNA sequencing from single polymerase molecules. *Science, 323*(5910), 133–138. https://doi.org/10.1126/science.1162986
18. Liaquat, F., Munis, M. F. H., Arif, S., et al. (2021). PacBio single-molecule long-read sequencing reveals genes tolerating manganese stress in schima superba saplings. *Frontiers in Genetics, 12*, 635043. https://doi.org/10.3389/fgene.2021.635043
19. Schadt, E. E., Turner, S., & Kasarskis, A. (2010). A window into third-generation sequencing. *Human Molecular Genetics, 19*(R2), R227–R240. https://doi.org/10.1093/hmg/ddq416
20. Korlach, J., & Turner, S. W. (2012). Going beyond five bases in DNA sequencing. *Current Opinion in Structural Biology, 22*(3), 251–261. https://doi.org/10.1016/j.sbi.2012.04.002
21. Wenger, A. M., Peluso, P., Rowell, W. J., et al. (2019). Accurate circular consensus long-read sequencing improves variant detection and assembly of a human genome. *Nature Biotechnology, 37*(10), 1155–1162. https://doi.org/10.1038/s41587-019-0217-9
22. Jain, M., Olsen, H. E., Paten, B., & Akeson, M. (2016). The Oxford Nanopore MinION: Delivery of nanopore sequencing to the genomics community. *Genome Biology, 17*(1), 239. https://doi.org/10.1186/s13059-016-1103-0.
23. Lu, H., Giordano, F., & Ning, Z. (2016). Oxford Nanopore MinION sequencing and genome assembly. *Genomics, Proteomics & Bioinformatics, 14*(5), 265–279. https://doi.org/10.1016/j.gpb.2016.05.004
24. Mikheyev, A. S., & Tin, M. M. (2014). A first look at the Oxford Nanopore MinION sequencer. *Molecular Ecology Resources, 14*(6), 1097–1102. https://doi.org/10.1111/1755-0998.12324
25. Deamer, D., Akeson, M., & Branton, D. (2016). Three decades of nanopore sequencing. *Nature Biotechnology, 34*(5), 518–524. https://doi.org/10.1038/nbt.3423
26. Ip, C. L. C., Loose, M., Tyson, J. R., et al. (2015). MinION analysis and reference consortium: Phase 1 data release and analysis. *F1000Res, 4*, 1075. https://doi.org/10.12688/f1000research.7201.1.
27. Levy, S., Sutton, G., Ng P. C., et al. (2007). The diploid genome sequence of an individual human. *PLoS Biology, 5*(10), e254. https://doi.org/10.1371/journal.pbio.0050254.
28. Ng, S. B., Buckingham, K. J., Lee, C., et al. (2010). Exome sequencing identifies the cause of a mendelian disorder. *Nature Genetics, 42*(1), 30–35. https://doi.org/10.1038/ng.499
29. Wang, Z., Gerstein, M., & Snyder, M. (2009). RNA-Seq: A revolutionary tool for transcriptomics. *Nature Reviews Genetics, 10*(1), 57–63. https://doi.org/10.1038/nrg2484
30. Tang, F., Barbacioru, C., Wang, Y., et al. (2009). MRNA-Seq whole-transcriptome analysis of a single cell. *Nature Methods, 6*(5), 377–382. https://doi.org/10.1038/nmeth.1315
31. Lister, R., Pelizzola, M., Dowen, R. H., et al. (2009). Human DNA methylomes at base resolution show widespread epigenomic differences. *Nature, 462*(7271), 315–322. https://doi.org/10.1038/nature08514
32. Buenrostro, J. D., Giresi, P. G., Zaba, L. C., Chang, H. Y., & Greenleaf, W. J. (2013). Transposition of native chromatin for fast and sensitive epigenomic profiling of open chromatin, DNA-binding proteins and nucleosome position. *Nature Methods, 10*(12), 1213–1218. https://doi.org/10.1038/nmeth.2688
33. Leipzig, J. (2017). A review of bioinformatic pipeline frameworks. *Briefings in Bioinformatics, 18*(3), 530–536. https://doi.org/10.1093/bib/bbw020
34. Wang, D. L., Zender, C. S., & Jenks, S. F. (2009). Efficient clustered server-side data analysis workflows using SWAMP. *Earth Science Informatic, 2*(3), 141–155. https://doi.org/10.1007/s12145-009-0021-z.
35. Yoo, J., Ha, I. C., Chang, G. T., Jung, K. S., Park, K., & Kim, Y. (2011). CNVAS: Copy number variation analysis system—the analysis tool for genomic alteration with a powerful visualization module. *BioChip Journal, 5*(3), 265–270. https://doi.org/10.1007/s13206-011-5311-0.
36. Goecks, J., Nekrutenko, A., & Taylor, J. (2010). The Galaxy T. Galaxy: A comprehensive approach for supporting accessible, reproducible, and transparent computational research in the life sciences. *Genome Biology, 11*(8), R86. https://doi.org/10.1186/gb-2010-11-8-r86.

37. Lahens, N. F., Kavakli, I. H., Zhang, R., et al. (2014). IVT-seq reveals extreme bias in RNA sequencing. *Genome Biology, 15*(6), R86. https://doi.org/10.1186/gb-2014-15-6-r86.
38. Adey, A., Morrison, H. G., Asan, et al. (2010). Rapid, low-input, low-bias construction of shotgun fragment libraries by high-density in vitro transposition. *Genome Biology, 11*(12), R119. https://doi.org/10.1186/gb-2010-11-12-r119.
39. van Dijk, E. L., Jaszczyszyn, Y., & Thermes, C. (2014). Library preparation methods for next-generation sequencing: Tone down the bias. *Experimental Cell Research, 322*(1), 12–20. https://doi.org/10.1016/j.yexcr.2014.01.008.
40. Wang, L., Wang, S., & Li, W. (2012). RSeQC: Quality control of RNA-seq experiments. *Bioinformatics, 28*(16), 2184–2185. https://doi.org/10.1093/bioinformatics/bts356
41. Kwong, J. C., McCallum, N., Sintchenko, V., & Howden, B. P. (2015). Whole genome sequencing in clinical and public health microbiology. *Pathology, 47*(3), 199–210. https://doi.org/10.1097/PAT.0000000000000235.
42. Flicek, P., & Birney, E. (2009). Sense from sequence reads: methods for alignment and assembly. *Nature Methods, 6*(11), S6–S12. https://doi.org/10.1038/nmeth.1376.
43. Li, H., & Durbin, R. (2009). Fast and accurate short read alignment with Burrows-Wheeler transform. *Bioinformatics, 25*(14), 1754–1760. https://doi.org/10.1093/bioinformatics/btp324
44. Alkan, C., Sajjadian, S., & Eichler, E. E. (2011). Limitations of next-generation genome sequence assembly. *Nature Methods, 8*(1), 61–65. https://doi.org/10.1038/nmeth.1527.
45. Rumble, S. M., Lacroute, P., Dalca, A. V., Fiume, M., Sidow, A., & Brudno, M. (2009). SHRiMP: Accurate mapping of short color-space reads. *PLoS Computational Biology., 5*(5), e1000386. https://doi.org/10.1371/journal.pcbi.1000386
46. Li, H., Handsaker, B., Wysoker, A., et al. (2009). The sequence alignment/map format and SAMtools. *Bioinformatics, 25*(16), 2078–2079. https://doi.org/10.1093/bioinformatics/btp352
47. Li, R., Yu, C., Li, Y., et al. (2009). SOAP2: An improved ultrafast tool for short read alignment. *Bioinformatics, 25*(15), 1966–1967. https://doi.org/10.1093/bioinformatics/btp336
48. Langmead, B., Trapnell, C., Pop, M., & Salzberg, S. L. (2009). Ultrafast and memory-efficient alignment of short DNA sequences to the human genome. *Genome Biology, 10*(3), R25. https://doi.org/10.1186/gb-2009-10-3-r25.
49. Robertson, G., Schein, J., Chiu, R., et al. (2010). De novo assembly and analysis of RNA-seq data. *Nature Methods, 7*(11), 909–912. https://doi.org/10.1038/nmeth.1517
50. Ng, P. C., & Kirkness, E. F. (2010). Whole genome sequencing. In M. R. Barnes, & G. Breen (Eds.), *Genetic variation: Methods and protocols* (pp. 215–226). Humana Press.
51. Myers, E. W. (1995). Toward simplifying and accurately formulating fragment assembly. *Journal of Computational Biology, 2*(2), 275–290. https://doi.org/10.1089/cmb.1995.2.275.
52. Pevzner, P. A., Tang, H., & Waterman, M. S. (2001). An Eulerian path approach to DNA fragment assembly. *Proceedings of the National Academy of Sciences, 98*(17), 9748–9753. https://doi.org/10.1073/pnas.171285098.
53. Myers, E. W., Sutton, G. G., Delcher, A. L., et al. (2000). A whole-genome assembly of *Drosophila. Science, 287*(5461), 2196–2204. https://doi.org/10.1126/science.287.5461.2196.
54. Jaffe, D. B., Butler, J., Gnerre, S., et al. (2003). Whole-genome sequence assembly for mammalian genomes: Arachne 2. *Genome Research, 13*(1), 91–96. https://doi.org/10.1101/gr.828403
55. Zerbino, D. R., & Birney, E. (2008). Velvet: Algorithms for de novo short read assembly using de Bruijn graphs. *Genome Research, 18*(5), 821–829. https://doi.org/10.1101/gr.074492.107
56. Luo, R., Liu, B., Xie, Y., et al. (2012). SOAPdenovo2: An empirically improved memory-efficient short-read de novo assembler. *Gigascience, 1*(1),18. https://doi.org/10.1186/2047-217x-1-18.
57. Simpson, J. T., Wong, K., Jackman, S. D., Schein, J. E., Jones, S. J., & Birol, I. (2009). ABySS: A parallel assembler for short read sequence data. *Genome Research, 19*(6), 1117–1123. https://doi.org/10.1101/gr.089532.108

58. Jiang, H., & Wong, W. H. (2008). SeqMap: Mapping massive amount of oligonucleotides to the genome. *Bioinformatics, 24*(20), 2395–2396. https://doi.org/10.1093/bioinformatics/btn429
59. Smith, A. D., Chung, W.-Y., Hodges, E., et al. (2009). Updates to the RMAP short-read mapping software. *Bioinformatics, 25*(21), 2841–2842. https://doi.org/10.1093/bioinformatics/btp533
60. Torri, F., Dinov, I. D., Zamanyan, A., et al. (2012). Next Generation Sequence Analysis and Computational Genomics Using Graphical Pipeline Workflows. *Genes, 3*(3), 545–575.
61. Yin, R., Kwoh, C. K., & Zheng, J. (2019). Whole genome sequencing analysis. In S. Ranganathan, M. Gribskov, K. Nakai, & C. Schönbach (Eds.), *Encyclopedia of bioinformatics and computational biology* (pp. 176–183). Academic Press.
62. Nielsen, R., Paul, J. S., Albrechtsen, A., & Song, Y. S. (2011). Genotype and SNP calling from next-generation sequencing data. *Nature Reviews Genetics, 12*(6), 443–451. https://doi.org/10.1038/nrg2986.
63. Hasan, M. S., Wu, X., & Zhang, L. (2015). Performance evaluation of indel calling tools using real short-read data. *Human Genomics, 9*(1), 20. https://doi.org/10.1186/s40246-015-0042-2.
64. Li, H. (2012). Exploring single-sample SNP and INDEL calling with whole-genome de novo assembly. *Bioinformatics, 28*(14), 1838–1844. https://doi.org/10.1093/bioinformatics/bts280
65. McKenna, A., Hanna, M., Banks, E., et al. (2010). The genome analysis toolkit: A MapReduce framework for analyzing next-generation DNA sequencing data. *Genome Research, 20*(9), 1297–1303. https://doi.org/10.1101/gr.107524.110
66. Li, R., Li, Y., Kristiansen, K., & Wang, J. (2008). SOAP: Short oligonucleotide alignment program. *Bioinformatics, 24*(5), 713–714. https://doi.org/10.1093/bioinformatics/btn025
67. Albers, C. A., Lunter, G., MacArthur, D. G., McVean, G., Ouwehand, W. H., & Durbin, R. (2011). Dindel: Accurate indel calls from short-read data. *Genome Research, 21*(6), 961–973. https://doi.org/10.1101/gr.112326.110
68. Fiegler, H., Redon, R., Andrews, D., et al. (2006). Accurate and reliable high-throughput detection of copy number variation in the human genome. *Genome Research, 16*(12), 1566–1574. https://doi.org/10.1101/gr.5630906
69. Alkan, C., Coe, B. P., & Eichler, E. E. (2011). Genome structural variation discovery and genotyping. *Nature Reviews Genetics, 12*(5), 363–376. https://doi.org/10.1038/nrg2958.
70. Chen, K., Wallis, J. W., McLellan, M. D., et al. (2009). BreakDancer: An algorithm for high-resolution mapping of genomic structural variation. *Nature Methods, 6*(9), 677–681. https://doi.org/10.1038/nmeth.1363.
71. Mimori, T., Nariai, N., Kojima, K., et al. (2013). iSVP: An integrated structural variant calling pipeline from high-throughput sequencing data. *BMC Systems Biology, 7*(6), S8. https://doi.org/10.1186/1752-0509-7-S6-S8.
72. Ye, K., Schulz, M. H., Long, Q., Apweiler, R., & Ning, Z. (2009). Pindel: A pattern growth approach to detect break points of large deletions and medium sized insertions from paired-end short reads. *Bioinformatics, 25*(21), 2865–2871. https://doi.org/10.1093/bioinformatics/btp394
73. Wong, K., Keane, T. M., Stalker, J., & Adams, D. J. (2010). Enhanced structural variant and breakpoint detection using SVMerge by integration of multiple detection methods and local assembly. *Genome Biology, 11*(12), R128. https://doi.org/10.1186/gb-2010-11-12-r128.
74. Kanehisa, M., & Goto, S. (2000). KEGG: Kyoto encyclopedia of genes and genomes. *Nucleic Acids Research, 28*(1), 27–30. https://doi.org/10.1093/nar/28.1.27
75. Primmer, C., Papakostas, S., Leder, E., Davis, M., & Ragan, M. (2013). Annotated genes and nonannotated genomes: Cross-species use of Gene Ontology in ecology and evolution research. *Molecular ecology., 22*(12), 3216–3241.
76. Albert, F. W., & Kruglyak, L. (2015). The role of regulatory variation in complex traits and disease. *Nature Reviews Genetics, 16*(4), 197–212. https://doi.org/10.1038/nrg3891.
77. Habegger, L., Balasubramanian, S., Chen, D. Z., et al. (2012). VAT: A computational framework to functionally annotate variants in personal genomes within a cloud-computing environment. *Bioinformatics, 28*(17), 2267–2269. https://doi.org/10.1093/bioinformatics/bts368

78. Lunter, G., & Goodson, M. (2011). Stampy: A statistical algorithm for sensitive and fast mapping of Illumina sequence reads. *Genome Research, 21*(6), 936–939. https://doi.org/10.1101/gr.111120.110
79. Carneiro, M. O., Russ, C., Ross, M. G., Gabriel, S. B., Nusbaum, C., & DePristo, M. A. (2012). Pacific biosciences sequencing technology for genotyping and variation discovery in human data. *BMC Genomics, 13*(1), 375. https://doi.org/10.1186/1471-2164-13-375.
80. Giardine, B., Riemer, C., Hardison, R. C., et al. (2005). Galaxy: A platform for interactive large-scale genome analysis. *Genome Research, 15*(10), 1451–1455. https://doi.org/10.1101/gr.4086505
81. Rex, D. E., Ma, J. Q., & Toga, A. W. (2003) The LONI pipeline processing environment. *NeuroImage, 19*(3), 1033–1048. https://doi.org/10.1016/S1053-8119(03)00185-X.
82. Oinn, T., Addis, M., Ferris, J., et al. (2004). Taverna: A tool for the composition and enactment of bioinformatics workflows. *Bioinformatics, 20*(17), 3045–3054. https://doi.org/10.1093/bioinformatics/bth361
83. Shao, Q., Jiang, Y., & Wu, J. (2014). Whole-genome sequencing and its application in the research and diagnoses of genetic diseases. *Yi Chuan, 36*(11), 1087–1098.
84. Lupski, J. R., Reid, J. G., Gonzaga-Jauregui, C., et al. (2010). Whole-genome sequencing in a patient with Charcot-Marie-Tooth neuropathy. *New England Journal of Medicine, 362*(13), 1181–1191. https://doi.org/10.1056/NEJMoa0908094
85. Roach, J. C., Glusman, G., Smit, A. F. A., et al. (2010). Analysis of genetic inheritance in a family quartet by whole-genome sequencing. *Science, 328*(5978), 636–639. https://doi.org/10.1126/science.1186802.
86. Hamosh, A., Scott, A. F., Amberger, J. S., Bocchini, C. A., & McKusick, V. A. (2005). Online Mendelian Inheritance in Man (OMIM), a knowledgebase of human genes and genetic disorders. *Nucleic Acids Research, 33*(Database issue), D514–D517. https://doi.org/10.1093/nar/gki033.
87. Barbieri, C. E., Baca, S. C., Lawrence, M. S., et al. (2012). Exome sequencing identifies recurrent SPOP, FOXA1 and MED12 mutations in prostate cancer. *Nature Genetics, 44*(6), 685–689. https://doi.org/10.1038/ng.2279.
88. Wang, K., Kan, J., Yuen, S. T., et al. (2011). Exome sequencing identifies frequent mutation of ARID1A in molecular subtypes of gastric cancer. *Nature Genetics, 43*(12), 1219–1223. https://doi.org/10.1038/ng.982.
89. Thompson, E. R., Doyle, M. A., Ryland, G. L., et al. (2012). Exome sequencing identifies rare deleterious mutations in DNA repair genes FANCC and BLM as potential breast cancer susceptibility alleles. *PLoS Genetics, 8*(9), e1002894. https://doi.org/10.1371/journal.pgen.1002894
90. Tsai, E. A., Shakbatyan, R., Evans, J., et al. (2016). Bioinformatics workflow for clinical whole genome sequencing at partners healthcare personalized medicine. *Journal of Personalized Medicine, 6*(1), 12.
91. Vassy, J. L., Lautenbach, D. M., McLaughlin, H. M., et al. (2014). The MedSeq project: A randomized trial of integrating whole genome sequencing into clinical medicine. *Trials, 15*, 85. https://doi.org/10.1186/1745-6215-15-85.
92. Krueger, F., & Andrews, S. R. (2011). Bismark: A flexible aligner and methylation caller for Bisulfite-Seq applications. *Bioinformatics, 27*(11), 1571–1572. https://doi.org/10.1093/bioinformatics/btr167
93. Xi, Y., & Li, W. (2009). BSMAP: Whole genome bisulfite sequence MAPping program. *BMC Bioinformatics, 10*(1), 232. https://doi.org/10.1186/1471-2105-10-232.
94. Akalin, A., Kormaksson, M., Li, S., et al. (2012). methylKit: A comprehensive R package for the analysis of genome-wide DNA methylation profiles. *Genome Biology, 13*(10), R87. https://doi.org/10.1186/gb-2012-13-10-r87.
95. Peters, T. J., Buckley, M. J., Statham, A. L., et al. (2015). De novo identification of differentially methylated regions in the human genome. *Epigenetics & Chromatin, 8*, 6. https://doi.org/10.1186/1756-8935-8-6

96. Wu, H., Xu, T., Feng, H., et al. (2015). Detection of differentially methylated regions from whole-genome bisulfite sequencing data without replicates. *Nucleic Acids Research, 43*(21), e141. https://doi.org/10.1093/nar/gkv715.
97. Zhang, Y., Liu, T., Meyer, C. A., et al. (2008). Model-based analysis of ChIP-Seq (MACS). *Genome Biology, 9*(9), R137. https://doi.org/10.1186/gb-2008-9-9-r137
98. Feng, J., Liu, T., Qin, B., Zhang, Y., & Liu, X. S. (2012). Identifying ChIP-seq enrichment using MACS. *Nature Protocols, 7*(9), 1728–1740. https://doi.org/10.1038/nprot.2012.101
99. Zang, C., Schones, D. E., Zeng, C., Cui, K., Zhao, K., & Peng, W. (2009). A clustering approach for identification of enriched domains from histone modification ChIP-Seq data. *Bioinformatics, 25*(15), 1952–1958. https://doi.org/10.1093/bioinformatics/btp340
100. Heinz, S., Benner, C., Spann, N., et al. (2010). Simple combinations of lineage-determining transcription factors prime cis-regulatory elements required for macrophage and B cell identities. *Molecular Cell, 38*(4), 576–589. https://doi.org/10.1016/j.molcel.2010.05.004
101. Yu, G., Wang, L. G., & He, Q. Y. (2015). ChIPseeker: An R/Bioconductor package for ChIP peak annotation, comparison and visualization. *Bioinformatics, 31*(14), 2382–2383. https://doi.org/10.1093/bioinformatics/btv145
102. Ou, J., Liu, H., Yu, J., et al. (2018). ATACseqQC: A bioconductor package for post-alignment quality assessment of ATAC-seq data. *BMC Genomics, 19*(1), 169. https://doi.org/10.1186/s12864-018-4559-3.
103. Ramírez, F., Dündar, F., Diehl, S., Grüning, B. A., & Manke, T. (2014). deepTools: A flexible platform for exploring deep-sequencing data. *Nucleic Acids Research, 42*(Web Server issue), W187–W191. https://doi.org/10.1093/nar/gku365.
104. Wu, D. Y., Bittencourt, D., Stallcup, M. R., & Siegmund, K. D. (2015). Identifying differential transcription factor binding in ChIP-seq. *Frontiers in Genetics, 6*, 169. https://doi.org/10.3389/fgene.2015.00169
105. Ross-Innes, C. S., Stark, R., Teschendorff, A. E., et al. (2012). Differential oestrogen receptor binding is associated with clinical outcome in breast cancer. *Nature, 481*(7381), 389–393. https://doi.org/10.1038/nature10730
106. Meyer, C. A., & Liu, X. S. (2014). Identifying and mitigating bias in next-generation sequencing methods for chromatin biology. *Nature Reviews Genetics, 15*(11), 709–721. https://doi.org/10.1038/nrg3788
107. He, H. H., Meyer, C. A., Shin, H., et al. (2010). Nucleosome dynamics define transcriptional enhancers. *Nature Genetics, 42*(4), 343–347. https://doi.org/10.1038/ng.545
108. Tarbell, E. D., & Liu, T. (2019). HMMRATAC: A hidden Markov ModeleR for ATAC-seq. *Nucleic Acids Research, 47*(16), e91–e91. https://doi.org/10.1093/nar/gkz533
109. Corces, M. R., Trevino, A. E., Hamilton, E. G., et al. (2017). An improved ATAC-seq protocol reduces background and enables interrogation of frozen tissues. *Nature Methods, 14*(10), 959–962. https://doi.org/10.1038/nmeth.4396
110. Bailey, T. L., & Machanick, P. (2012). Inferring direct DNA binding from ChIP-seq. *Nucleic Acids Research, 40*(17), e128. https://doi.org/10.1093/nar/gks433.
111. Bailey, T. L., Johnson, J., Grant, C. E., & Noble, W. S. (2015). The MEME suite. *Nucleic Acids Research, 43*(W1), W39-49. https://doi.org/10.1093/nar/gkv416
112. Robinson, J. T., Thorvaldsdóttir, H., Winckler, W., et al. (2011). Integrative genomics viewer. *Nature Biotechnology, 29*(1), 24–26. https://doi.org/10.1038/nbt.1754
113. Zhou, X., Maricque, B., Xie, M., et al. (2011). The human epigenome browser at Washington University. *Nature Methods, 8*(12), 989–990. https://doi.org/10.1038/nmeth.1772
114. Halachev, K., Bast, H., Albrecht, F., Lengauer, T., & Bock C. (2012). EpiExplorer: Live exploration and global analysis of large epigenomic datasets. *Genome Biology, 13*(10), R96. https://doi.org/10.1186/gb-2012-13-10-r96.
115. Kundaje, A., Meuleman, W., Ernst, J., et al. (2015). Integrative analysis of 111 reference human epigenomes. *Nature, 518*(7539), 317–330. https://doi.org/10.1038/nature14248.

116. Quinlan, A. R., & Hall, I. M. (2010). BEDTools: A flexible suite of utilities for comparing genomic features. *Bioinformatics, 26*(6), 841–842. https://doi.org/10.1093/bioinformatics/btq033
117. Huber, W., Carey, V. J., Gentleman, R., et al. (2015). Orchestrating high-throughput genomic analysis with bioconductor. *Nature Methods, 12*(2), 115–121. https://doi.org/10.1038/nmeth.3252

Metagenomics and Microbiome Analysis

Vivek Kumar, Rohit Shukla, Surabhi Gangani, Himanshu Shekhar, and Vaishali Shukla

Abstract Microbiome analysis have emerged as powerful tool for studying the composition, diversity, and function of microbial communities in diverse environments. It involves the high-throughput sequencing and analysis of DNA extracted directly from the samples, allowing for the comprehensive characterization of microbial genomes without the need for cultivation. Microbiome analysis focuses on the study of microbial communities inhabiting specific host-associated or environmental niches, such as the human gut, soil, ocean, or air. This chapter will provides an overview of metagenomics and microbiome analysis, highlighting their methodologies, applications, and challenges. Key steps in metagenomic and microbiome analysis include sample collection, DNA extraction, sequencing, bioinformatics analysis, and interpretation of results. Metagenomics and microbiome analysis have diverse applications across various fields, including environmental science, human health, agriculture, and biotechnology. Despite their transformative potential, metagenomics and microbiome analysis face challenges such as data complexity, standardization issues, and ethical considerations. Addressing these challenges requires interdisciplinary collaboration, methodological advancements, and the development of robust computational tools. Overall, metagenomics and microbiome analysis continue to drive discoveries in microbiology, ecology, and biomedicine, offering insights into the complex microbial world and its role in shaping ecosystems and human health.

V. Kumar (✉)
Molecular Biology Department, Lab MicroCare Diagnostics India, New Delhi 110063, India
e-mail: vivek.kumar2012@vitalum.ac.in

V. Kumar · R. Shukla · S. Gangani
USF Center for Microbiome Research, Institute of Microbiomes, University of South Florida Morsani College of Medicine, Tampa, FL 33612, USA

Center of Excellence for Aging and Brain Repair, University of South Florida Morsani College of Medicine, Tampa, FL, USA

H. Shekhar
Biotechnology Division, CSIR-IHBT, Palampur 176061, Himachal Pradesh, India

V. Shukla
Department of Molecular Medicine, Sanjay Gandhi Post Graduate Institute of Medical Sciences, Lucknow 226014, Uttar Pradesh, India

© The Author(s), under exclusive license to Springer Nature Singapore Pte Ltd. 2024
A. Chaudhary et al. (eds.), *Unraveling New Frontiers and Advances in Bioinformatics*,
https://doi.org/10.1007/978-981-97-7123-3_4

Keywords Microbiome · 16 s rRNA sequencing · Metagenomics · FMT · NGS

1 Introduction

The utilization of metagenomics, a potent method in microbiology and microbial ecology, has completely transformed our comprehension of microbial populations and their functions in many habitats [1]. A thorough examination of microbial communities may be done directly from environmental samples using metagenomics, in contrast to conventional microbiological techniques which depend on growing individual microbes in the lab [2]. Metagenomics reveals the astounding diversity, functional potential, and ecological dynamics of microbial communities across many environments, from soil and seas to human beings, by sequencing and analyzing the aggregate genetic material inside these specimens [3].

The field of metagenomics began to take shape in the early years of the twenty-first century, when advanced sequencing methods like next-generation sequencing (NGS) made it easier to sequence DNA isolated from complicated environmental and biological samples. The investigation of microbial communities at a never-before-seen size and resolution is now possible because to this technical advances. Since then, genomic research has grown essential in many areas, such as biotechnology, the agricultural sector, ecological science, and human health.

Metagenomics has given environmental and biological microbiologists a great deal of information into the composition and activities of microbial communities in both naturally occurring and artificially induced settings [4]. The complex interactions between microbes and their surroundings, which promote nutrient cycling, pollutant degradation, and ecosystem resilience, have been made clear by metagenomic investigations. Moreover, metagenomics has aided in the identification of new microbes, enzymes, and metabolic pathways that may find use in sustainable agriculture, biological remediation, and the generation of bioenergy [5].

Metagenomics has provided insights into the intricate microbial communities that reside within the human body, referred to as the human microbiome, with implications for human health [6]. Metagenomic research has clarified the significance of microbes in health and disease by describing the microbiomes of many bodily locations, such as the gut, skin, oral cavity, and urogenital tract. Changes in the human microbiota's makeup and activity have been connected to several illnesses, including diabetes, obesity, inflammatory bowel disease, and even neurological diseases. Understanding the processes underlying these relationships and creating microbiome-based diagnostics and treatments are possible with the help of metagenomics [7].

Metagenomics has the potential to be revolutionary, but it also comes with special difficulties and concerns. These include biases in sample processing, pipelines in bioinformatics, and data analysis. To advance the field and maximize the value of metagenomic techniques, continuing efforts include standardizing methodologies, benchmarking tools, and integrating multi-omics data.

The ideas, uses, and developments of metagenomics will be covered in this chapter with an emphasis on important research and foundational works that have influenced the field. Our goal is to provide an overview of the state of metagenomics today and its potential to address important issues in ecology, biomedicine, and microbiology by combining the most recent research findings and methodological advancements [4].

2 Metagenomics and Microbiome Analysis Process

The metagenomics process involves several key steps, starting with sample collection, where scientists acquire biological or environmental samples from a specific species or habitat. After collecting samples, whole genome DNA is extracted, and the library is prepared by fragmenting DNA, adapter ligation, and amplification steps. Further, cutting-edge sequencing methods are utilized where the produced library is subjected to high-throughput sequencing, thus producing millions of short DNA sequences that describe the genetic makeup of the microbial population. Currently, metagenomics or microbiome analysis relies on two major approaches of sequencing, including whole-genome shotgun sequencing and 16 s rRNA sequencing. The goal of WGS metagenomics is to sequence every genome found in a sample to analyze the functionality and diversification of the microbial community under study. Moreover, the metagenomic process retrieves entire genetic constituents of the specimen, thus making it feasible to define the whole variety of a habitat, including archaea, bacteria, eukaryotes, viruses, and plasmids [8]. Conversely, 16 s rRNA based studies elucidate the richness and diversity of certain taxonomic groupings found in a specimen by sequencing a gene-specific region. The primary marker genes utilized in the ecology of microbes are the internal transcribed spacer (ITS) region (which are employed to define the makeup of the fungus communities), the 16S rRNA gene (which is utilized to analyze the existence of microorganisms and archaea) [9], and the 18S rRNA (which is utilized to indicate the existence of eukaryotic) [10, 11].

Following sequencing, the enormous volumes of data generated by sequencing are processed and interpreted using bioinformatics methods. To find patterns and interactions throughout the microbial population, this involves several processes including quality control, read trimming, taxonomic categorization, functional annotation, and statistical analysis. Ultimately, the metagenomic analysis results are analyzed considering the study objectives and hypotheses, offering insightful information on the relationships between microbes and hosts, biogeochemical cycles, microbial diversity, and microbial ecology.

2.1 Sample Collection

In metagenomics or microbiome investigations, gathering samples is an essential step since it directly affects the accurate representation and quality of the microbial population profiles that are acquired. An outline of the metagenomics sample collection procedure is provided below.

2.1.1 Selection of Samples

The first step in a metagenomics study is selecting the location of specimen collection. This might include any milieu that supports microbial populations, including soil, water, sediment, air, plant surfaces, animal intestines, and even harsh settings like deep-sea hydrothermal vents or hot springs. The purpose of the study and the microbial communities of interest influence the sample site selection.

2.1.2 Preparation and Sterilization of Equipment

In order to prevent contaminants, all containers and equipment must be sanitized prior to sample collection. This usually entails employing techniques like autoclaving, UV irradiation, or chemical disinfection to sterilize instruments like sampling scoops, core samplers, syringes, or collecting bottles.

2.1.3 Sample Collection

- Soil Sampling: Soil samples can be gathered with sterilized scoops or trowel for surface soil collection, or by employing coring equipment to get cores at various depths. To measure spatial heterogeneity, many samples may be gathered from various places throughout the same site.
- Water Sampling: The samples may be obtained by pumps, samples, or bottles, based on where and how deep we are going. During obtaining, caution has to be exercised to prevent contaminants from the air or surrounding environment.
- Fecal Sampling: The feces collection instruction sheet is given to participants. This usually entails transferring a tiny bit of feces into the collecting container using a clean, disposable spoon or spatula. Urine or toilet water contamination of the specimen should be avoided at all costs. The feces specimen should be frozen as soon as feasible to protect the DNA of the microbial population if fast processing is not possible.

2.2 *DNA Extraction*

In metagenomics, DNA isolation is crucial process for straightaway investigation of microbial populations from environmental samples. Extracting DNA from every type of microorganism in the sample—including bacteria, viruses, fungi, protists, and archaea—is the aim of DNA isolation. There are several commercial kits available for the isolation of DNA from different types of metagenomic samples; every kit is based on the same technology. The list of commercially available kits is summarized in Table 1. In brief, samples are homogenized, and cells are lysed using chemical treatment or mechanical means. Further, samples are centrifuged to rid out cellular debris and contaminants. Next, DNA is purified using phenol–chloroform or silica-based spin column-based, magnetic bead-based, or by precipitating with ethanol or isopropanol. Every method has its own advantages and disadvantages, and the selection of the method is based on the type of samples, throughput, and downstream applications [12].

3 Sequencing Process

The sequencing process for metagenomics or microbiome analysis starts with wet lab process which includes DNA isolation, quality control, fragmentation, library preparation, and qubit analysis, followed by the library pooling and sequencing. The library preparation method also depends upon the researcher's choice, currently two main approaches of library preparation are followed for metagenomics or microbiome analysis. The first method is based on 16 s rRNA, where the targeted region is PCR amplified and further pooled for sequencing [9]. Conversely, in whole genome shotgun sequencing method, the DNA is fragmented, followed by library construction, amplification, library pooling and sequencing [8]. In this portion we will briefly

Table 1 Commercially available kits for DNA isolation

Commercially available kit	Company	Type of samples
QIAamp fast DNA stool mini kit	Qiagen	Stool sample
QIAamp DNA microbiome kit	Qiagen	Human body fluid
QIAamp power fecal DNA kit	Qiagen	Body fluid
DNeasy blood and tissue kit	Qiagen	Blood and tissue
DNeasy power soil pro kit	Qiagen	Soil
PureLink microbiome DNA purification kit	Invitrogen	Wide variety of samples (Soil, Stool, Body fluid etc.)
Monarch HMW DNA extraction kit	NEB	Wide variety of samples (Soil, Stool, Body fluid etc.)
Soil gnomic DNA isolation kit	LSBio	Soil kit

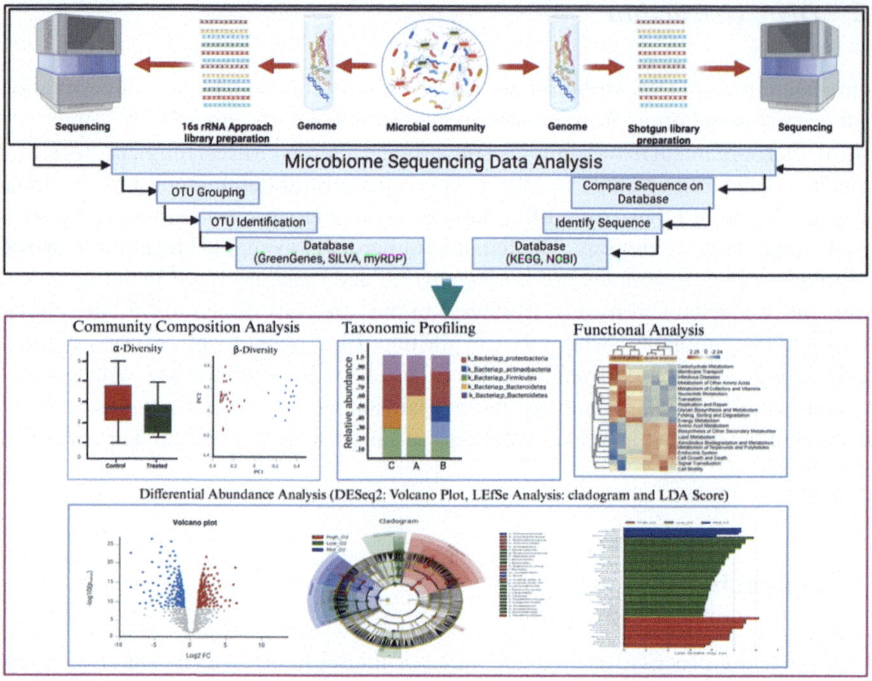

Fig. 1 Showing workflow of 16 s rRNA and shotgun sequencing approaches of metagenomics

look into the both library preparation and sequencing method. A comprehensive flow chart of sequencing and computational analysis is given in Fig. 1.

3.1 Shotgun Sequencing Methodology

- DNA fragmentation: In order to facilitate sequencing, the collected DNA must usually be broken up into smaller fragments. DNA fragmentation is done by physical or enzymatic methods. During the enzymatic method, the DNA is subjected to DNase I or transposases and incubated for a specific period at 37 °C. The selection of an enzyme totally depends upon the application. Further, the enzyme is deactivated by chelating agents or by using heat deactivation and purified for further application. In the physical process, DNA is subjected to either sonication, hydrodynamic shearing, or nebulization. The size of DNA fragments can be controlled by the intensity and duration of sonication.
- Library Construction: Fragmented DNA is further subjected to library construction. In brief, fragmented DNA is ligated with an adapter at both ends. Adapters consist of special i5 and i7 index sequences for later use in sequencing. Further,

these ligated DNA products are shorted and selected based on the size, and only the desired DNA length is selected for the library.

- Amplification: Further, the shorted DNA fragment library is enriched with short PCR amplification, and moreover, specialized sequences are added for further sequencing steps.
- Sequencing: Further, enriched libraries are pooled and loaded in the cartridges for high-throughput sequencing using a choice of a sequencing platform. After sequencing the data is converted in FASTA format and stored in Illumina base space or local drive. These files are further used for downstream analysis.

3.2 16 s rRNA Targeted Sequencing

- PCR enrichment of Specific Region: The objective of 16S rRNA sequencing is to enrich and sequence a particular portion of the 16S rRNA gene, which is mostly conserved in microbes and archaea but also includes variable segments which can be utilized for taxonomical segregation.
- Targeted Amplification: During this process, targeted primers are designed to variable segments of the 16 s rRNA gene. Mostly, primers are highly specific to conserved segments flanked by the variable segment, which allows the selective amplification of the target region from the genomic DNA or cDNA.
- Library Construction: Similar to the shotgun method, the amplified DNA is subjected to adaptor ligation, which contains the unique molecular identifiers to determine the PCR duplicates during the sequencing process.
- Pooling and Sequencing: After library construction, all samples are pooled together and further loaded on the choice of sequencing platform.

4 Sequencing Data Analysis

Once the sequencing is completed, the major hurdle is to analyze the billions of reads that come from the sequencing of multiple samples. Therefore, it is necessary to refine the data to extract nucleotide sequences that support the hypothesis and provide meaningful and relevant information. Various software tools have been developed to analyze large datasets, each designed to perform specific functions [13]. In this section, we will briefly understand the flow of metagenomics data analysis, software, and tools used.

4.1 Quality Control and Preprocessing of Raw Sequences

Post-sequencing, the raw sequencing reads are processed for quality control, leading to the removal of low-quality reads and adapter sequences. The reads are subjected to a trimming and filtering process; sequences meeting the specific criteria, including GC percentage and length, are only kept for further analysis (called high-quality reads), and other low-quality reads are discarded for subsequent analysis. There are several freely available tools for this process that can perform trimming and filtering of raw sequencing; these include Trimmomatic, FastqCleaner, Cutadapt, EA-Tools, Trim Galore, and AdaptorRemoval [14].

4.2 Assembly Preparation

- Genome assembly is the process of reconstructing a genome from the short DNA reads obtained after the sequencing process [15]. There are several algorithms available for genome assembly, such as de novo assembly and comparative assembly. The available short reads were assembled and aligned against a sequence, and then a genome assembly was reconstructed and used for further analysis. The assembly preparation workflow have been shown in Fig. 2.

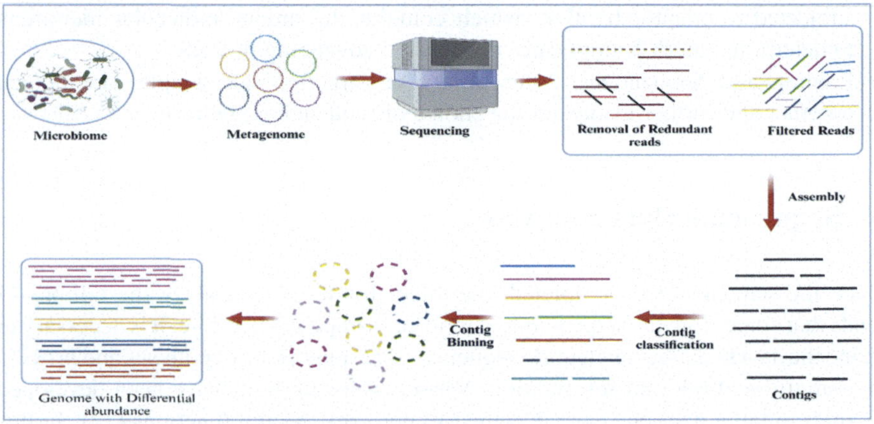

Fig. 2 Represent the process of assembly preparation

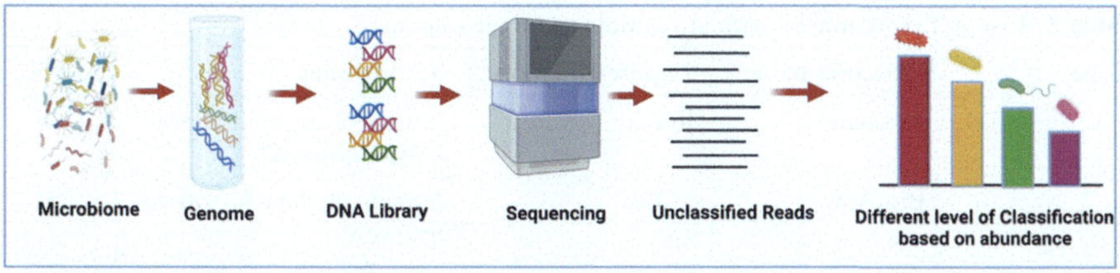

Fig. 3 Represents the process of taxonomical classification of unclassified reads. After sequencing, the unclassified reads are assigned with labels and further classified to particular taxa, and their abundance is calculated and further represented as a graph

4.3 *Taxonomy Analysis*

In metagenomics, taxonomy characterization refers to the identification and quantification of the various kinds of microorganisms that exist in a particular biological specimen using the genetic material (DNA or RNA) that has been recovered from the specimen. The fundamental process of metagenomics analysis is the taxonomical categorization of the corresponding sequence reads and their grouping into OTUs (operational taxonomic units). This process involves assigning taxonomical labels to each sequenced read or contigs by comparing them with microbial genomic sequences present on the reference sequence database NCBI RefSeq, SILVA [16], GreenGenes [17], unite, RDP [18], and EzBioCloud [19]. The most often used pipelines for taxonomic analysis are QIIME2 [20], MetaPhlAn 4.0, Kraken 2.0, and Mothur [21]. However, these pipelines necessitate the deployment of extra tools for subsequent analysis, and most importantly, these pipelines also require proficiency in bioinformatics programming knowledge. The process of taxonomic analysis is depicted in Fig. 3.

Recently, a great deal of algorithms have been created with the goal of quickly and accurately categorizing vast volumes of sequencing data to established taxa. These tools are often termed as classifier [22]. These classifiers function on two different approaches known as taxonomic binning and taxonomic profiling. The taxonomic binning approach classifies each sequence read to reference taxa, while the profiler tells us about the relative abundances of taxa in given data without classifying the individual reads. These classifier tools are broadly divided into three groups: (1) DNA-DNA classification, (2) Protein, and (3) Marker classification [23]. The available tools in each category are tabulated in Table 2.

4.4 *Community Composition Analysis*

Characterizing community composition in metagenomics involves the identification and quantification of microbial taxa present in a sample. This is typically achieved

Table 2 List of Taxonomic classification tools and their source

Type	Classifier name	Required memory	Tools source	Refs.
	Kraken2	36 Gb	https://github.com/DerrickWood/kraken2	[24]
	Bracken	<1 Gb	https://ccb.jhu.edu/software/bracken/	[25]
	Centrifuge	20 Gb	https://github.com/DaehwanKimLab/centrifuge	[26]
	PathSeq	140 Gb	https://github.com/ChandraPedamallu/PathSeq	[27]
Protein	DIAMOND	Variable	https://github.com/bbuchfink/diamond	[28]
	Kaiju	25 Gb	https://github.com/bioinformatics-centre/kaiju	[29]
	MMSeqs2	85 Gb	https://github.com/soedinglab/MMseqs2-App	[30]
Marker based	mOTUs2	2 Gb	https://github.com/motu-tool/mOTUs	[31]
	MetaPhlAn4	2 Gb	https://github.com/biobakery/MetaPhlAn4	[32]

through taxonomic classification of sequencing reads or contigs, followed by abundance profiling to determine the relative abundance of different taxa within the microbial community. This analysis provides insights into the diversity, abundance, and structure of microbial populations, facilitating the understanding of ecosystem dynamics, biogeochemical processes, and associations with health or disease. After taxonomic classification, alpha and beta diversity are measured, stating the diversity of microbial community [33]. Alpha Diversity measures the diversity of microbial community within the samples. During this analysis, parameters such as species diversity and distribution of abundance among taxa are considered. While beta diversity measures the differences in microbial community composition between samples. Tools such as QIIME2, MetaPhlAn4 R script, Mothur, and Phyloseq are frequently used for the analysis of alpha and beta diversity [21, 34, 35].

4.5 Differential Abundance Analysis

Differential abundance analysis is a statistical methodology performed for determining the microbial taxa or functional traits that are highly abundant or reduced between various categories of samples. For taxonomical data, tools such as DESeq2, edgeR, or ALDEx2 are used to identify the differential abundance, while for functional data, LEfSe (Linear discriminant analysis Effect Size), DEseq2 or ANCOM (analysis of Composition of Microbiome) are used to measure the differential abundant features. After statistical significance analysis, effect size estimation is also performed to evaluate the magnitude of differences in abundance between sample groups and is represented in terms of log-fold change, fold change, or effect size coefficients, revealing the biological significance of differential abundance [36, 37]. These analyses play an important part in microbiome research by revealing the key taxa or traits that play key roles in particular phenotypes, conditions, or treatment processes. It sheds light on microbial signatures, assists in discovering putative markers, and highlights the biological processes underpinning the dynamic nature of microbial communities.

4.6 Functional Analysis

Metagenomics functional analysis facilitates the characterization of the function potential of microbial communities present in the environmental species. In contrast to taxonomy analysis, which deals with determining the taxonomic makeup of microorganisms, functional analysis attempts to comprehend the metabolic processes and biological roles that are encoded in the whole genomes of community microbes [38]. Metagenomic functional analysis can reveal better knowledge of the metabolic capabilities and functionality of microbial communities, as well as their involvement in different ecosystems, environmental processes, and host-microbe interactions.

The functional analysis starts with sequence annotation, where raw sequence data (reads/contigs) are aligned against the known database using BLAST, HMMER, or DIAMOND. This step will reveal the genes, protein-coding sequences, and other functional elements. Further, functional profiling is performed to measure the abundant genes, pathways, and biological features present within the metagenomic dataset. Next, the process involves pathway reconstruction, where metagenomic data are reconstructed to predict the metabolic pathways and metabolic potential of microbial communities. The most popular tools, including HUMAnN, MEGAN, or PICRUSt, are most frequently used to decipher the pathways and their function from the metagenomics dataset. During functional annotation, reference databases such as KEGG (Kyoto Encyclopedia of Genes and Genomes), COG (Clusters of Orthologous Groups), Pfam, or SEED are frequently used for functional annotation.

5 Application of Metagenomics

Metagenomics applications are distributed in different fields, which include clinical metagenomics, biotechnology, agriculture, and environmental studies. The application of metagenomics in different fields and the recent studies are given in the section below.

5.1 *Clinical Metagenomics*

Recent clinical metagenomics studies are focused on identifying the pathogenic as well as therapeutic microbes in different diseases. Moreover, tons of studies are focused on FMT (fecal microbiota transplantation) based therapeutic studies in different diseases. The recent finding reveals the association of intestinal microbiome in several diseases, including non-alcoholic fatty liver [39], autoimmune disease [40], and tumours [41].

5.2 *Environmental Metagenomics*

Metagenomics enables the comprehensive study of microbial communities in environmental samples, facilitating environmental monitoring, bioremediation, and assessment of ecosystem health. Moreover, recent metagenomics studies on environmental samples also provide insights into the microbial process involved in the biogeochemical cycling of elements such as carbon, nitrogen, and Sulphur [42].

6 Challenges and Limitations of Metagenomics

The metagenomics and microbiome analysis has several limitations, which start from sample acquisition to the bioinformatics analysis and interpretation of results. These challenges require interdisciplinary collaboration, standardization of methodologies, improved computational algorithms, and ethical frameworks to ensure the responsible and ethical conduct of metagenomics and microbiome research. Some crucial challenges and limitations of metagenomics and microbiome analysis are discussed below.

- Sample contamination: There is always a potential chance of sample contamination during collection, handling, and processing, leading to erroneous interpretations of microbial community composition [43].
- Data analysis complexity: Metagenomic data analysis requires sophisticated computational tools and bioinformatics expertise, posing challenges for researchers without access to computational resources or specialized training [44].
- Taxonomic and Functional Annotation Accuracy: Taxonomic and functional annotation of metagenomic data may suffer from inaccuracies and biases, affecting the reliability of downstream analyses and interpretations [45].
- Standardization and Reproducibility: Lack of standardized protocols and methodologies for sample collection, DNA extraction, sequencing, and data analysis hinders reproducibility and comparability of results across studies [46].

- Host-Microbiome Interactions: Understanding the complex interactions between host genetics, environmental factors, and the microbiome presents challenges in deciphering causal relationships and identifying biomarkers of disease.
- Ethical Legal Considerations: Ethical dilemmas regarding informed consent, data privacy, and ownership of microbiome data pose challenges for researchers, healthcare providers, and policymakers [47].

7 Conclusion

In conclusion, metagenomics and microbiome analysis are effective methods for researching microbial communities in a variety of settings, including the human body, soils, seas, and more. Metagenomics offers insights into the taxonomic makeup, functional potential, and ecological dynamics of microbial ecosystems by examining genetic material that has been extracted directly from environmental samples. Likewise, microbiome analysis provides insight into the intricate relationships that exist between microorganisms and their hosts, illuminating the function of the microbiome in human health, illness, and other domains.

The applications of metagenomics and microbiome analysis are vast and continue to expand rapidly. These methods are essential to environmental microbiology in order to comprehend biogeochemical cycling, ecosystem processes, and reactions to shifts in the environment. Research on the microbiome has potential applications in human health, including illness prevention, personalized medicine, and microbiome-based therapies. Furthermore, metagenomics in agriculture provides chances to improve crop yield, soil health, and sustainable food production methods. Metagenomics and microbiome analysis will become more crucial as technology and computational tools develop to tackle urgent problems in science, health, and sustainability. But along with these developments come difficulties, such as standardization, data integration, and ethical issues. Interdisciplinary cooperation, creative approaches, and conscientious management of microbial resources are all necessary to meet these problems.

In summary, metagenomics and microbiome analysis have revolutionized our understanding of microbial life on Earth, offering unprecedented insights into the hidden world of microbes and their myriad roles in shaping ecosystems, human health, and the environment. By harnessing the power of metagenomics, we are poised to unlock new frontiers in biology, medicine, and biotechnology, with the potential to improve human health, protect the environment, and advance our understanding of life's diversity and complexity.

References

1. Hiraoka, S., Yang, C. C., & Iwasaki, W. (2016). Metagenomics and bioinformatics in microbial ecology: Current status and beyond. *Microbes and Environments, 31*(3), 204–212. https://doi.org/10.1264/jsme2.ME16024
2. Prayogo, F. A., Budiharjo, A., Kusumaningrum, H. P., Wijanarka, W., Suprihadi, A., & Nurhayati, N. (2020). Metagenomic applications in exploration and development of novel enzymes from nature: A review. *Journal, Genetic Engineering & Biotechnology, 18*(1), 39. https://doi.org/10.1186/s43141-020-00043-9
3. Nwachukwu, B. C., & Babalola, O. O. (2022). Metagenomics: A tool for exploring key microbiome with the potentials for improving sustainable agriculture. *Front Sustain Food Syst., 6*, 886987. https://doi.org/10.3389/fsufs.2022.886987
4. Zhang, L., Chen, F., Zeng, Z., et al. (2021). Advances in metagenomics and its application in environmental microorganisms. *Frontiers in Microbiology, 12*, 766364. https://doi.org/10.3389/fmicb.2021.766364
5. Garg, D., Patel, N., Rawat, A., & Rosado, A. S. (2024). Cutting edge tools in the field of soil microbiology. *Current Research in Microbial Sciences, 6*, 100226. https://doi.org/10.1016/j.crmicr.2024.100226
6. Martín, R., Miquel, S., Langella, P., & Bermúdez-Humarán, L. G. (2014). The role of metagenomics in understanding the human microbiome in health and disease. *Virulence, 5*(3), 413–423. https://doi.org/10.4161/viru.27864
7. Afzaal, M., Saeed, F., Shah, Y. A., et al. (2022). Human gut microbiota in health and disease: Unveiling the relationship. *Frontiers in Microbiology, 13*, 999001. https://doi.org/10.3389/fmicb.2022.999001
8. Kim, C., Pongpanich, M., & Porntaveetus, T. (2024). Unraveling metagenomics through long-read sequencing: A comprehensive review. *Journal of Translational Medicine, 22*(1), 111. https://doi.org/10.1186/s12967-024-04917-1
9. Case, R. J., Boucher, Y., Dahllöf, I., Holmström, C., Doolittle, W. F., & Kjelleberg, S. (2007). Use of 16S rRNA and *rpoB* genes as molecular markers for microbial ecology studies. *Applied and Environment Microbiology, 73*(1), 278–288. https://doi.org/10.1128/AEM.01177-06
10. Scholz, M. B., Lo, C. C., & Chain, P. S. G. (2012). Next generation sequencing and bioinformatic bottlenecks: The current state of metagenomic data analysis. *Current Opinion in Biotechnology, 23*(1), 9–15. https://doi.org/10.1016/j.copbio.2011.11.013
11. Lindahl, B. D., Nilsson, R. H., Tedersoo, L., et al. (2013). Fungal community analysis by high-throughput sequencing of amplified markers–a user's guide. *New Phytologist, 199*(1), 288–299. https://doi.org/10.1111/nph.12243
12. Greathouse, K. L., Sinha, R., & Vogtmann, E. (2019). DNA extraction for human microbiome studies: The issue of standardization. *Genome Biology, 20*(1), 212. https://doi.org/10.1186/s13059-019-1843-8
13. Navgire, G. S., Goel, N., Sawhney, G., et al. (2022). Analysis and Interpretation of metagenomics data: An approach. *Biological Procedures Online, 24*(1), 18. https://doi.org/10.1186/s12575-022-00179-7
14. Lindgreen, S. (2012). AdapterRemoval: Easy cleaning of next-generation sequencing reads. *BMC Research Notes, 5*, 337. https://doi.org/10.1186/1756-0500-5-337
15. Reich, J. G., Drabsch, H., & Däumler, A. (1984). On the statistical assessment of similarities in DNA sequences. *Nucleic Acids Research, 12*(13), 5529–5543. https://doi.org/10.1093/nar/12.13.5529
16. Yilmaz, P., Parfrey, L. W., Yarza, P., et al. (2014). The SILVA and "All-species Living Tree Project (LTP)" taxonomic frameworks. *Nucleic Acids Research, 42*(D1), D643–D648. https://doi.org/10.1093/nar/gkt1209
17. DeSantis, T. Z., Hugenholtz, P., Larsen, N., et al. (2006). Greengenes, a chimera-checked 16S rRNA gene database and workbench compatible with ARB. *Applied and Environment Microbiology, 72*(7), 5069–5072. https://doi.org/10.1128/AEM.03006-05

18. Cole, J. R. (2004). The Ribosomal Database Project (RDP-II): Sequences and tools for high-throughput rRNA analysis. *Nucleic Acids Research, 33*(Database issue), D294–D296. https://doi.org/10.1093/nar/gki038.
19. Yoon, S. H., Ha, S. M., Kwon, S., et al. (2017). Introducing EzBioCloud: A taxonomically united database of 16S rRNA gene sequences and whole-genome assemblies. *International Journal of Systematic and Evolutionary Microbiology, 67*(5), 1613–1617. https://doi.org/10.1099/ijsem.0.001755
20. Caporaso, J. G., Kuczynski, J., Stombaugh, J., et al. (2010). QIIME allows analysis of high-throughput community sequencing data. *Nature Methods, 7*(5), 335–336. https://doi.org/10.1038/nmeth.f.303
21. Schloss, P. D., Westcott, S. L., Ryabin, T., et al. Introducing Mothur: Open-source, platform-independent, community-supported software for describing and comparing microbial communities. *Applied and Environmental Microbiology, 75*(23), 7537–7541. https://doi.org/10.1128/AEM.01541-09.
22. Wu, Y. W., Simmons, B. A., & Singer, S. W. (2016). MaxBin 2.0: An automated binning algorithm to recover genomes from multiple metagenomic datasets. *Bioinformatics, 32*(4), 605–607. https://doi.org/10.1093/bioinformatics/btv638.
23. Edgar, R. C. (2018). Updating the 97% identity threshold for 16S ribosomal RNA OTUs. *Bioinformatics, 34*(14), 2371–2375. https://doi.org/10.1093/bioinformatics/bty113
24. Wood, D. E., & Salzberg, S. L. (2014). Kraken: Ultrafast metagenomic sequence classification using exact alignments. *Genome Biology, 15*(3), R46. https://doi.org/10.1186/gb-2014-15-3-r46
25. Lu, J., Breitwieser, F. P., Thielen, P., & Salzberg, S. L. (2017). Bracken: Estimating species abundance in metagenomics data. *PeerJ Computer Science, 3*, e104. https://doi.org/10.7717/peerj-cs.104
26. Morgulis, A., Coulouris, G., Raytselis, Y., Madden, T. L., Agarwala, R., & Schäffer, A. A. (2008). Database indexing for production MegaBLAST searches. *Bioinformatics, 24*(16), 1757–1764. https://doi.org/10.1093/bioinformatics/btn322
27. Walker, M. A., Pedamallu, C. S., Ojesina, A. I., et al. (2018). GATK PathSeq: A customizable computational tool for the discovery and identification of microbial sequences in libraries from eukaryotic hosts. *Bioinformatics, 34*(24), 4287–4289. https://doi.org/10.1093/bioinformatics/bty501
28. Buchfink, B., Xie, C., & Huson, D. H. (2015). Fast and sensitive protein alignment using DIAMOND. *Nature Methods, 12*(1), 59–60. https://doi.org/10.1038/nmeth.3176
29. Menzel, P., Ng, K. L., & Krogh, A. (2016). Fast and sensitive taxonomic classification for metagenomics with Kaiju. *Nature Communications, 7*, 11257. https://doi.org/10.1038/ncomms11257
30. Steinegger, M., & Söding, J. (2017). MMseqs2 enables sensitive protein sequence searching for the analysis of massive data sets. *Nature Biotechnology, 35*(11), 1026–1028. https://doi.org/10.1038/nbt.3988
31. Milanese, A., Mende, D. R., Paoli, L., et al. (2019). Microbial abundance, activity and population genomic profiling with mOTUs2. *Nature Communications, 10*(1), 1014. https://doi.org/10.1038/s41467-019-08844-4
32. Truong, D. T., Franzosa, E. A., Tickle, T. L., et al. (2015). MetaPhlAn2 for enhanced metagenomic taxonomic profiling. *Nature Methods, 12*(10), 902–903. https://doi.org/10.1038/nmeth.3589
33. Jankowski, P., Gan, J., Le, T., et al. (2022). Metagenomic community composition and resistome analysis in a full-scale cold climate wastewater treatment plant. *Environmental Microbiome, 17*(1), 3. https://doi.org/10.1186/s40793-022-00398-1
34. Bolyen, E., Rideout, J. R., Dillon, M. R., et al. (2018). QIIME 2: Reproducible, interactive, scalable, and extensible microbiome data science. https://doi.org/10.7287/peerj.preprints.27295v2. (Published online December 3, 2018)
35. McMurdie, P. J., & Holmes, S. (2013). phyloseq: An R package for reproducible interactive analysis and graphics of microbiome census data. *PLoS ONE, 8*(4), e61217. https://doi.org/10.1371/journal.pone.0061217. (M. Watson (Ed.))

36. Love, M. I., Huber, W., & Anders, S. (2014). Moderated estimation of fold change and dispersion for RNA-seq data with DESeq2. *Genome Biology, 15*(12), 550. https://doi.org/10.1186/s13059-014-0550-8
37. Segata, N., Izard, J., Waldron, L., et al. (2011). Metagenomic biomarker discovery and explanation. *Genome Biology, 12*(6), R60. https://doi.org/10.1186/gb-2011-12-6-r60
38. Gilbert, J. A., Field, D., Swift, P., et al. (2010). The taxonomic and functional diversity of microbes at a temperate coastal site: A 'Multi-Omic' study of seasonal and diel temporal variation. *PLoS ONE, 5*(11), e15545. https://doi.org/10.1371/journal.pone.0015545. (F. Rodriguez-Valera (Ed.))
39. Aron-Wisnewsky, J., Vigliotti, C., Witjes, J., et al. (2020). Gut microbiota and human NAFLD: Disentangling microbial signatures from metabolic disorders. *Nature Reviews Gastroenterology & Hepatology, 17*(5), 279–297. https://doi.org/10.1038/s41575-020-0269-9
40. Svoboda, E. (2021). Gut feeling yields evidence of microbial involvement in autoimmunity. *Nature, 595*(7867), S54–S55. https://doi.org/10.1038/d41586-021-01837-8
41. Andrews, M. C., Duong, C. P. M., Gopalakrishnan, V., et al. (2021). Gut microbiota signatures are associated with toxicity to combined CTLA-4 and PD-1 blockade. *Nature Medicine, 27*(8), 1432–1441. https://doi.org/10.1038/s41591-021-01406-6
42. Zhao, A., Lu, Y., Li, Q., Li, T., & Zhao, J. (2023). Metagenomics reveals the diversity and role of surface-water microbes in biogeochemical cycles in lakes at different terrain ladders. *Frontiers in Environmental Science, 1*, 1121775. https://doi.org/10.3389/fenvs.2023.1121775
43. Cheek, M., Onana, J. M., Chapman, H. M. (2021). The montane trees of the Cameroon Highlands, West-Central Africa, with *Deinbollia onanae* sp. nov. (Sapindaceae), a new primate-dispersed, endangered species. *PeerJ, 9*, e11036. https://doi.org/10.7717/peerj.11036.
44. Pasolli, E., Asnicar, F., Manara, S., et al. (2019). Extensive unexplored human microbiome diversity revealed by over 150,000 genomes from metagenomes spanning age, geography, and lifestyle. *Cell, 176*(3), 649-662.e20. https://doi.org/10.1016/j.cell.2019.01.001
45. Egbebiyi, T. S., Crespo, O., Lennard, C., et al. (2020). Investigating the potential impact of 1.5, 2 and 3 °C global warming levels on crop suitability and planting season over West Africa. *PeerJ, 8*, e8851. https://doi.org/10.7717/peerj.8851.
46. David, M. M. (2021). The role of the microbiome in autism: All that we know about all that we don't know. *mSystems, 6*(2), e00234–21. https://doi.org/10.1128/mSystems.00234-21.
47. Dhariwal, A., Chong, J., Habib, S., King, I. L., Agellon, L. B., & Xia, J. (2017). MicrobiomeAnalyst: A web-based tool for comprehensive statistical, visual and meta-analysis of microbiome data. *Nucleic Acids Research, 45*(W1), W180–W188. https://doi.org/10.1093/nar/gkx295

Opportunities and Prospects of Artificial Intelligence in Plant Genomics

Sona Charles, A. Subeesh, V. G. Dhanya, and V. M. Malathi

Abstract The global food production system is currently impacted by several challenges, and hence, the adoption of modern technologies for crop improvement and sustainable agriculture is the need of the hour. The realm of *omic* technologies has revolutionized agricultural research by exploring high throughput genomic and transcriptomic data to identify suitable candidate genes for developing favorable agronomic and physiological traits. With a rapid burst of "biological big data", the use of artificial intelligence methods has been vital in extracting remarkable patterns contributing to predictions of biomarker genes from genomic data, genomic selection (GS) and marker-assisted selection (MAS), gene structure and regulatory element prediction and also in processes like host–pathogen interaction prediction. We review the prospects of machine learning techniques in dealing with complex biological datasets with predictive potential in various fields, including plant genomics, as well as the successful applications thereof.

Keywords Deep learning · Gene prediction · Machine learning · Transcriptomics · Neural networks

---

All authors contributed equally to the manuscript.

S. Charles (✉)
ICAR-Indian Institute of Spices Research, Calicut 673012, Kerala, India
e-mail: Sona.Charles@icar.gov.in

A. Subeesh
ICAR-Central Institute of Agricultural Engineering (CIAE), Bhopal, Madhya Pradesh, India

V. G. Dhanya
ICAR-Sugarcane Breeding Institute (Regional Station), Kannur, Kerala, India

V. M. Malathi
ICAR-Indian Institute of Millets Research (IIMR), Hyderabad 500030, India

© The Author(s), under exclusive license to Springer Nature Singapore Pte Ltd. 2024
A. Chaudhary et al. (eds.), *Unraveling New Frontiers and Advances in Bioinformatics*,
https://doi.org/10.1007/978-981-97-7123-3_5

1 Introduction

The global food production system is under a daunting and challenging phase, owing to rampant climate change, dwindling groundwater supply, increasing input costs, and pest and disease incidences. Further, with the appalling incidence of the COVID-19 pandemic, the agriculture supply chain has been severely cringed, sending waves of shock to the future of food security [1]. Though, even in the past, the agricultural system has shown crest trough graphs, signaling the process of continuous evolution, recent incidents of severe fallouts need to be dealt with with caution. Significant innovations will hence be needed to keep the pace of food production in tune with these global challenges [2]. But the one obvious question here is how to produce sufficient quantities of food for the burgeoning human population, utilizing the available land and water resources as well as the presently accessible state-of-the-art technologies. Throughout the history of agriculture, the scientific utilization of agronomic and physiological traits has paved the way for the development of crop-specific entities, favoring the domestication of crops that are most desirable in terms of production and utilization [3]. Citing the example of developing "rust resistance under the guidance of man" in the 20th century, major selection pressure had been deployed for rust resistance breeding on relevant populations of pathogens, leading to the development of novel pathogenicity profiles capable of overcoming the challenge of rust incidence [4]. Moving forward, the science of plant breeding has been fortunately revolutionized through the technological innovations referred to as '*omics*', for the comprehensive characterization of the traits of concern. Genomics, the information regarding genetic material (the DNA), is the core of crop improvement today, for the identifications of phenotypic variations and novel traits, for the characterizations of physiological processes, etc. The allied technologies viz. phenomics, proteomics, metabolomics, transcriptomics, and epigenomics are primarily aimed at the phenotypes, proteins, metabolites, mRNA, and in the detection of methylation profiles, respectively [5]. Compared to single-omic technology, multiomics offers the capability to analyze and understand the hidden layer of information surrounding different kinds of environmental stresses. Short-read technology for sequencing has made large genome data sets easily and readily accessible for the evaluation of structural variations in germline, RNA sequence expression, and DNA methylation, which in today's advanced stage is used to integrate with the single cell-level genomic processes [6]. With the availability of whole genome sequences, genome editing strategies also became popular, enabling the manipulation of crop genomes, thereby making them adaptable to various environmental stresses. The deployment of CRISPR/Cas9 into the stream has indeed spearheaded the developments for fast and precise modifications of genomes thus becoming a promising tool for the rapid shifting of knowledge from the lab to the land [7]. This, indeed, was a milestone in the arena of agriculture. Nevertheless, these advances in biological sciences are generating huge amounts of data sets, which have become hard and nearly impossible for humankind to process, analyze, and evaluate. And this demands the need to integrate technologies, including artificial intelligence, so that the targets of farming can be achieved at the desired pace.

On one hand, plant breeders and agriculture scientists are bringing in clear pictures for the fine understanding of plant behaviour under diverse climatic conditions thus adding to the bulk of data to be processed [8]. Several multi-national companies have already invested millions of dollars in developing such extensive technologies that can leverage these types of data sets, critically evaluating the type of soil, varieties of seed, weather data, requirement of agricultural inputs, forecast of pest and disease incidences etc. aiding farmers in planning their crop season [9]. In the area of precision agriculture, countries like Romania are currently using technologies such as environmental sensors that predict the variability in weather as early as possible. Thus, due to the complexity of the data type, novel architectures, algorithms and frameworks, several obstacles are to be handled in extracting the valuable and hidden information in the data sets. The recent research on utilizing the capabilities of artificial intelligence, machine learning, and deep learning for analyzing, processing, planning, learning, reasoning, and to bring in intelligent decisions using these sorts of data sets are a boon to the scientists who are deeply dwelled in this challenge.

Increasingly, machine learning technologies have been used in predicting biomarker genes from genomic data, genomic selection (GS) and marker-assisted selection (MAS), gene structure and regulatory element prediction, and processes like host–pathogen interaction prediction. The world's most powerful supercomputer, Summit, is said to have the potential to handle nearly 27,000 GPUs of data sets, thus showing the light of a bright future [10]. The development of AI-induced spatial extraction methods has thus the opportunity to integrate the results of multi-omics technologies with the data collected from manual phenotyping, thus paving the way for climate-resilient agriculture [11].

In this review, we have provided a gentle introduction to the basic machine learning workflow commencing from the different types of data, pre-processing methods, development and validation of various types of machine learning models. Further the recent advances and progress in the application of machine learning techniques in plant genomics research is elaborated based on research and review articles from literature databases such as PubMed, Scopus, Web of Science, ScienceDirect etc.

2 Machine Learning Process Workflow

Artificial intelligence research, and in particular, the advances in machine learning and deep learning, have led to breakthrough innovations in genomics, pathology, agronomy, and other related fields [12]. Machine learning is a rapidly growing area of computer science that focuses on two interrelated questions: (1) How can we develop a computer system/program that automatically improves through experience? (2) What are the fundamental laws (statistical, computational, etc.) that govern all learning systems? [13]. In a traditional programming approach, a developer writes a series of instructions that tells a computer to do certain processing or transformation

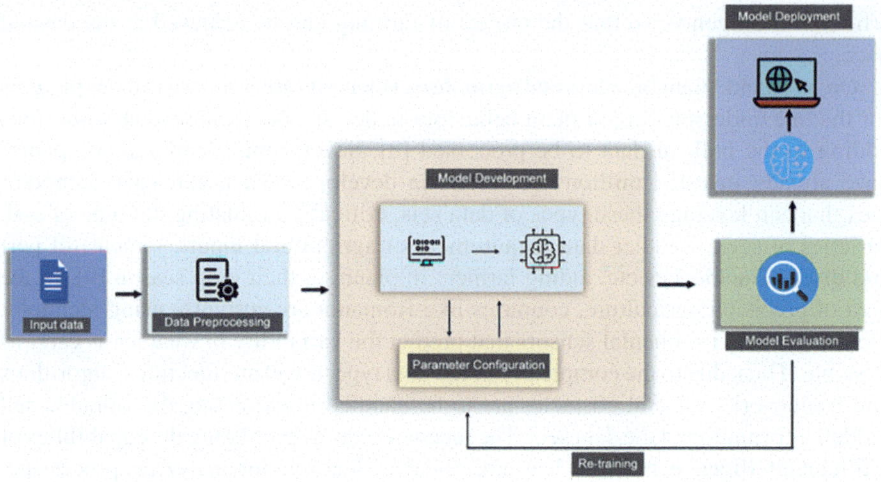

Fig. 1 Machine learning workflow

of input data into desired output. The instructions can be conditional statements (IF-ELSE structure), control statements, etc. Machine learning automates this process of producing desired outcomes with little or no human input and takes actions based on past observations.

Machine learning refers to a computer system or program's ability to acquire and integrate knowledge through a wide range of observations and learn new knowledge without being programmed explicitly. They include programs of data-driven inference rules that make automated pattern recognition in high dimensional-complex data [14]. Some of the patterns in huge datasets may be hidden, and hence, finding those hidden patterns can add significant improvements to the strategic decision-making of any organization. This is typically performed by leveraging computational and statistical algorithms to turn the empirical data into usable intelligent models that can generalize the data with utmost precision [15, 16] A general configuration of the ML system is presented in Fig. 1.

A typical machine learning system starts with the input dataset. Labeled or unlabeled training data is collected from various sources and fed into the machine learning system. This data is consumed by the machine learning algorithms to learn related patterns, and the availability of data is considered the key aspect in developing a data-driven machine learning system.

2.1 Types of Data

There are several types of real-world data viz. structured, semi-structured or unstructured.

2.1.1 Structured Data

A structured data is highly organized and can be easily analyzed and processed by computers. The data is represented in a standardized format, like tables in relational databases and can be easily analyzed by using analytical tools. For instance, SQL relational tables in RDBMS, consisting of tables with rows and columns constitute typical structured data. Structured data storage minimizes data redundancy and makes efficient utilization of memory. This type of data is more interrelated and less flexible.

2.1.2 Semi-Structured Data

Semi-structured data is a type of data with some definite characteristics. However, it does not possess a fixed format like relational databases. Metadata or semantic tags make it more manageable, but still, they contain some variability and inconsistency. The key advantage of semi-structured data is that it is more flexible and easier to scale compared to structured data.

2.1.3 Unstructured Data

Unstructured data is not arranged according to a fixed data schema and cannot be stored in a traditional RDBMS. The majority of the data generated by organizations, IoT, social media, etc., are unstructured, with increasing volume. Unstructured data are usually stored in NoSQL databases, data lakes, and data warehouses. The data's Lack of structure makes it cumbersome to process the data using conventional analytic tools.

However, the advent of AI and machine learning algorithms are currently being utilized to process this data.

2.2 Data Pre-processing

The input data are likely imperfect, containing inconsistencies and redundancies, making them inappropriate for direct application into the machine learning algorithms. After the successful application of the data preprocessing operations [17], the resultant dataset obtained can be regarded as a reliable and suitable source for a machine learning algorithm. One main assumption made by the machine learning and data mining techniques is that the data is complete. However, the presence of missing values is very common in data collection processes 17 and 18 and leads to false conclusions, creating severe difficulties for practitioners. Data imputation techniques like mean/mode/median imputation, arbitrary value imputations, random sample imputations, maximum likelihood procedures, etc. are used to fill the missing values. Identifying the most relevant features from a huge high-dimensional dataset with several features is another main challenge for practitioners in solving complex machine learning problems. Feature selection provides an effective way to solve this problem by providing irrelevant and redundant data that can reduce processing time and improve learning performance [18]. Along with feature selection, space transformation techniques are also one of the data pre-processing approaches. Space transformation generates a new set of features combining the original ones, reducing the dimensionality of the data. Some of the approaches include factor analysis and PCA [19]. In transformation processes, discretization transforms continuous variables using discrete intervals, whereas normalization performs a scaling to the distributions so that it can be represented in a smaller range [20, 21]. In many real-world datasets, there are significant difference between the probabilities of different target classes (in classification problem). This imbalance in classes is addressed by resampling the data to balance the class distribution [22]. Under-sampling decreases the amount of majority target class samples and oversampling increases the amount of minority class samples [23]. Once the data is pre-processed, it can be plugged directly into the machine learning algorithms.

2.3 *Machine Learning Model Development*

In machine learning model development, a set of supervised (for labelled data), unsupervised (for unlabelled data), semi supervised (both labelled and unlabelled data) and reinforcement learning algorithms are available to choose from depending on the nature of input and outcome required. In supervised algorithms, the model gets trained on the labelled training data and the aim of a supervised learning algorithm is to find a best mapping function to map the input variables to the output variable. Two main areas where supervised machine learning approaches used are regression and classification problems [24, 25]. The classification problems predict/classify a discrete value and try to find the decision boundary that divides the data into different classes. Regression finds a function for predicting continuous real values instead of classes or discrete values. Unsupervised learning works without explicit instructions on the description of the target variable on an unlabeled dataset. The algorithm finds the structure and hidden patterns in the input data on its own without requiring the output labels. Clustering and association rule mining are some of the examples of unsupervised learning. Semi-supervised learning uses a small portion of labeled data and a large number of unlabelled data to train a model. This reduces expenses in datacollection and manual annotation, reducing data preparation time. As compared to unsupervised learning, reinforcement learning is different in terms of goals. In reinforcement learning, an agent learns in an interactive environment by trial and error using a feedback from its own actions and experiences. Reinforcement learning uses rewards and punishment for positive and negative actions of the agent respectively [14]. The objective is to find a suitable action model that maximizes the total cumulative award of the agent. Reinforcement learning approaches are widely used in computer games, automation systems, chat bots, etc.

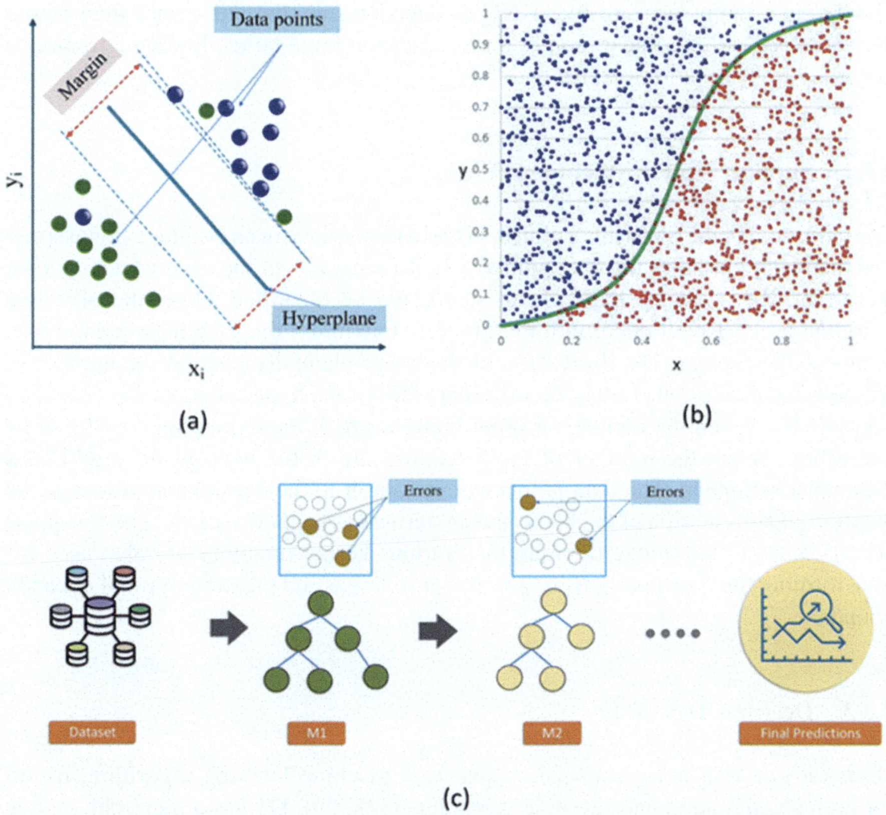

Fig. 2 Machine learning models **a** SVM **b** Logistic regression **c** Gradient boosting model

It is worth noting that no single algorithm works for all the problems. In genomics, majority of the problems falls into the supervised classification problem. Figure 2 represents the various machine learning models.

2.3.1 Logistic Regression

Though the logistic regression approach can be extended to more than two classes however it is often used for binary classification. A logistic regression model is very similar to a linear regression model except that the former uses a more sophisticated function known as 'sigmoid' or 'logistic' function instead of a linear function. Logistic regression is a process of modelling the probability of a discrete outcome, given an input variable. The mathematics of logistic regression rely on the concept of the odds of the event, which indicates the probability of an event occurring divided by the probability of event not occurring [26]. The interpretability of logistic regression is not very straight forward like KNN or decision tree, but still easier than neural networks. Due to the linear, non-complex decision boundaries, linear regression is generally less prone to overfitting.

2.3.2 Support Vector Machine (SVM)

The main goal of the SVM algorithm is to find a hyperplane in an N-dimensional space that distinctly classifies the data points [27]. To separate out the class of data points, there are many possible hyperplanes that could be identified. Here, the objective is to find a plane with maximum margin. i.e., maximum distance between the data points of the classes. The dimension of the hyper plane depends on the number of features and data points falling on different sides of the hyper plane and is classified accordingly. When the number of input features are 2, the hyperplane used will be just a line. When the number of input features are 3, the hyperplane would be a 2-dimensional plane. The data points that are closer to the hyperplane influence the position and orientation of the plane and are termed as support vectors. These support vectors helps to maximize the margins. Various kernel functions are also used for transforming the data points from non-linear to linear and creating optimal decision boundary.

2.3.3 Decision Tree (DT)

Decision tree is a non-parametric supervised machine learning algorithm, useful for both classification and regression problems [28, 29]. DT has a hierarchical, tree structure with root node, internal nodes, leaves and branches. The goal of decision tree is to create a model which can be used to predict the class of the target variable by learning simple decision rules inferred from the training data. There are different ways to select the best feature at each node. Information gain and Gini impurity are two best methods adopted for deciding the splitting criterion for the decision tree model. These approaches evaluate the quality of each condition and how well it can classify the samples into a class. ID3, C4.5, CART are some of the popular decision tree algorithm. Decision trees are comparatively easier to interpret and requires less data pre-processing.

2.3.4 Random Forest (RF)

Random forest, as the name indicates, contains large number of individual decision trees that operate as an ensemble. Each individual tree in the forest makes a prediction for a classification problem, and the class with the most votes becomes the final prediction of the model. For a regression problem, the decision made by the individual trees are averaged to obtain the final prediction. In this approach, a random subset of features are generated, and this ensures low correlation among the decision trees

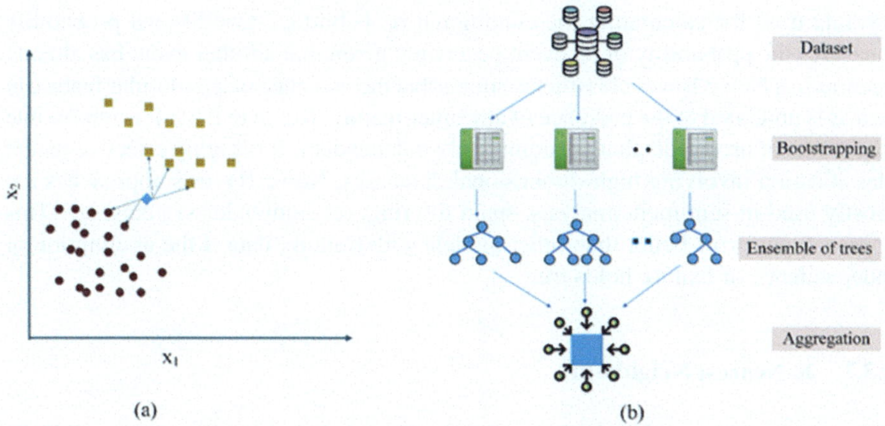

Fig. 3 Machine learning models **a** KNN **b** Random forest

developed (Fig. 3). Random data points are drawn from the training dataset with replacement, and the process is known as 'bootstrap sampling'. RF classifiers tend to outperform most other classification models in terms of performance without any overfitting. RF is more robust to noise and does not require feature scaling [30]. RF is harder to interpret compared to decision trees but it easier to tune the hyper-parameter and obtain accurate predictions.

2.3.5 Gradient Boosting Model

Boosting algorithms are supervised learning algorithms that are mostly used by researchers to increase the level of accuracy in the models. Gradient boosting develops a strong learner in the form of a combination of weak prediction models [31]. The approach relies on the perception that the next model in the sequence will minimize the error present in the previous model. The performance of the models are boosted by assigning higher importance to the samples that are incorrectly classified. Due to the sequential connection, gradient boosting models are relatively slow to learn but provide the best performances. The gradient boosting approach is less prone to overfitting and has an implicit feature selection method. Advanced boosting models provide better cache optimization and distributed computing, ensuring optimum utilization of resources.

2.3.6 Naive Bayes

Naïve Bayes is one of the simplest and most effective classification approaches that helps in developing fast and accurate machine learning models [32]. The algorithm works on the basis of Baye's theorem. Baye's theorem is a simple mathematical formula used for calculating the conditional probabilities. Conditional probability measures the probability of an event occurring, given that another event has already occurred. A Naïve Bayes classifier assumes that the presence of a particular feature in a class is unrelated to the presence of any other feature. But in real life it is impossible to get a set of predictors that are completely independent. It is highly effective in text classification involving high-dimensional databases. Naive Bayes's approaches are mostly used in sentiment analysis, spam filtering, recommender systems, etc. This approach performs better than other models with training data if the assumption of independence of feature holds true.

2.3.7 K-Nearest Neighbours

K-nearest neighbours (KNN) is an instance-based, lazy-learning approach. KNN does not focus on building an internal model, rather it stores all data points in an N-dimensional space [33, 34]. A new data point is classified based on the similarity measures (Euclidean distance, Manhattan distance, etc.). Being a non-parametric algorithm, KNN does not make any assumptions on the underlying data. The value of K in KNN defines how many neighbors are to be taken into consideration to determine the classification of a data point. Identifying the K value is a balancing act, as different values may cause overfitting or underfitting. The performance of the model heavily depends on the data quality, and outliers can impact the overall accuracy of the model. KNN is effective in solving both classification and regression problems (Fig. 3). A major drawback of KNN is its slow processing as the size of the data increases.

2.3.8 Deep Learning Models

Machine learning algorithms play a crucial role in cognitive computing, but many of these algorithms require hand crafting of features, which is often a tedious process. Advanced machine learning approaches like deep learning algorithms are becoming more and more popular as they can automatically extract and learn feature representations [35]. Deep learning systems can learn and interpret complex data, including unstructured and unlabelled data. Different deep learning architectures like ANN, CNN, and RNN have found applications in the field of computer vision, audio/speech recognition, machine translation, bioinformatics, and applications in agriculture. Artificial neural networks are biologically inspired computational models, and like the structure of the human brain, the ANN comprises of neurons in a complex and non-linear fashion. The ANN consists of input, hidden, and output layers, and the neurons are connected to each other by weighted links (Fig. 4). The input layer nodes pass the input information to the hidden layer, and at the hidden layer, most of the computations are done. The nodes get fired using activation functions and output gets produced at the output layer. The ANN gets trained by a back propagation algorithm and requires huge amount of data for training.

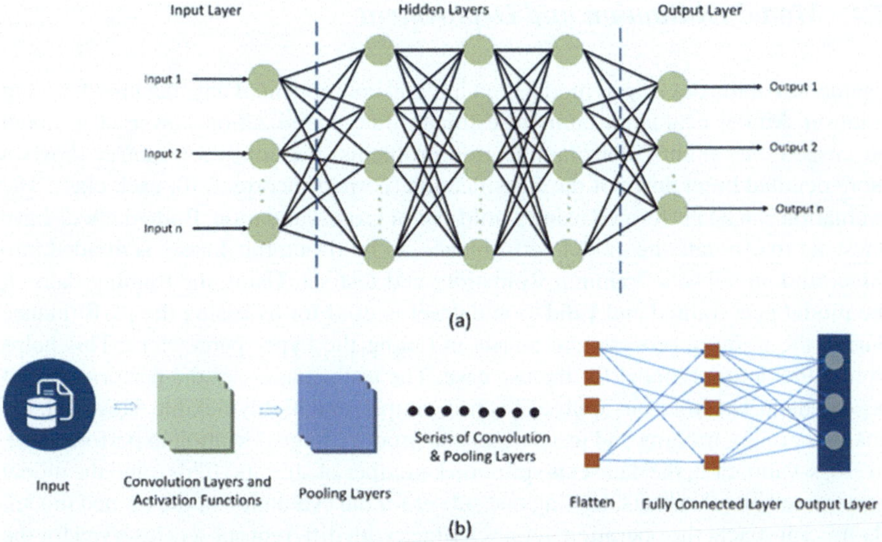

Fig. 4 Deep learning models **a** Simple artificial neural network **b** CNN

CNN (Convolutional Neural Network) is an integral part of modern computer vision systems for achieving object detection, image classification and segmentation tasks. CNN can act as an image recognition tool for machines. CNN is different from ANN as it is structured differently in multi-dimensional layers with height, width and depth parameters. The model can extract each and every portion of the input image and assigns weights for each neuron based on the significant role of the receptive field [36].

The key components of the CNN model are convolution, pooling, and fully connected layers (Fig. 4). The convolution operation uses several filters to extract features from the input image dataset and generates a feature map. The pooling operation is called subsampling, that reduces the dimensionality of the feature map generated by the convolution operation [37]. Max pooling and average pooling are the commonly used pooling operations in CNN. The fully connected layer acts as the last few layers of the ANN and is a feed-forward neural network. The input to the fully connected layer is the input generated from the final convolutional or pooling layer and it is then flattened before feeding to the fully connected layer.

Apart from CNN, there are other DL architectures such as autoencoder, generative models and recurrent neural networks (RNN).

2.4 Model Evaluation and Deployment

Evaluation of the developed model is an integral component of any machine-learning solution. Model evaluation aims to estimate the generalization power of a model on unseen data points. For classification problems, the confusion matrix shows a more detailed breakdown of the classification (correct/incorrect) for each class. The evaluation can be performed using a hold-out or cross-validation. Both methods have a test set to evaluate the model performance. In hold-out, the dataset is divided into three random subsets: training, validation, and test set. Using the training dataset, the model gets trained and validation dataset is used for assessing the performance during the training process and assists in tuning the hyper parameters. This helps to find the optimal model for the use case. The test set assesses the performance of unseen data. The hold-out approach is fast, simple, and highly flexible. However, the changes among training and test dataset can produce huge deviation in performance. In cross-validation, the dataset is split into k number of subsets (folds), and the model gets trained on all subsets, leaving one subset for the evaluation of the trained model. By this approach, the operation iterates k times with different subsets reserved for the testing process each time. Evaluation with cross-validation is more efficient as every observation is used for both training and testing, thus avoiding potential overfitting problems.

Model evaluation metrics quantify the model performance, and the metrics depend on the type of machine learning algorithm. The performance measures used to evaluate machine learning classifier models are given in Table 1. Deployment of a machine learning model includes the integration of a developed ML model into an existing production environment which can accept input parameters and return the output that can be used in making practical decisions. Most of the projects deploy the developed machine learning models as an on-demand decision-making service. Some modern applications deploy the ML models in edge and mobile devices. Models deployed in cloud services can save computation power and provide scalability. Mobile/web applications leverage web services to provide cheaper and near real-time predictions. The model can be easily made available to the public through API calls.

3 Applications of Artificial Intelligence in Plant Genomics

Technological advances in the genomic era have paved the way for the generation of complex multidimensional datasets. The overwhelming flood of massive datasets contributed to the increasing applications of artificial intelligence-based methods such as machine learning and deep learning in biological sciences. The data derived from genomic sciences are characterized by high dimensionality and uncertainty

Table 1 Performance evaluation metrices of machine learning models

Performance measure	Definition	Formula/plot
Accuracy	The fraction correct classifications, the model got right. i.e., number of correct predictions made by the model divided by the total number of predictions made	$\text{Accuracy} = \frac{\text{TP}+\text{TN}}{\text{TP}+\text{TN}+\text{FP}+\text{N}}$
Precision	The total number of correct positive classifications made (true positives) divided by the total number of positive classifications made by the model (true positives and false positives)	$\text{Precision} = \frac{\text{TP}}{\text{TP}+\text{FP}}$
Recall/ sensitivity	The total number of correct positive classifications made (true positives) divided by the number of positive class data points present in the data (true positives and false negatives)	$\text{Recall} = \frac{\text{TP}}{\text{TP}+\text{FN}}$
Specificity	Specificity evaluates a model's ability to predict true negatives of each available category	$\text{Specificity} = \frac{\text{TN}}{\text{TN}+\text{FP}}$
F1 Score	F1-Score gives a combined idea about precision and recall metrics. It is calculated as the harmonic mean of precision and recall	$\text{F1 Score} = \frac{2*(\text{Precision}*\text{Recall})}{\text{Precision}+\text{Recall}}$
AUC-ROC	The ROC (Receiver Operator Characteristics) is a probability curve plots between TPR (True positive rate) and FPR (False positive rate). AUC (Area Under the Curve) measures the ability of the classifier to distinguish between the classes	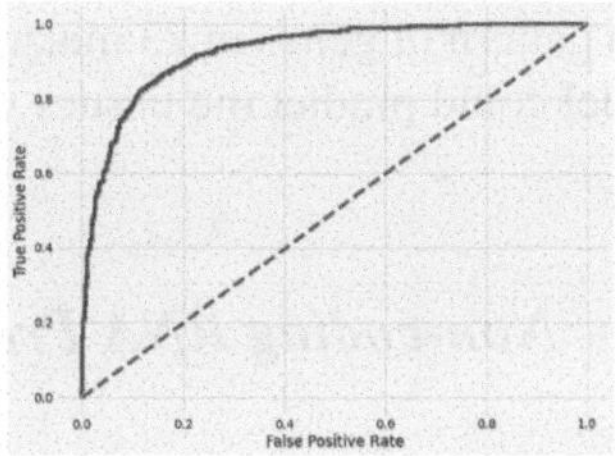

(for example, transcriptomic datasets), which makes it challenging to apply traditional algorithms to extract meaningful hidden information. The broad areas in which machine learning approaches have been extensively applied are briefed below.

3.1 Gene Expression Profiling

The advent of modern sequencing techniques and the enormous generation of high throughput genomic data has delivered a holistic approach to molecular biotechnology. The application of artificial intelligence to achieve an in-depth understanding of *omics* datasets has catered to the investigations for determining candidate genes in various cellular conditions. Computational methods, in combination with experimental data, have huge potential to identify hidden causative elements in a specific biological process. However, the application of machine learning in plant sciences is limited, most of which is trained on the genomic data from the model plant *Arabidopsis thaliana.* NGS technologies have opened up investigative areas of research, such as the identification of novel markers and the discovery of molecules that connect the DNA and phenotypic expression, such as non-coding RNAs and epigenetic variations [38].

Naïve Bayes algorithm, in combination with correlation-based feature selection, identified 128 essential genes in drought tolerance in rice with an accuracy of 93% [39]. The cell population in the roots of *Arabidopsis thaliana* are highly heterogeneous which signifies the potential of identifying marker genes. Machine learning methods, SVM, KNN, RF, and NN were applied to scRNA-seq datasets of root cells to identify marker genes for the classification of ten root cell types in *Arabidopsis thaliana* [40]. Functional characterization of the FBX family of genes in Arabidopsis thaliana was performed by comparing the gene structure, phylogeny functions, and gene expression patterns collected from publicly available databases of known and unknown FBX genes. A neural network-based model was constructed to predict the activity of FBX in Arabidopsis [41]. Meta-analysis of microarray-based studies in combination with machine learning using recursive SVM has been efficiently able to classify abiotic and biotic stress-related genes in rice [42]. Gene expression varies with tissues and conditions. A machine learning model which combines Markov models and a Bayesian network was applied to predict the tissue-specific expression pattern of genes in 23 maize tissues using promoter and protein sequences. The model could predict the tissues with an accuracy of 95% [43].

3.2 Non-coding RNA Prediction

Genomic data is heterogeneous as well as voluminous. Artificial intelligence-based algorithms have the potential to interrogate heterogeneous multidimensional data to uncover patterns through rule-based approaches. Identification of the functional regions of the genome is integral part of annotation. Several machine learning algorithms have been used to predict genomic elements such as miRNA, lncRNAs, polyA sites etc. Broad areas of application and common bioinformatics tools used for the purpose is illustrated in Table 2. MiRLocator is a machine learning-based miRNA prediction tool that uses sequence and structure-based features for prediction. The tool implements RF, SVM, NB, DT, and kNN in Python [44]. Position-specific dinucleotide density variation near miRNA target sites was exploited as features for implementing Support Vector Regression to predict miRNA targets [45]. Prediction of miRNAs responsive to abiotic stress using SVM was developed using pseudo-K-tuple nucleotide compositional feature. The algorithm was implemented on the web through the server ASRmiRNA [46]. Sequence and structural features were combined to develop an SVM classification model to identify pre-miRNAs and

Table 2 Bioinformatics tools for gene structure and structural elements prediction

Application	Tools	Algorithm
Gene structure prediction	Augustus GeneID GRAIL GENSCAN	Markov models Dynamic programming Neural networks Markov models
Splice junctions	SplicePredictor	Bayesian statistics
Transcription start sites	TSSPlant	Artificial Neural Network
Promoter elements	DeePromoter	CNN, LSTM

miRNAs with an accuracy of 90%. The algorithm also identified 522 novel miRNAs from *Solanum lycopersicum* [47]. MaturePred is a tool that can predict mature plant miRNAs from within pre-miRNAs. The tool uses position-specific features, stability-related features, triplet structure features, and energy features to train an SVM model for prediction [48].

3.3 Plant-Pathogen Interaction Prediction

In the current scenario of changing climate, technological interventions in agriculture is aimed at increasing the yield and productivity of crops. The interaction of plants with the microbes, both beneficial and hostile, plays a vital role in plant growth, and the balance is essential for productivity. Advances in genomics have opened up new arenas to decipher the complex mechanisms underlying host–pathogen interactions, including understanding the signaling pathways that are activated in response to the interaction.

Dissection of plant-pathogen interaction involves the prediction of the regulatory network of genes, the prediction of effectors, and genomic selection for resistance breeding [49]. Transcriptomic datasets can help to unveil the underlying regulatory network in host–pathogen interaction. Prediction of plant-associated bacteria was performed using an RF algorithm and was developed into a web server PIFAR (Plant-bacteria Interaction Factors Resource [50]. Plant pathogens localize effectors and other proteins into the apoplast as well as within the host cells. Localization of plant proteins to the apoplast is predicted using machine learning and is implemented through the tool Apoplast-P [51]. Fungal pathogens are known to cause devastating effects on plants. The infection is triggered by effector molecules produced by the pathogens to facilitate infection [52]. One of the first prediction systems for effectors, EffectorP uses features such as length, molecular weight, charge, and amino acid content of cysteine, serine, and tryptophan to distinguish effectors from no effector molecules. In combination with gene expression data from plants, the tool has been efficient in predicting valid candidates [53]. Emerging artificial intelligence method, CNN was used to classify effectors and non-effectors in pathogens, especially bacteria, fungi, and oomycetes in deepredeff. The algorithm has been deployed as a R package where any sequence can be used for prediction [54]. NOD-like receptors play a major role in the plant's innate immunity. They possess diverse sequence motifs, which are identified using a combination of 17 machine learning-based predictors and is available as a web server NLRexpress [55].

3.4 Gene Function Prediction

Identifying and characterizing genomic regions from large-scale genome assemblies is crucial in obtaining high-quality genomic annotations for reproducible research. The Gene Ontology Meta Annotator for Plants (GOMAP) is an annotation pipleline specially designed for plants to extract the ontology terms. GOMAP has been successfully used to annotate four inbred lines of maize [56]. Seqping is a computational tool that combines a self-training HMM model and transcriptomic data for gene structure prediction [57]. Identifying protein functions by exploiting sequence similarity with proteins whose functions are known in combination with gene ontology has been implemented in DeepGOPlus [58].

3.5 Plant Breeding

The goal of any breeding experiment is to correlate the expression of a complex trait as a function of data at the level of gene, phenotype or environment. The application of machine learning in breeding is to identify Quantitative Trait Loci (QTL) which are locations in the genome that are associated with a trait of interest. Other outcomes of QTL mapping are to characterize the genomic structure and estimate the combined effects of all the loci as well as overlapping loci. These problems come under variable selection, estimation and prediction [59].

Using the marker information available, machine learning enables the prediction of phenotypic patterns for new genotypes, Random forest for the prediction of unknown phenotypes has been extensively reviewed by Chen and Ishwaran [60]. Meta-analysis is a method of choice for combining loci from various studies for QTL mapping [61]. The commonly applied genomic selection algorithm ridge regression best linear unbiased predictor (rrBLUP) was compared with multilayer perceptron (MLP) and convolutional neural network (CNN) in spring wheat nested association mapping datasets collected over three years for five traits. The study revealed that deep learning enables better prediction compared to rrBLUP [62]. Though the genomic selection is best associated with the genomic best linear unbiased prediction (GBLUP), other models used for breeding applications include the least absolute shrinkage and selection operator (LASSO) and Bayesian models. MetaQRL [63], solQTL [64] and RASQUAL [65] are some of the bioinformatics tools available for identifying and visualizing QTL effects. By using genotypic data the prediction of unobserved phenotypes can be done in a time and cost effective manner.

4 Conclusion

Artificial intelligence as a data driven technology solution has taken over all facets of crop improvement, and genomics is no exception to it. The massive flood of big data in plant genomics demands the use of advanced computational algorithms to excavate patterns and derive meaningful insights. The massive flood of big data has now become challenging in addressing omics-based technology upgradation in agriculture, where AI is proving a supportive ecosystem for the same. The wealth of sequencing data has made statistical and computational methods indispensable for addressing a plethora of problems in crop improvement and biotechnology.

Integrated analysis of high throughput datasets has enabled the integration of multi-omic data for the identification of regulatory elements in the genome, development of improved breeding strategies, host–pathogen interaction studies, and gene expression profiling. Data-driven technologies are capable of analyzing real-time inputs and are capable of providing inferences in real-time and thereby reducing latency limitations.

Conflict of Interest Authors have no conflict of interest.

References

1. Outlaw, J. L., Fischer, B. L., Anderson, D. P., Klose, S. L., Ribera, L. A., Raulston, J. M., et al. (2020). COVID-19 Impact on Texas Production Agriculture. *Agricultural and Food Policy Center Texas A&M University,* 879.
2. Andersen, M. A., Alston, J. M., Pardey, P. G., & Smith, A. (2018). A century of U.S. farm productivity growth: A surge then a slowdown. *American Journal of Agricultural Economics, 100*(4), 1072–1090. https://doi.org/10.1093/ajae/aay023.
3. Thrall, P. H., Bever, J. D., & Burdon, J. J. (2010). Evolutionary change in agriculture: The past, present and future. *Evolutionary Applications, 3*(5–6), 405–408. https://doi.org/10.1111/j.1752-4571.2010.00155.x
4. Johnson, T. (1961). Man-guided evolution in plant rusts: Through his modification of the host plants of the cereal rusts, man is also modifying the rusts. *Science, 133*(3450), 357–362. https://doi.org/10.1126/science.133.3450.357
5. Rasheed, A., Hao, Y., Xia, X., et al. (2017). Crop breeding chips and genotyping platforms: Progress, challenges, and perspectives. *Molecular Plant, 10*(8), 1047–1064. https://doi.org/10.1016/j.molp.2017.06.008
6. Leonavicius, K., Nainys, J., Kuciauskas, D., & Mazutis, L. (2019). Multi-omics at single-cell resolution: Comparison of experimental and data fusion approaches. *Current Opinion in Biotechnology, 55*, 159–166. https://doi.org/10.1016/j.copbio.2018.09.012

7. Scheben, A., Wolter, F., Batley, J., Puchta, H., & Edwards, D. (2017). Towards CRISPR/Cas crops–bringing together genomics and genome editing. *New Phytologist, 216*(3), 682–698. https://doi.org/10.1111/nph.14702
8. Varshney, R. K., Thudi, M., Roorkiwal, M., et al. (2019). Resequencing of 429 chickpea accessions from 45 countries provides insights into genome diversity, domestication and agronomic traits. *Nature Genetics, 51*(5), 857–864. https://doi.org/10.1038/s41588-019-0401-3
9. Shaw, J., Rudzicz, F., Jamieson, T., & Goldfarb, A. (2019). Artificial intelligence and the implementation challenge. *Journal of Medical Internet Research, 21*(7), e13659. https://doi.org/10.2196/13659.(Published2019July10)
10. Streich, J., Romero, J., Gazolla, J. G. F. M., et al. (2020). Can exascale computing and explainable artificial intelligence applied to plant biology deliver on the United Nations sustainable development goals? *Current Opinion in Biotechnology, 61*, 217–225. https://doi.org/10.1016/j.copbio.2020.01.010
11. Bolger, A. M., Poorter, H., Dumschott, K., et al. (2019). Computational aspects underlying genome to phenome analysis in plants. *The Plant Journal, 97*(1), 182–198. https://doi.org/10.1111/tpj.14179
12. Belyadi, H., & Haghighat, A. (2021). Chapter 1-introduction to machine learning and Python. In H. Belyadi, & A. Haghighat (Eds.), *Machine learning guide for oil and gas using Python* (pp. 1–55). Gulf Professional Publishing. https://doi.org/10.1016/B978-0-12-821929-4.00006-8.
13. Jordan, M. I., & Mitchell, T. M. (2015). Machine learning: Trends, perspectives, and prospects. *Science, 349*(6245), 255–260. https://doi.org/10.1126/science.aaa8415
14. El Bouchefry, K., & de Souza, R. S. (2020). Learning in big data: Introduction to machine learning. In P. Škoda, & F. Adam (Eds.), *Knowledge discovery in big data from astronomy and earth observation* (pp. 225–249). https://doi.org/10.1016/B978-0-12-819154-5.00023-0.
15. Edgar, T. W., & Manz, D. O. (2017). Chapter 6-machine learning. In T. W. Edgar, & D. O. Manz (Eds.), *Research methods for cyber security* (pp. 153–173). Syngress. https://doi.org/10.1016/B978-0-12-805349-2.00006-6.
16. Wolf, B. P. (2009). Chapter 7-machine learning. In B. P. Woolf (Ed.), *Building intelligent interactive tutors* (pp. 221–297). San Francisco: Morgan Kaufmann. https://doi.org/10.1016/B978-0-12-373594-2.00007-1.
17. García, S., Ramírez-Gallego, S., Luengo, J., Benítez, J. M., & Herrera, F. (2016). Big data preprocessing: Methods and prospects. *Big Data Analytics, 1*(1), 9. https://doi.org/10.1186/s41044-016-0014-0
18. Cai, J., Luo, J., Wang, S., & Yang, S. (2018). Feature selection in machine learning: A new perspective. *Neurocomputing, 300*, 70–79. https://doi.org/10.1016/j.neucom.2017.11.077
19. Gündoğdu, Y., Karabağlı, P., Alptekin, H., Şahin, M., & Kılıç, H. Ş. (2019). Comparison of performances of Principal Component Analysis (PCA) and Factor Analysis (FA) methods on the identification of cancerous and healthy colon tissues. *International Journal of Mass Spectrometry, 445*, 116204. https://doi.org/10.1016/j.ijms.2019.116204
20. Obaid, H. S., Dheyab, S. A., & Sabry, S. S. (2019). The impact of data pre-processing techniques and dimensionality reduction on the accuracy of machine learning. In 2019 *9th Annual Information Technology, Electromechanical Engineering and Microelectronics Conference (IEMECON)*, pp. 279–283. https://doi.org/10.1109/IEMECONX.2019.8877011.
21. Singh, D., & Singh, B. (2020). Investigating the impact of data normalization on classification performance. *Applied Soft Computing, 97*, 105524. https://doi.org/10.1016/j.asoc.2019.105524
22. Wang, H., & Wang, S. (2010). Mining incomplete survey data through classification. *Knowledge and Information Systems, 24*(2), 221–233. https://doi.org/10.1007/s10115-009-0245-8
23. Mohammed, R., Rawashdeh, J., & Abdullah, M. (2020). Machine learning with oversampling and undersampling techniques: Overview study and experimental results. In 2020 *11th International Conference on Information and Communication Systems (ICICS)*, pp. 243–248. https://doi.org/10.1109/ICICS49469.2020.239556.
24. Kotsiantis, S. B., Zaharakis, I. D., & Pintelas, P. E. (2006). Machine learning: A review of classification and combining techniques. *Artificial Intelligence Review, 26*, 159–190.

25. Matloff, N. (2017). Statistical regression and classification: From linear models to machine learning. *Chapman and Hall/CRC*. https://doi.org/10.1201/9781315119588
26. Bartosik, A., & Whittingham, H. (2021). Chapter 7-evaluating safety and toxicity. In S. K. Ashenden (Ed.), *The era of artificial intelligence, machine learning, and data science in the pharmaceutical industry* (pp. 119–137). Academic Press. https://doi.org/10.1016/B978-0-12-820045-2.00008-8.
27. Chandra, M. A., & Bedi, S. S. (2021). Survey on SVM and their application in image classification. *International Journal of Information Technology, 13*(5), 1–11. https://doi.org/10.1007/s41870-017-0080-1
28. Myles, A. J., Feudale, R. N., Liu, Y., Woody, N. A., & Brown, S. D. (2004). An introduction to decision tree modeling. *Journal of Chemometrics, 18*(6), 275–285. https://doi.org/10.1002/cem.873
29. Song, Y. Y., & Lu Y. (2012). Decision tree methods: Applications for classification and prediction. *Shanghai Arch Psychiatry, 27*(2), 130–135. https://doi.org/10.11919/j.issn.1002-0829.215044.
30. Misra, S., & Li, H. (2020). Chapter 9-noninvasive fracture characterization based on the classification of sonic wave travel times. In S. Misra, H. Li, & J. He (Eds.), Machine learning for subsurface characterization (pp. 243–287). Gulf Professional Publishing. https://doi.org/10.1016/B978-0-12-817736-5.00009-0.
31. Natekin, A., & Knoll, A. (2013). Gradient boosting machines, a tutorial. *Front Neurorobotics*, 7. https://doi.org/10.3389/fnbot.2013.00021.
32. Flach, P. A., & Lachiche, N. (2004). Naive Bayesian classification of structured data. *Machine Learning, 57*(3), 233–269. https://doi.org/10.1023/B:MACH.0000039778.69032.ab
33. Guo, G., Wang, H., Bell, D., Bi, Y., & Greer, K. (2003). KNN model-based approach in classification. In R. Meersman, Z. Tari, & D. C. Schmidt (Eds.), *On the move to meaningful internet systems* 2003: *CoopIS, DOA, and ODBASE* (pp. 986–996). Springer. https://doi.org/10.1007/978-3-540-39964-3_62.
34. Zhang, M. L., & Zhou, Z. H. (2007). ML-KNN: A lazy learning approach to multi-label learning. *Pattern Recognition, 40*(7), 2038–2048. https://doi.org/10.1016/j.patcog.2006.12.019
35. Gudivada, V. N. (2016). Chapter 1-cognitive computing: concepts, architectures, systems, and applications. In V. N. Gudivada, V. V. Raghavan, V. Govindaraju, C. R. Rao (Eds.), *Handbook of statistics. Cognitive computing: Theory and applications* (Vol. 35, pp. 3–38). Elsevier. https://doi.org/10.1016/bs.host.2016.07.004.
36. Nisha, S. S., & Meeral, M. N. (2021). Applications of deep learning in biomedical engineering. In *Handbook of deep learning in biomedical engineering* (pp. 245–270). Academic Press.
37. Zhu, W., Ma, Y., Zhou, Y., Benton, M., & Romagnoli, J. (2018). Deep learning based soft sensor and its application on a pyrolysis reactor for compositions predictions of gas phase components. In M. R. Eden, M. G. Ierapetritou, & G. P. Towler (Eds.), *Computer aided chemical engineering, 13 international symposium on process systems engineering* (*PSE*2018) (pp. 2245–2250). Elsevier. https://doi.org/10.1016/B978-0-444-64241-7.50369-4.
38. Xu, G., Lyu, J., Li, Q., Liu, H., Wang, D., Zhang, M., Springer, N. M., Ross-Ibarra, J., & Yang, J. (2020). Evolutionary and functional genomics of DNA methylation in maize domestication and improvement. *Nature Communications, 11*(1), 5539. https://doi.org/10.1038/s41467-020-19333-4.PMID:33139747;PMCID:PMC7606521
39. Thanmalagan, R. R., Roy, A., Jayaprakash, A., & Lakshmi, P. T. V. (2022). Comprehensive meta-analysis and machine learning approaches identified the role of novel drought specific genes in Oryza sativa. *Plant Gene, 32*, 100382. https://doi.org/10.1016/j.plgene.2022.100382
40. Yan, H., Lee, J., Song, Q., Li, Q., Schiefelbein, J., Zhao, B., et al. (2022). Identification of new marker genes from plant single-cell RNA-seq data using interpretable machine learning methods. *New Phytologist, 234*(4), 1507–1520. https://doi.org/10.1111/nph.18053. Epub 2022 Mar 26. PMID: 35211979; PMCID: PMC9314150.
41. Li, Y., Yapa, M. M., & Hua, Z. A. (2021). Machine learning approach to prioritizing functionally active F-box members in Arabidopsis thaliana. *Frontiers in Plant Science, 12*, 639253. https://doi.org/10.3389/fpls.2021.639253.PMID:34122469;PMCID:PMC8192846

42. Shaik, R., & Ramakrishna, W. (2014). Machine learning approaches distinguish multiple stress conditions using stress-responsive genes and identify candidate genes for broad resistance in rice. *Plant Physiology, 164*(1), 481–495. https://doi.org/10.1104/pp.113.225862. Epub 2013 Nov 14. PMID: 24235132; PMCID: PMC3875824.
43. Cho, K. T., Sen, T. Z., & Andorf, C. M. (2022). Predicting tissue-specific mRNA and protein abundance in maize: A machine learning approach. *Frontiers in Artificial Intelligence, 26*(5), 830170. https://doi.org/10.3389/frai.2022.830170.PMID:35719692;PMCID:PMC9204276
44. Cui, H., Zhai, J., & Ma, C. (2015). MiRLocator: Machine learning-based prediction of mature MicroRNAs within plant pre-miRNA sequences. *PLoS ONE, 10*(11), e0142753. https://doi.org/10.1371/journal.pone.0142753.PMID:26558614;PMCID:PMC4641693
45. Jha, A., & Shankar, R. (2011). Employing machine learning for reliable miRNA target identification in plants. *BMC Genomics, 12*, 636.
46. Meher, P. K., Begam, S., Sah, T. K., Gupta, A., Kumar, A., Kumar, U., et al. (2022). ASRmiRNA: Abiotic stress-responsive miRNA prediction in plants by using machine learning algorithms with pseudo K-tuple nucleotide compositional features. *International Journal of Molecular Sciences, 23*(3), 1612. https://doi.org/10.3390/ijms23031612. PMID: 35163534.
47. Meng, J., Liu, D., Sun, C., & Luan, Y. (2014). Prediction of plant pre-microRNAs and their microRNAs in genome-scale sequences using structure-sequence features and support vector machine. *BMC Bioinformatics, 15*(1), 423. https://doi.org/10.1186/s12859-014-0423-x.PMID: 25547126;PMCID:PMC4310204
48. Xuan, P., Guo, M., Huang, Y., Li, W., & Huang, Y. (2011). MaturePred: Efficient identification of microRNAs within novel plant pre-miRNAs. *PLoS One, 26*(11), e27422. https://doi.org/10.1371/journal.pone.0027422. Epub 2011 Nov 16. PMID: 22110646; PMCID: PMC3217989.
49. Sperschneider, J. (2020). Machine learning in plant–pathogen interactions: Empowering biological predictions from field scale to genome scale. *New Phytologist, 1*(228), 35–41.
50. Martínez-García, P. M., López-Solanilla, E., Ramos, C., & Rodríguez-Palenzuela, P. (2016). Prediction of bacterial associations with plants using a supervised machine-learning approach. *Environmental Microbiology, 18*(12), 4847–4861. https://doi.org/10.1111/1462-2920.13389. Epub 2016 Jun 27 PMID: 27234490.
51. Sperschneider, J., Dodds, P. N., Singh, K. B., & Taylor, J. M. (2018). ApoplastP: Prediction of effectors and plant proteins in the apoplast using machine learning. *New Phytologist, 217*, 1764–1778.
52. Kamoun, S. (2006). A catalogue of the effector secretome of plant pathogenic oomycetes. *Annual Review of Phytopathology, 44*, 41–60.
53. Peyraud, R., Dubiella, U., Barbacci, A., Genin, S., Raffaele, S., & Roby, D. (2017). Advances on plant-pathogen interactions from molecular toward systems biology perspectives. *The Plant Journal, 90*, 720–737.
54. Kristianingsih, R., & MacLean, D. (2021). Accurate plant pathogen effector protein classification ab initio with deepredeff: An ensemble of convolutional neural networks. *BMC Bioinformatics, 22*, 372.
55. Martin, E. C., Spiridon, L., Goverse, A., & Petrescu, A. J. (2022). NLRexpress-A bundle of machine learning motif predictors-Reveals motif stability underlying plant Nod-like receptors diversity. *Frontiers in Plant Science, 13*, 975888.
56. Wimalanathan, K., & Lawrence-Dill, C. J. (2021). Gene ontology meta annotator for plants (GOMAP). *Plant Methods, 17*, 54.
57. Chan, K. L., Rosli, R., Tatarinova, T. V., Hogan, M., Firdaus-Raih, M., & Leslie Low, E. T. (2017). Seqping: Gene prediction pipeline for plant genomes using self-training gene models and transcriptomic data. *BMC Bioinformatics, 18*(Suppl 1), 1–7.
58. Kulmanov, M., Zhapa-Camacho, F., & Hoehndorf, R. (2021). DeepGOWeb: Fast and accurate protein function prediction on the (Semantic) web. *Nucleic Acids Research, 49*, 40–146.
59. Van Dijk, A. D. J., Kootstra, G., Kruijer, W., & de Ridder, D. (2020). Machine learning in plant science and plant breeding. *iScience, 24*, 101890.
60. Chen, X., & Ishwaran, H. (2012). Random forests for genomic data analysis. *Genomics, 99*, 323–329.

61. Mace, E. S., & Jordan, D. R. (2011). Integrating sorghum whole genome sequence information with a compendium of sorghum QTL studies reveals uneven distribution of QTL and of gene-rich regions with significant implications for crop improvement. *Theoretical and Applied Genetics, 123*, 169–191.
62. Sandhu, K. S., Lozada, D. N., Zhang, Z., Pumphrey, M. O., & Carter, A. H. (2021). Deep learning for predicting complex traits in spring wheat breeding program. *Frontiers in Plant Science, 11*, 613325.
63. Veyrieras, J. B., Goffinet, B., & Charcosset, A. (2007). MetaQTL: A package of new computational methods for the meta-analysis of QTL mapping experiments. *BMC Bioinformatics, 8*, 49.
64. Tecle, I. Y., Menda, N., Buels, R. M., van der Knaap, E., & Mueller, L. A. (2010). SolQTL: A tool for QT analysis, visualization and linking to genomes at SGN database. *BMC Bioinformatics, 11*, 525.
65. Kumasaka, N., Knights, A. J., & Gaffney, D. J. (2016). Fine-mapping cellular QTLs with RASQUAL and ATAC-seq. *Nature Genetics, 48*, 206–213. *Erratum in: Nature Genetics, 48*, 473.

Immunoinformatics and Vaccine Development

Shruti Shukla and Ashutosh Mani

Abstract Immunoinformatics, a dynamic field bridging immunology and bioinformatics, plays a vital role in vaccine development. This interdisciplinary approach integrates immunological principles with computational biology and artificial intelligence (AI) techniques to accelerate vaccine discovery and design for infectious diseases, cancer, and emerging pathogens, offering promising solutions to global health challenges. This chapter explores recent advancements, existing challenges, and future prospects, highlighting the transformative impact of immunoinformatics and AI on biomedical research and public health initiatives.

Keywords Immunoinformatics · Computational biology · Artificial intelligence · Immunology · Bioinformatics

1 Introduction

The world's health has significantly improved as a result of vaccine use. It has improved the quality of life for both humans and animals, lowered treatment expenses, and saved a number of lives [1]. The discovery of immunological phenomena and the advancement of bioinformatics tools for epitope prediction have paved the way for a fresh perspective on vaccine formulation [2]. Immunoinformatics is a multidisciplinary field that uses computational techniques to research the immune system by combining immunology and bioinformatics [3]. It is essential for locating and creating viable vaccine candidates in the context of vaccine development [4, 5]. Research in immunology produces enormous amounts of data. Furthermore, the need to gather, organize, and assess this data has grown due to large-scale pathogen screening, proteomics and genomics projects, and pathogen-host interactions this is how Immunoinformatics was created [6].

S. Shukla · A. Mani (✉)
Department of Biotechnology, Motilal Nehru National Institute of Technology, Prayagraj, Uttar Pradesh, India
e-mail: amani@mnnit.ac.in

© The Author(s), under exclusive license to Springer Nature Singapore Pte Ltd. 2024
A. Chaudhary et al. (eds.), *Unraveling New Frontiers and Advances in Bioinformatics*,
https://doi.org/10.1007/978-981-97-7123-3_6

It applies biological, mathematical, statistical, and computational knowledge and techniques to precisely store and evaluate immune system and function data [7], and handles the diversity of evidence by utilizing approaches that cut across several bioinformatics fields, such as database creation and upkeep [8], the usage and characterization of structural and functional markers, along with the development and implementation of predictive methodologies [1, 7]. Due to the intricate nature of vertebrate immune systems, the diverse range of diseases and environmental antigens they encounter, and the multitude of regulatory pathways involved, a substantial volume of data will be necessary to gain a comprehensive understanding of the workings of the human immune system [9]. Many of the aims are usually unachievable because of the sophistication of the immune system and the pathogenic antigen. Nonetheless, computational vaccinology has simplified, improved, and focused vaccine design research [10, 11].

2 Immunological Memory

Immunological memory pertains to the immune system's ability to identify infections it has previously faced and react appropriately when encountering them again [12]. When a pathogen or its associated antigens enter the body for the first time, whether through self-infection or immunization, it initiates a cascade of immune system responses [13]. There are immune cells that form a "memory" of the invader during this first interaction. Re-exposure to the virus will trigger a more robust and swift immune response, enabling the body to effectively eliminate the pathogen without experiencing severe sickness or disease development [14].

3 How is Immunological Memory Established?

The cells within the adaptive immune system, known for coordinating a highly specific and intricate immune response to eradicate invading cells, are tasked with generating immunological memory [15]. It is essential that this specificity be maintained in order to target just foreign antigen molecules rather than host-specific ones. As shown in Fig. 1 the primary actors in this mechanism are specialized white blood cells known as T and B lymphocytes [16]. These cells generate antibodies and effectively eliminate pathogen-infected cells through targeted attacks [9].When a virus is reintroduced, immunological memory guarantees a speedier and more potent immune response [14].

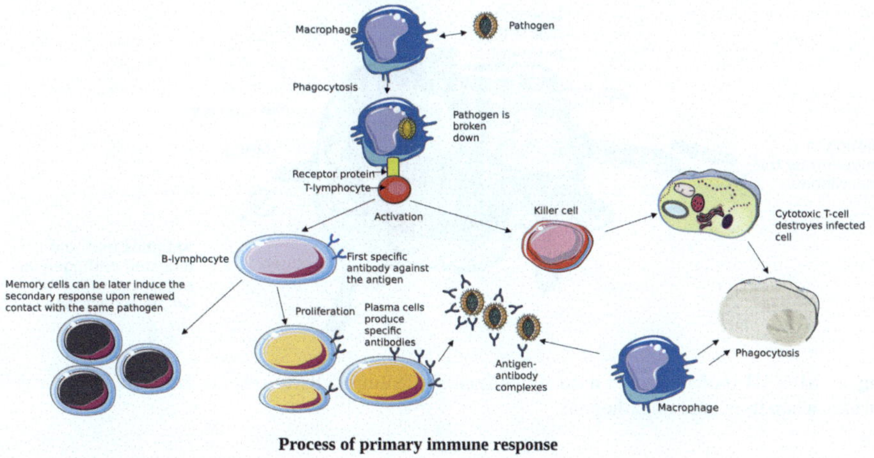

Fig. 1 The activation, differentiation, and proliferation of T and B cells constitute the fundamental immunological response

3.1 B Cell

Primary response: When an antigen (a foreign material) is encountered by a naïve B cell, it undergoes activation and differentiation into plasma cells, which release antibodies that are specific to that antigen [17].

Memory B cell formation: Some B cells differentiate into memory B cells alongside their production of plasma cells. These cells are different from plasma cells in that they live longer and remain in the body for longer periods of time—years or even decades [18]. Memory T cells support immune response coordination and regulation, while memory B cells are principally responsible for generating antibodies [13].

Secondary response: Memory B cells possess the capacity to rapidly recognize and respond to the same antigen upon re-encounter. Upon activation, memory B cells promptly differentiate into plasma cells, which causes a quicker and more robust generation of antibodies. Greater protection against reinfection is offered by this secondary response, which is typified by higher antibody titers and faster pathogen clearance [19].

Adaptive immunity encompasses both antibody-mediated and cell-mediated responses. In the antibody-mediated response, B-lymphocytes produce immunoglobulins. The response mediated by T-cells is termed cell-mediated. T-lymphocytes exist in two classes: helper T-cells (CD4+) and cytotoxic T-cells (CD8+) [20].

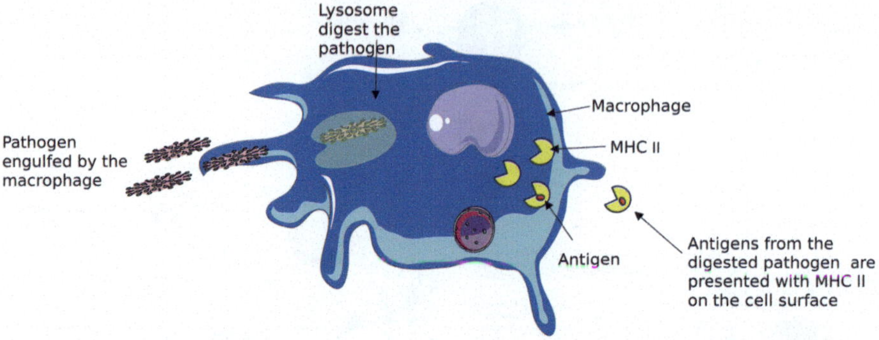

Fig. 2 MHC II molecules are used by antigen-presenting cells to display antigens on their cell surface when they take in pathogens

3.2 T Cell

Primary response: When native T lymphocytes come into contact with antigen-presenting cells (APCs) that have processed and displayed antigens using major histocompatibility complex (MHC) molecules, they become activated [21, 22]. This interaction leads to the activation and differentiation of T cells into effector T cells, including cytotoxic T cells (CD8+) and helper T cells (CD4+) [23]. As shown in Fig. 2.

Memory T cell formation: Certain effector T cells can develop into memory T cells during the first immunological response, just like memory B cells can. These cells have a longer life.

Secondary response: Re-exposure to the antigen causes memory T cells to quickly identify it and expand clonally, producing a sizable population of effector T cells [12].

By delivering safe variants of pathogens to induce the development of memory B and T cells and offer long-term protection against infections in the future, Vaccinations take advantage of the idea of immunological memory [19].

Bioinformatics approaches are used to identify putative epitopes for vaccine development. Each method has the capacity to screen protein sequences, identify supertype motifs and MHC binding aggregates, and may be applied in immunological investigations on human populations with diverse genetic compositions as well as in the development of vaccines based on epitopes There are numerous databases and Tools (Table 1) that provide in-depth details on different kinds of immunological research. Exploring these databases may actually pique your interest in the immunization of illnesses associated with infections that are resurfacing and evolving.

Table 1 Immunological databases, tools and their applications

Epitope prediction			References
NetCTL	Anticipates MHC class I binding and proteasomal cleavage sites, as well as CTL (Cytotoxic T Lymphocyte) epitopes	https://services.healthtech.dtu.dk/services/NetCTL-1.2/	[24]
Jenner-Predict	Use host–pathogen interactions to predict protein vaccine candidates (PVCs)	http://117.211.115.67/vaccine/home.html	[25]
PanRV	Find viable vaccine candidates in the microbial pangenome using the pangenome-reverse vaccinology technique	https://sourceforge.net/projects/panrv2/	[26]
NetMHCpan	Forecasts the The process of peptides attaching to a specific sequence of MHC molecules	http://www.cbs.dtu.dk/services/NetMHCpan-4.1/	[24]
IEDB	Offers a thorough resource for epitope analysis and prediction	https://www.iedb.org/	[27]
NERVE	For the selection of the most suitable vaccine candidates from the entire bacterial pathogen proteomes	http://www.bio.unipd.it/molbinfo	[28]
VacSol	A high throughput in silico method that uses subtractive reverse vaccinology to identify possible treatment targets in prokaryotic infections	https://sourceforge.net/projects/vacsol/	[29]
Antigenicity prediction			
Vaxijen	Predicts a protein sequence's antigenic characteristics	https://www.ddg-pharmfac.net/vaxijen/VaxiJen/VaxiJen.html	[30]
SVMTriP	Uses support vector machines to predict if a protein sequence is antigenic, and forecast linear antigenic epitopes	http://sysbio.unl.edu/SVMTriP/	[31]
Structure prediction			
SWISS-MODEL	For protein structure homology modeling	https://swissmodel.expasy.org/	[32]

(continued)

Table 1 (continued)

Epitope prediction			References
Phyre2	Generates 3D protein structures by homology modeling	http://www.sbg.bio.ic.ac.uk/phyre2/html/page.cgi?id=index	[33]
C-IMMSIM	Enable immune stimulation in silico prediction using the Celada-Seiden model	https://kraken.iac.rm.cnr.it/C-iMMSIM/	[14]
Ifnepitope	Forecast, recognize, and create the peptides that trigger the IFN-Y (gamma) cytokine	https://webs.iiitd.edu.in/ragbava/ifnepitope/	[34]
Molecular docking			
AutoDock Vina	Carries out molecular docking to anticipate binding interactions between small ligands and proteins	https://vina.scripps.edu/	[9]
PatchDock	Identify molecular interactions in docked complexes	http://bioinfo3d.cs.tau.ac.il/PatchDock/	[35]
Immunogenicity prediction			
Immune Epitope Database (IEDB)	Provides resources for epitope analysis and immunogenicity prediction	http://tools.iedb.org/immunogenicity/	[36]
BCPred	Predicts B cell epitopes	https://webs.iiitd.edu.in/raghava/bcepred/bcepred_submission.html	[37]
Population coverage analysis			
IEDB Population Coverage	Evaluates the extent to which various ethnic groups' populations contain possible epitopes	http://tools.iedb.org/population/	[5]
Allergenicity prediction			
AllerTOP	Predicts if a protein will cause allergies	https://www.ddg-pharmfac.net/AllerTOP/	[38]

(continued)

Table 1 (continued)

Epitope prediction			References
SDAP	Protein Structure Database for Allergenic Substances	https://fermi.utmb.edu/SDAP/AllergenAI/AllergenAI.html	[39]
T-cell epitope mapping			
TEPITOPE	Maps protein sequences to T-cell epitopes	http://www.biokdd.fudan.edu.cn/Service/TEPITOPEpan/	[40]
SYFPEITHI	Offers a T-cell epitope database	http://www.syfpeithi.de/bin/MHCServer.dll/EpitopePrediction.htm	[41]
Vaccine design			
Vaxign, Vaxitop	Identify protective antigens for the creation of vaccines and determines possible vaccination targets	https://violinet.org/vaxign/vaxitop/index.php	[30]
IMGT	Focuses on T cell receptors, immunoglobulins or antibodies, the major histocompatibility complex (MHC) found in humans and other vertebrate species, and related proteins across all immune systems. IMGT comprises five databases, such as IMGT/3Dstructure-DB, IMGT/GENE-DB, and IMGT/LIGM-DB	https://www.imgt.org/	[27]
AbMiner	A tool for antibody exploration	http://discover.nci.nih.gov/abminer	[42]
HIV Molecular Immunology	Comprises a categorized and searchable compilation of T-cell epitopes and antibodies	https://www.hiv.lanl.gov/content/immunology/	[43]
ExactAntigen	A repository containing 22,000 monoclonal antibody products	http://www.exactantigen.com/	[1]

(continued)

Table 1 (continued)

Epitope prediction			References
EPIMHC	A database containing naturally processed MHC-restricted peptide ligands and epitopes tailored for personalized computational vaccinology	http://bio.med.ucm.es/epimhc/	[44]
SNPBinder	A Tissue-Specific Minor Histocompatibility Antigen Prediction System	http://www.sipep.org/	[45]
Bcipep	A database containing experimentally determined linear B-cell epitopes exhibiting varying levels of immunogenicity	http://www.imtech.res.in/raghava/bcipep	[37]
MotifScan	Identify every known motif that appears within a sequence	https://myhits.sib.swiss/cgi-bin/motif_scan	[46]

4 Steps of Vaccine Development Using Immunoinformatics

Combining computational methods with experimental validation is one of the crucial elements in the Immunoinformatics-based vaccine development process [47]. An outline of the standard steps in the immunoinformatics-based vaccine development process is explained in Fig. 3.

(1) **Retrieval and analysis of the sequence of the target protein:**

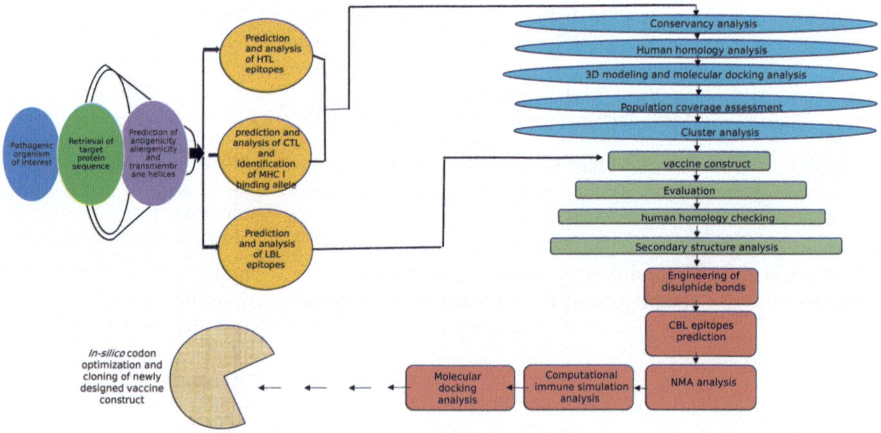

Fig. 3 In silico development of vaccine: Accelerating Identification, Design, and Optimization through Computational Tools and Techniques

The amino acid sequences of target proteins from desired pathogens are available in numerous protein databases, including UniProt [48] and the National Centre for Biotechnology Information [34, 49]. The VaxiJen v2.0 web server [30] is utilized for predicting antigenicity, while the AllergenFP v1.0 server [38] identifies allergenicity. Additionally, the TMHMM v2.0 server [50] is employed to forecast the presence of transmembrane (TM) helices in the target protein. Finally, amino acid sequences that are highly antigenic, non-allergic, and have a lower TM helicase content are chosen for additional analysis [21].

(2) **Epitope prediction:**

Short amino acid sequences known as epitopes are identified by the immune system and cause an immunological reaction. By employing epitopes, vaccines can be created to specifically target parts of a pathogen, resulting in a more potent and focused immune response [51].

(a) **CTL (Cytotoxic T Lymphocyte)Epitope**: Target protein CTL epitopes can be found using the NetCTL v1.2 service [24, 36, 52]. Additionally, the MHC I binding module available in the IEDB (Immune Epitope Database) server aids in identifying the MHC I binding allele associated with each identified CTL epitope.

(b) **HTL (Helper T Lymphocyte) Epitopes**: The MHC II binding tool provided by the IEDB server [53] can be utilized to predict Helper T Lymphocyte (HTL) epitopes within a provided protein sequence.

(c) **LBL epitope**: Using ABCpred [18], LBL epitopes can be predicted within a given protein sequence set, with a sensitivity, specificity, and accuracy threshold set at 100.

(3) Epitope selection:

The chosen HTL, CTL, and LBL epitopes can then be examined for immunogenicity, toxicity, and allergenicity using the IEDB MHC I Immunogenicity tool of the IEDB server, ToxinPred server [54], AllerTOP v2.0 [38], and VaxiJen v2.0 [30], respectively [55].

(a) **Cytokine-inducing Capacity Analysis**: Interferon-gamma (IFNγ) is essential for certain innate immune responses against microbial infections, as it activates natural killer cells and macrophages [46]. Using IFNepitope server [21], a strong IFN-γ producing MHC II binding helper T cell epitopes can be predicted.

(b) **Conservancy Analysis**: The MHC I and MHC II epitopes that were previously identified can have their degree of conservation across antigens evaluated using the epitope conservancy analysis tool that is accessible on the IEDB server [56].

(c) **Human Homology Analysis**: To find the epitope homology to the human proteome, using the NCBI database's BLAST (Basic Local Alignment Search Tool) module [57]. To create a powerful vaccination, homologous epitopes within the human proteome must be found. This is because human epitopes that are identical to one another may prevent the body from mounting a strong enough defense [58].

(d) **Molecular Docking and Structural Modelling**: Using software like PyMOL [59] or UCSF Chimera [60], create three-dimensional models of molecules. Then, employ molecular docking tools such as AutoDock [61] or SwissDock [62] to simulate the interaction between the target protein and ligands, predicting their binding affinity and potential interactions [63].

(e) **Population Coverage Assessment**: There are regional and ethnic variations worldwide in the expression and distribution of class I and class II HLA alleles. Evaluating population coverage is crucial for developing a peptide vaccination strategy based on epitopes that effectively target diverse populations [64]. The IEDB server's population coverage tool facilitates the assessment of population coverage for selected CTL and HTL epitopes [56].

(f) **Cluster Analysis**: The genes encoding both classes of MHC molecules in humans exhibit great variability, encompassing hundreds of thousands of variants in HLA antigens. Using MHC clustering analysis and the MHCcluster 2.0 server [65], MHC I and II molecules with comparable binding affinities can be identified.

(4) **Vaccine Construct:**

Assessment of the Recently Built Vaccine Candidate: The ProtParam tool (ExPASy proteomic server) [66] can be used to evaluate the physicochemical characteristics of the formulated vaccine, such as theoretical pH (pI), amino acid composition, total number of positively and negatively charged residues, aliphatic index (AI), and grand average of hydrophilicity (GRAVY). To elicit a robust immune response, a newly developed vaccine component needs to demonstrate extreme antigenicity, high solubility, and non-allergenicity. The sensitivity to allergies A vaccine's solubility can be examined using the Protein-Sol server-like SOLpro program [67].

The PSIBLAST algorithm's BLASTp module in the NCBI database allows for the assessment of the vaccine candidate's relative similarity to the human proteome, reducing the possibility of an autoimmune reaction. Furthermore, the secondary structure and peptide configuration of the completed vaccine can be examined by utilizing the PSIPRED v4.0 [68] SOPMA programs [69]. Disulfide bonds within a protein are vital for preserving its tertiary/quaternary structure, stability, and dynamic interactions [70]. After refinement, the vaccine construct can be submitted for disulfide engineering using the Disulfide by Design v2.12 server [71]. Analysis of Normal Modes To comprehend a protein complex's spontaneous functional mobility in its internal (dihedral) coordinates, NMA (Normal Mode Analysis) is crucial [72]. The intended vaccine candidate's normal mode can be examined using the iMODS server [11, 73, 74].

(5) **B-cell epitope prediction**: The IEDB analysis resource's ElliPro: Antibody Epitope Prediction tool [75] can be used to predict the CBL epitopes inside the developed vaccine design [18, 76].

(6) **Molecular docking**: Perform molecular docking analysis using tools like AutoDock [61] or SwissDock [62] to simulate the interaction between molecules, predicting binding affinities and potential interactions.
(7) **MD (Molecular Dynamics) Simulation**: Conduct molecular dynamics simulations using software such as GROMACS [77] or NAMD to simulate the motion and behaviour of atoms and molecules over time. Apply force fields and algorithms to model the interactions between particles accurately. Analyse trajectories and structural changes to gain insights into the dynamic behaviour of the system under study [78].
(8) **Performing in silico codon optimization and in silico cloning of the developed vaccine construct**: Utilizing computational methods to optimize codon usage and virtually clone the designed vaccine construct, often employing tools like GeneOptimizer [79] for codon optimization and in silico cloning [80].

The successful creation of a vaccine depends on cooperation between molecular biologists, immunologists, immunoinformatic professionals, and other experts throughout these processes [81]. The creation of vaccines is sped up and the chance of finding viable candidates is raised when computational techniques are combined with experimental validation [82].

5 Limitations of Immunoinformatic in Vaccine Development

Despite being a great resource for vaccine research, immunoinformatic tools have certain drawbacks when it comes to vaccine design and development.

1. Wet lab research is necessary in immunoinformatic to generate raw data for vaccine development.
2. The predictions made through the immunoinformatic approach do not constitute formal proof of concepts; they cannot replace conventional experimental research methods, which involve hypothesis testing.
3. The sophistication of the immunoinformatic tools utilized during vaccine development affects the accuracy of the output data. Consequently, the conclusions drawn from the analysis may also be inaccurate.
4. In the immunoinformatics approach to vaccine generation, proteins, linear, and discontinuous epitopes are utilized, whereas other macromolecules like polysaccharides are not employed in vaccine production.
5. It has been demonstrated that under certain circumstances, antigen residues can form epitopes, influencing prediction algorithms. Therefore, it is advisable for an immunoinformatic tool to incorporate improved prediction techniques.

6 Conclusion and Prospects for Future Research

Immunoinformatics has become a potent tool in the vaccine development process, providing creative ways to speed up the identification and creation of new vaccines. By combining immunological expertise, bioinformatics tools, and computer algorithms, scientists have achieved amazing precision in immune response simulation, antigenic epitope prediction, and vaccine candidate optimization. The combination of experimental techniques and immunoinformatics has greatly advanced vaccine design, enabling the creation of vaccinations against a variety of diseases, such as bacteria, viruses, and parasites. Translating immunoinformatics-driven discoveries into clinically viable vaccine candidates will require the joint efforts of immunologists, computational biologists, and vaccine scientists. It will take a multidisciplinary team and strong experimental validation to overcome vaccine development obstacles and ensure the timely delivery of safe and effective vaccinations to combat present and emerging disease threats.

7 Summary

Immunoinformatics integrates computational tools with immunology to accelerate vaccine development. By predicting epitopes and antigenicity, it streamlines candidate selection and optimization. Immunological memory ensures swift and potent immune responses upon pathogen re-encounter. Computational tools aid in epitope prediction, population coverage analysis, and vaccine design. Despite limitations, immunoinformatics tools and techniques have been helpful in accelerating vaccine research and development.

Acknowledgements The Authors acknowledge the Department of Biotechnology, MNNIT, Allahabad, for providing the necessary support. All the authors have conceptualized, edited and have helped in writing the chapter.

Conflict of Interest There are no conflicts of interest to declare by the authors.

References

1. Excler, J. L., Saville, M., Berkley, S., & Kim, J. H. (2021). Vaccine development for emerging infectious diseases. *Nature Medicine, 27*(4), 591–600. https://doi.org/10.1038/s41591-021-01301-0
2. Hegde, N. R., Gauthami, S., Sampath Kumar, H. M., & Bayry, J. (2018). The use of databases, data mining and immunoinformatics in vaccinology: Where are we? *Expert Opinion on Drug Discovery, 13*(2), 117–130. https://doi.org/10.1080/17460441.2018.1413088

3. García-Machorro, J., Ramírez-Salinas, G. L., Martinez-Archundia, M., & Correa-Basurto, J. (2022). The advantage of using immunoinformatic tools on vaccine design and development for coronavirus. *Vaccines, 10*(11), 1844. https://doi.org/10.3390/vaccines10111844
4. Abass, O. A., Timofeev, V. I., Sarkar, B., et al. (2022). Immunoinformatics analysis to design novel epitope based vaccine candidate targeting the glycoprotein and nucleoprotein of Lassa mammarenavirus (LASMV) using strains from Nigeria. *Journal of Biomolecular Structure & Dynamics, 40*(16), 7283–7302. https://doi.org/10.1080/07391102.2021.1896387
5. Rapin, N., Lund, O., Bernaschi, M., & Castiglione, F. (2015). Computational immunology meets bioinformatics: The use of prediction tools for molecular binding in the simulation of the immune system. *PLoS ONE, 5*(4), e9862. https://doi.org/10.1371/journal.pone.0009862. (V. Brusic (Ed.))
6. Dey, J., Mahapatra, S. R., Raj, T. K., et al. (2022). Designing a novel multi-epitope vaccine to evoke a robust immune response against pathogenic multidrug-resistant Enterococcus faecium bacterium. *Gut Pathogens, 14*(1), 21. https://doi.org/10.1186/s13099-022-00495-z
7. He, Y., Rappuoli, R., De Groot, A. S., & Chen, R. T. (2010). Emerging vaccine informatics. *Journal of Biomedicine & Biotechnology, 2010*, 1–26. https://doi.org/10.1155/2010/218590
8. Rai, S., Shukla, S., Scotti, L., & Mani, A. (2023). Drug repurposing against novel therapeutic targets in plasmodium falciparum for malaria: The computational perspective. *Current Medicinal Chemistry,* 31. https://doi.org/10.2174/0929867331666230807151708.
9. Kardani, K., Bolhassani, A., & Namvar, A. (2020). An overview of *in silico* vaccine design against different pathogens and cancer. *Expert Review of Vaccines, 19*(8), 699–726. https://doi.org/10.1080/14760584.2020.1794832
10. Oli, A. N., Obialor, W. O., Ifeanyichukwu, M. O., et al. (2020). Immunoinformatics and vaccine development: An overview. *ImmunoTargets and Therapy, 9*, 13–30. https://doi.org/10.2147/ITT.S241064
11. Rauch, S., Jasny, E., Schmidt, K. E., & Petsch, B. (2018). New vaccine technologies to combat outbreak situations. *Frontiers in Immunology, 9*, 1963. https://doi.org/10.3389/fimmu.2018.01963
12. Ratajczak, W., Niedźwiedzka-Rystwej, P., Tokarz-Deptuła, B., & Deptuła, W. (2018). Immunological memory cells. *Central European Journal of Immunology, 43*(2), 194–203. https://doi.org/10.5114/ceji.2018.77390
13. Marshall, J. S., Warrington, R., Watson, W., & Kim, H. L. (2018). An introduction to immunology and immunopathology. *Allergy, Asthma and Clinical Immunology, 14*(S2), 49. https://doi.org/10.1186/s13223-018-0278-1
14. Sanchez-Trincado, J. L., Gomez-Perosanz, M., & Reche, P. A. (2017). Fundamentals and methods for T- and B-cell epitope prediction. *Journal of Immunology Research, 2017*, 2680160. https://doi.org/10.1155/2017/2680160
15. Bugya, Z., Prechl, J., Szénási, T., Nemes, É., Bácsi, A., & Koncz, G. (2021). Multiple levels of immunological memory and their association with vaccination. *Vaccines, 9*(2), 174. https://doi.org/10.3390/vaccines9020174
16. Nielsen, M., Andreatta, M., Peters, B., & Buus, S. (2020). Immunoinformatics: Predicting peptide–MHC binding. *Annual Review of Biomedical Data Science, 3*(1), 191–215. https://doi.org/10.1146/annurev-biodatasci-021920-100259
17. Palm, A. K. E., & Henry, C. (2019). Remembrance of things past: Long-term B cell memory after infection and vaccination. *Frontiers in Immunology, 10*, 1787. https://doi.org/10.3389/fimmu.2019.01787
18. Saha, S., & Raghava, G. P. S. (2007). Prediction methods for B-cell epitopes. In D. R. Flower (Ed.), *Immunoinformatics*. (Vol. 409, pp. 387–394). Methods in molecular biology. Humana Press. https://doi.org/10.1007/978-1-60327-118-9_29.
19. Santosa, E. K., & Sun, J. C. (2023). Cardinal features of immune memory in innate lymphocytes. *Nature Immunology, 24*(11), 1803–1812. https://doi.org/10.1038/s41590-023-01607-w
20. Raskov, H., Orhan, A., Christensen, J. P., & Gögenur, I. (2021). Cytotoxic CD8+ T cells in cancer and cancer immunotherapy. *British Journal of Cancer, 124*(2), 359–367. https://doi.org/10.1038/s41416-020-01048-4

21. Dhanda, S. K., Vir, P., & Raghava, G. P. S. (2013). Designing of interferon-gamma inducing MHC class-II binders. *Biology Direct, 8*, 30. https://doi.org/10.1186/1745-6150-8-30
22. Calis, J. J. A., Maybeno, M., Greenbaum, J. A., et al. (2013). Properties of MHC class I presented peptides that enhance immunogenicity. *PLoS Computational Biology, 9*(10), e1003266. https://doi.org/10.1371/journal.pcbi.1003266
23. Andersen, M. H., Schrama, D., Thor Straten, P., & Becker, J. C. (2006). Cytotoxic T cells. *The Journal of Investigative Dermatology, 126*(1), 32–41. https://doi.org/10.1038/sj.jid.5700001
24. Reynisson, B., Alvarez, B., Paul, S., Peters, B., & Nielsen, M. (2020). NetMHCpan-4.1 and NetMHCIIpan-4.0: Improved predictions of MHC antigen presentation by concurrent motif deconvolution and integration of MS MHC eluted ligand data. *Nucleic Acids Research, 48*(W1), W449–W454. https://doi.org/10.1093/nar/gkaa379.
25. Jaiswal, V., Chanumolu, S. K., Gupta, A., Chauhan, R. S., & Rout, C. (2013). Jenner-predict server: Prediction of protein vaccine candidates (PVCs) in bacteria based on host-pathogen interactions. *BMC Bioinformatics, 14*(1), 211. https://doi.org/10.1186/1471-2105-14-211
26. Naz, K., Naz, A., Ashraf, S. T., et al. (2019). PanRV: Pangenome-reverse vaccinology approach for identifications of potential vaccine candidates in microbial pangenome. *BMC Bioinformatics, 20*(1), 123. https://doi.org/10.1186/s12859-019-2713-9
27. Lefranc, M. P., Giudicelli, V., Ginestoux, C., et al. (2009). IMGT(R), the international ImMunoGeneTics information system(R). *Nucleic Acids Research, 37*(Database), D1006–D1012. https://doi.org/10.1093/nar/gkn838.
28. Vivona, S., Bernante, F., & Filippini, F. (2006). NERVE: New enhanced reverse vaccinology environment. *BMC Biotechnology, 6*(1), 35. https://doi.org/10.1186/1472-6750-6-35
29. Rizwan, M., Naz, A., Ahmad, J., et al. (2017). VacSol: A high throughput in silico pipeline to predict potential therapeutic targets in prokaryotic pathogens using subtractive reverse vaccinology. *BMC Bioinformatics, 18*(1), 106. https://doi.org/10.1186/s12859-017-1540-0
30. Doytchinova, I. A., & Flower, D. R. (2007). VaxiJen: A server for prediction of protective antigens, tumour antigens and subunit vaccines. *BMC Bioinformatics, 8*(1), 4. https://doi.org/10.1186/1471-2105-8-4
31. Yao, B., Zhang, L., Liang, S., & Zhang, C. (2012). SVMTriP: A method to predict antigenic epitopes using support vector machine to integrate tri-peptide similarity and propensity. *PLoS ONE, 7*(9), e45152. https://doi.org/10.1371/journal.pone.0045152. (A. Porollo (Ed.))
32. Schwede, T., Kopp, J., Guex, N., & Peitsch, M. C. (2003). SWISS-MODEL: An automated protein homology-modeling server. *Nucleic Acids Research, 31*(13), 3381–3385. https://doi.org/10.1093/nar/gkg520
33. Kelley, L. A., Mezulis, S., Yates, C. M., Wass, M. N., & Sternberg, M. J. E. (2015). The Phyre2 web portal for protein modeling, prediction and analysis. *Nature Protocols, 10*(6), 845–858. https://doi.org/10.1038/nprot.2015.053
34. Shawan, M. M. A. K., Sharma, A. R., Halder, S. K., et al. (2023). Advances in computational and bioinformatics tools and databases for designing and developing a multi-epitope-based peptide vaccine. *International Journal of Peptide Research and Therapeutics, 29*(4), 60. https://doi.org/10.1007/s10989-023-10535-0
35. Schneidman-Duhovny, D., Inbar, Y., Nussinov, R., & Wolfson, H. J. (2005). PatchDock and SymmDock: Servers for rigid and symmetric docking. *Nucleic Acids Research, 33*(Web Server issue), W363–367. https://doi.org/10.1093/nar/gki481.
36. Pappalardo, F., Flower, D., Russo, G., Pennisi, M., & Motta, S. (2015). Computational modelling approaches to vaccinology. *Pharmacological Research, 92*, 40–45. https://doi.org/10.1016/j.phrs.2014.08.006
37. Saha, S., Bhasin, M., & Raghava, G. P. (2005). Bcipep: A database of B-cell epitopes. *BMC Genomics, 6*(1), 79. https://doi.org/10.1186/1471-2164-6-79
38. Dimitrov, I., Bangov, I., Flower, D. R., & Doytchinova, I. (2014). AllerTOP vol 2—a server for in silico prediction of allergens. *Journal of Molecular Modeling, 20*(6), 2278. https://doi.org/10.1007/s00894-014-2278-5
39. Negi, S. S., Schein, C. H., & Braun, W. (2023). The updated structural database of allergenic proteins (SDAP 2.0) provides 3D models for allergens and incorporated bioinformatics tools.

Journal of Allergy and Clinical Immunology: Global, 2(4), 100162. https://doi.org/10.1016/j.jacig.2023.100162.
40. Zhang, L., Chen, Y., Wong, H. S., Zhou, S., Mamitsuka, H., & Zhu, S. (2012). TEPITOPEpan: Extending TEPITOPE for peptide binding prediction covering over 700 HLA-DR molecules. *PLoS ONE, 7*(2), e30483. https://doi.org/10.1371/journal.pone.0030483. (C. Schönbach (Ed.))
41. Rammensee, H. G., Bachmann, J., Emmerich, N. P. N., Bachor, O. A., & Stevanović, S. (1999). SYFPEITHI: Database for MHC ligands and peptide motifs. *Immunogenetics, 50*(3–4), 213–219. https://doi.org/10.1007/s002510050595
42. Major, S. M., Nishizuka, S., Morita, D., et al. (2006). AbMiner: A bioinformatic resource on available monoclonal antibodies and corresponding gene identifiers for genomic, proteomic, and immunologic studies. *BMC Bioinformatics, 7*, 192. https://doi.org/10.1186/1471-2105-7-192
43. Pandey, R. K., Ojha, R., Aathmanathan, V. S., Krishnan, M., & Prajapati, V. K. (2018). Immunoinformatics approaches to design a novel multi-epitope subunit vaccine against HIV infection. *Vaccine, 36*(17), 2262–2272. https://doi.org/10.1016/j.vaccine.2018.03.042
44. Reche, P. A., Zhang, H., Glutting, J. P., & Reinherz, E. L. (2005). EPIMHC: A curated database of MHC-binding peptides for customized computational vaccinology. *Bioinformatics, 21*(9), 2140–2141. https://doi.org/10.1093/bioinformatics/bti269
45. Kumar, S., & Hasija, Y. (2022). Immunoinformatics tools: A boon in vaccine development against Covid-19. In *2022 IEEE Delhi Section Conference (DELCON)* (pp. 1–9). IEEE. https://doi.org/10.1109/DELCON54057.2022.9753152.
46. Sanami, S., Rafieian-Kopaei, M., Dehkordi, K. A., et al. (2022). In silico design of a multi-epitope vaccine against HPV16/18. *BMC Bioinformatics, 23*(1), 311. https://doi.org/10.1186/s12859-022-04784-x
47. Joon, S., Singla, R. K., & Shen, B. (2022). Vaccines and immunoinformatics for vaccine design. In B. Shen (Ed.), *Translational informatics* (Vol. 1368, pp. 95–110). Advances in experimental medicine and biology. Springer Nature Singapore. https://doi.org/10.1007/978-981-16-8969-7_5.
48. UniProt Consortium. (2008). The universal protein resource (UniProt). *Nucleic Acids Research, 36*(Database issue), D190–195. https://doi.org/10.1093/nar/gkm895.
49. NCBI Resource Coordinators. (2012). Database resources of the National Center for Biotechnology Information. *Nucleic Acids Research, 41*(D1), D8–D20. https://doi.org/10.1093/nar/gks1189
50. Krogh, A., Larsson, B., Von Heijne, G., & Sonnhammer, E. L. L. (2001). Predicting transmembrane protein topology with a hidden Markov model: Application to complete genomes. *Journal of Molecular Biology, 305*(3), 567–580. https://doi.org/10.1006/jmbi.2000.4315. (F. Cohen (Ed.))
51. Tong, J. C., Tan, T. W., & Ranganathan, S. (2006). Methods and protocols for prediction of immunogenic epitopes. *Briefings in Bioinformatics, 8*(2), 96–108. https://doi.org/10.1093/bib/bbl038
52. Larsen, M. V., Lundegaard, C., Lamberth, K., Buus, S., Lund, O., & Nielsen, M. (2007). Large-scale validation of methods for cytotoxic T-lymphocyte epitope prediction. *BMC Bioinformatics, 8*, 424. https://doi.org/10.1186/1471-2105-8-424
53. Fleri, W., Paul, S., Dhanda, S.K., et al. (2017). The immune epitope database and analysis resource in epitope discovery and synthetic vaccine design. *Frontiers in immunology,* 8. https://doi.org/10.3389/fimmu.2017.00278.
54. Gupta, S., Kapoor, P., Chaudhary, K., et al. (2013). *In Silico* approach for predicting toxicity of peptides and proteins. *PLoS ONE*. 8(9):e73957. https://doi.org/10.1371/journal.pone.0073957. (R. L. Patterson (Ed.))
55. Shiragannavar, S., & Madagi, S. (2022). *In Silico* vaccine design tools. In Y. Desheva (Ed.), *Vaccine development*. IntechOpen. https://doi.org/10.5772/intechopen.100180.
56. Bui, H. H., Sidney, J., Dinh, K., Southwood, S., Newman, M. J., & Sette, A. (2006). Predicting population coverage of T-cell epitope-based diagnostics and vaccines. *BMC Bioinformatics, 7*, 153. https://doi.org/10.1186/1471-2105-7-153

57. McGinnis, S., & Madden, T. L. (2004). BLAST: At the core of a powerful and diverse set of sequence analysis tools. *Nucleic Acids Research, 32*(Web Server issue), W20–25. https://doi.org/10.1093/nar/gkh435.
58. Pearson, W. R. (2013). An introduction to sequence similarity ("Homology") searching. *Current Protocols in Bioinformatics, 42*(1). https://doi.org/10.1002/0471250953.bi0301s42.
59. Rosignoli, S., & Paiardini, A. (2022). Boosting the full potential of PyMOL with structural biology plugins. *Biomolecules, 12*(12), 1764. https://doi.org/10.3390/biom12121764
60. Goddard, T. D., Huang, C. C., & Ferrin, T. E. (2005). Software extensions to UCSF chimera for interactive visualization of large molecular assemblies. *Structure, 13*(3), 473–482. https://doi.org/10.1016/j.str.2005.01.006
61. Cosconati, S., Forli, S., Perryman, A. L., Harris, R., Goodsell, D. S., & Olson, A. J. (2010). Virtual screening with AutoDock: Theory and practice. *Expert Opinion on Drug Discovery, 5*(6), 597–607. https://doi.org/10.1517/17460441.2010.484460
62. Grosdidier, A., Zoete, V., & Michielin, O. (2011). SwissDock, a protein-small molecule docking web service based on EADock DSS. *Nucleic Acids Research, 39*(suppl), W270–W277. https://doi.org/10.1093/nar/gkr366
63. Arora, N., Keshri, A. K., Kaur, R., Rawat, S. S., & Prasad, A. (2022). Immunoinformatic approaches for vaccine designing for pathogens with unclear pathogenesis. In S. Thomas (Ed.), *Vaccine design* (Vol. 2412, pp. 425–437). Methods in molecular biology. Springer US. https://doi.org/10.1007/978-1-0716-1892-9_22.
64. Misra, N., Panda, P. K., Shah, K., Sukla, L. B., & Chaubey, P. (2011). Population coverage analysis of T-cell epitopes of Neisseria meningitidis serogroup B from Iron acquisition proteins for vaccine design. *Bioinformation, 6*(7), 255–261. https://doi.org/10.6026/97320630006255
65. Thomsen, M., Lundegaard, C., Buus, S., Lund, O., & Nielsen, M. (2013). MHCcluster, a method for functional clustering of MHC molecules. *Immunogenetics, 65*(9), 655–665. https://doi.org/10.1007/s00251-013-0714-9
66. Gasteiger, E., Gattiker, A., Hoogland, C., Ivanyi, I., Appel, R. D., & Bairoch, A. (2003). ExPASy: The proteomics server for in-depth protein knowledge and analysis. *Nucleic Acids Research, 31*(13), 3784–3788. https://doi.org/10.1093/nar/gkg563
67. Magnan, C. N., Randall, A., & Baldi, P. (2009). SOLpro: Accurate sequence-based prediction of protein solubility. *Bioinformatics, 25*(17), 2200–2207. https://doi.org/10.1093/bioinformatics/btp386
68. Buchan, D. W. A., & Jones, D. T. (2019). The PSIPRED protein analysis workbench: 20 years on. *Nucleic Acids Research, 47*(W1), W402–W407. https://doi.org/10.1093/nar/gkz297
69. Geourjon, C., & Deléage, G. (1995). SOPMA: Significant improvements in protein secondary structure prediction by consensus prediction from multiple alignments. *Bioinformatics, 11*(6), 681–684. https://doi.org/10.1093/bioinformatics/11.6.681.
70. Rawat, S. S., Keshri, A. K., Kaur, R., & Prasad, A. (2023). Immunoinformatics approaches for vaccine design: A fast and secure strategy for successful vaccine development. *Vaccines, 11*(2), 221. https://doi.org/10.3390/vaccines11020221
71. Craig, D. B., & Dombkowski, A. A. (2013). Disulfide by design 2.0: A web-based tool for disulfide engineering in proteins. *BMC Bioinformatics, 14*(1), 346. https://doi.org/10.1186/1471-2105-14-346.
72. Bauer, J. A., Pavlović, J., & Bauerová-Hlinková, V. (2019). Normal mode analysis as a routine part of a structural investigation. *Molecules (Basel Switzerland), 24*(18), 3293. https://doi.org/10.3390/molecules24183293
73. López-Blanco, J. R., Aliaga, J. I., Quintana-Ortí, E. S., & Chacón, P. (2014). iMODS: Internal coordinates normal mode analysis server. *Nucleic Acids Research, 42*(Web Server issue), W271–276. https://doi.org/10.1093/nar/gku339.
74. Usmani, S. S., Kumar, R., Bhalla, S., Kumar, V., & Raghava, G. P. S. (2018). *In Silico* tools and databases for designing peptide-based vaccine and drugs. In *Advances in protein chemistry and structural biology* (Vol. 112, pp. 221–263). Elsevier. https://doi.org/10.1016/bs.apcsb.2018.01.006.

75. Ponomarenko, J., Bui, H. H., Li, W., et al. (2008). ElliPro: A new structure-based tool for the prediction of antibody epitopes. *BMC Bioinformatics, 9*, 514. https://doi.org/10.1186/1471-2105-9-514
76. El-Manzalawy, Y., Dobbs, D., & Honavar, V. (2008). Predicting linear B-cell epitopes using string kernels. *J Mol Recognit JMR., 21*(4), 243–255. https://doi.org/10.1002/jmr.893
77. Vieira, I. H. P., Botelho, E. B., De Souza Gomes, T. J., Kist, R., Caceres, R. A., & Zanchi, F. B. (2023). Visual dynamics: A WEB application for molecular dynamics simulation using GROMACS. *BMC Bioinformatics, 24*(1), 107. https://doi.org/10.1186/s12859-023-05234-y
78. Durrant, J. D., & McCammon, J. A. (2011). Molecular dynamics simulations and drug discovery. *BMC Biology, 9*(1), 71. https://doi.org/10.1186/1741-7007-9-71
79. Raab, D., Graf, M., Notka, F., Schödl, T., & Wagner, R. (2010). The GeneOptimizer algorithm: Using a sliding window approach to cope with the vast sequence space in multiparameter DNA sequence optimization. *Systems and Synthetic Biology, 4*(3), 215–225. https://doi.org/10.1007/s11693-010-9062-3
80. Mauro, V. P., & Chappell, S. A. (2014). A critical analysis of codon optimization in human therapeutics. *Trends in Molecular Medicine, 20*(11), 604–613. https://doi.org/10.1016/j.molmed.2014.09.003
81. Awasthi, A., Sharma, G., & Agrawal, P. (2022). Computational approaches for vaccine designing. In *Bioinformatics* (pp. 317–335) Elsevier. https://doi.org/10.1016/B978-0-323-89775-4.00011-0.
82. Arya, H., & Bhatt, T. K. (2021). Role of bioinformatics in subunit vaccine design. In *Molecular docking for computer-aided drug design* (pp. 425–439). Elsevier. https://doi.org/10.1016/B978-0-12-822312-3.00013-8.

Machine Learning and Artificial Intelligence in Bioinformatics

Shruti Shukla, Brijesh Singh, and Ashutosh Mani

Abstract The integration of Machine Learning (ML) and Artificial Intelligence (AI) in bioinformatics has revolutionized the field by enabling the efficient analysis and interpretation of biological data. This chapter explores the significant impact of ML and AI in various bioinformatics applications, including drug discovery, personalized medicine, genomic analysis, and disease research, and discusses the utilization of ML algorithms such as neural networks, convolutional neural networks, random forests, and support vector machines to address challenges in genetic disease studies, cancer genetics, and personalized medicine. Additionally, it outlines AI's contributions to biomedical image analysis, text mining, and literature analysis, facilitating biomarker discovery and therapeutic target identification. Overall, the integration of ML and AI enhances data analysis capabilities, leading to significant advancements in biomedical research, healthcare, and drug development within the field of bioinformatics.

Keywords Machine learning · Artificial intelligence · Drug discovery · Bioinformatics · Neural networks · Convolutional neural networks · Random forests · Support vector machines

1 Introduction

Artificial intelligence (AI) is a multidisciplinary field that has transformed numerous domains, including bioinformatics, which focuses on analyzing and interpreting biological data efficiently [1]. The impact of AI on bioinformatics has been substantial, with various AI algorithms and machine learning techniques being applied to handle large volumes of biological data effectively [2]. These algorithms have shown promise in drug discovery, lead compound identification, molecular dynamics simulation, and the prediction and analysis of biological structures, as well as in designing

S. Shukla · B. Singh · A. Mani (✉)
Department of Biotechnology, Motilal Nehru National Institute of Technology, Prayagraj, Uttar Pradesh, India
e-mail: amani@mnnit.ac.in

© The Author(s), under exclusive license to Springer Nature Singapore Pte Ltd. 2024
A. Chaudhary et al. (eds.), *Unraveling New Frontiers and Advances in Bioinformatics*,
https://doi.org/10.1007/978-981-97-7123-3_7

personalized medicine. The integration of bioinformatics and AI has the potential to address challenges in genetic disease studies, cancer genetics, and personalized medicine [3].

Machine learning, a branch of artificial intelligence (AI), is cantered on crafting algorithms and statistical models to empower computers in executing tasks devoid of direct programming commands [4, 5]. These algorithms acquire insights and connections from data, empowering them to make predictions, decisions, or identifications with novel input [6, 7].

At its essence, machine learning comprises the following fundamental elements.

DATA: Machine learning relies on data for learning, which can encompass examples, features, or pertinent observations.

Model: A model, derived from the data, serves as a mathematical representation capturing patterns or rules, enabling the algorithm to make predictions or decisions [8].

Learning Algorithm: Serving as the driving force behind model construction, the learning algorithm iteratively adjusts parameters to enhance performance, minimizing errors or maximizing efficacy.

Training: Training involves exposing the machine learning algorithm to labelled or unlabelled data, facilitating the generation of a model. Through this process, the algorithm refines its parameters, enhancing its performance.

Testing and Evaluation: Following training, the model undergoes testing on a distinct dataset to assess its performance and generalization capability. This evaluation gauges the model's proficiency in making predictions or decisions on unseen data [9].

Prediction or Inference: Once trained and evaluated, the model can be deployed to make predictions, decisions, or identifications on new input data.

Machine learning techniques can be broadly classified into three primary categories:

(1) **Supervised Learning**: Supervised learning algorithms aim to understand and model the connections and interdependencies between the predicted output and input features. Through this understanding, they can forecast output values for new data by applying learned relationships from prior datasets [10]. In machine learning it requires external guidance for learning, models are trained using labelled datasets, and their performance is evaluated using test data [11]. As shown in Fig. 1 the Supervised learning can be categorized into two types of problems: classification and regression [12].

Common algorithms in this category include:

- **k-nearest neighbor (KNN)** [13]
- **Naive Bayes** [14]
- **Decision Trees** [15]
- **Linear Regression** [16]

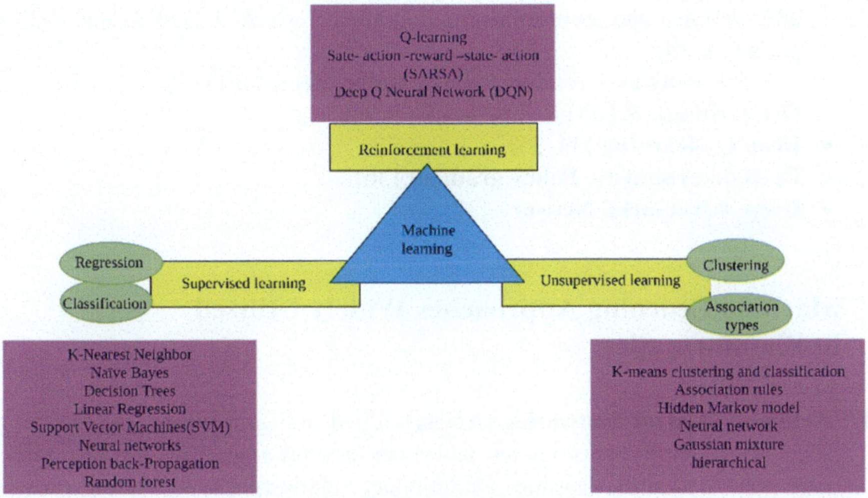

Fig. 1 Classification of machine learning techniques

- **Support Vector Machines (SVM)** [17]
- **Neural Networks** [18]
- **Random Forest** [19]
- **Perception back Propagation** [20].
- **Unsupervised Learning**: Unsupervised learning is a branch of machine learning where the model learns from unlabelled data without external guidance, hence its name [21]. Models are trained using unclassified datasets, aiming to derive insights from the data without predefined outputs. This approach tackles Association and Clustering problems, categorized into **Clustering** and **Association types** [22, 23].

  Common algorithms in this realm include:
- **k-means clustering and classification** [24]
- **The Apriori algorithm** [25]
- **Principal component analysis** [26].
- **Reinforcement Learning**: In Reinforcement Learning, an agent engages with its environment by taking actions and learns through feedback. This feedback comes in the form of rewards, where positive rewards are given for good actions and negative rewards (penalties) for bad ones known as the reinforcement signal enabling it to refine strategies or policies for decision-making [27]. The agent operates without external supervision. Various algorithms address this challenge, with all solutions falling under the category of Reinforcement Learning algorithms [28]. In this problem domain, an agent must decide the best action based on its current state, forming a Markov Decision Process when repeated. Reinforcement learning typically involves these steps: observing input states, using decision-making functions to select actions, receiving rewards or reinforcement from the environment

after actions, and storing information about rewards linked to state-action pairs [29, 30].

Commonly used reinforcement learning algorithms [31]:

- **Q-Learning,** [32, 33]
- **Deep Q –learning** [34, 35]
- **Deep deterministic Policy gradient** [36]
- **Deep Adversarial Network** [37].

2 Machine Learning Approaches Widely Utilized in Bioinformatics

1. **Artificial Neural Networks (ANNs)**: Used for comparing and aligning sequences, identifying promoters and genes, interpreting expression data, inferring gene networks, learning evolutionary relationships, predicting protein structure, and molecular design [20].
2. **Convolutional Neural Networks (CNNs)**: Applied for analysing genomic data, where they leverage hierarchical patterns and shared weights to automatically extract features. Phylogenetic CNNs (Ph-CNNs) are specifically designed for classifying metagenomics data by incorporating phylogenetic information and utilizing convolutional filters [38].
3. **Random Forest (RF)**: are utilized in various bioinformatic applications. RF constructs an ensemble of decision trees and outputs the average prediction of individual trees. This technique, a modification of bootstrap aggregating, is suitable for classification or regression tasks. RF offers advantages such as internal estimation of generalization error, generation of proximities for imputing missing values, and ease of implementation in parallel. Moreover, RF provides measures of variable importance, enables missing value imputation, visualization, outlier detection, and unsupervised learning [19].
4. **Support Vector Machines (SVM)**: SVM, a supervised learning algorithm, is commonly applied to classification tasks. It aims to establish a hyperplane or decision boundary to separate datasets into distinct classes. The crucial data points defining this boundary are termed support vectors, giving the algorithm its name. SVM finds application in various real-world scenarios like face detection, image classification, and drug discovery [17, 39].
5. **Linear Regression**: The linear regression algorithm reveals a linear link between a dependent (y) variable and one or more independent (x) variables, hence its name. By depicting this linear relationship, it determines how the dependent variable's value shifts based on changes in the independent variable(s) [40].
6. **The Apriori algorithm**: an unsupervised learning technique, addresses association problems by leveraging frequent item sets to generate rules. It's tailored for transactional databases. By analysing association rules, it gauges the strength of connection between objects. Apriori employs breadth-first search and Hash

Tree for efficient itemset calculation, iteratively processing large datasets to identify frequent itemsets [25].

7. **K-Nearest Neighbors (KNN)**: applicable to classification and regression tasks. It operates by comparing similarities between new and existing data points, assigning the new data to the most similar categories. The algorithm assesses similarity using a distance function, placing the new case in the class with the closest resemblance [13, 15].
8. **Principal component analysis**: PCA, an unsupervised learning method, focuses on dimensionality reduction, particularly for datasets with correlated features [41]. Through a statistical process, it transforms correlated features into linearly uncorrelated ones via orthogonal transformation. It's widely used in exploratory data analysis and predictive modelling. PCA prioritizes attributes with high variance, as it signifies effective class separation, facilitating dimensionality reduction [42, 43].
9. **k-means clustering and classification:** K-Means Clustering is a method in Unsupervised Learning that organizes unlabelled data into distinct clusters. The parameter K determines the number of clusters. Through iterative steps, data points are grouped based on similarity, aiming to minimize distances between points and their cluster centroids. This approach autonomously identifies patterns within data without requiring prior training [44, 45].
10. **Naive Bayes**: The Naïve Bayes algorithm, a supervised learning technique based on Bayes' theorem, tackles classification tasks. Its "naïve" assumption is that features are independent, like identifying fruit solely by colour, shape, and taste. This method is widely applied in text classification with large datasets. Known for its simplicity and speed, Naïve Bayes is a probabilistic classifier, making predictions based on object probabilities [14].
11. **Hidden Markov model (HMMs)**: Employed for profiling and converting sequence alignments, describing ecological phenomena, and modelling sequential data evolving over time [46].

Artificial intelligence (AI) significantly contributes to the progress of bioinformatics, an interdisciplinary field merging biology, computer science, and information technology for analyzing biological data [2, 47]. The use of of AI in various fields of bioinformatics is shown in Fig. 2.

- **Data Analysis and Interpretation**: Bioinformatics handles extensive biological data like genomic sequences, protein structures, and gene expressions. AI, such as machine learning and deep learning, analyses these datasets, extracting meaningful insights and patterns not easily discernible through traditional methods [48].
- **Genomic Sequence Analysis**: AI algorithms scrutinize genomic sequences to detect genes, regulatory elements, and disease-related genetic variations. They predict gene functions, classify sequences, and identify relevant sequence motifs [8].

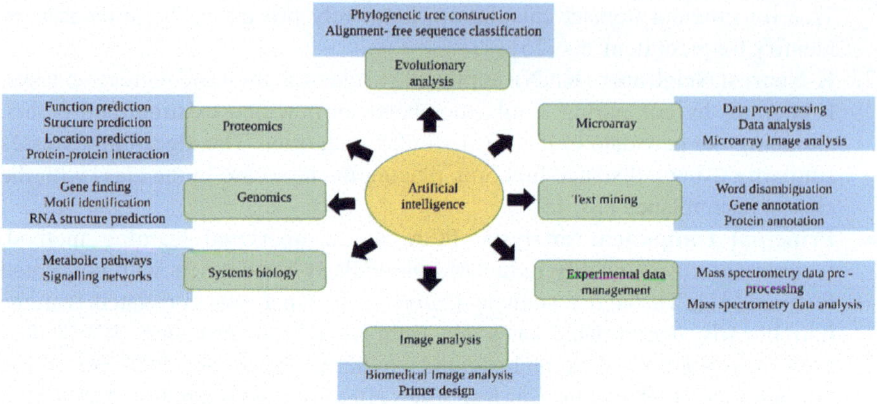

Fig. 2 Application of machine learning methods in various domains

- **Protein Structure Prediction**: AI, particularly deep learning, forecasts protein three-dimensional structures from amino acid sequences, essential for understanding functions, interactions, and drug targeting [49].
- **Drug Discovery and Design**: AI expedites drug discovery by predicting drug-biological target interactions. Machine learning models analyse chemical structures, assays, and clinical data to identify potential drugs, optimize properties, and predict toxicity [50, 51].
- **Biological Network Analysis**: AI models biological networks, unveiling relationships among genes, proteins, and metabolites. They elucidate biological processes and disease mechanisms through network analysis [52].
- **Personalized Medicine and Clinical Support**: AI integrates genomic, clinical, and imaging data for personalized treatment strategies. Models predict disease risk, treatment response, and patient outcomes, improving clinical decisions and healthcare [29, 53].
- **Biomedical Image Analysis**: AI, including convolutional neural networks, analyzes biomedical images, aiding in disease diagnosis and prognosis by detecting abnormalities and segmenting structures [37, 38].
- **Microarrays**: help to analyses gene expressions in microarrays, enabling gene classification and clustering [30, 39].
- **Text Mining and Literature Analysis**: AI techniques like natural language processing extract knowledge from biomedical literature and databases, facilitating discovery and curation of biomarkers and therapeutic targets [42, 54].

Overall, AI empowers bioinformatics by facilitating diverse data analysis, leading to biomedical breakthroughs in research [47], healthcare, and drug development [55]. Its ongoing advancements hold promise for addressing complex disease challenges [56].

3 Summary

Artificial intelligence has revolutionized bioinformatics by enabling efficient analysis of biological data through machine learning algorithms. These algorithms have been instrumental in drug discovery, molecular dynamics simulation, and personalized medicine design leading to significant advancements in understanding disease mechanisms and improving patient outcomes.

Acknowledgements The Authors acknowledge the Department of Biotechnology, MNNIT, Allahabad, for providing the necessary support for conducting the study smoothly. All the authors have conceptualized, edited, and helped in writing the manuscript.

Conflict of Interest There are no conflicts of interest to declare by the authors.

References

1. Arora, S., Chettri, S., Percha, V., Kumar, D., & Latwal, M. (2024). Artificial intelligence: A virtual chemist for natural product drug discovery. *Journal of Biomolecular Structure & Dynamics, 42*(7), 3826–3835. https://doi.org/10.1080/07391102.2023.2216295
2. Auslander, N., Gussow, A. B., & Koonin, E. V. (2021). Incorporating machine learning into established bioinformatics frameworks. *International Journal of Molecular Sciences, 22*(6), 2903. https://doi.org/10.3390/ijms22062903
3. Rawlings, C. J., Fox, J. P. (1994). Artificial intelligence in molecular biology: A review and assessment. *Philos Trans R Soc Lond B Biol Sci, 344*(1310), 353–362; discussion 362–363. https://doi.org/10.1098/rstb.1994.0074
4. Inza, I., Calvo, B., Armañanzas, R., Bengoetxea, E., Larrañaga, P., & Lozano, J. A. (2010). Machine learning: An indispensable tool in bioinformatics. In R. Matthiesen (Ed.), *Bioinformatics methods in clinical research* (Vol. 593, pp. 25–48). Methods in molecular biology. Humana Press. https://doi.org/10.1007/978-1-60327-194-3_2.
5. Bhaskar, H., Hoyle, D. C., & Singh, S. (2006). Machine learning in bioinformatics: A brief survey and recommendations for practitioners. *Computers in Biology and Medicine, 36*(10), 1104–1125. https://doi.org/10.1016/j.compbiomed.2005.09.002
6. Duc, T. L., Leiva, R. G., Casari, P., & Östberg, P. O. (2020). Machine learning methods for reliable resource provisioning in edge-cloud computing: A survey. *ACM Computing Surveys, 52*(5), 1–39. https://doi.org/10.1145/3341145
7. Kumar, I., Singh, S. P., & Shivam. (2022). Machine learning in bioinformatics. In *Bioinformatics* (pp. 443–456). Elsevier. https://doi.org/10.1016/B978-0-323-89775-4.00020-1.
8. Li, W., Wen, Y., Wang, K., et al. (2024). Developing a machine learning model for accurate nucleoside hydrogels prediction based on descriptors. *Nature Communications, 15*(1), 2603. https://doi.org/10.1038/s41467-024-46866-9
9. Aldahiri, A., Alrashed, B., & Hussain, W. (2021). Trends in using IoT with machine learning in health prediction system. *Forecasting, 3*(1), 181–206. https://doi.org/10.3390/forecast3010012
10. Olcay, B., Ozdemir, G. D., Ozdemir, M. A., Ercan, U. K., Guren, O., & Karaman, O. (2024). Prediction of the synergistic effect of antimicrobial peptides and antimicrobial agents via supervised machine learning. *BMC Biomedical Engineering, 6*(1), 1. https://doi.org/10.1186/s42490-024-00075-z
11. Che, D., Liu, Q., Rasheed, K., & Tao, X. (2011). Decision tree and ensemble learning algorithms with their applications in bioinformatics. In H. R. Arabnia, & Q. N. Tran (Eds.), *Software tools*

and algorithms for biological systems (Vol. 696, pp. 191–199). Advances in experimental medicine and biology. Springer New York. https://doi.org/10.1007/978-1-4419-7046-6_19.
12. Kampezidou, S. I., Tikayat Ray, A., Bhat, A. P., Pinon Fischer, O. J., & Mavris, D. N. (2024). Fundamental components and principles of supervised machine learning workflows with numerical and categorical data. *Engineering, 5*(1), 384–416. https://doi.org/10.3390/eng5010021
13. Yao, Z., & Ruzzo, W. L. (2006). A regression-based K nearest neighbor algorithm for gene function prediction from heterogeneous data. *BMC Bioinformatics, 7*(S1), S11. https://doi.org/10.1186/1471-2105-7-S1-S11
14. Murakami, Y., & Mizuguchi, K. (2010). Applying the Naïve Bayes classifier with kernel density estimation to the prediction of protein–protein interaction sites. *Bioinformatics, 26*(15), 1841–1848. https://doi.org/10.1093/bioinformatics/btq302
15. Karalis G. (2020). Decision trees and applications. In P. Vlamos (Ed.), *GeNeDis 2018* (Vol. 1194. Pp. 239–242). Advances in experimental medicine and biology. Springer International Publishing. https://doi.org/10.1007/978-3-030-32622-7_21.
16. Priya, K. S. (2021). Linear regression algorithm in machine learning through MATLAB. *Ijraset Journal For Research in Applied Science and Engineering Technology, 9*(12), 989–995. https://doi.org/10.22214/ijraset.2021.39410.
17. Yang, Z. R. (2004). Biological applications of support vector machines. *Briefings in Bioinformatics, 5*(4), 328–338. https://doi.org/10.1093/bib/5.4.328
18. Yang, Z. R. (2010). Neural networks. In: O. Carugo, & F. Eisenhaber (Eds.), *Data mining techniques for the life sciences* (Vol. 609, pp. 197–222). Methods in molecular biology. Humana Press. https://doi.org/10.1007/978-1-60327-241-4_12.
19. Ashari, Z. E., & Broschat, S. L. (2019). t-Tree and t-Forest: Decision tree and random forest algorithms including the relevance factor with applications in bioinformatics. In *2019 IEEE International Conference on Bioinformatics and Biomedicine (BIBM)* (pp. 2779–2783). IEEE. https://doi.org/10.1109/BIBM47256.2019.8983065.
20. Asriani, E. R. (2020). Application of artificial neural network with backpropagation algorithm for estimating leaf area. *IOP Conference Series: Earth and Environmental Science, 599*(1), 012046. https://doi.org/10.1088/1755-1315/599/1/012046.
21. Sohail, A., & Arif, F. (2020). Supervised and unsupervised algorithms for bioinformatics and data science. *Progress in Biophysics and Molecular Biology, 151*, 14–22. https://doi.org/10.1016/j.pbiomolbio.2019.11.012
22. *Data Mining: A Knowledge Discovery Approach.* (2007). Springer.
23. Gulzar, K., Ayoob Memon, M., Mohsin, S. M., Aslam, S., Akber, S. M. A., & Nadeem, M. A. (2023). An efficient healthcare data mining approach using Apriori algorithm: A case study of eye disorders in young adults. *Information, 14*(4), 203. https://doi.org/10.3390/info14040203
24. Nayak, G. S., Puttamadappa, C., & Davide, O. (2010). Classification of bio optical signals using K-means clustering for detection of skin pathology. *International Journal of Computers and Applications, 1*(2), 112–116. https://doi.org/10.5120/32-141
25. Chen, M., & Yin, Z. (2022). Classification of cardiotocography based on the Apriori algorithm and multi-model ensemble classifier. *Frontiers in Cell and Developmental Biology, 10*, 888859. https://doi.org/10.3389/fcell.2022.888859
26. Yao, F., Coquery, J., & Lê Cao, K. A. (2012). Independent principal component analysis for biologically meaningful dimension reduction of large biological data sets. *BMC Bioinformatics, 13*(1), 24. https://doi.org/10.1186/1471-2105-13-24
27. Wang, Y. C., & Usher, J. M. (2007). A reinforcement learning approach for developing routing policies in multi-agent production scheduling. *International Journal of Advanced Manufacturing Technology, 33*(3–4), 323–333. https://doi.org/10.1007/s00170-006-0465-y
28. Mahmud, M., Kaiser, M. S., Hussain, A., & Vassanelli, S. (2017). Applications of deep learning and reinforcement learning to biological data. https://doi.org/10.48550/ARXIV.1711.03985. (Published online 2017)
29. Yu, C., Liu, J., & Nemati, S. (2019). Reinforcement learning in healthcare: A survey. https://doi.org/10.48550/ARXIV.1908.08796. (Published online 2019)

30. Karami, M., Alizadehsani, R., Khadijeh, et al. (2023). Revolutionizing genomics with reinforcement learning techniques. https://doi.org/10.48550/ARXIV.2302.13268. (Published online 2023)
31. Chadi, M. A., & Mousannif, H. (2023). Understanding reinforcement learning algorithms: The progress from basic Q-learning to proximal policy optimization. https://doi.org/10.48550/ARXIV.2304.00026. (Published online 2023)
32. Neves, M., Vieira, M., & Neto, P. (2021). A study on a Q-Learning algorithm application to a manufacturing assembly problem. *Journal of Manufacturing Systems, 59*, 426–440. https://doi.org/10.1016/j.jmsy.2021.02.014
33. Watkins, C. J. C. H., & Dayan, P. (1992). Q-learning. *Machine Learning, 8*(3–4), 279–292. https://doi.org/10.1007/BF00992698
34. Andrianov, A. M., Nikolaev, G. I., Shuldov, N. A., Bosko, I. P., Anischenko, A. I., & Tuzikov, A. V. (2022). Application of deep learning and molecular modeling to identify small drug-like compounds as potential HIV-1 entry inhibitors. *Journal of Biomolecular Structure & Dynamics, 40*(16), 7555–7573. https://doi.org/10.1080/07391102.2021.1905559
35. Varga, B., Kulcsar, B., & Chehreghani, M. H. (2022). Deep Q-learning: A robust control approach. https://doi.org/10.48550/ARXIV.2201.08610. (Published online 2022)
36. Sumiea, E. H., AbdulKadir, S. J., Al-Selwi, S. M., et al. (2023). Deep deterministic policy Gradient algorithm: A systematic review. https://doi.org/10.21203/rs.3.rs-3544387/v1. (Published online November 6, 2023)
37. Zhang, Y., Yang, L., Chen, J., Fredericksen, M., Hughes, D. P., & Chen, D. Z. (2017). Deep adversarial networks for biomedical image segmentation utilizing unannotated images. In M. Descoteaux, L. Maier-Hein, A. Franz, P. Jannin, D. L. Collins, & S. Duchesne (Eds.), *Medical image computing and computer assisted intervention−MICCAI 2017* (Vol. 10435, pp. 408–416). Lecture notes in computer science. Springer International Publishing. https://doi.org/10.1007/978-3-319-66179-7_47.
38. Pastore, V. P., Ciranni, M., Bianco, S., Fung, J. C., Murino, V., & Odone, F. (2023). Efficient unsupervised learning of biological images with compressed deep features. *Image and Vision Computing, 137*, 104764. https://doi.org/10.1016/j.imavis.2023.104764
39. Zhang, J., Lee, R., & Wang, Y. J. (2003). Support vector machine classifications for microarray expression data set. In *Proceedings Fifth International Conference on Computational Intelligence and Multimedia Applications. ICCIMA 2003* (pp. 67–71). IEEE Computer Society. https://doi.org/10.1109/ICCIMA.2003.1238102.
40. Kim, S. J., Bae, S. J., & Jang, M. W. (2022). Linear regression machine learning algorithms for estimating reference evapotranspiration using limited climate data. *Sustainability, 14*(18), 11674. https://doi.org/10.3390/su141811674
41. Tsuyuzaki, K., Sato, H., Sato, K., & Nikaido, I. (2020). Benchmarking principal component analysis for large-scale single-cell RNA-sequencing. *Genome Biology, 21*(1), 9. https://doi.org/10.1186/s13059-019-1900-3
42. Elhaik, E. (2022). Principal Component Analyses (PCA)-based findings in population genetic studies are highly biased and must be reevaluated. *Science and Reports, 12*(1), 14683. https://doi.org/10.1038/s41598-022-14395-4
43. Kitao, A. (2022). Principal component analysis and related methods for investigating the dynamics of biological macromolecules. *J, 5*(2), 298–317. https://doi.org/10.3390/j5020021.
44. Ahmed, M., Seraj, R., & Islam, S. M. S. (2020). The k-means algorithm: A comprehensive survey and performance evaluation. *Electronics, 9*(8), 1295. https://doi.org/10.3390/electronics9081295
45. Wu, F. X. (2008). Genetic weighted k-means algorithm for clustering large-scale gene expression data. *BMC Bioinformatics, 9*(S6), S12. https://doi.org/10.1186/1471-2105-9-S6-S12
46. Krogh, A., Larsson, B., Von Heijne, G., & Sonnhammer, E. L. L. (2001). Predicting transmembrane protein topology with a hidden Markov model: Application to complete genomes. *Journal of Molecular Biology, 305*(3), 567–580. https://doi.org/10.1006/jmbi.2000.4315. (F. Cohen (Ed.))

47. Saifi, I., Bhat, B. A., Hamdani, S. S., et al. (2023). Artificial intelligence and cheminformatics tools: A contribution to the drug development and chemical science. *Journal of Biomolecular Structure and Dynamics,* 1–19. https://doi.org/10.1080/07391102.2023.2234039. (Published online July 11)
48. Karim, Md. R., Islam, T., Beyan, O., et al. (2022). Explainable AI for bioinformatics: Methods, tools, and applications. https://doi.org/10.48550/ARXIV.2212.13261. (Published online 2022)
49. Liu, S., Liu, C., & Deng, L. (2018). Machine learning approaches for protein-protein interaction hot spot prediction: Progress and comparative assessment. *Molecules, 23*(10), 2535. https://doi.org/10.3390/molecules23102535
50. Larrañaga, P., Calvo, B., Santana, R., et al. (2006). Machine learning in bioinformatics. *Briefings in Bioinformatics, 7*(1), 86–112. https://doi.org/10.1093/bib/bbk007
51. Rai, S., Shukla, S., Scotti, L., & Mani, A. (2023). Drug repurposing against novel therapeutic targets in plasmodium falciparum for malaria: The computational perspective. *Current Medicinal Chemistry,* 31. https://doi.org/10.2174/0929867331666230807151708.
52. Rabie Ahmed, E. H. A. E. (2024). Clinical applications of machine learning in the diagnosis, classification and prediction of heart failure. *Journal Electrical System, 20*(3s), 2160–2170. https://doi.org/10.52783/jes.1815.
53. Parmar, U. P. S., Surico, P. L., Singh, R. B., et al. (2024). Artificial Intelligence (AI) for early diagnosis of retinal diseases. *Medicina (Mexico), 60*(4), 527. https://doi.org/10.3390/medicina60040527
54. Wang, Z., Wang, D., Li, C., Xu, Y., Li, H., & Bao, Z. (2018). Deep reinforcement learning of cell movement in the early stage of *C.elegans* embryogenesis. *Bioinformatics (Oxford, England), 34*(18), 3169–3177. https://doi.org/10.1093/bioinformatics/bty323.
55. Zheng, P., Wang, S., Wang, X., & Zeng, X. (2022). Editorial: Artificial intelligence in bioinformatics and drug repurposing: Methods and applications. *Frontiers in Genetics, 13*, 870795. https://doi.org/10.3389/fgene.2022.870795
56. Kaushal, K., Sarma, P., Rana, S. V., Medhi, B., & Naithani, M. (2022). Emerging role of artificial intelligence in therapeutics for COVID-19: A systematic review. *Journal of Biomolecular Structure & Dynamics, 40*(10), 4750–4765. https://doi.org/10.1080/07391102.2020.1855250

Structural Bioinformatics and Protein Structure Prediction

Kavita Patel and Ashutosh Mani

Abstract Structural bioinformatics is a rapidly growing field and is essential in understanding the three-dimensional structure of biological macromolecules like proteins. This chapter thoroughly introduces structural bioinformatics and how it is used to predict protein structures. The chapter begins with an introduction to the fundamentals of protein structure determination, covering experimental methods like nuclear magnetic resonance (NMR) spectroscopy, X-ray crystallography, and cryo-electron microscopy (cryo-EM). It then explores the difficulties of using experimental methods and the necessity of computational approaches in protein structure prediction, like homology modeling, ab initio model, and threading. The focus is on utilizing databases, machine learning techniques, and bioinformatics tools in concert to improve prediction accuracy. We look at new developments, including the prediction of protein–protein interactions and the effects of genetic variants on the structure and function of proteins. We also address future directions, emphasizing the multidisciplinary nature of the discipline and its implications for drug discovery and personalized medicine. These include the development of novel computational tools and the integration of multi-omics data.

Keywords Template based modelling · Template free modelling · Homology modelling · Ab-initio modelling · Threading

1 Introduction

The dynamic and interdisciplinary area of structural bioinformatics uses concepts from computational biology, bioinformatics, and biology to analyze and predict the three-dimensional (3D) structures of biological macromolecules, with a primary focus on proteins [1]. The basic building blocks of life, proteins are involved in almost every biological function, including molecular recognition, gene regulation,

---

K. Patel · A. Mani (✉)
Department of Biotechnology, Motilal Nehru National Institute of Technology, Prayagraj, Uttar Pradesh 211004, India
e-mail: amani@mnnit.ac.in

© The Author(s), under exclusive license to Springer Nature Singapore Pte Ltd. 2024
A. Chaudhary et al. (eds.), *Unraveling New Frontiers and Advances in Bioinformatics*,
https://doi.org/10.1007/978-981-97-7123-3_8

signal transduction, enzyme catalysis, and signal transduction. Clarifying the roles, relationships, and methods of action of proteins requires an understanding of their three-dimensional structures [2].

The combination of computational approaches for modeling and predicting protein structures with experimental methods for determining protein structures gave rise to the discipline of structural bioinformatics. Atomic resolution protein structure visualization and characterization have been made possible by the development of experimental methods, including cryo-electron microscopy (cryo-EM) [3], nuclear magnetic resonance (NMR) spectroscopy [4], and X-ray crystallography [5]. Proteins are crystallized in X-ray crystallography, and the electron density and three-dimensional structure of the protein are inferred by examining the diffraction patterns of X-rays going through the crystals. In order to ascertain a protein's structure in solution, NMR spectroscopy analyses the interactions between atomic nuclei in the protein. Protein structures are visualized using electron microscopy in cryo-EM, which determines high-resolution structures of massive macromolecular complexes by freezing the proteins in vitreous ice.

However, protein structure identification by experimental approaches is frequently labor-intensive, time-consuming, and not always possible for all proteins, particularly membrane-bound or multi-domain proteins. Furthermore, due to protein flexibility and conformational changes, experimental methods may not always provide high-resolution structures or may yield inconclusive results. Because of this, computational tools [6–9] are becoming essential for predicting protein structures in situations when experimental approaches are impractical or yield conflicting results.

There are two main approaches for the structure prediction of the proteins, first one is template based modelling and second one is template free modelling (Fig. 1). Template based modelling methods constructs model by assessing and identifying the structural frameworks of existing proteins from PDB [10] which is called homologous templates while template free modelling methods predict protein without using global template structures. Template based modelling methods accuracy depends upon the quality of alignments and sequence identities between target protein and template, which is also dependent on the evolutionary distances between template and query. For proteins with sequence identities >30–50%, the models have ~85% of the core regions to the native structure, however, the modelling accuracy also decreases as the sequence identities drops <30% due to lack of significant templates [11, 12].

Template free modelling methods have been used traditionally to model those proteins which have no homologous templates identified from the PDB. Template free modelling method based on the knowledge based and physics based energy functions and use extensive sampling methods to model protein structure, referred as de novo modelling or ab initio modelling approaches [13, 14]. This method has not been used historically as its accuracy is low as compare to template based modelling but recently, the gap between these two approaches has been filled with the use of deep learning to build protein structure models [15].

The field of structure prediction has gained significant advancements recently due to the availability of high-throughput experimental data, advancement in computational algorithms and methods, and integration of resources and bioinformatics

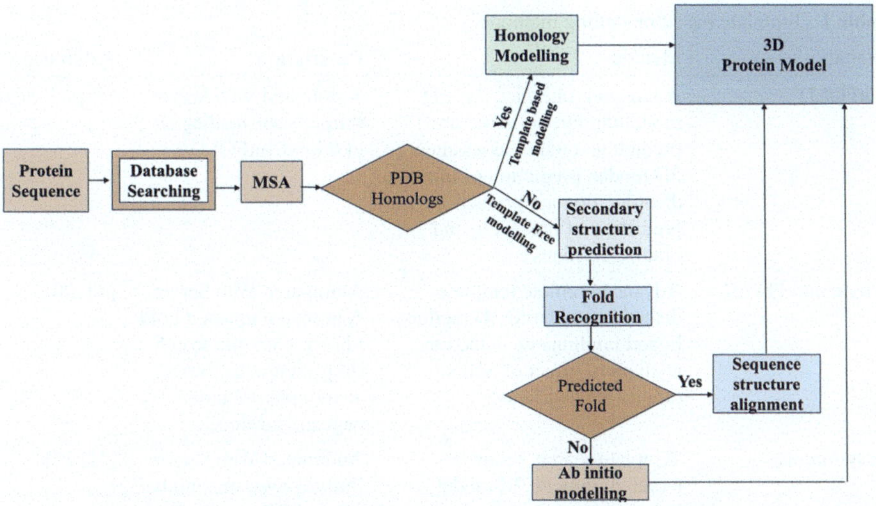

Fig. 1 Decision-making diagram for the process of protein structure prediction

tools. Structural bioinformatics advances biological research and the development of new therapeutics by clarifying protein structures and providing insights into their relationships, roles, and activities in both health and disease.

2 Template Based Modelling

Template based modelling or comparative modelling is a computational approach used for the protein structure prediction from the known structures of homologous protein, which is referred as templates used to predict the target protein structure [16]. The principle underly in template based modelling is that it assume that proteins which have similar sequences share similar functions and structures. Template based modelling method (Table 1) aims to transfer the structural information from the templates to target protein by aligning the sequences of target protein with known protein structure sequences, which leads to 3D structure prediction.

2.1 Homology Modelling

Homology modelling is used for 3D structure predictions of proteins based on its amino acid sequences and the known structure of homologous proteins. This method is based on the hypothesis for the proteins which shares high sequence homology, also will be structurally similar. So, when the sequence similarity is high $\geq$30% between

Table 1 Template based modelling methods

Name	Method	Description	References
IntFOLD	A integrate interface for: 3D modelling/tertiary structure prediction, quality assessment of 3D model, prediction of intrinsic disorder, domain prediction, prediction of protein ligand binding residues	Automated Web Server (http://www.reading.ac.uk/bioinf/IntFOLD/)	[17]
Phyre and Phyre2	Advanced remote template detection and build 3D models, ligand binding site detection, analysis of effect of SNPs (amino acid variants)	Automated Web Server, Automated updated Fold library, Genome search (http://www.sbg.bio.ic.ac.uk/~phyre2/html/page.cgi?id=index)	[18–20]
ESyPred3D	Template search, template-target alignment, 3d-model building, model evaluation	Automated Web Server (https://www.unamur.be/sciences/biologie/urbm/bioinfo/esypred/)	[21, 22]
FoldX	A integrated method for: Total folding energy calculations, build model, repair PDB, analyse complex, Print Networks, and stability of protein	Command line tool and downloadable program (http://foldxsuite.crg.eu/)	[23]
Biskit	Protein structure prediction, analysis of molecular dynamics trajectories, plateform for integrated programs and algorithms DSSP, T-Coffee, Fold-X, TM-Align and MODELLER	Python platform for structural bioinformatics (https://biskit.pasteur.fr/)	[24]
RaptorX	Template based protein structure modelling	Automated Web Server (http://raptorx6.uchicago.edu/)	[25–27]
MODELLER	Template search, target template alignment, comparative modelling, model evaluation	Software package based on Python and Fortran (https://salilab.org/modeller/)	[28]
HHpred	Detection of templates, alignment, 3D modelling	Interactive Web Server (https://toolkit.tuebingen.mpg.de/tools/hhpred)	[29]
ROSETTA	Protein modelling, and analysis, ligand docking	Software program (https://www.rosettacommons.org/software)	[30]
Yasara	Template detection, alignment, ligand and oligomers modelling, model fragment hybridizations. simulation	(Software program https://www.yasara.org/)	[31]

(continued)

Table 1 (continued)

Name	Method	Description	References
MOE (Molecular Operating Environment)	Template search, model build, visualization, simulation	Software program (https://www.chemcomp.com/Products.htm)	[32]
SWISS-MODEL	Template search, user defined target- template alignment, 3D structure prediction	Automated Web Server (https://swissmodel.expasy.org/)	[33]
BHAGEERATH-H	Combination of homology methods and ab initio folding	Automated Web Server (http://www.scfbio-iitd.res.in/bhageerath/bhageerath_h.jsp)	[34, 35]

target and templates, this method is used. This makes it remarkably easier to find the high similarity templates by aligning them. The identification of structurally conserved regions among the structures corresponding to templates is a crucial step in homology modelling, it determine by calculating the C-α distance matrix of each structure and small sections of this matrix get compared to find the small peptide with low RMSD (root mean square deviations) of related structures [36, 37].

Simple sequence-sequence alignment algorithm like Smith–Waterman algorithm [38] (for local alignment) and Needleman–Wunsch algorithm [39] (for global alignment) is used to find homologous sequence or conserved regions but its relatively slow dynamic programming methods, BLAST [40] software is used for rapid sequence-sequence alignment.

The target protein 3D structure is built by transferring the atom coordinates from the corresponding regions of template structures. The target sequence is superimposed onto the template structure during this procedure, maintaining the structural features that are conserved.

Template-target alignment shows that some blocks in the target match the template's structurally conserved regions. Gaps appear between the aligned model and the template sequence during template-target alignment. It is necessary to identify and incorporate the structural fold of the gap residues, or loop, between the two conserved core regions. This require modifications of backbone, orientation or conformational changes are absent from the normal secondary structural elements. As a result, it is safe to make all of the insertions and deletions inside the alignment form loops, turns, and helixes.

Two approaches mainly used for loop modelling, knowledge based and energy based approach. In knowledge based approach, it look for a loop region's structure with endpoints that resemble known structures, and then we sandwich the loop structure's coordinates between two cores. Mostly molecular modelling programs like Swiss model, MODELLER [41], insight support knowledge based approach for loop modelling. In energy based approach, molecular dynamics or Monte Carlo simulation techniques [42] are used to produce the loop form most accurately, and the energy function is utilised to evaluate the loop quality. It is possible to alter the energy function to produce a loop structure that fits the core more snugly [43, 44].

High-precision side-chain rotamer prediction, mainly depends on the rotamers and their packaging, requires the correct backbone [45]. The primary method for handling a tangle of this type involves repeatedly predicting rotamers, followed by backbone shifts and new backbone rotamers, until the process converges. By using this technique, the number of rotamer predictions and energy minimization stages are decreased. These approaches should be used for the whole protein structure since they are applicable to both loop modelling and model optimization [46, 47].

Two approaches (1) Quantum force fields [48] (2) Self- Parameterizing force fields [49], are used for the better optimization of a model which can be achieved by calculating force field. Recent developments in computational biology have made it possible to apply quantum chemical techniques to provide a more precise interpretation of the charge distribution across the protein molecule. Initialise the force field with certain parameters, adjust it, reduce model energy, and store the updated force field if model quality improved; otherwise, revert to the original force field value. By using this method, the force field during energy minimization can be more accurately directed in the desired direction.

Molecular dynamics simulation can be used to optimise protein models. It generates the true folding dynamics of the protein by sampling the trajectory of the protein's motions over a 10-fs time span [50]. As a result, it is anticipated that as the simulation runs, the model will resemble real structure [51].

2.2 *Threading*

Threading also known as fold recognition method, is a computational method used in tertiary structure prediction of proteins by identifying the most similar fold or structural arrangement for target protein sequence. This method predicts the 3D structure when the similarity between target and template sequence is very low which is $\leq 30\%$ [52]. Understanding the link between sequences, structure, and function is still incomplete. Understanding the link between sequences, structure, and function is still incomplete. This method begins with the generation of structural template or protein folds [53] by identifying the distantly homologous templates [54].

The hypothesis behind finding the 3D structure by aligning the sequences by PSI-BLAST [55] which is extension of BLAST methodology. PSI BLAST perform multiple sequence alignment of the sequences which it get by BLAST alignment. The alignment is subsequently transformed into a position-specific score matrix (PSSM), which records the amino acid trends at every MSA position. An method similar to BLAST is used to iteratively search through a sequence database a predefined number of times using the PSSM rather than the query sequence. The profile, or PSSM, is changed to reflect the sequences found in the previous round following each step. In order to find more distantly related proteins, PSI-BLAST's methodology involves iteratively searching a database utilising profiles, which describe more about the sequence space compatible with a given protein fold. A template structure is expressed as a descriptor string that characterises the structural environment in the 3D profile approach. Three primary environment classifications exist: (1) the portion of the lateral chains covered by polar atoms, (2) the area of the lateral chains buried by other protein atoms, and (3) secondary local structure. The environmental class of each residue in the folded protein structure is represented by an ID string in this instance, which describes a 3D protein structure. Here, the profile of a template structure is generated using the 3D-1D score table. This technique is also referred to as sequence-structure since it represents both the target and the template as strings. The template structure is represented as a string of environmental classes, while the target protein is represented as a string of amino acids. Using a dynamic programming approach, the fit score between the template and target environment classes is determined.

Profile Hidden Markov Models (HMMs) can also be used to represent sequence profiles in addition to PSSMs. Utilising profile HMMs has the benefit of applying substitution probabilities and position-specific gap penalties, which more accurately reflect the underlying sequence distribution [56, 57].

TASSER is used to generate 3D model of distantly related homology protein target, which extracts generally the contiguous fragments from the aligned regions of threading. And for the unaligned regions, it uses a lattice-based template-free modelling approach. Aside from using constraints from the aligned region, TASSER also uses different knowledge-based energy functions, which is important for protein folding (e.g., secondary structure formation, hydrogen bonding, side-chain contact formation) Monte Carlo Simulations and produce large number of reduced models. Based on structural similarity, low-energy decoys are grouped, and the biggest cluster centroid is chosen for further full-atom refining. The lowest energy structure is then chosen and put through full-atom refining.

3 Template Free Modelling

This computational approach used in protein structure prediction which does not rely on the known structures of homologous proteins as templates. As an alternative, TFM uses only the target protein's amino acid sequence to predict protein structures de novo. This method is especially useful for predicting novel folds or in the absence of closely related homologous proteins with known structures. There are following different techniques (Table 2) used in this approach.

3.1 Ab Initio Methods

This method creates a protein model based only on sequence data because structural folds or their counterparts are not readily available. The physicochemical concept

Table 2 Template free modelling methods

Name	Method	Description	References
D-QUARK	Orientation-guided folding/deep learning distance, method for ab-initio protein structure predictions	Automated web server (https://zhanggroup.org/QUARK/)	[58]
D-I-TASSER	Hydrogen bond network-guided folding/orientation/template and deep learning distance/	Automated web server (https://zhanggroup.org/D-I-TASSER/)	[59]
Rosetta	Orientation-guided folding/Deep learning distance/	Automated web server Robetta Server (https://robetta.bakerlab.org) trRosetta Server: (https://yanglab.nankai.edu.cn/trRosetta/)	[30]
AlphaFold2	End-to-end deep learning-based protein model prediction	Automated web server (https://alphafold.ebi.ac.uk/)	[60]
FragFold	Based on fragment assembly to resolve 3D structure prediction	https://github.com/swanss/FragFold	[53]
PSICOV	Distance based deep learning model prediction	Downloadable program (https://maayanlab.cloud/datasets2tools/landing/tool/PSICOV)	[61]
GREMLIN	Distance based deep learning model prediction	Automated web server (https://gremlin.bakerlab.org)	[62, 63]
CCMpred	Based on Markov model contact based prediction	Open source software (https://github.com/soedinglab/CCMpred)	[64, 65]
NeBcon	Protein prediction on contact map prediction using neural netwrok	Online server and standalone package (http://zhanglab.ccmb.med.umich.edu/NeBcon/)	[66]
ResPRE	Protein prediction on contact map prediction using coupling precision matrix	Online server (https://zhanglab.ccmb.med.umich.edu/ResPRE)	[67]
DMPfold	Deep learning-based prediction	Automated web server (https://github.com/psipred/DMPfold)	[68]

(continued)

Table 2 (continued)

Name	Method	Description	References
RaptorX-Contact	Distance-based deep learning model prediction	(http://raptorx6.uchicago.edu/ComplexContact/)	[25, 27]
TripletRes	Distance-based deep learning model prediction	Automated web server (https://zhanggroup.org/TripletRes/)	[69]

pertaining to the nature of proteins can be comprehended through the ab initio method. Compared to other approaches of structure prediction, ab initio modelling has a low accuracy [70]. Protein structure can be constructed by figuring out the configurational space of the atoms in amino acids if the target sequence is not having any structural similarities with structures present in the database. This approach makes use of an understanding of several physics, chemistry, and mathematics concepts. The computation is simplified by using reduced protein representations. In certain models, a residue can represent only two locations like backbone and side chain [71]. Some employ several sites, such as a side link and heavy backbone atoms. Hydrophobic interactions are recognised to be the primary factor in protein folding, and an empirical energy function may be used to calculate these interactions. Three criteria need to be determined in order to make the ab initio prediction: (1) proteins representation reduction; (2) interaction's potential energy function; and (3) a method to explore the conformational space.

Simulated annealing is used to search the configuration space of fragment structures. A step is performed to substitute the present configuration with the torsion angles of a neighbour chosen at random at a randomly chosen position. Motions that bring two atoms closer together by 2.5 Å are eliminated, while other motions are assessed. Since most structure prediction algorithms now in use rely on information gained from experimentally predicted structures, they are not very helpful in delving into the fundamentals of protein folding. Template-free approaches take into account both the fundamentals of protein folding and their practical application. The development of template-free approaches may more accurately represent the prediction of the technical and theoretical level of protein structure than template-based methods since they are based on information from existing structures. ROSETTA A template-free method developed by the David Baker Lab that assembles a full structure based on fragments of 3–9 residues from PDB is one of the effective ab initio modelling approaches [72, 73]. The fragments are chosen according to the degree of similarity between the predicted and known secondary structure, much like in template-based approaches. The annealing search strategy used in the Monte Carlo approach simulates the assembly process. The fragments utilised in QUARK range in size from one to twenty residues, and Monte Carlo replica-exchange simulation is employed to simulate the assembly process while following the guidance of a knowledge-based atom-level force field. Fragment assembly is also the foundation of many other techniques, such as Scrape, PROFESY, FRAGFOLD, etc. The primary difference between these methods and the template-based methods is that the first do not rely on any global structural blueprint, while the latter do not take advantage of structural similarities or homology between the target and the proteins that the fragments originate from. For template-free approaches, it is better at modelling the target of new folds. However, modelling proteins with a length of more than 150 residues remains a significant difficulty for template-free approaches due to the high computational need and low force field accuracy. Recently, contact map prediction using a co-evolution method has demonstrated success in breaking through this length restriction of ab initio structure folding.

3.2 Protein Structure Prediction by Physics Based Energy Minimization

Physics based energy minimization is a computational approach to predict the 3D structure of the proteins by using the principle of physics and chemistry. This approach describes the interactions between atoms in a protein molecule using mathematical models called force fields. The protein's native or physiologically active conformation may be found by minimising the potential energy of the protein structure, which is the most stable conformation. The choice of a suitable force field is the first stage in the energy minimization method of protein structure prediction. Protein molecules' atomic interactions' strength and geometry are described by characteristics found in force fields. In order to replicate experimental observables, these parameters are obtained through the use of empirical fitting, quantum mechanical calculations, and experimental data. A force field is a mathematical model that uses atom-to-atom interactions to represent the potential energy of a protein structure. CHARMM [74–76], AMBER [77–79], and GROMOS [80] are common force fields used in protein structure prediction. The process of creating the protein's first structure usually involves the use of techniques like threading, homology modelling, or fragment-based assembly, this serves as a starting point for energy minimization. Using the force field that has been chosen, the potential energy of the protein structure is computed. Bond stretching, dihedral angle rotation, angle bending, electrostatic interactions and van der Waals interactions are some of the energy factors that affect the potential energy [9]. A multidimensional surface with each dimension representing a degree of freedom in the protein conformation can be used to represent the potential energy of a protein structure. The goal of energy minimization is to locate the global minimum, or the most stable protein conformation, on this potential energy surface. Molecular dynamics simulations, conjugate gradient, and steepest descent are common optimization algorithms for energy minimization. By sampling various conformations and calculating their potential energies, energy minimization investigates the protein's conformational space. Until a termination condition is satisfied, such as hitting a predetermined energy threshold or having the optimisation algorithm converge, the energy minimization process keeps going. The final protein structure is acquired when optimisation is finished [81].

3.3 Protein Structure Prediction by Fragment Assembly Method

Fragment Assembly (FA) techniques for protein structure prediction are computational approaches that assemble small peptide fragments into a whole 3D protein structure [82]. Fragment Assembly techniques create structures de novo, based only on the amino acid sequence of the target protein, in contrast to Template-Based Modelling (TBM), which uses templates of known structures of homologous proteins. FA techniques allow the assembly of these fragments into a global protein structure by taking use of the fact that local structural motifs, such secondary structure elements, are frequently conserved among proteins.

In Fragment Assembly, a library of short peptide fragments generated from known protein structures is created as the initial step. These segments usually consist of three to nine amino acids and exhibit local structural motifs such as turns, β-strands, and α-helices. Fragment libraries can be produced using computational techniques like ab initio structure prediction or using experimental structures found in databases like the Protein Data Bank (PDB). Proteins often possess local structural motifs like α-helices, β-strands, and turns, which is why fragment assembly methods take use of this characteristic. Global protein structures can be produced by the assembly of small peptide fragments that represent these patterns. Based on projected secondary structure and sequence similarity, appropriate fragments are chosen from the fragment library for a specified target protein sequence. For assembly, only the fragments that align with the target protein's predicted secondary structure and sequence are retained.

Fragment Assembly methods search through the enormous conformational space of protein structures using combinatorial search strategies. Assembly algorithms find the best possible arrangement of fragments that takes energy and structural constraints into account by methodically merging and assessing the compatibility of each fragment. The chosen fragments are combined repeatedly by the assembly algorithm to provide a pool of potential structures. Based on their projected sequence and secondary structure, fragments are overlapped and aligned. The assemblies that develop are evaluated for geometric compatibility and energy scores.

The fragment assembly-generated candidate structures are refined to maximise their geometry and eliminate steric conflicts [14, 53]. Energy minimization, loop modelling, and molecular dynamics simulations are examples of refinement approaches which is used to enhance the accuracy and quality of the predicted protein structures. Using geometric and energy parameters, Fragment Assembly techniques optimise the arrangement of fragments. Protein structure is optimised by aligning and overlapping fragments to reduce steric conflicts and enhance favourable interactions, such hydrogen bonds and van der Waals contacts.

3.4 Protein Structure Prediction by Rapid Gradient Descent Based Folding Methods

One effective template free modelling method is fragment assembly, but the downside is that it depends on the total length of the protein, the simulations might take hours or even days to complete. Consequently, it is preferable to create techniques that can quickly produce structures. Gradient descent-based folding techniques can be used to accomplish this.

A drawback of these methods is that they could be more likely to trap in local minima than to identify global minimum conformation of the energy distribution. This is especially true in cases like protein folding when the energy landscape is complicated. Deep learning has recently been used to reliably anticipate pair-wise spatial constraints [83], such as inter-residue distances, which can smooth the energy landscape and enable gradient-based algorithms [84] to fold protein structures appropriately. The initial AlphaFold iteration in CASP13 used a gradient descent-based folding technique to attain state-of-the-art performance. Moreover, the most recent version of the Rosetta modelling programme, trRosetta, rapidly folds protein structures using an L-BFGS gradient descent technique and shows that appropriate deep learning-based constraints [83] may enable high-accuracy predictions even with quick simulations.

3.5 Protein Structure Prediction by Deep Learning

Deep Learning (DL) protein structure prediction is an advanced technique that uses artificial intelligence methods to predict the 3D structure of proteins based only on their amino acid sequences. Protein structure prediction challenges are suitable for deep learning models, especially neural networks, which have shown impressive capacities in identifying complex patterns and correlations in biological data. Preparing training data is the initial stage in Deep Learning's prediction of protein structure. A dataset of protein sequences matched to their experimentally established structures must be assembled; these may be found in sources such as the Protein Data Bank (PDB). The dataset is divide into training, testing, and validation sets. In order to capture local and global sequence properties pertinent to protein structure, Deep Learning models develop hierarchical representations of protein sequences [85]. Deep Learning models automatically learn to extract meaningful representations by training on large data sets [86], which helps with accurate structure prediction. Numerical representations of protein sequences that capture significant sequence properties are called feature vectors. Embedding methods, in which amino acids are mapped to continuous vector spaces, and one-hot encoding, in which each amino acid is represented as a binary vector, are common encoding approaches. Deep Learning models can capture the complex correlations between sequence and structure in proteins because they are very good at recognising patterns in large, complicated datasets. Deep Learning models may generalise to accurately anticipate the structures of previously unknown proteins by learning from a variety of examples in the training data. By predicting their structures and deducing their activities from structural data, deep learning algorithms help with the functional annotation of proteins.

Recurrent neural networks (RNNs), transformer models, convolutional neural networks (CNNs), and other concepts made up of several layers of linked neurons are common components of Deep Learning models for protein structure prediction [87]. Processing the encoded protein sequences and predicting their matching three-dimensional structures is the architecture's intended function. By predicting the structures of designed proteins with certain activities or features, deep learning-based protein structure prediction aids in the design and engineering of proteins. Stochastic gradient descent (SGD) or Adam are two optimisation methods that are used to train the Deep Learning model on the provided dataset. By modifying the neural network's parameters (weights and biases) in response to the difference between the predicted and real structures, the model gains the ability to map input protein sequences to the relevant 3D structures during training. Without the need for manual feature creation or intermediary representations, end-to-end learning is made possible by deep learning models, which allow raw input data (protein sequences) to be directly mapped to output predictions (3D structures). Complex sequence-structure interactions may be seamlessly integrated with this end-to-end technique, which also speeds the prediction process. Utilising an independent validation dataset, the trained Deep Learning model's prediction accuracy of protein structures is assessed. The global distance test total score (GDT-TS), certain geometric and energy-based parameters, and the root-mean-square deviation (RMSD) from experimental constructions are examples of evaluation metrics. Model evaluation aids in determining the projected structures' accuracy and reliability. The trained Deep Learning model may be used to forecast the three-dimensional structures of novel protein sequences once it has been verified. The protein's spatial organisation in three dimensions is represented by the model, which takes as input the amino acid sequence of a protein and outputs projected coordinates for its atoms.

4 Model Evaluation and Validation

It is necessary to assess the final projected model to ensure that its structural characteristics are in line with the physicochemical laws. This entails examining irregularities in bond lengths, close to contacts, φ–ψ angles, and other related factors. Implicitly accounting for these stereochemical features is another technique to assess the quality of modelled protein. Using statistical profiles of interaction energy and spatial features compiled from structures identified through experimentation, this technique identifies errors. By comparing the statistical parameters with the developed model, it is able to ascertain whether sections of a sequence appear to be folded normally and which do not. If structural differences are discovered, the area is considered to have errors and requires more refinement. A collection of programmes called SAVES server allows you to verify the model's correctness by simply uploading the expected structure. One such programme that can verify general physical properties including bond angles, bond lengths, chirality, and bond angles is called Procheck.

The model's parameters are compared to those obtained from precisely defined high-resolution structures. The programme highlights the areas that require more inspection or refinement if it finds any odd traits. Another thorough protein analysis site that verifies a protein model's chemical accuracy is WHATIF. It performs a variety of tasks, such as planarity checking, proline puckering, collisions with anomalous bond angles, symmetry axes (close contacts), and bond lengths. Additionally, it permits the creation of Ramachandran plots to evaluate the model's quality.

A web server called Atomic Non-Local Environment Assessment (ANOLEA) employs the statistical evaluation method. It computes the energy of atomic interactions inside a protein chain and contrasts the resultant interaction energy values with data gathered from an X-ray structural database of proteins. It may be an indication that the appropriate area has not been accurately modelled if the energy terms of some regions differ notably from those of the typical crystal structures. Typically, 5.0 is the threshold for unfavourable residues. Regions containing mistakes are defined as residues with scores more than 5.0.

Another server that use the statistical method is Verify3D. It makes use of a pre-calculated database made up of 18 environmental profiles derived from high-resolution protein structures that are based on solvent exposure and secondary structures. The solvent exposure and secondary structure propensity of each residue are computed to evaluate the quality of a modelled protein. The solvent exposure and secondary structure propensity of each residue are computed to evaluate the quality of a protein model. A residue is given either a high or low score depending on whether its properties fit into one of the profiles. The outcome is a two-dimensional graph that shows how well each protein structural residue folds. Normally, the threshold value is zero. A residue is said to have an unfavourable environment if its score is less than zero.

Using various verification programmes may yield different evaluation findings. Verify3D determines that the protein's C-terminus has low-quality residues, despite ANOLEA declaring the model's full-length protein chain to be favourable. It's a good idea to employ a variety of verification methods and determine which ones are in agreement with one another because there isn't a single approach that is obviously better than the others. It's also critical to remember that the evaluation tests carried out by these programmes only assess the stereochemical correctness; they do not assess the model's accuracy, which may or may not have any bearing on biology. Based on the Ramachandran plot of amino acid residues, several methods forecast the correctness of the anticipated model. It is a 2D scatter plot that displays the torsion angles of every residue of an amino acid in a protein. Based on established protein structures, the map indicates both permitted and prohibited areas of the angles. The quality assessment of a novel protein model is aided by this plot.

5 Applications

A vast quantity of data is being integrated more easily thanks to the availability of protein 3D (3-dimensional structure) structures and other structural analysis tools. This information may be valuable to investigate other avenues for strengthening our understanding of protein function and structure in the future. A protein's 3D structure offers greater information about its binding site and other functionally significant areas, information that may be used in the development of new drugs. To begin the process of creating drugs based on structure, a target structure must be available. This structure also directs the modifications that lead molecules undergo. Better knowledge about the residues involved in the interaction can be obtained from a protein's 3D complex structure with a ligand. The mechanism of a drug's pharmacological activity, binding affinity, and lead modification is explained by the interaction of a receptor with ligand (small molecule). A protein's natural structure, which is necessary for the protein to function normally, is destroyed when an amino acid is mutated, which is another way that computational modelling explains why this occurs. Through the representation of structural alterations in the mutant target protein that result in a loss of appropriate drug binding or interaction, it can also help explain the mechanism of drug resistance. To maintain stability, a variety of forces, including Van der Waals, hydrophobic interaction, electrostatic interaction, and hydrogen bonding interaction between protein–ligand complexes. The process of modelling the intermolecular linkages within the protein–ligand complex is complicated since there are several degrees of freedom and little data regarding the influence of water on binding.

6 Conclusion

Molecular modelling has emerged as a major and fundamental method for the experts working in the drug creation field. Molecular modelling helps to understand the relevant physicochemical characteristics of proteins by revealing their three-dimensional structures. The protein modelling effectively combines theoretical scientific concepts, computational science techniques, and experimental data to reveal a macromolecule's structural and biological characteristics. The most effective technology for prediction of protein structure depend on the kind of problem that has to be solved. The fundamental strategies and the most current developments in protein structure prediction techniques have been covered in this chapter. A necessary emphasis has also been placed on the kinds of mistakes that might occur and compound while working with protein models. The accuracy of predicted protein structure is a crucial step as structure-based drug design depends on protein structure.

Since X-ray crystallography and NMR spectroscopy are time-consuming methods that don't seem to be viable for determining the structure of each particular protein, the creation of a highly accurate and exact modelling tools need to be the evolved.

Acknowledgements The authors acknowledge the Department of Biotechnology, MNNIT, Allahabad, for supporting this study smoothly. All authors have edited, conceptualised, and helped write the manuscript.

Conflict of Interest "The authors declare that they have no conflicts of interest in this work."

References

1. Kumar, A., & Chordia, N. (2017). Role of Bioinformatics in Biotechnology. *Resaerch and Reviews in BioSciences, 12*(1). Retrieved May 6, 2024, from https://www.tsijournals.com/abstract/role-of-bioinformatics-in-biotechnology-13046.html.
2. Khan, R. H., Siddiqi, M. K., & Salahuddin, P. (2017). *Protein structure and function.*
3. Cheng, Y. (2015). Single-particle cryo-EM at crystallographic resolution. *Cell, 161*(3), 450–457. https://doi.org/10.1016/j.cell.2015.03.049
4. Cavanagh, J. (1996). *Protein NMR spectroscopy: Principles and practice.* Academic Press.
5. Methods of Biochemical Analysis. Retrieved May 6, 2024, from https://onlinelibrary.wiley.com/doi/book/, https://doi.org/10.1002/9780470110584#page=15.
6. A method to identify protein sequences that fold into a known three-dimensional structure. Science. Retrieved May 6, 2024, from https://www.science.org/doi/abs/, https://doi.org/10.1126/science.1853201.
7. Folding of polypeptide chains in proteins: A proposed mechanism for folding. PNAS. Retrieved May 6, 2024, from https://www.pnas.org/doi/abs/, https://doi.org/10.1073/pnas.68.9.2293.
8. Levitt, M., & Warshel, A. (1975). Computer simulation of protein folding. *Nature, 253*(5494), 694–698. https://doi.org/10.1038/253694a0
9. McCammon, J. A., Gelin, B. R., & Karplus, M. (1977). Dynamics of folded proteins. *Nature, 267*(5612), 585–590. https://doi.org/10.1038/267585a0
10. The protein structure prediction problem could be solved using the current PDB library. PNAS. Retrieved May 6, 2024, from https://www.pnas.org/doi/abs/, https://doi.org/10.1073/pnas.0407152101.
11. Kryshtafovych, A., Monastyrskyy, B., Fidelis, K., Moult, J., Schwede, T., & Tramontano, A. (2018). Evaluation of the template-based modeling in CASP12. *Proteins: Structure, Function, and Bioinformatics, 86*(S1), 321–334. https://doi.org/10.1002/prot.25425
12. Protein structure prediction and structural genomics. Science. Retrieved May 6, 2024, from https://www.science.org/doi/abs/, https://doi.org/10.1126/science.1065659.
13. Simons, K. T., Kooperberg, C., Huang, E., & Baker, D. (1997). Assembly of protein tertiary structures from fragments with similar local sequences using simulated annealing and bayesian scoring functions1. *Journal of Molecular Biology, 268*(1), 209–225. https://doi.org/10.1006/jmbi.1997.0959
14. Xu, D., & Zhang, Y. (2012). Ab initio protein structure assembly using continuous structure fragments and optimized knowledge-based force field. *Proteins: Structure, Function, and Bioinformatics, 80*(7), 1715–1735. https://doi.org/10.1002/prot.24065
15. "It will change everything": DeepMind's AI makes gigantic leap in solving protein structures-document-gale academic OneFile. Retrieved May 6, 2024, from https://go.gale.com/ps/i.do?id=GALE%7CA649545694&sid=googleScholar&v=2.1&it=r&linkaccess=abs&issn=00280836&p=AONE&sw=w&userGroupName=anon%7E51452b6e&aty=open-web-entry.

16. Browne, W. J., North, A. C. T., Phillips, D. C., Brew, K., Vanaman, T. C., & Hill, R. L. (1969). A possible three-dimensional structure of bovine α-lactalbumin based on that of hen's egg-white lysozyme. *Journal of Molecular Biology, 42*(1), 65–86. https://doi.org/10.1016/0022-2836(69)90487-2
17. McGuffin, L. J., Adiyaman, R., Maghrabi, A. H. A., et al. (2019). IntFOLD: An integrated web resource for high performance protein structure and function prediction. *Nucleic Acids Research, 47*(W1), W408–W413. https://doi.org/10.1093/nar/gkz322
18. Kelley, L. A., Mezulis, S., Yates, C. M., Wass, M. N., & Sternberg, M. J. (2015). The Phyre2 web portal for protein modelling, prediction and analysis. *Nature Protocols, 10*(6), 845–858. https://doi.org/10.1038/nprot.2015.053
19. PHYRE Protein Fold Recognition Server. Retrieved May 1, 2024, from http://www.sbg.bio.ic.ac.uk/phyre2/html/help.cgi?id=help/interpret_intensive.
20. The Phyre2 web portal for protein modelling, prediction and analysis-PMC. Retrieved May 1, 2024, from https://www.ncbi.nlm.nih.gov/pmc/articles/PMC5298202/.
21. ESyPred3D submitting form. Retrieved May 1, 2024, from https://www.unamur.be/sciences/biologie/urbm/bioinfo/esypred/.
22. Lambert, C., Léonard, N., De Bolle, X., & Depiereux, E. (2002). ESyPred3D: Prediction of proteins 3D structures. *Bioinformatics, 18*(9), 1250–1256. https://doi.org/10.1093/bioinformatics/18.9.1250
23. Buß, O., Rudat, J., & Ochsenreither, K. (2018). FoldX as protein engineering tool: Better than random based approaches? *Computational and Structural Biotechnology Journal, 16*, 25–33. https://doi.org/10.1016/j.csbj.2018.01.002
24. Welcome to Biskit!—Biskit: Python for structural bioinformatics. Retrieved April 17, 2024, from http://biskit.pasteur.fr.
25. Källberg, M., Wang, H., Wang, S., et al. (2012). Template-based protein structure modeling using the RaptorX web server. *Nature Protocols, 7*(8), 1511–1522. https://doi.org/10.1038/nprot.2012.085
26. RaptorX. Retrieved April 17, 2024, from https://bio.tools/raptorx.
27. RaptorX-complex contact: A protein complex contact map prediction server. Retrieved May 6, 2024, from http://raptorx6.uchicago.edu/ComplexContact/.
28. About MODELLER. Retrieved May 6, 2024, from https://salilab.org/modeller/.
29. Söding, J., Biegert, A., Lupas, A. N. (2005). The HHpred interactive server for protein homology detection and structure prediction. *Nucleic Acids Research, 33*(Web Server issue), W244–W248. https://doi.org/10.1093/nar/gki408.
30. The Rosetta Software. RosettaCommons. Retrieved May 1, 2024, from https://www.rosettacommons.org/software.
31. Land, H., & Humble, M. S. (2018). YASARA: A tool to obtain structural guidance in biocatalytic investigations. *Methods in Molecular Biology Clifton NJ, 1685*, 43–67. https://doi.org/10.1007/978-1-4939-7366-8_4
32. Wang, Y., Cheng, X., Shan, Q., et al. (2014). Simultaneous editing of three homoeoalleles in hexaploid bread wheat confers heritable resistance to powdery mildew. *Nature Biotechnology, 32*(9), 947–951. https://doi.org/10.1038/nbt.2969
33. Schwede, T., Kopp, J., Guex, N., & Peitsch, M. C. (2003). SWISS-MODEL: An automated protein homology-modeling server. *Nucleic Acids Research, 31*(13), 3381–3385.
34. Bhageerath-H. Retrieved May 6, 2024, from http://www.scfbio-iitd.res.in/bhageerath/bhageerath_h.jsp.
35. Bhageerath-H: A homology/ab initio hybrid server for predicting tertiary structures of monomeric soluble proteins-PMC. Retrieved May 6, 2024, from https://www.ncbi.nlm.nih.gov/pmc/articles/PMC4290660/.
36. Krieger, E., Nabuurs, S. B., & Vriend, G. (2003). Homology modeling. *Methods of Biochemical Analysis, 44*, 509–523. https://doi.org/10.1002/0471721204.ch25
37. Rodriguez, R., Chinea, G., Lopez, N., Pons, T., & Vriend, G. (1998). Homology modeling, model and software evaluation: Three related resources. *Bioinformatics (Oxford, England), 14*(6), 523–528. https://doi.org/10.1093/bioinformatics/14.6.523

38. Smith, T. F., & Waterman, M. S. (1981). Identification of common molecular subsequences. *Journal of Molecular Biology, 147*(1), 195–197. https://doi.org/10.1016/0022-2836(81)90087-5
39. Needleman, S. B., & Wunsch, C. D. (1970). A general method applicable to the search for similarities in the amino acid sequence of two proteins. *Journal of Molecular Biology, 48*(3), 443–453. https://doi.org/10.1016/0022-2836(70)90057-4
40. Altschul, S. F., Gish, W., Miller, W., Myers, E. W., & Lipman, D. J. (1990). Basic local alignment search tool. *Journal of Molecular Biology, 215*(3), 403–410. https://doi.org/10.1016/S0022-2836(05)80360-2
41. Sali, A., & Blundell, T. L. (1993). Comparative protein modelling by satisfaction of spatial restraints. *Journal of Molecular Biology, 234*(3), 779–815. https://doi.org/10.1006/jmbi.1993.1626
42. Fiser, A., Do, R. K., & Sali, A. (2000). Modeling of loops in protein structures. *Protein Science Publication Protein Society, 9*(9), 1753–1773. https://doi.org/10.1110/ps.9.9.1753
43. Sánchez, R., & Sali, A. (1997). Advances in comparative protein-structure modelling. *Current Opinion in Structural Biology, 7*(2), 206–214. https://doi.org/10.1016/s0959-440x(97)80027-9
44. Tappura, K. (2001). Influence of rotational energy barriers to the conformational search of protein loops in molecular dynamics and ranking the conformations. *Proteins, 44*(3), 167–179. https://doi.org/10.1002/prot.1082
45. Scouras, A. D., & Daggett, V. (2011). The Dynameomics rotamer library: Amino acid side chain conformations and dynamics from comprehensive molecular dynamics simulations in water. *Protein Science Publication Protein Society, 20*(2), 341–352. https://doi.org/10.1002/pro.565
46. Hintze, B. J., Lewis, S. M., Richardson, J. S., & Richardson, D. C. (2016). Molprobity's ultimate rotamer-library distributions for model validation. *Proteins, 84*(9), 1177–1189. https://doi.org/10.1002/prot.25039
47. Hansen, D. F., & Kay, L. E. (2011). Determining valine side-chain rotamer conformations in proteins from methyl 13C chemical shifts: Application to the 360 kDa half-proteasome. *Journal of the American Chemical Society, 133*(21), 8272–8281. https://doi.org/10.1021/ja2014532
48. Liu, H., Elstner, M., Kaxiras, E., Frauenheim, T., Hermans, J., & Yang, W. (2001). Quantum mechanics simulation of protein dynamics on long timescale. *Proteins, 44*(4), 484–489. https://doi.org/10.1002/prot.1114
49. Krieger, E., Koraimann, G., & Vriend, G. (2002). Increasing the precision of comparative models with YASARA NOVA–a self-parameterizing force field. *Proteins, 47*(3), 393–402. https://doi.org/10.1002/prot.10104
50. Adcock, S. A., & McCammon, J. A. (2006). Molecular dynamics: Survey of methods for simulating the activity of proteins. *Chemical Reviews, 106*(5), 1589–1615. https://doi.org/10.1021/cr040426m
51. Hospital, A., Goñi, J. R., Orozco, M., & Gelpí, J. L. (2015). Molecular dynamics simulations: Advances and applications. *Advances and Applications in Bioinformatics and Chemistry (AABC), 8*, 37–47. https://doi.org/10.2147/AABC.S70333
52. Hendlich, M., Lackner, P., Weitckus, S., et al. (1990). Identification of native protein folds amongst a large number of incorrect models. The calculation of low energy conformations from potentials of mean force. *Journal of Molecular Biology, 216*(1):167–180. https://doi.org/10.1016/S0022-2836(05)80068-3.
53. Jones, D. T. (2001). Predicting novel protein folds by using FRAGFOLD. *Proteins* (Suppl 5), 127–132. https://doi.org/10.1002/prot.1171.
54. Jaroszewski, L., Rychlewski, L., Zhang, B., & Godzik, A. (1998). Fold prediction by a hierarchy of sequence, threading, and modeling methods. *Protein Science Publication Protein Society, 7*(6), 1431–1440.
55. Altschul, S. F., Madden, T. L., Schäffer, A. A., et al. (1997). Gapped BLAST and PSI-BLAST: A new generation of protein database search programs. *Nucleic Acids Research, 25*(17), 3389–3402. https://doi.org/10.1093/nar/25.17.3389

56. Krogh, A., Brown, M., Mian, I. S., Sjölander, K., & Haussler, D. (1994). Hidden Markov models in computational biology: Applications to protein modeling. *Journal of Molecular Biology, 235*(5), 1501–1531. https://doi.org/10.1006/jmbi.1994.1104
57. Söding, J. (2005). Protein homology detection by HMM–HMM comparison. *Bioinformatics, 21*(7), 951–960. https://doi.org/10.1093/bioinformatics/bti125
58. De Novo Protein Structure Prediction by QUARK. Retrieved May 6, 2024, from https://zhanggroup.org/QUARK/.
59. D-I-TASSER: deep learning-based protein structure prediction. Retrieved May 6, 2024, from https://zhanggroup.org/D-I-TASSER/.
60. Google-deepmind/alphafold. Published online May 5, 2024. Retrieved May 6, 2024, from https://github.com/google-deepmind/alphafold.
61. Jones, D. T., Buchan, D. W. A., Cozzetto, D., & Pontil, M. (2012). PSICOV: Precise structural contact prediction using sparse inverse covariance estimation on large multiple sequence alignments. *Bioinformatics, 28*(2), 184–190. https://doi.org/10.1093/bioinformatics/btr638
62. O S. Sokrypton/GREMLIN. Published online December 26, 2023. Retrieved May 6, 2024, from https://github.com/sokrypton/GREMLIN.
63. Pearce, R., & Zhang, Y. (2021). Toward the solution of the protein structure prediction problem. *Journal of Biological Chemistry, 297*(1), 100870. https://doi.org/10.1016/j.jbc.2021.100870
64. CCMpred–fast and precise prediction of protein residue-residue contacts from correlated mutations-PubMed. Retrieved May 6, 2024, from https://pubmed.ncbi.nlm.nih.gov/25064567/.
65. Soedinglab/CCMpred. Published online April 9, 2024. Retrieved May 6, 2024, from https://github.com/soedinglab/CCMpred.
66. NeBcon: protein contact map prediction using neural network training coupled with naïve Bayes classifiers. Bioinformatics. Oxford Academic. Retrieved May 6, 2024, from https://academic.oup.com/bioinformatics/article/33/15/2296/3091850.
67. Li, Y., Hu, J., Zhang, C., Yu, D. J., & Zhang, Y. (2019). ResPRE: High-accuracy protein contact prediction by coupling precision matrix with deep residual neural networks. *Bioinformatics, 35*(22), 4647–4655. https://doi.org/10.1093/bioinformatics/btz291
68. Greener, J. G., Kandathil, S. M., & Jones, D. T. (2019). Deep learning extends de novo protein modelling coverage of genomes using iteratively predicted structural constraints. *Nature Communications, 10*(1), 3977. https://doi.org/10.1038/s41467-019-11994-0
69. TripletRes: contact map prediction based on a triplet of coevolutionary features and deep residual neural networks. Retrieved May 6, 2024, from https://zhanggroup.org/TripletRes/.
70. Simons, K. T., Bonneau, R., Ruczinski, I., & Baker, D. (1999). Ab initio protein structure prediction of CASP III targets using ROSETTA. *Proteins,* (Suppl 3), 171–176. https://doi.org/10.1002/(sici)1097-0134(1999)37:3+<171::aid-prot21>3.3.co;2-q.
71. Cohen, M., Potapov, V., & Schreiber, G. (2009). Four distances between pairs of amino acids provide a precise description of their interaction. *PLoS Computational Biology, 5*(8), e1000470. https://doi.org/10.1371/journal.pcbi.1000470
72. Han, K. F., & Baker, D. (1995). Recurring local sequence motifs in proteins. *Journal of Molecular Biology, 251*(1), 176–187. https://doi.org/10.1006/jmbi.1995.0424
73. Shortle, D., Simons, K. T., & Baker, D. (1998). Clustering of low-energy conformations near the native structures of small proteins. *Proceedings of the National Academy of Sciences USA, 95*(19), 11158–11162. https://doi.org/10.1073/pnas.95.19.11158
74. All-atom empirical potential for molecular modeling and dynamics studies of proteins. The *Journal of Physical Chemistry B*. Retrieved May 6, 2024, from https://pubs.acs.org/doi/abs/, https://doi.org/10.1021/jp973084f.
75. Brooks, B. R., et al. (1983). CHARMM: A program for macromolecular energy, minimization, and dynamics calculations. Journal of Computational Chemistry. Retrieved May 6, 2024, from https://onlinelibrary.wiley.com/doi/abs/, https://doi.org/10.1002/jcc.540040211. (Wiley Online Library)
76. Neria, E., Fischer, S., & Karplus, M. (1996). Simulation of activation free energies in molecular systems. *The Journal of Chemical Physics, 105*(5), 1902–1921. https://doi.org/10.1063/1.472061

77. Cornell, W. D., Cieplak, P., Bayly, C. I., et al. (1995, 1996). A second generation force field for the simulation of proteins, nucleic acids, and organic molecules. *Journal of the American Chemical Society, 117*, 5179−5197. *Journal of the American Chemical Society, 118*(9), 2309–2309. https://doi.org/10.1021/ja955032e.
78. Duan, Y., & Kollman, P. A. (1998). Pathways to a protein folding intermediate observed in a 1-microsecond simulation in aqueous solution. *Science, 282*(5389), 740–744. https://doi.org/10.1126/science.282.5389.740
79. Weiner, S. J., Kollman, P. A., Case, D. A., et al. (1984). A new force field for molecular mechanical simulation of nucleic acids and proteins. *ACS Publications*. https://doi.org/10.1021/ja00315a051
80. Scott, W. R. P., Hünenberger, P. H., Tironi, I. G., et al. (1999). The GROMOS biomolecular simulation program package. *Journal of Physical Chemistry A, 103*(19), 3596–3607. https://doi.org/10.1021/jp984217f
81. Atomic-level characterization of the structural dynamics of proteins. Science. Retrieved May 6, 2024, from https://www.science.org/doi/abs/, https://doi.org/10.1126/science.1187409.
82. Bowie, J. U., & Eisenberg, D. (1994). An evolutionary approach to folding small alpha-helical proteins that uses sequence information and an empirical guiding fitness function. *Proceedings of the National Academy of Sciences, 91*(10), 4436–4440. https://doi.org/10.1073/pnas.91.10.4436
83. Improved protein structure prediction using predicted interresidue orientations. PNAS. Retrieved May 6, 2024, from https://www.pnas.org/doi/abs/, https://doi.org/10.1073/pnas.1914677117.
84. Senior, A. W., Evans, R., Jumper, J., et al. (2020). Improved protein structure prediction using potentials from deep learning. *Nature, 577*(7792), 706–710. https://doi.org/10.1038/s41586-019-1923-7
85. Pearce, R., & Zhang, Y. (2021). Deep learning techniques have significantly impacted protein structure prediction and protein design. *Current Opinion in Structural Biology, 68*, 194–207. https://doi.org/10.1016/j.sbi.2021.01.007
86. Jumper, J., Evans, R., Pritzel, A., et al. (2021). Highly accurate protein structure prediction with AlphaFold. *Nature, 596*(7873), 583–589. https://doi.org/10.1038/s41586-021-03819-2
87. Callaway, E. (2020). "It will change everything": DeepMind's AI makes gigantic leap in solving protein structures. *Nature, 588*(7837), 203–205.

Computational Approaches for Structure-Assisted Drug Discovery and Repurposing

Anand Kumar Pandey and Shalja Verma

Abstract The drug discovery process is highly challenging due to the excess cost, time, and resource requirements. From identifying potential molecules to their utilization in clinics as drugs, molecules face a variety of validations and many counterparts land up to rejection in between the process. Many potent molecules seem effective in treating the pathologies of the disease but are discarded due to their toxic nature in *in-vitro* and *in-vivo* conditions. Repurposing of known already available drugs can provide strong candidates that have already gone through toxicity tests and will thus reduce the time for the drug to be implemented as a treatment for a new disease. Structural information about the disease target and the potential lead molecules facilitates the identification of significant interactions that can convey regulation of the target by inhibiting it. Computational tools and approaches have immensely supported and accelerated the process of drug discovery by providing effective predictions based on structural information. The lead computational approach comprising identification of pharmacophores, physiochemical and biological activities, drug target interaction and binding affinity, and analysis of the binding stability, together lead to the identification of a potent druglike molecule. Further, different AI/ML-based models have come forth to assist drug design on the basis of structure and to increase the pace of the process without compromising accuracy. Therefore, this piece of work elaborates on the significance of different computational approaches, and recent advancements with a lead focus on the software, tools, and models developed for use in drug discovery.

Keywords Drug discovery and repurposing · Pharmacophore modeling · QSAR · Molecular docking · MD (Molecular Dynamic) simulation

A. K. Pandey (✉) · S. Verma
Department of Biotechnology Engineering, Institute of Engineering and Technology, Bundelkhand University, Jhansi 284128, Uttar Pradesh, India
e-mail: akpandey.bme@gmail.com

S. Verma
Department of Biosciences and Bioengineering, Indian Institute of Technology, Roorkee 247667, Uttarakhand, India

© The Author(s), under exclusive license to Springer Nature Singapore Pte Ltd. 2024
A. Chaudhary et al. (eds.), *Unraveling New Frontiers and Advances in Bioinformatics*,
https://doi.org/10.1007/978-981-97-7123-3_9

1 Introduction

The continuous evolution and ability to adapt to diverse environmental conditions are giving rise to new pathogens now and then [1]. Also, the change in lifestyle has led to several lifestyle diseases including stroke, obesity, metabolic syndrome, diabetes, cancers, heart disease, obstructive chronic pulmonary diseases, etc. The increasing variety of diseases in today's scenario has consequently increased global death cases and demands for effective treatments [2]. The identification of high-efficacy drugs has been challenging due to the excess cost and duration of time associated with their development. Also, the evolving causative agents in the case of infectious diseases and the high complexity of associated pathologies in degenerative diseases have made the scenario far more adverse than the estimation for the development of drugs [1, 3]. Not only the effectiveness of a drug in a particular disease is required but its side effects and toxicity against undesired targets are of great concern and stand out to be the major reason for the rejection of a drug-like molecule in clinical trials [4]. Drug repurposing of already-known drugs having no proven toxicity or side effects can accelerate the procedure of investigation of drugs for a certain disease by providing effective treatment and can mediate further research for efficacy enhancement by contributing structural scaffolds [5].

Structure-based approaches are being a landmark in drug discovery field and have contributed to a variety of potential compounds of which several have reached the clinics. Structural information about the disease targets can be utilized to know the chemistry of interactions of the target protein with the potential compounds and can reveal the significant interactions essential for further improvement of the compound to be developed into a drug [6]. The functional groups present in the structure of the compound convey specific electronic and steric properties to the compound and mediate its chemical interactions defining its affinity for the target protein [7]. Also, compounds with similar structural composition possess similar biological properties, thus such an approach can be beneficial for drug repurposing of known drugs. Ligand-based approaches or algorithms are based on coverage of the chemical space of known drug molecules [8].

Computational approaches have accelerated the stride of structure and ligand based drug discovery (Fig. 1). High-throughput virtual screening using huge libraries of chemicals to inhibit the target proteins mediates the investigation of huge chemical space and leads to the disclosure of the best counterpart. Homology modeling approach for target structure prediction, pharmacophore modeling for identification of effective functional groups in ligands, molecular docking for studying the interactions of the ligands with effective proteins, molecular dynamic (MD) simulation for evaluating the stability of interaction, and free binding energy analysis for calculating the affinity of the ligand and its binding stability, have been among the most frequent approaches for drug discovery [9–11]. The quantitative structure–activity relationship (QSAR) scheme is often utilized to develop models that reveal the association between structural properties and the biological activities (significant for treatment of

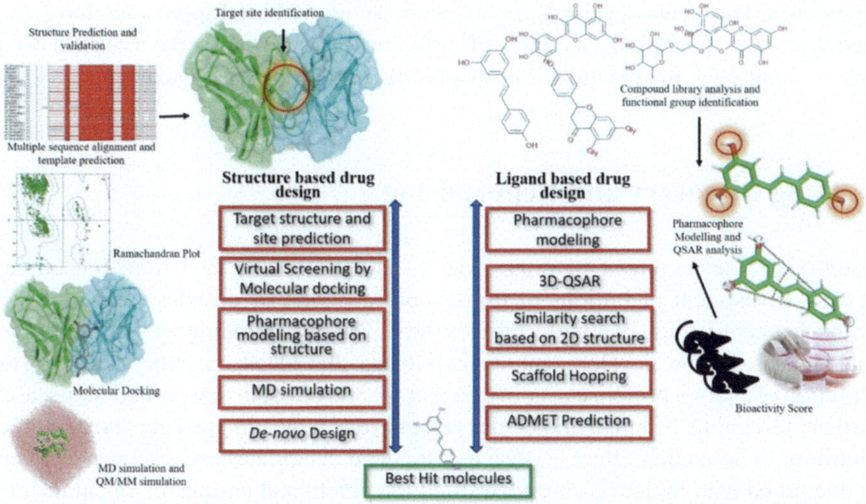

Fig. 1 Different computational approaches and methods used for drug design

some disease) of chemical compounds. These models of QSAR are used for the identification and classification of novel compounds which possess a required biological activity [12].

Artificial intelligence has made a highly significant contribution in diverse areas of drug discovery and repurposing. Screening of numerous libraries containing diverse chemical spaces, prediction of toxicity, physiochemical properties, bioactivity, affinity, and efficacy of druglike molecules, designing of novel drug scaffolds, and prediction of the drug-target interaction network, are the lead applications of AI being investigated to hasten the procedure of drug discovery [13]. Numerous researchers have generated several deep learning models that can be utilized to identify potential drug-like molecules utilizing structural information of the target or vice versa [14]. Artificial neural networks (ANNs) comprise sophisticated interconnected computing elements that involve perceptrons that have analogies to human neurons and mimic the electrical impulse transmission of the brain. These networks contain nodes that receive separate inputs either singly or in a multi-linked manner and convert these inputs into outputs using different algorithms for problem-solving. Different types of ANNs include, multilayer perceptron networks (MPN), recurrent neural networks (RNN), and convolutional neural networks (CNN) that use unsupervised or supervised learning procedures [15]. Graph convolutional neural network (GCNN) approach is effective in the estimation of drug-target interactions and has benefitted society by providing several promising drug candidates [16].

Therefore, computational methods have incredibly enhanced the drug discovery process and have provided cost-effective solutions for the challenges associated with the drug development. The present chapter elaborates on the process of drug discovery and repurposing with a major focus on structure-assisted computational approaches. The recent developments associated with AI-based approaches have also been discussed. Eventually, a review of different software and online tools has been done to emphasize the practical aspects of structure-based drug discovery.

2 Drug Discovery and Repurposing

Traditionally development of drugs was a capital-intensive and time-consuming process. Investment of billions of dollars and a period of decades were required for the characterization of a molecule as a drug. The initial finding on a disease from academia leads to the development of a hypothesis that states that either activation or inhibition of a target protein can convey a therapeutic effect in that particular disease. Further, screening of small molecules against the target to mediate the required inhibitory or activation effect is done and the lead candidates are discovered. Then the identified lead molecules are forwarded for preclinical and then clinical testing eventually landing in the market. The most exhaustive and laborious stage of drug discovery is to identify the lead molecule against an effective target, having the least toxicity and high bioavailability. Screening of all the molecules via *in-vitro* assay and *ex-vivo* assays will result in the consumption of a huge amount of time and resources [17, 18]. Structure-based approaches can potentiate the lead candidate identification and reduce time and resource consumption. Information about the structural aspects of targets and the molecules present in chemical libraries support the identification of best-fit molecules into the target cavity of desired enzymes. Analysis of structures of target enzymes and millions of chemical molecules present in diverse drug libraries is impossible without using computational support [19]. Computational algorithms have accelerated the pace of drug discovery and have reduced the cost by decreasing the number of experiments to a great extent. Different steps in drug discovery including identification of the target, screening of chemical libraries, synthesis of drugs, clinical testing of potential drug molecules, assessment of risk associated with the drug or its toxicities, and development of the relationship between molecular structures and their efficacy for generating novel potential molecules are utilizing the power of computational algorithms [20].

Drug repurposing is re-tasking, repositioning, reprofiling, recycling, and redirecting of already-known drugs. It includes the identification of a new pharmacological target for an old drug or the application of the recently developed drug for treating a disease apart from the original disease for which it is being used in clinics [21]. The approach of drug repurposing can reduce the monetary inputs, time required, and the failure risk associated with the drug. In recent scenario repurposing drugs is a frequently utilized approach and corresponds to the approval of around 1/3rd of the new drugs which accounts for about 25% of pharma revenue annually. It is documented that among the FDA-approved vaccines and drugs, 30% are repositioned drugs [5, 22]. Various repurposed drugs currently in use for their new target disease are aspirin, sildenafil, minoxidil, methotrexate, and valproic acid [22]. Aspirin, a well-known painkiller has been repurposed for cardiovascular disorders [23]. Sildenafil established as a treatment for angina and hypertension is now used for treating erectile dysfunction [24]. Minoxidil is an exceptional drug that has been repurposed as well as reformulated. As an oral formulation, it acts as a treatment for hypertension but it has hair growth as its side effect which promoted its utilization as a topical formulation in treating androgenetic alopecia [25]. Methotrexate, a known chemotherapeutic agent and immunosuppressant has been repurposed for the treatment of cardiovascular disease and as an antiaggregatory agent that can be effective in protein aggregation-associated diseases like Parkinson's or Alzheimer's disease [26]. Valproic acid is an established treatment for neurological disorders and epilepsy and has been repurposed for certain cancers as inhibits histone deacetylase modifying the epigenetic changes, thus affecting the gene expression in differentiation, cell cycle, and apoptosis [27]. Thus, by increasing the application of known drugs in different diseases, drug repurposing increases the therapeutic value of drugs. The strategy of repurposing utilizes computational as well as experimental approaches for investigation of new target of already known drugs. Computational structural approaches are frequently utilized to identify new targets for known drugs based on the similarity in active site architecture of the new target with that of known target proteins. Also, the ligand-based approach which emphasize on the similarity of the chemical architecture of a drug with the scaffold of active binder of the new target has been used for repurposing of drugs [21].

Thus, the upcoming section focuses on different computational approaches with a lead focus on the applied aspects of drug discovery and repurposing.

3 Computational Approaches for Structure-Based Discovery of Drugs and Repurposing

Computational approaches for drug design leadingly focus on the structure of target proteins or the structure or chemical features of ligands and thus follow the two major approaches namely structure-based drug design and ligand-based drug design (Fig. 1) [28]. In structure-based drug design the features of the active site of the target protein are considered for the determination of an effective ligand that can interact with the target protein while in ligand-based drug design, the ligands and their activities are utilized to identify high-efficacy ligands that can favorably interact with the protein structures. Both these drug design approaches utilize different algorithms to fulfill the purpose [29]. To mediate structure-based drug design, several methodologies are used of which *de-novo* designing, molecular docking, fragment-based designing of drug, and pharmacophore modeling based on protein structures are the lead methodologies [30]. For ligand-based drug design, QSAR and pharmacophore modeling are the effective methodologies utilized to identify the potential ligand. Based on the availability of the information regarding target protein structure and its binding interaction with ligands different strategies for drug design can be used [31].

After the generation of lead compounds, several rules for the evaluation of drug-likeness including Lipinski's, Ghose, Veber, Egan, and Muegge rules, and bioavailability score, are used. Lipinski's rule considers less than 5 H-bond donors, less than 10 H-bond acceptors, molecular weight (MW) lower than 500, and logP value (partition coefficient of octanol–water) lesser than 5 [32]. The Ghose rule ensures that the MW should be between 160 to 480 Da, logP between −0.4 to 5.6, molecular refractivity in the range of 40–130, and the number of atoms should lie between 20 to 70 [33]. Veber rule elaborates that the rotatable bonds should be less than 10 and the total polar surface area (TPSA) value should be lesser than 140 [34]. Further according to Egan rule the logP value of 5.88 and TPSA of 131.6 is considerable. The Muegge rule follows that the MW should be in the range of 200–600, logP between −2 to 5, TPSA 150, aromatic rings lesser than 7, number of carbon atom more than 4, heteroatoms greater than 1, rotatable bonds less than 15, hydrogen bond acceptors lesser than 10 and hydrogen bond donors lesser than 5. The bioavailability (F) score provides a probability value when $F > 10\%$ in rats [35].

Currently, artificial intelligence (AI) has made an immense contribution to the computer-aided drug design field. Numerous studies have highlighted the significance of AI in drug designing, development, and repurposing. Pan et al. [36], reviewed the general procedure of 4 steps being followed by most of the studies for drug repurposing using AI. Firstly, data sources containing high-quality data on proteins, compounds, and diseases are needed. Secondly, significant feature vectors are generated using graphs, text, or sequence representations. Thirdly, deep learning models are built and evaluated. Eventually, the binding affinity of the drug for a target, interactions between drug and target, and the association between drug and disease is predicted [36]. Rajput et al. [37], developed a drug repurposing method for *SARS-CoV-2* treatment using different approaches of machine learning (ML) including Support Vector Machine (SVM), k-Nearest Neighbour, deep learning, Random Forest (RF), and artificial neural network. They used the DrugRepV library to prepare a dataset of anticorona drugs and extracted structural or chemical descriptors and fingerprints. Most significant features were identified by utilizing the algorithm of recursive feature selection and then the selected descriptors were utilized for the development of prediction models with 0.6–0.9 Pearson's correlation coefficients. The prediction from the developed model along with molecular docking gave metergoline, verteporfin, rescinnamine, alatrofloxacin, telotristat ethyl, and leuprolide as effective inhibitors of spike-ACE-2 receptor complex [37]. Jin et al. [38], in their investigation of combination therapy against SARS-CoV-2, developed a deep learning model that can predict robustly when limited data for drug combinations is available. This model possessed a neural network (NN)-based architecture that learns drug-to-drug synergy and drug-target interactions. Two major modules of this model include drug to target interactions and target to disease association. Both interaction data of drug targets and the antiviral activity of single agents were utilized along with small in-reach datasets of combinations of drugs. Also, the information related to biological aspects has been incorporated in this model and thus the model performed significantly better in predicting synergy compared to other methods. Eventually, reserpine and remdesivir and IQ-1S and remdesivir are the combinations that revealed strong synergy in in-vitro experiments [38]. Pan et al. [39], developed a framework based on deep learning for prediction and identification of drugs repurposed for Alzheimer's disease. Also, their framework can be used for the investigation of combinations of drugs for different diseases. The features of multiple drugs and targets were used to build drug-target pairs. These interacting pairs and the associations between them were used for the drug discovery network development namely AI-DrugNet based on artificial intelligence with robust prediction performance [39]. Rao et al. [40], developed a framework for repurposing small molecules combining AI/ML and target prediction based on chemical similarity along with cross-species transcriptomics information. This methodology for repurposing integrates 8 methods for the prediction of targets, of which 3 include ML. A dataset of 2766 drugs that were FDA-approved was used for targeting different therapeutic classes and 27,371 interactions with 2013 off-target proteins were identified. Related to the dataset drugs, around 150,620 drugs were identified as having similar structures. The maximum number of interactions predicted were for receptors coupled with G-proteins, and kinases. Further, sixty three percent of the off-target interactions were confirmed by experiments done in *in-vitro* conditions of which around 4000 had less than 100 nM IC50 for drugs approved by FDA while 1661 interactions displayed less than 10 nM IC50 for drugs approved by FDA [40].

Lenselink et al. [41], performed a comparative analysis of different AI/ML algorithms on ChEMBL dataset using standard metrics to evaluate the potential of different methods in drug discovery. RF, Deep Neural Networks, SVM, Naïve Bayes, and Logistic Regression were the algorithms assessed utilizing the methods of proteochemometric and QSAR. Out of these methods Deep Neural Networks with proteochemometric method was observed to be exceptionally effective [41]. Feng et al. [42], performed their analysis on opioid use disorder for drug repurposing. ML-based rapid compound screening of the compounds of DrugBank database was implemented. Data on inhibitors of major four opioid receptors was collected and ML advanced predictors of affinity of binding were used that fuse a 2D traditional fingerprint and two molecular fingerprints based on natural language processing with the algorithm of gradient boosting decision tree. By utilizing these predictors systematic analysis of DrugBank compound's binding affinities was done for 4 main opioid receptors. Moreover, the outcomes of this analysis were subjected to ADMET analysis to identify the repurposing ability of DrugBank compounds for inhibition of considered receptors [42]. Rusinko et al. [43], developed a platform based on AI/ML namely AIDDISON that integrates generative models to predict ADMET properties, chemical library search, and molecular docking to accelerate discovery of drugs. It can generate thousands of new molecules and screen available molecules based on 2D chemical similarity, and pharmacophores search and can mediate de novo designing of molecules utilizing generative models or by user input directly. Filtering molecules based on their properties is a critical step followed by molecular docking or alignment of molecules based on their shape complementarity. Eventually, the synthesizability of the molecules can be checked and necessary reagents can be predicted using SYNTHIA software [43].

4 Databases, Software, and Tools Required for Drug Discovery and Repurposing

Diverse varieties of drug libraries, databases, software, and tools are being utilized to explore the chemical space which can be exploited for the identification of drugs against the potent targets of different diseases. The lead databases and libraries include significant information about genes and proteins responsible for mediating diseased conditions, protein structures, different chemical compounds, their physio-chemical, and biological activities, FDA-approved drugs, and repurposed drugs. The software based on pharmacophore modeling, QSAR, molecular modeling, molecular docking, and MD simulation approaches are frequently used in discovery and repurposing of drugs. To analyze the properties of chemical compounds or drugs, their bioavailability and bioactivity values, and model the structure of protein and specific functional groups, etc., an extensive range of algorithms and tools have been developed and are currently in active use supporting the exhaustive process of drug discovery [44]. The subsequent subsection elaborates in detail on the databases, algorithms, procedures, approaches, and associated tools and software utilized for the development of drugs.

4.1 Databases and Drug Libraries

The different open-source databases of genes, genomes, proteins, and drugs are available which provide information related to disease targets and the chemical compounds thus promotes drug discovery. GenBank NIH database (https://www.ncbi.nlm.nih.gov/genbank/) of nucleotide sequences is associated with the International Nucleotide Sequence Database that contains data from DNA Databank of Japan (DDBJ). GenBank, National Center for Biotechnology Information (NCBI) and European Nucleotide Archive exchange their information daily to provide accurate and updated information [45]. The Gene (https://www.ncbi.nlm.nih.gov/gene) database of the NIH integrates information which is species specific. This database of genes provides information about Reference Sequences (RefSeqs), nomenclature, pathways, maps, phenotypes, variations, and links to specific phenotypes, genomes, and locus-associated global resources [46]. Online Mendelian Inheritance in Man (OMIM) is an authoritative and inclusive collection of genes of humans and their genetic phenotypes. It focuses on genes of humans and their associated genetic disorders (https://www.omim.org/) [47]. Kyoto Encyclopedia of Genes and Genomes (KEGG) (https://www.genome.jp/kegg/) provides information about the utilities and functions of biological systems like ecosystems, organisms, and cells from molecular sets of data obtained by sequencing genomes and other experiments utilizing high throughput technologies [48]. KEGG GENES database (https://www.genome.jp/kegg/genes.html) contains information about the genes and proteins of viruses and cellular organisms obtained from GenBank and NCBI annotated by KEGG as KEGG orthology. This collection in addition contains the KEGG collection of characterized function of proteins from the literature published to date. The sequences of proteins and RNA of all the entries of GENES are computed by SSDB and KO assignment is done by KOALA tools [49]. To procure information about the different mutations responsible for mediating different inherited diseases in humans, the Human Gene Mutation Database (https://www.hgmd.cf.ac.uk/ac/index.php) can be useful in collecting all known lesions of genes [50]. Another database DisGeNET database (https://www.disgenet.org/) contains a huge collection of human disease-associated genes and their variants. It integrates data from different repositories including scientific repositories, GWAS catalogues, and animal models. The data present in this database is annotated homogenously with ontologies that are community-driven and controlled vocabularies. In addition, many original metrics to support the prioritization of the phenotype-genotype relationship have been provided [51]. Gene ontology https://geneontology.org/ database provides a computational comprehensive biological systems model, extending from the molecular level to the organism level across the species in the life tree. It is the largest information source related to genes and their functions and is readable by both machines and humans, thus providing a base for computational analysis of genetics and molecular biology large-scale experiments [52]. GeneDB (https://www.sanger.ac.uk/tool/genedb/) is a genome database consisting data of sequences and annotations of pathogenic organisms [53]. UniProt is a complete database of sequence and function-based information on proteins that supports drug discovery by elucidating information about various drug targets of different diseases [54]. The Cancer Genome Atlas Program (TCGA) (https://www.cancer.gov/ccg/research/genome-sequencing/tcga) is a landmark program for cancer genomics and it is characterized molecularly over 20,000 cancers and matches samples of normal 33 cancer types. It is a joint effort of the National Human Genome Research Institute (NHGRI) and NCI which was initiated in 2006. To date, TCGA generated data from genomics, transcriptomics, epigenomics, and proteomics accounting for 2.5 petabytes of data. This data not only encourages diagnosis and prevention of cancer but also provides information regarding essential targets to develop treatments [55]. GenAge database (https://genomics.senescence.info/genes/index.html) contains information about the genes mediating aging. It is divided into genes associated with aging and longevity of organisms like worms, yeast, mice, flies, etc., and can be beneficial for the anti-aging therapies development. The human aging-related genes section contains few genes that are directly associated with human aging and some obtained from developed model organisms. Hence, human genes contain wide information and are precisely annotated. Also, this database is curated for high quality by experts manually [56]. MGI is a resource for providing mouse biological, genomic, and integrated genetic data that facilitate human disease and health studies [57]. Gene Expression Database (GDX) (https://www.informatics.jax.org/mgihome/GXD/aboutGXD.shtml) contains information on gene expression procured from MGI. It is focused on the expression of endogenous genes regulated during development in mouse used in laboratories. It integrates and stores different expression data types and makes them accessible at no cost for comprehensive analysis in different formats [58]. dbSNP (database Single Nucleotide Polymorphism) (https://www.ncbi.nlm.nih.gov/snp/) of NCBI is a database of genetic variation. It classifies variations in the sequence of nucleotide into following percentage composition and types, in the database: (1). 99.77% of single nucleotide substitution, (2). 0.21% of small deletion or insertion polymorphisms, (3). 0.02% of invariant sequence regions, (4). 0.001% accounts for microsatellite repeats, (5). <0.001% are named variants and lesser than 0.001% of heterozygous uncharacterized assays. The major scope of this database lies in neutral polymorphisms as well as clinical mutations that can cause disease [59]. DDBJ (https://www.ddbj.nig.ac.jp/index-e.html) provides analysis and sharing services for data from researchers of advanced science and life sciences [60]. The Catalog of human genes and transcripts also called CHESS (http://ccb.jhu.edu/chess) is a database of sequences from experiments of deep RNA sequencing. These sequences are assembled using Genotype-Tissue Expression project as a source. This database consists of 42,611 genes out of which 20,352 genes code proteins, 22,259 are those genes that do not code proteins, and 323,258 are total transcripts [61]. Gene Expression Omnibus database (GEO) is a database of information associated with expression of genes and functional genomics. It was developed in 2000 as a resource for studies related to gene expression. It has evolved along with changing technologies and accepts data from several other applications including genome methylation data, interaction data of genome protein, and chromatin structure data [62]. RCSB Protein Data Bank (PDB) provides access to 3D structures which have been experimentally determined. It also contains structures generated using Alphafold and ModelArchive. In addition tools for visualization, exploration, and analysis of 3D structures are also available in PDB database. This is the most frequently utilized database for procuring the 3D structures of the drug targets to mediate computational drug design [63].

All these databases mentioned above play vital roles in identifying potential drug targets and obtaining their structural and functional information.

NCATS (https://ncats.nih.gov/) database provides data related to the development, examination, and application of therapeutics and diagnostics for different diseases and conditions. Major approaches include understanding the similarity among diseases to develop different treatments at one time, building models that predict better reactions to treatment, and clinical trial enhancement to take into account the full diversity of the population. Furthermore, NCATS also collaborates with several government agencies, organizations for patient support, industry, and academia [64]. ReFrame (https://reframedb.org/) is a library focused on drug repurposing as it contains 12000 chemical compounds assembled from GVK Excelra GoStar, the Citeline Pharmaprojects and Clarivate Integrity databases [65]. ChemEMBL database (https://www.ebi.ac.uk/chembl/) contains information of small molecules that are bioactive and have the drug-like properties. It combines bioactive, genomic, and chemical data to identify effective drugs [66]. Phenol-Explorer database (http://phenol-explorer.eu/) is a food polyphenols database that contains content values of 35000 for around 500 different polyphenols obtained from about 400 different types of foods. All the information is derived from the collection of greater than 60,000 content values procured from around 1300 publications. Each publication was evaluated critically before utilizing its data for the database. All the information is available to be downloaded for the users [67]. Natural products online, a project that is open-source supports storing, searching, and analyzing natural products. Its present database version hosts COCONUT, i.e., collection of open natural products, constituting the best annotated natural product resource at no cost and restriction [68]. FooDB (www.foodb.ca) is the world's most inclusive and largest resource of constituents of foods, and their biochemistry. It provides details about both micronutrients and macronutrients which include several food constituents like aroma, flavor, taste, colour, and texture. Each entry of FooDB consists of 100 data files containing detailed information about biochemical, compositional, and physiological aspects including information related to nomenclature, structure, description, chemical class, source physiochemical data, aroma, effect on physiology, taste, concentration in source food, and health effects [69]. ZINC20 (https://zinc.docking.org/) database of compounds that are commercially available for the purpose of virtual screening. It contains 230 million compounds that can be used directly for docking in 3D confirmation. It also contains 750 million which can be searched for the identification of analogs in a minute [70]. MolPort (https://www.molport.com/) is a commercial database containing a stock of 5 million screening compounds and 1 billion made-to-order compounds. This database is accessible by registered users on request and the best compounds identified in screening can be directly ordered for the synthesis [71]. Selleckchem.com drug library (https://www.selleckchem.com/screening/fda-approved-drug-library.html) contains a collection of compound libraries including FDA-approved drug libraries and drugs approved by pharmacopeia of different countries [72]. LIT-PCBA (https://drugdesign.unistra.fr/LIT-PCBA/) a dataset for computational or virtual screening and ML has been developed using 15 sets of targets, and 7761 active molecules and 382674 inactive molecules using the PubChem Bioassay data [73]. All the above-mentioned collections of compounds can be directly utilized and associated physiochemical information can be accessed for virtual screening against diverse drug targets.

4.2 Drug Target Structure Prediction

The protein 3D structure is associated with its specific function and interactions with other molecules in the organism. It provides molecular-level details about the different interacting partners that mediate the associated pathways. Though the number of sequences of proteins known is large only a part of these sequences has been structurally characterized revealing the exact 3D experimental structures [74]. Recently, around 248,805,733 sequences of diverse proteins are available in the database of UniProtKB while only 217,705 experimental structures are available as of April 2024 in the PDB [54, 63]. For designing drugs utilizing a structure-based approach, the availability of target protein structure becomes obligatory. As the determination of experimental structures is highly laborious and requires specific skills and resources, the prediction of structures using computational tools provides an effective alternative. Using computational algorithms like homology modeling for predicting the drug target proteins is a frequently used approach to increase the pace of the process of drug discovery [63, 75].

Homology modeling is a wellknown method to build 3D structures of proteins utilizing known experimental structures of homologous protein families as their templates. For a target protein sequence, for which the structure is to be predicted, sequence alignment is performed with the sequences of the experimental structures present in the library. The structure having an amino acid sequence with maximum identity (>30%) with the target sequence, is then considered as a template for prediction of structure. The reliability of the structure modeled using the template is established by considering different model quality assessment parameters including the Ramachandra outliers, clashes, steric hindrance, etc. [76]. Several tools for the prediction of structure based on homology modeling algorithms and assessment of predicted structure quality have been developed. In case of a lack of availability of homologous structures, de novo methods for structure prediction can be used for the generation of models of protein without using a homologous template [77]. Currently, several artificial intelligence-based algorithms have been developed for predition of the structures of proteins and these predictions are of exceptional accuracy even for those proteins for which the homologous or template experimental structures are lacking. Though these tools provide structures of high accuracy of most of the proteins still prediction of some protein structures has not been as accurate as expected and homology modeling plays an effective role in this scenario if homologous experimental structures are available [78]. Table 1 summarizes the various tools that can be applied for structure prediction of proteins and its assessment.

4.3 Pharmacophore Modeling

A pharmacophore consists of a collection of electronic and steric features, important for supramolecular-interactions with the structure of a specific protein target and to mediate a significant biological response. Pharmacophore describes the interactions present between the ligand and the receptor and identification of interactions is called pharmacophore modeling. Models based on pharmacophores can be utilized for lead identification, virtual screening optimization, and the de novo drug design along with drug designing against multiple targets. The first software for the screening of 3D pharmacophores was developed in 1977 by Gund [96]. Two approaches namely receptor-based or ligand-based can be used for modeling pharmacophore. To date, several programs for pharmacophore modeling have been reported (Table 2). The accuracy of software following the ligand-based approach depends upon their efficiency in handling the flexibility of the ligand and its alignment, thus requiring a set of active ligands that are pharmacologically active while the receptor-based approach focuses on the isolation of pharmacophoric features by analyzing complex structures of receptor and ligand [31]. In conditions, when no ligand for a target protein is known, hotspots or highly probable sites of ligand binding on the receptor can be identified. MD simulations of considered target proteins in complex with drug-like

Table 1 List of structure prediction software and web servers along with their basic algorthims

Webserver/software	Links	Features and algorithms	References
ModWeb/ MODELLER	https://modbase.compbio.ucsf.edu/modweb/	Homology-based structure modeling	[79]
SWISS-MODEL	https://swissmodel.expasy.org/	Homology-based structure modeling	[80]
Phyre2	http://www.sbg.bio.ic.ac.uk/phyre2/html/page.cgi?id=index	Recognition of Protein homology or analogy	[81]
I-TASSER	https://zhanggroup.org/I-TASSER/	Utilize multiple threading approach	[82]
HHpred	https://toolkit.tuebingen.mpg.de/tools/hhpred	Detction of homology and prediction of structure by HMM–HMM comparison	[83]
PSI-BLAST–based secondary structure prediction (PSIPRED)	http://bioinf.cs.ucl.ac.uk/psipred/	GenTHREADER for secondary structure (ss) prediction and MEMSAT-2 for transmembrane topology prediction	[84]
Robetta	https://robetta.bakerlab.org/	Uses deep learning based RoseTTAFold and TrRosetta methods for structure modeling. Modeling of complexes of multiple chains can be done using RoseTTAFold	[85]
RaptorX	http://raptorx6.uchicago.edu/ContactMap/	Predicts properties of local structures, disordered regions, distance and contact map, inter-residue orientation, solvent accessibility, functional annotation, as well as the tertiary structure and binding sites of protein by utilizing an ultra-deep convolutional neural network	[86]
AlphaFold-2	https://colab.research.google.com/github/sokrypton/ColabFold/blob/main/AlphaFold2.ipynb#scrollTo=G4yBrceuFbf3	Uses ML approach incorporating biological and physical knowledge about the structure of protein using multi-sequence alignments, into deep learning algorithm design	[87]

(continued)

Table 1 (continued)

Webserver/software	Links	Features and algorithms	References
CABS-flex	https://biocomp.chem.uw.edu.pl/CABSflex2/submit	It is a Python package for modeling flexibility of structures of the protein utilizing CABS coarse-grained model of protein	88
SAVES v6.0	https://saves.mbi.ucla.edu/	Structure assessment using PROVE, ERRAT, PROCHECK, WHATCHECK and the VERIFY 3D algorithms	[89–93]
ProteinTools—Toolkit for Analyzing Protein Structures	https://proteintools.uni-bayreuth.de/	Predicts networks of hydrogen bonds, cluster of hydrophobic interactions, contact maps, charge segregation and the salt bridges	[94]
MolProbity	http://molprobity.biochem.duke.edu/	It assesses the quality of structure by using the algorithms of ProCheck, WhatIf, and Oops, which validates on the basis of Ramachandran-plot or Ramachandran outliers and the sidechain rotamer criteria. It uses library of backbone geometry which is conformation-dependent for validation	[95]
RFdiffusion	https://github.com/RosettaCommons/RFdiffusion	De novo protein design based on fine-tuned RoseTTAFold structure prediction network	[77]

molecules are being used to assess the hotspot of ligands while maintaining the flexibility of the target protein [97]. Also, several algorithms for the prediction of active sites of target proteins can be used to predict the highly probable site on the receptor (Table 2). The fingerprint of a 2D pharmacophore is a kind of binary code possessing pharmacophore properties. These fingerprints contain fragments of molecules that can be applied with multiple models based on artificial intelligence (Table 2).

4.4 Quantitative Conformational Relationships or QSAR

The origin of the QSAR approach dates back to 1962 when Hansch et al. reported that partition coefficient and Hammett functions would be significant for establishing a relationship between activity and structure. In past 60 years, QSAR has advanced from classification or the regression analysis of some compounds to AI/ML-based approach that has the ability to extract physical, chemical, and biological functions associated with massive data sets of molecules having complex structures [129]. QSAR has now been established as an important drug discovery component. It allows high accuracy, low-cost activity, and property prediction along with the virtual

Table 2 List of software, web servers and AI models developed for pharmacophore modeling

Webserver/software	Links	References
Pharmit	https://pharmit.csb.pitt.edu/	[98]
PharmaGist	https://bioinfo3d.cs.tau.ac.il/PharmaGist/php.php	[99]
Pharmmaker	http://prody.csb.pitt.edu/pharmmaker/	[100]
LigandScout	https://docs.inteligand.com/ligandscout/	[101]
MOE-Chemical Computing Group	https://www.chemcomp.com/index.htm	
Hypogen algorithm of Discovery Studio	https://www.3ds.com/products/biovia/discovery-studio	
MolAlign	http://carbon.structbio.vanderbilt.edu/index.php/servers/molalign	[102]
ZINC-Pharmer	http://zincpharmer.csb.pitt.edu/	[103]
ELIXIR:A	https://github.com/sfernando-BAEN/ELIXIR-A/releases	[96]
GBPM: GRID-based pharmacophore model		[104]
PyRod	https://github.com/schallerdavid/pyrod	[105]
Phase (Schrodinger)	https://www.schrodinger.com/platform/products/phase/#:~:text=Phase%20is%20an%20intuitive%20pharmacophore,known%20to%20have%20biological%20activity	
IChem-Shaper2	http://bioinfo-pharma.u-strasbg.fr/labwebsite/download.html	[106]
SiteAlign	http://bioinfo-pharma.u-strasbg.fr/labwebsite/download.html	[107]
Apo2ph4	https://github.com/molinfo-vienna/apo2ph4	[108]
Pocket v.2	http://www.pkumdl.cn:8080/pocketv2.html	[109]
Perturbation-Theory and Machine Learning (PTML) Model		[110]
Pharmacoprint	https://github.com/lstruski/Pharmacoprint	[111]
GALAHAD: 1		[112]
Q-SiteFinder	http://www.bioinformatics.leeds.ac.uk/qsitefinder	[113]
ICM Pocket-Finder	https://www.molsoft.com/icmpocketfinder.html	[114]
CavityPlus	http://www.pkumdl.cn:8000/cavityplus/index.php#/computation	[115]
CRAFT	http://pitools.niper.ac.in/CRAFT	[116]
webPDBinder	http://cbm.bio.uniroma2.it/pdbinder/	[117]
SuperStar	https://www.ccdc.cam.ac.uk/solutions/software/superstar/	[118]
COACH	https://zhanggroup.org/COACH/	[119]
CavBench	https://github.com/mosqueteer/CavBench	[120]
Meta Pocket	http://projects.biotec.tu-dresden.de/metapocket/	[121]
CavitySpace	http://www.pkumdl.cn:8000/cavityspace/	[122]

(continued)

Table 2 (continued)

Webserver/software	Links	References
PUResNet	https://github.com/jivankandel/PUResNet	[123]
GRID	https://www.moldiscovery.com/software/grid/	[124]
POCASA	https://altair.sci.hokudai.ac.jp/g6/Research/POCASA_e.html https://g6altair.sci.hokudai.ac.jp/g6/service/pocasa/	[125]
ET Viewer	http://evolution.lichtargelab.org/ETviewer	[126]
CASTp	http://sts.bioe.uic.edu/castp/index.html?1bxw	[127]
ConCavity	https://compbio.cs.princeton.edu/concavity/	[128]

screening of active hits on the basis of structures from chemical libraries consisting of millions of candidate molecules. Classical approaches to QSAR are based on mathematical models and establish associations between biological activities and different molecular descriptors [130]. The lead molecular descriptors considered for QSAR are 2D graphs, fingerprints, or representations while the lead biological activities include binding free energies, ADMET (Absorption-Distribution-Metabolism-Excretion-Toxicity) profiling, ligand–protein kinetic parameters of molecules similar in function and topology. Incorporation of AI/ML in QSAR dates early for processing huge amount of multi-dimensional information by developing robust models that utilize non-linear relationships between structure and function [131]. Deep neural networks and RF algorithms have been the most robust algorithms for developing the QSAR models [132]. As most ML models lack interpretability recent focus has shifted towards explainable AI models to encourage human intervention. Informative explanations provided with mathematical associations enhance understandability, avoid biasedness and false positive outcomes and link the community of ML with the scientific community. Explainable QSAR models have been programmed to encourage transparency and accuracy of outcomes and eventually potentiate the exhaustive drug discovery process [133].

Numerous tools and software for QSAR modeling have been established based on diverse algorithms. The following Table 3 enlists the most potent tools available for use.

4.5 Molecular Docking

Molecular docking is a tool that predicts different modes of binding, and the interactions between proteins and ligands in three dimensions. It also reveals the affinity of binding of ligands for respective proteins [141]. The usage of this technique age back to the 1980s when it was first discovered. Initially, molecular docking focused on the rigid view of interactions between protein and ligand due to limited computational infrastructure but now flexible modes of docking have come into play to provide

Table 3 Software, web servers and AI-based models developed for QSAR analysis

Webserver/ software	Links	References
QPHAR	https://github.com/StefanKohlbacher/QuantParmacophore	[134]
3D-QSAR	https://www.3d-qsar.com/	[135]
QSAR Toolbox	https://qsartoolbox.org/	
MOZART	https://github.com/mpperez3/MOZART	[136]
QSAR-ready	https://github.com/NIEHS/QSAR-ready	[137]
QECD QSAR Toolbox	https://www.oecd.org/chemicalsafety/risk-assessment/oecd-qsar-toolbox.htm#Download_qsar_application_toolbox	
QSAR:Co	https://sites.google.com/view/qsar-co	[138]
Web4D:QSAR	https://github.com/rougeth/Web-4D-QSAR	[139]
TargetNet	http://targetnet.scbdd.com/calcnet/index/	[140]
MOE-QSAR	https://www.chemcomp.com/Products.htm	

more effective and efficient interactions mimicking the biological systems [142]. Successful methods of dockings investigate effectively high dimensional spaces and utilize scoring functions that rank correctly the docking candidates. Docking can screen compound libraries, rank them, and predict the binding modes and efficiencies of ligands to inhibit target proteins. The binding of ligands for specific receptors can be well-defined by variables like x-y-z translations indicating the ligand position, and Euler angles, quaternion, or axis angle indicating orientations [143]. In the case of flexible docking, the variables like rotatable bond torsional angles can also be considered. All these variables indicate the degree of freedom as one in a space search of multiple dimensions. Rigid docking due to the small search space is faster than flexible docking but lacks accuracy. The scoring functions of docking for ranking binding modes of candidates include functions based on knowledge or force fields. While the methods for search are classified as stochastic and systematic. Stochastic methods make random iterative changes to state variables till the criterion defined by the user is met but the systemic method searches the whole space at intervals defined previously. The methods for search can also be categorized into local and global search methods. The local methods investigate the minimum local or nearest energy to the present conformation while the global method investigates the minimum global energy in the search space defined. Hybrid methods incorporating both local and global search have been reported to improve performance [144]. The most widely used software for molecular docking “AutoDock4” utilizes global, local as well as hybrid algorithms including Monte Carlo, simulated annealing, pattern search, genetic algorithm and Lamarckian GA [145]. In addition to flexible and rigid docking a semi-flexible mode of docking is also used. In flexible mode both the ligand and amino acids side chains of proteins are taken as flexible and thus follow the induce-fit mechanism. The semi-flexible mode of docking considers ligands as flexible but the protein remains rigid while in rigid docking both the ligand or any small molecue and protein remain rigid [146].

The affinity of the binding of the ligand to respective protein is influenced by several forces that mediate the docking interactions. The most significant forces that play a role in interactions include electrostatic forces like dipole–dipole, charge-charge, and charge-dipole, electrodynamic forces like van der Waals, steric forces, forces related to solvent and modification of ligand conformations [147]. These forces mediate the calculation or estimation of the scoring function which eventually evaluates the binding energy. The net binding energy of docking is a result of energies associated with the above-mentioned forces and includes hydrogen bond, total internal, torsional free, electrostatic, unbound system's, repulsion, and dispersion energy types. The constant of dissociation is utilized to specify the binding energy as Gibbs' free energy [148].

In the current scenario, AI/ML has potentiated the conventional algorithms of molecular docking with the aim to increase accuracy, docking speed and reduce the computational cost. Convolutional neural networks have been utilized as scoring functions to improve the accuracy of docking [149]. Further, the utilization of deep QSAR models which are trained using the scores of dockings promotes the iterative removal of unfavorable molecules thus promoting accurate and fast docking of billions of molecules against different targets [150].

Being an essential part of drug discovery molecular docking captures an eminent position in the field and hence numerous software depending on diverse methodologies have been developed (Table 4). Molecular docking has also been applied to study the interactions between two proteins as well as interactions of ligands at the interface of the considered proteins to generate protein–protein interaction inhibitors [151]. Thus, specific tools for molecular docking purpose have been mentioned in Table 4, out of which some are freely available in the public domain.

4.6 MD Simulation and Quantum Mechanics/molecular Mechanics (QM/MM) Simulations

A biomolecule structure at the atomic level is utmost helpful to understand the function of the macromolecule and the mechanism it follows to mediate the concerned function. The atoms of biomolecules are constantly in motion and their function and interactions with other molecules depend on their dynamics. To visualize the dynamics and motions of biomolecules at the atomic level, computational simulations are the most effective alternative [162].

MD simulation provides a prediction of the movement of every atom in the protein or the concerned molecular system during the simulation trajectory time utilizing the information from general physics model which governs the interactions at inter-atomic level. Simulations can display different processes occurring in biomolecules like ligand binding, change in conformation, folding of proteins, and disclosing the atomic positions at the temporal resolution of femtosecond. These simulations can

Table 4 Molecular docking software, web servers and AI/ML models

Webserver/software	Links	References
DUD•E	https://dude.docking.org/generate	[152]
Surflex-Dock	https://www.computabio.com/applications-of-surflex-dock-software.html	
Deep Docking	https://github.com/vibudh2209/D2	[150]
ROSIE Ligand Docking Server	https://rosie.graylab.jhu.edu/ligand_docking	[153]
GNINA-1.0	https://github.com/gnina/gnina	[154]
SwissDock	http://www.swissdock.ch/	[155]
ZDOCK	https://zdock.umassmed.edu/	[156]
pyDOCK	https://life.bsc.es/pid/pydock/	[157]
ClusPro-2.0	https://cluspro.bu.edu/login.php	[158]
HEX	https://hex.loria.fr/#:~:text=Hex%20is%20an%20interactive%20protein,read%20small%2Dmolecule%20SDF%20files	[159]
AutoDock Vina	https://vina.scripps.edu/	[160]
PyRx	https://pyrx.sourceforge.io/	[161]

also predict the response of biomolecules at the level of atoms to any change like phosphorylation, protonation, mutation, and removal or addition of ligands [162, 163]. The forces at every atom in different conditions act differently and convey different motions at the atomic level. These forces in simulations are estimated using molecular mechanics force field models which fit into calculations of quantum mechanics and to experimental measurement up to certain levels. Typically, a force field contains terms such as Coulombic or electrostatic interactions among atoms, spring-like terms for each covalent bond that model the length preferred, and the terms for arresting interatomic interactions of other types [164]. From the 1st protein MD simulation performed in the 1970s, which fetched the Nobel Prize in 2013 in Chemistry, to date the algorithms, force fields, and software have improved immensely. Furthermore, in the classical form of simulations, formation and breaking of covalent bonds are not observed while Quantum mechanics/molecular mechanics (QM/MM) simulations that treat some part of the system with calculations at the quantum and another part at the molecular level can be utilized to model interactions that possess variations in covalent bonds or can also model reactions which are mediated by light absorption [165].

A wide variety of software, tools, and models for MD simulation and QM/MM simulation have been developed and are listed in Table 5.

Table 5 Molecular dynamic (MD) simulation and QM/MM software, web servers and models

Webserver/software	Links	References
Visual Dynamics	https://visualdynamics.fiocruz.br/en-US/login	[166]
MDWeb	https://mmb.irbbarcelona.org/MDWeb/	[167]
SiBioLead	https://sibiolead.com/	
GRID-MARKETS	https://www.pharma.gridmarkets.com/gromacs	
CABS flex	https://biocomp.chem.uw.edu.pl/CABSflex2/submit	[88]
GOMACS	https://www.gromacs.org/	[168]
AMBER Tools	https://ambermd.org/	[169]
NAMD	https://www.ks.uiuc.edu/Research/namd/	[170]
Desmond-Schrödinger	https://www.schrodinger.com/platform/products/desmond/	
Making it rain	https://github.com/pablo-arantes/making-it-rain	[171]
HTMD	https://software.acellera.com/htmd/index.html	[172]
LAMMPS	https://www.lammps.org/#gsc.tab=0	[173]
QMMM 2018	https://comp.chem.umn.edu/qmmm/	
ONIOM	https://gaussian.com/oniom/	
CP2K-QM/MM	https://manual.cp2k.org/trunk/methods/qm_mm/builtin.html	
pDynamo	https://sites.google.com/site/pdynamomodeling/home?authuser=0	[174]
ChemShell	https://chemshell.org/	[175]
GAMESS	https://www.charmm-gui.org/charmmdoc/gamess.html	[176]

5 Summary

Computers have made incredible contributions for the acceleration of the process of the drug discovery. Repurposing, the known or currently in-use drug for a particular disease, to treat some other disease will in turn minimize the time, labour and cost needed for toxicological studies. The availability of diverse gene, protein, and chemical databases increases the ease and pace of identification of disease targets and drug molecules. Computational approaches including molecular modeling, pharmacophore modeling, active site or hotspots prediction, QSAR, molecular docking, MD simulation, and quantum mechanics/molecular dynamics simulations are frequently utilized for lead molecule identification, optimization, and validation in the process of drug discovery. The AI/ML based algorithms have been utilized in different aspects to increase the accuracy and reduce computational costs associated with these approaches to speed up drug discovery and to make it more economical without compromising with accuracy and efficacy of the identified treatment. The present chapter provides comprehensive information on diverse computational approaches and algorithms used for drug identification and provides a resource of different tools and software available to encourage the use of available computational resources by the biologist with ease for drug discovery against a diverse range of diseases.

Conflict of Interest There is no conflict of interest among authors.

References

1. Bagre, A., Patel, P. R., Naqvi, S., Jain, K. (2022). Emerging concerns of infectious diseases and drug delivery challenges. *Nanotheranostics for Treatment and Diagnosis of Infectious Diseases*, 1–23. https://doi.org/10.1016/B978-0-323-91201-3.00013-X.
2. Damle, S. G. (2018). Big tobacco, big food, big alcohol "lifestyle diseases." *Contemporary Clinical Dentistry, 9*(3), 327–329. https://doi.org/10.4103/ccd.ccd_580_18
3. Cummings, J., Montes, A., Kamboj, S., & Cacho, J. F. (2022). The role of basket trials in drug development for neurodegenerative disorders. *Alzheimer's Research & Therapy, 14*(1), 73. https://doi.org/10.1186/s13195-022-01015-6
4. Sun, D., Gao, W., Hu, H., & Zhou, S. (2022). Why 90% of clinical drug development fails and how to improve it? *Acta Pharmaceutica Sinica B, 12*(7), 3049–3062. https://doi.org/10.1016/j.apsb.2022.02.002
5. Pillaiyar, T., Meenakshisundaram, S., Manickam, M., & Sankaranarayanan, M. (2020). A medicinal chemistry perspective of drug repositioning: Recent advances and challenges in drug discovery. *European Journal of Medicinal Chemistry, 195*, 112275. https://doi.org/10.1016/j.ejmech.2020.112275
6. Batool, M., Ahmad, B., & Choi, S. (2019). A structure-based drug discovery paradigm. *International Journal of Molecular Sciences, 20*(11), 2783. https://doi.org/10.3390/ijms20112783
7. Bissantz, C., Kuhn, B., & Stahl, M. (2010). A medicinal chemist's guide to molecular interactions. *Journal of Medicinal Chemistry, 53*(14), 5061–5084. https://doi.org/10.1021/jm100112j
8. Choudhury, C., Arul Murugan, N., & Priyakumar, U. D. (2022). Structure-based drug repurposing: Traditional and advanced AI/ML-aided methods. *Drug Discovery Today, 27*(7), 1847–1861. https://doi.org/10.1016/j.drudis.2022.03.006
9. Opo, F. A. D. M., Rahman, M. M., Ahammad, F., Ahmed, I., Bhuiyan, M. A., & Asiri, A. M. (2021). Structure based pharmacophore modeling, virtual screening, molecular docking and ADMET approaches for identification of natural anti-cancer agents targeting XIAP protein. *Science and Reports, 11*(1), 4049. https://doi.org/10.1038/s41598-021-83626-x
10. Yadav, D. K., Kumar, S., Teli, M. K., & Kim, M.-H. (2020). Ligand-based pharmacophore modeling and docking studies on vitamin D receptor inhibitors. *Journal of Cellular Biochemistry, 121*(7), 3570–3583. https://doi.org/10.1002/jcb.29640
11. Ali, A., Mir, G. J., Ayaz, A., et al. (2023). In silico analysis and molecular docking studies of natural compounds of Withania somnifera against bovine NLRP9. *Journal of Molecular Modeling, 29*(6), 171. https://doi.org/10.1007/s00894-023-05570-z
12. Kwon, S., Bae, H., Jo, J., & Yoon, S. (2019). Comprehensive ensemble in QSAR prediction for drug discovery. *BMC Bioinformatics, 20*(1), 521. https://doi.org/10.1186/s12859-019-3135-4
13. Chen, W., Liu, X., Zhang, S., & Chen, S. (2023). Artificial intelligence for drug discovery: Resources, methods, and applications. *Molecular Therapy-Nucleic Acids, 31*, 691–702. https://doi.org/10.1016/j.omtn.2023.02.019
14. Askr, H., Elgeldawi, E., Aboul Ella, H., Elshaier, Y. A. M. M., Gomaa, M. M., & Hassanien, A. E. (2023). Deep learning in drug discovery: An integrative review and future challenges. *Artificial Intelligence Review, 56*(7), 5975–6037. https://doi.org/10.1007/s10462-022-10306-1

15. Scotti, L., Ishiki, H., Mendonça Júnior, F. J. B., da Silva, M. S., & Scotti, M. T. (2015). Artificial neural network methods applied to drug discovery for neglected diseases. *Combinatorial Chemistry & High Throughput Screening, 18*(8), 819–829. https://doi.org/10.2174/1386207318666150803141219
16. Sun, M., Zhao, S., Gilvary, C., Elemento, O., Zhou, J., & Wang, F. (2020). Graph convolutional networks for computational drug development and discovery. *Briefings in Bioinformatics, 21*(3), 919–935. https://doi.org/10.1093/bib/bbz042
17. Hughes, J. P., Rees, S., Kalindjian, S. B., & Philpott, K. L. (2011). Principles of early drug discovery. *British Journal of Pharmacology, 162*(6), 1239–1249. https://doi.org/10.1111/j.1476-5381.2010.01127.x
18. Mohs, R. C., & Greig, N. H. (2017). Drug discovery and development: Role of basic biological research. *Alzheimers Dement (NY)., 3*(4), 651–657. https://doi.org/10.1016/j.trci.2017.10.005
19. Singh, S., Malik, B. K., & Sharma, D. K. (2006). Molecular drug targets and structure based drug design: A holistic approach. *Bioinformation, 1*(8), 314–320. https://doi.org/10.6026/97320630001314
20. Sadybekov, A. V., & Katritch, V. (2023). Computational approaches streamlining drug discovery. *Nature, 616*(7958), 673–685. https://doi.org/10.1038/s41586-023-05905-z
21. Kulkarni, V. S., Alagarsamy, V., Solomon, V. R., Jose, P. A., & Murugesan, S. (2023). Drug repurposing: An effective tool in modern drug discovery. *Russian Journal of Bioorganic Chemistry, 49*(2), 157–166. https://doi.org/10.1134/S1068162023020139
22. Rudrapal, M., Khairnar, S. J., Jadhav, A. G. (2020). Drug Repurposing (DR): an emerging approach in drug discovery. In *Intechopen eBooks*. https://doi.org/10.5772/intechopen.93193.
23. Hennekens, C. H., Dyken, M. L., & Fuster, V. (1997). Aspirin as a therapeutic agent in cardiovascular disease: A statement for healthcare professionals from the American Heart Association. *Circulation, 96*(8), 2751–2753. https://doi.org/10.1161/01.cir.96.8.2751
24. Ghofrani, H. A., Osterloh, I. H., & Grimminger, F. (2006). Sildenafil: From angina to erectile dysfunction to pulmonary hypertension and beyond. *Nature Reviews Drug Discovery, 5*(8), 689–702. https://doi.org/10.1038/nrd2030
25. Suchonwanit, P., Thammarucha, S., & Leerunyakul, K. (2019). Minoxidil and its use in hair disorders: A review. *Drug Design, Development and Therapy, 13*, 2777–2786. https://doi.org/10.2147/DDDT.S214907
26. Westlake, S. L., Colebatch, A. N., Baird, J., et al. (2010). The effect of methotrexate on cardiovascular disease in patients with rheumatoid arthritis: A systematic literature review. *Rheumatology (Oxford English), 49*(2), 295–307. https://doi.org/10.1093/rheumatology/kep366
27. Wawruszak, A., Halasa, M., Okon, E., Kukula-Koch, W., & Stepulak, A. (2021). Valproic acid and breast cancer: State of the art in 2021. *Cancers, 13*(14), 3409. https://doi.org/10.3390/cancers13143409
28. Hung, C.-L., & Chen, C.-C. (2014). Computational approaches for drug discovery. *Drug Development Research, 75*(6), 412–418. https://doi.org/10.1002/ddr.21222
29. Ferreira, L. G., Dos Santos, R. N., Oliva, G., & Andricopulo, A. D. (2015). Molecular docking and structure-based drug design strategies. *Molecules, 20*(7), 13384–13421. https://doi.org/10.3390/molecules200713384
30. Tang, Y., Moretti, R., & Meiler, J. (2024). Recent advances in automated structure-based de novo drug design. *Journal of Chemical Information and Modeling, 64*(6), 1794–1805. https://doi.org/10.1021/acs.jcim.4c00247
31. Giordano, D., Biancaniello, C., Argenio, M. A., & Facchiano, A. (2022). Drug design by pharmacophore and virtual screening approach. *Pharmaceuticals (Basel, Switzerland), 15*(5), 646. https://doi.org/10.3390/ph15050646
32. Lipinski, C. A., Lombardo, F., Dominy, B. W., & Feeney, P. J. (2001). Experimental and computational approaches to estimate solubility and permeability in drug discovery and development settings. *Advanced Drug Delivery Reviews, 46*(1–3), 3–26. https://doi.org/10.1016/s0169-409x(00)00129-0

33. Bickerton, G. R., Paolini, G. V., Besnard, J., Muresan, S., & Hopkins, A. L. (2012). Quantifying the chemical beauty of drugs. *Nature Chemistry, 4*(2), 90–98. https://doi.org/10.1038/nchem.1243
34. Veber, D. F., Johnson, S. R., Cheng, H.-Y., Smith, B. R., Ward, K. W., & Kopple, K. D. (2002). Molecular properties that influence the oral bioavailability of drug candidates. *Journal of Medicinal Chemistry, 45*(12), 2615–2623. https://doi.org/10.1021/jm020017n
35. Wardana, A. P., Abdjan, M. I., Aminah, N. S., et al. (2022). 3,4,3′-Tri-O-methylellagic acid as an anticancer agent: In vitro and in silico studies. *RSC Advances, 12*(46), 29884–29891. https://doi.org/10.1039/d2ra05246f
36. Pan, X., Lin, X., Cao, D., et al. (2022). Deep learning for drug repurposing: methods, databases, and applications. *Wiley Interdisciplinary Reviews: Computational Molecular Science, 12*(4). https://doi.org/10.1002/wcms.1597.
37. Rajput, A., Thakur, A., Mukhopadhyay, A., et al. (2021). Prediction of repurposed drugs for coronaviruses using artificial intelligence and machine learning. *Computational and Structural Biotechnology Journal, 19*, 3133–3148. https://doi.org/10.1016/j.csbj.2021.05.037
38. Jin, W., Stokes, J. M., Eastman, R. T., et al. (2021). Deep learning identifies synergistic drug combinations for treating COVID-19. *Proceedings of the National Academy of Sciences, 118*(39), e2105070118. https://doi.org/10.1073/pnas.2105070118
39. Pan, X., Yun, J., Coban Akdemir, Z. H., et al. (2023). AI-DrugNet: A network-based deep learning model for drug repurposing and combination therapy in neurological disorders. *Computational and Structural Biotechnology Journal, 21*, 1533–1542. https://doi.org/10.1016/j.csbj.2023.02.004
40. Rao, M., McDuffie, E., & Sachs, C. (2023). Artificial intelligence/machine learning-driven small molecule repurposing via off-target prediction and transcriptomics. *Toxics, 11*(10), 875. https://doi.org/10.3390/toxics11100875
41. Lenselink, E. B., Ten Dijke, N., Bongers, B., et al. (2017). Beyond the hype: Deep neural networks outperform established methods using a ChEMBL bioactivity benchmark set. *Journal of Cheminformatics, 9*(1), 45. https://doi.org/10.1186/s13321-017-0232-0
42. Feng, H., Jiang, J., & Wei, G.-W. (2023). Machine-learning repurposing of DrugBank compounds for opioid use disorder. *Computers in Biology and Medicine, 160*, 106921. https://doi.org/10.1016/j.compbiomed.2023.106921
43. Rusinko, A., Rezaei, M., Friedrich, L., Buchstaller, H.-P., Kuhn, D., & Ghogare, A. (2024). AIDDISON: Empowering drug discovery with AI/ML and CADD tools in a secure, web-based SaaS platform. *Journal of Chemical Information and Modeling, 64*(1), 3–8. https://doi.org/10.1021/acs.jcim.3c01016
44. Xia, X. (2017). Bioinformatics and drug discovery. *Current Topics in Medicinal Chemistry, 17*(15), 1709–1726. https://doi.org/10.2174/1568026617666161116143440
45. Benson, D. A., Cavanaugh, M., Clark, K., et al. (2013). GenBank. *Nucleic Acids Research, 41(database issue)*, D36–D42. https://doi.org/10.1093/nar/gks1195.
46. Brown, G. R., Hem, V., Katz, K. S., et al. (2015). Gene: a gene-centered information resource at NCBI. *Nucleic Acids Research, 43(database issue)*, D36–D42. https://doi.org/10.1093/nar/gku1055.
47. Amberger, J. S., Bocchini, C. A., Schiettecatte, F., Scott, A. F., & Hamosh, A. (2015). OMIM.org: Online Mendelian Inheritance in Man (OMIM®), an online catalog of human genes and genetic disorders. OMIM.org. *Nucleic Acids Research, 43(database issue)*, D789–D798. https://doi.org/10.1093/nar/gku1205.
48. Kanehisa, M., & Goto, S. (2000). KEGG: Kyoto encyclopedia of genes and genomes. *Nucleic Acids Research, 28*(1), 27–30. https://doi.org/10.1093/nar/28.1.27
49. Kanehisa, M., Sato, Y., Kawashima, M., Furumichi, M., & Tanabe, M. (2016). KEGG as a reference resource for gene and protein annotation. *Nucleic Acids Research, 44*(D1), D457–D462. https://doi.org/10.1093/nar/gkv1070
50. Stenson, P. D., Mort, M., Ball, E. V., et al. (2017). The Human Gene Mutation Database: Towards a comprehensive repository of inherited mutation data for medical research, genetic diagnosis and next-generation sequencing studies. *Human Genetics, 136*(6), 665–677. https://doi.org/10.1007/s00439-017-1779-6

51. Piñero, J., Bravo, À., Queralt-Rosinach, N., et al. (2017). DisGeNET: A comprehensive platform integrating information on human disease-associated genes and variants. *Nucleic Acids Research, 45*(D1), D833–D839. https://doi.org/10.1093/nar/gkw943
52. Gene Ontology Consortium, Aleksander, S. A., Balhoff, J., et al. (2023). The Gene Ontology knowledgebase in 2023. *Genetics, 224*(1), iyad031. https://doi.org/10.1093/genetics/iyad031.
53. Logan-Klumpler, F. J., De Silva, N., Boehme, U., et al. (2012). GeneDB—an annotation database for pathogens. *Nucleic Acids Research, 40(database issue)*, D98–D108. https://doi.org/10.1093/nar/gkr1032.
54. Apweiler, R., Bairoch, A., Wu, C. H., et al. (2004). UniProt: the Universal Protein knowledgebase. *Nucleic Acids Research, 32(database issue)*, D115–D119. https://doi.org/10.1093/nar/gkh131.
55. Tomczak, K., Czerwińska, P., & Wiznerowicz, M. (2015). The Cancer Genome Atlas (TCGA): An immeasurable source of knowledge. *Contemporary Oncology (Poznan), 19*(1A), A68–A77. https://doi.org/10.5114/wo.2014.47136
56. Tacutu, R., Thornton, D., Johnson, E., et al. (2018). Human ageing genomic resources: New and updated databases. *Nucleic Acids Research, 46*(D1), D1083–D1090. https://doi.org/10.1093/nar/gkx1042
57. Eppig, J. T. (2017). Mouse genome informatics (MGI) resource: Genetic, genomic, and biological knowledgebase for the laboratory mouse. *ILAR Journal, 58*(1), 17–41. https://doi.org/10.1093/ilar/ilx013
58. Hill, D. P., Begley, D. A., Finger, J. H., et al. (2004). The mouse Gene Expression Database (GXD): updates and enhancements. *Nucleic Acids Research, 32(database issue)*, D568–D571. https://doi.org/10.1093/nar/gkh069.
59. Sherry, S. T., Ward, M. H., Kholodov, M., et al. (2001). DbSNP: The NCBI database of genetic variation. *Nucleic Acids Research, 29*(1), 308–311. https://doi.org/10.1093/nar/29.1.308
60. Tanizawa, Y., Fujisawa, T., Kodama, Y., et al. (2023). DNA Data Bank of Japan (DDBJ) update report 2022. *Nucleic Acids Research, 51*(D1), D101–D105. https://doi.org/10.1093/nar/gkac1083
61. Pertea, M., Shumate, A., Pertea, G., et al. (2018). CHESS: A new human gene catalog curated from thousands of large-scale RNA sequencing experiments reveals extensive transcriptional noise. *Genome Biology, 19*(1), 208. https://doi.org/10.1186/s13059-018-1590-2
62. Clough, E., & Barrett, T. (2016). The gene expression omnibus database. *Methods in Molecular Biology, 1418*, 93–110. https://doi.org/10.1007/978-1-4939-3578-9_5
63. Bittrich, S., Bhikadiya, C., Bi, C., et al. (2023). RCSB Protein Data Bank: Efficient searching and simultaneous access to one million computed structure models alongside the PDB structures enabled by architectural advances. *Journal of Molecular Biology, 435*(14), 167994. https://doi.org/10.1016/j.jmb.2023.167994
64. Brimacombe, K. R., Zhao, T., Eastman, R. T., et al. (2020). An OpenData portal to share COVID-19 drug repurposing data in real time. *bioRxiv*. https://doi.org/10.1101/2020.06.04.135046.
65. Janes, J., Young, M. E., Chen, E., et al. (2018). The ReFRAME library as a comprehensive drug repurposing library and its application to the treatment of cryptosporidiosis. *Proceedings of the National Academy of Sciences, 115*(42), 10750–10755. https://doi.org/10.1073/pnas.1810137115
66. Zdrazil, B., Felix, E., Hunter, F., et al. (2024). The ChEMBL Database in 2023: A drug discovery platform spanning multiple bioactivity data types and time periods. *Nucleic Acids Research, 52*(D1), D1180–D1192. https://doi.org/10.1093/nar/gkad1004
67. Neveu, V., Perez-Jiménez, J., Vos, F., et al. (2010). Phenol-explorer: an online comprehensive database on polyphenol contents in foods. *Database (Oxford), 2010*, bap024. https://doi.org/10.1093/database/bap024.
68. Sorokina, M., Merseburger, P., Rajan, K., Yirik, M. A., & Steinbeck, C. (2021). COCONUT online: Collection of open natural products database. *Journal of Cheminformatics, 13*(1), 2. https://doi.org/10.1186/s13321-020-00478-9
69. FOODB. Retrieved April 14, 2024, from http://www.foodb.ca.

70. Irwin, J. J., Tang, K. G., Young, J., et al. (2020). ZINC20-A free ultralarge-scale chemical database for ligand discovery. *Journal of Chemical Information and Modeling, 60*(12), 6065–6073. https://doi.org/10.1021/acs.jcim.0c00675
71. MolPort. Retrieved April 14, 2024, from https://www.molport.com/.
72. Selleckchem.com. Retrieved April 14, 2024, from https://www.selleckchem.com/screening/fda-approved-drug-library.html.
73. Tran-Nguyen, V.-K., Jacquemard, C., & Rognan, D. (2020). LIT-PCBA: An unbiased data set for machine learning and virtual screening. *Journal of Chemical Information and Modeling, 60*(9), 4263–4273. https://doi.org/10.1021/acs.jcim.0c00155
74. Morris, R., Black, K. A., & Stollar, E. J. (2022). Uncovering protein function: From classification to complexes. *Essays in Biochemistry, 66*(3), 255–285. https://doi.org/10.1042/EBC20200108
75. Hameduh, T., Haddad, Y., Adam, V., & Heger, Z. (2020). Homology modeling in the time of collective and artificial intelligence. *Computational and Structural Biotechnology Journal, 18*, 3494–3506. https://doi.org/10.1016/j.csbj.2020.11.007
76. Fiser, A. (2010). Template-based protein structure modeling. *Methods Molecular Biology (Clifton NJ), 673*, 73–94. https://doi.org/10.1007/978-1-60761-842-3_6.
77. Watson, J. L., Juergens, D., Bennett, N. R., et al. (2023). De novo design of protein structure and function with RFdiffusion. *Nature, 620*(7976), 1089–1100. https://doi.org/10.1038/s41586-023-06415-8
78. Jambrich, M. A., Tusnady, G. E., & Dobson, L. (2023). How AlphaFold2 shaped the structural coverage of the human transmembrane proteome. *Science and Reports, 13*(1), 20283. https://doi.org/10.1038/s41598-023-47204-7
79. Webb, B., & Sali, A. (2021). Protein structure modeling with MODELLER. *Methods Molecular Biology (Clifton NJ), 2199*, 239–255. https://doi.org/10.1007/978-1-0716-0892-0_14
80. Bordoli, L., Kiefer, F., Arnold, K., Benkert, P., Battey, J., & Schwede, T. (2009). Protein structure homology modeling using Swiss-MODEL workspace. *Nature Protocols, 4*(1), 1–13. https://doi.org/10.1038/nprot.2008.197
81. Kelley, L. A., Mezulis, S., Yates, C. M., Wass, M. N., & Sternberg, M. J. E. (2015). The Phyre2 web portal for protein modeling, prediction and analysis. *Nature Protocols, 10*(6), 845–858. https://doi.org/10.1038/nprot.2015.053
82. Zhou, X., Zheng, W., Li, Y., et al. (2022). I-TASSER-MTD: A deep-learning-based platform for multi-domain protein structure and function prediction. *Nature Protocols, 17*(10), 2326–2353. https://doi.org/10.1038/s41596-022-00728-0
83. Zimmermann, L., Stephens, A., Nam, S.-Z., et al. (2018). A completely reimplemented MPI bioinformatics toolkit with a new HHpred server at its core. *Journal of Molecular Biology, 430*(15), 2237–2243. https://doi.org/10.1016/j.jmb.2017.12.007
84. Buchan, D. W. A., & Jones, D. T. (2019). The PSIPRED Protein Analysis Workbench: 20 years on. *Nucleic Acids Research, 47*(W1), W402–W407. https://doi.org/10.1093/nar/gkz297
85. Baek, M., DiMaio, F., Anishchenko, I., et al. (2021). Accurate prediction of protein structures and interactions using a three-track neural network. *Science, 373*(6557), 871–876. https://doi.org/10.1126/science.abj8754
86. Xu, J., Mcpartlon, M., & Li, J. (2021). Improved protein structure prediction by deep learning irrespective of co-evolution information. *Nature Machine Intelligence, 3*, 601–609. https://doi.org/10.1038/s42256-021-00348-5
87. Jumper, J., Evans, R., Pritzel, A., et al. (2021). Highly accurate protein structure prediction with AlphaFold. *Nature, 596*(7873), 583–589. https://doi.org/10.1038/s41586-021-03819-2
88. Kurcinski, M., Oleniecki, T., Ciemny, M. P., Kuriata, A., Kolinski, A., & Kmiecik, S. (2019). CABS-flex standalone: A simulation environment for fast modeling of protein flexibility. *Bioinformatics, 35*(4), 694–695. https://doi.org/10.1093/bioinformatics/bty685
89. Colovos, C., & Yeates, T. O. (1993). Verification of protein structures: Patterns of nonbonded atomic interactions. *Protein Science, 2*(9), 1511–1519. https://doi.org/10.1002/pro.5560020916

90. Bowie, J. U., Lüthy, R., & Eisenberg, D. (1991). A method to identify protein sequences that fold into a known three-dimensional structure. *Science, 253*(5016), 164–170. https://doi.org/10.1126/science.1853201
91. Lüthy, R., Bowie, J. U., & Eisenberg, D. (1992). Assessment of protein models with three-dimensional profiles. *Nature, 356*(6364), 83–85. https://doi.org/10.1038/356083a0
92. Pontius, J., Richelle, J., & Wodak, S. J. (1996). Deviations from standard atomic volumes as a quality measure for protein crystal structures. *Journal of Molecular Biology, 264*(1), 121–136. https://doi.org/10.1006/jmbi.1996.0628
93. Laskowski, R. A., MacArthur, M. W., Moss, D. S., & Thornton, J. M. (1993). PROCHECK: A program to check the stereochemical quality of protein structures. *Journal of Applied Crystallography, 26*(2), 283–291. https://doi.org/10.1107/S0021889892009944
94. Ferruz, N., Schmidt, S., & Höcker, B. (2021). ProteinTools: A toolkit to analyze protein structures. *Nucleic Acids Research, 49*(W1), W559–W566. https://doi.org/10.1093/nar/gkab375
95. Williams, C. J., Headd, J. J., Moriarty, N. W., et al. (2018). MolProbity: More and better reference data for improved all-atom structure validation. *Protein Science, 27*(1), 293–315. https://doi.org/10.1002/pro.3330
96. Wang, H., Mulgaonkar, N., Pérez, L. M., & Fernando, S. (2022). ELIXIR-A: An interactive visualization tool for multi-target pharmacophore refinement. *ACS Omega, 7*(15), 12707–12715. https://doi.org/10.1021/acsomega.1c07144
97. Hall, D. R., Kozakov, D., Whitty, A., & Vajda, S. (2015). Lessons from hot spot analysis for fragment-based drug discovery. *Trends in Pharmacological Sciences, 36*(11), 724–736. https://doi.org/10.1016/j.tips.2015.08.003
98. Koes, D. R., & Camacho, C. J. (2011). Pharmer: Efficient and exact pharmacophore search. *Journal of Chemical Information and Modeling, 51*(6), 1307–1314. https://doi.org/10.1021/ci200097m
99. Schneidman-Duhovny, D., Dror, O., Inbar, Y., Nussinov, R., & Wolfson, H. J. (2008). PharmaGist: a webserver for ligand-based pharmacophore detection. *Nucleic Acids Research, 36*(Web Server issue), W223–W228. https://doi.org/10.1093/nar/gkn187.
100. Lee, J. Y., Krieger, J. M., Li, H., & Bahar, I. (2020). Pharmmaker: Pharmacophore modeling and hit identification based on druggability simulations. *Protein Science, 29*(1), 76–86. https://doi.org/10.1002/pro.3732
101. Wolber, G., & Langer, T. (2005). LigandScout: 3-D pharmacophores derived from protein-bound ligands and their use as virtual screening filters. *Journal of Chemical Information and Modeling, 45*(1), 160–169. https://doi.org/10.1021/ci049885e
102. Brown, B. P., Mendenhall, J., & Meiler, J. (2019). BCL::MolAlign: Three-dimensional small molecule alignment for pharmacophore mapping. *Journal of Chemical Information and Modeling, 59*(2), 689–701. https://doi.org/10.1021/acs.jcim.9b00020
103. Koes DR, Camacho CJ. (2012). ZINCPharmer: pharmacophore search of the zinc database. *Nucleic Acids Research, 40*(Web Server issue), W409–W414. https://doi.org/10.1093/nar/gks378.
104. Ortuso, F., Langer, T., & Alcaro, S. (2006). GBPM: GRID-based pharmacophore model: Concept and application studies to protein-protein recognition. *Bioinformatics, 22*(12), 1449–1455. https://doi.org/10.1093/bioinformatics/btl115
105. Schaller, D., Pach, S., & Wolber, G. (2019). PyRod: Tracing water molecules in Molecular Dynamics simulations. *Journal of Chemical Information and Modeling, 59*(6), 2818–2829. https://doi.org/10.1021/acs.jcim.9b00281
106. Tran-Nguyen, V.-K., Da Silva, F., Bret, G., & Rognan, D. (2019). All in one: Cavity detection, druggability estimate, cavity-based pharmacophore perception, and virtual screening. *Journal of Chemical Information and Modeling, 59*(1), 573–585. https://doi.org/10.1021/acs.jcim.8b00684
107. Schalon, C., Surgand, J.-S., Kellenberger, E., & Rognan, D. (2008). A simple and fuzzy method to align and compare druggable ligand-binding sites. *Proteins, 71*(4), 1755–1778. https://doi.org/10.1002/prot.21858

108. Heider, J., Kilian, J., Garifulina, A., Hering, S., Langer, T., & Seidel, T. (2023). Apo2ph4: A versatile workflow for the generation of receptor-based pharmacophore models for virtual screening. *Journal of Chemical Information and Modeling, 63*(1), 101–110. https://doi.org/10.1021/acs.jcim.2c00814
109. Chen, J., & Lai, L. (2006). Pocket vol 2: Further developments on receptor-based pharmacophore modeling. *Journal of Chemical Information and Modeling, 46*(6), 2684–2691. https://doi.org/10.1021/ci600246s
110. Simón-Vidal, L., García-Calvo, O., Oteo, U., et al. (2018). Perturbation-theory and machine learning (PTML) model for high-throughput screening of Parham reactions: Experimental and theoretical studies. *Journal of Chemical Information and Modeling, 58*(7), 1384–1396. https://doi.org/10.1021/acs.jcim.8b00286
111. Warszycki, D., Struski, Ł, Śmieja, M., Kafel, R., & Kurczab, R. (2021). Pharmacoprint: A combination of a pharmacophore fingerprint and artificial intelligence as a tool for computer-aided drug design. *Journal of Chemical Information and Modeling, 61*(10), 5054–5065. https://doi.org/10.1021/acs.jcim.1c00589
112. Richmond, N. J., Abrams, C. A., Wolohan, P. R. N., Abrahamian, E., Willett, P., Clark, R. D. (2006). GALAHAD: 1. pharmacophore identification by hypermolecular alignment of ligands in 3D. *Journal of Computer-Aided Molecular Design, 20*(9), 567–587. https://doi.org/10.1007/s10822-006-9082-y.
113. Laurie, A. T. R., & Jackson, R. M. (2005). Q-SiteFinder: An energy-based method for the prediction of protein-ligand binding sites. *Bioinformatics, 21*(9), 1908–1916. https://doi.org/10.1093/bioinformatics/bti315
114. An, J., Totrov, M., & Abagyan, R. (2005). Pocketome via comprehensive identification and classification of ligand binding envelopes. *Molecular and Cellular Proteomics, 4*(6), 752–761. https://doi.org/10.1074/mcp.M400159-MCP200
115. Xu, Y., Wang, S., Hu, Q., et al. (2018). CavityPlus: A web server for protein cavity detection with pharmacophore modelling, allosteric site identification and covalent ligand binding ability prediction. *Nucleic Acids Research, 46*(W1), W374–W379. https://doi.org/10.1093/nar/gky380
116. Gahlawat, A., Singh, A., Sandhu, H., & Garg, P. (2024). CRAFT: A web-integrated cavity prediction tool based on flow transfer algorithm. *Journal of Cheminformatics, 16*(1), 12. https://doi.org/10.1186/s13321-024-00803-6
117. Bianchi, V., Mangone, I., Ferrè, F., Helmer-Citterich, M., & Ausiello, G. (2013). webPDBinder: a server for the identification of ligand binding sites on protein structures. *Nucleic Acids Research, 41*(Web Server issue), W308–W313. https://doi.org/10.1093/nar/gkt457.
118. Boer, D. R., Kroon, J., Cole, J. C., Smith, B., & Verdonk, M. L. (2001). SuperStar: Comparison of CSD and PDB-based interaction fields as a basis for the prediction of protein-ligand interactions. *Journal of Molecular Biology, 312*(1), 275–287. https://doi.org/10.1006/jmbi.2001.4901
119. Yang, J., Roy, A., & Zhang, Y. (2013). Protein-ligand binding site recognition using complementary binding-specific substructure comparison and sequence profile alignment. *Bioinformatics, 29*(20), 2588–2595. https://doi.org/10.1093/bioinformatics/btt447
120. Dias, S., Simões, T., Fernandes, F., et al. (2019). CavBench: A benchmark for protein cavity detection methods. *PLoS ONE, 14*(10), e0223596. https://doi.org/10.1371/journal.pone.0223596
121. Zhang, Z., Li, Y., Lin, B., Schroeder, M., & Huang, B. (2011). Identification of cavities on protein surface using multiple computational approaches for drug binding site prediction. *Bioinformatics, 27*(15), 2083–2088. https://doi.org/10.1093/bioinformatics/btr331
122. Wang, S., Lin, H., Huang, Z., et al. (2022). CavitySpace: A database of potential ligand binding sites in the human proteome. *Biomolecules, 12*(7), 967. https://doi.org/10.3390/biom12070967
123. Kandel, J., Tayara, H., & Chong, K. T. (2021). PUResNet: Prediction of protein-ligand binding sites using deep residual neural network. *Journal of Cheminformatics, 13*(1), 65. https://doi.org/10.1186/s13321-021-00547-7

124. Goodford, P. J. (1985). A computational procedure for determining energetically favorable binding sites on biologically important macromolecules. *Journal of Medicinal Chemistry, 28*(7), 849–857. https://doi.org/10.1021/jm00145a002
125. Yu, J., Zhou, Y., Tanaka, I., & Yao, M. (2010). Roll: A new algorithm for the detection of protein pockets and cavities with a rolling probe sphere. *Bioinformatics, 26*(1), 46–52. https://doi.org/10.1093/bioinformatics/btp599
126. Morgan, D. H., Kristensen, D. M., Mittelman, D., & Lichtarge, O. (2006). ET viewer: An application for predicting and visualizing functional sites in protein structures. *Bioinformatics, 22*(16), 2049–2050. https://doi.org/10.1093/bioinformatics/btl285
127. Tian, W., Chen, C., Lei, X., Zhao, J., & Liang, J. (2018). CASTp 3.0: computed atlas of surface topography of proteins. *Nucleic Acids Research, 46*(W1), W363–W367. https://doi.org/10.1093/nar/gky473.
128. Capra, J. A., Laskowski, R. A., Thornton, J. M., Singh, M., & Funkhouser, T. A. (2009). Predicting protein ligand binding sites by combining evolutionary sequence conservation and 3D structure. *PLoS Computational Biology, 5*(12), e1000585. https://doi.org/10.1371/journal.pcbi.1000585
129. Cherkasov, A., Muratov, E. N., Fourches, D., et al. (2014). QSAR modeling: Where have you been? Where are you going to? *Journal of Medicinal Chemistry, 57*(12), 4977–5010. https://doi.org/10.1021/jm4004285
130. Soares, T. A., Nunes-Alves, A., Mazzolari, A., Ruggiu, F., Wei, G.-W., & Merz, K. (2022). The (Re)-evolution of quantitative structure-activity relationship (QSAR) studies propelled by the surge of machine learning methods. *Journal of Chemical Information and Modeling, 62*(22), 5317–5320. https://doi.org/10.1021/acs.jcim.2c01422
131. Mao, J., Akhtar, J., Zhang, X., et al. (2021). Comprehensive strategies of machine-learning-based quantitative structure-activity relationship models. *iScience 24*(9), 103052. https://doi.org/10.1016/j.isci.2021.103052.
132. Niazi, S. K., & Mariam, Z. (2023). Recent advances in machine-learning-based chemoinformatics: A comprehensive review. *International Journal of Molecular Sciences, 24*(14), 11488. https://doi.org/10.3390/ijms241411488
133. Jiménez-Luna, J., Grisoni, F., & Schneider, G. (2020). Drug discovery with explainable artificial intelligence. *Nature Machine Intelligence, 2*(10), 573–584. https://doi.org/10.1038/s42256-020-00236-4
134. Kohlbacher, S. M., Langer, T., & Seidel, T. (2021). QPHAR: Quantitative pharmacophore activity relationship: Method and validation. *Journal of Cheminformatics, 13*(1), 57. https://doi.org/10.1186/s13321-021-00537-9
135. Ragno, R. (2019). www.3d-qsar.com: A web portal that brings 3-D QSAR to all electronic devices-the Py-CoMFA web application as tool to build models from pre-aligned datasets. *Journal of Computer-Aided Molecular Design, 33*(9), 855–864. https://doi.org/10.1007/s10822-019-00231-x.
136. Concu, R., Cordeiro, M. N. D. S., Pérez-Pérez, M., & Fdez-Riverola, F. (2023). MOZART, a QSAR multi-target web-based tool to predict multiple drug-enzyme interactions. *Molecules, 28*(3), 1182. https://doi.org/10.3390/molecules28031182
137. Mansouri, K., Moreira-Filho, J. T., Lowe, C. N., et al. (2024). Free and open-source QSAR-ready workflow for automated standardization of chemical structures in support of QSAR modeling. *Journal of Cheminformatics, 16*(1), 19. https://doi.org/10.1186/s13321-024-00814-3
138. Ambure, P., Halder, A. K., González Díaz, H., & Cordeiro, M. N. D. S. (2019). QSAR-Co: An open source software for developing robust multitasking or multitarget classification-based QSAR models. *Journal of Chemical Information and Modeling, 59*(6), 2538–2544. https://doi.org/10.1021/acs.jcim.9b00295
139. Ataide Martins, J. P., Rougeth de Oliveira, M. A., & Oliveira de Queiroz, M. S. (2018). Web-4D-QSAR: A web-based application to generate 4D-QSAR descriptors. *Journal of Computational Chemistry, 39*(15), 917–924. https://doi.org/10.1002/jcc.25166

140. Yao, Z.-J., Dong, J., Che, Y.-J., et al. (2016). TargetNet: A web service for predicting potential drug-target interaction profiling via multi-target SAR models. *Journal of Computer-Aided Molecular Design, 30*(5), 413–424. https://doi.org/10.1007/s10822-016-9915-2
141. Agu, P. C., Afiukwa, C. A., Orji, O. U., et al. (2023). Molecular docking as a tool for the discovery of molecular targets of nutraceuticals in diseases management. *Science and Reports, 13*(1), 13398. https://doi.org/10.1038/s41598-023-40160-2
142. Stanzione, F., Giangreco, I., & Cole, J. C. (2021). Use of molecular docking computational tools in drug discovery. *Progress in Medicinal Chemistry, 60*, 273–343. https://doi.org/10.1016/bs.pmch.2021.01.004
143. Torres, P. H. M., Sodero, A. C. R., Jofily, P., & Silva-Jr, F. P., Jr. (2019). Key topics in molecular docking for drug design. *International Journal of Molecular Sciences, 20*(18), 4574. https://doi.org/10.3390/ijms20184574
144. Morris, G. M., & Lim-Wilby, M. (2008). Molecular docking. *Methods in Molecular Biology, 443*, 365–382. https://doi.org/10.1007/978-1-59745-177-2_19
145. Morris, G. M., Huey, R., Lindstrom, W., et al. (2009). AutoDock4 and AutoDockTools4: Automated docking with selective receptor flexibility. *Journal of Computational Chemistry, 30*(16), 2785–2791. https://doi.org/10.1002/jcc.21256
146. Mohanty, M., Mohanty, P. S. (2023). Molecular docking in organic, inorganic, and hybrid systems: a tutorial review. *Monatshefte für Chemie-Chemical Monthly*, 1–25. https://doi.org/10.1007/s00706-023-03076-1. (Advance online publication)
147. Pantsar, T., & Poso, A. (2018). Binding affinity via docking: Fact and fiction. *Molecules, 23*(8), 1899. https://doi.org/10.3390/molecules23081899
148. Meng, X.-Y., Zhang, H.-X., Mezei, M., & Cui, M. (2011). Molecular docking: A powerful approach for structure-based drug discovery. *Current Computer-Aided Drug Design, 7*(2), 146–157. https://doi.org/10.2174/157340911795677602
149. Ragoza, M., Hochuli, J., Idrobo, E., Sunseri, J., Koes, D. R. (2017). Protein–ligand scoring with convolutional neural networks. *Journal of Chemical Information and Modelling, 57*(4), 942–957. https://doi.org/10.1021/acs.jcim.6b00740.
150. Gentile, F., Agrawal, V., Hsing, M., et al. (2020). Deep docking: A deep learning platform for augmentation of structure based drug discovery. *ACS Central Science, 6*(6), 939–949. https://doi.org/10.1021/acscentsci.0c00229
151. Vakser, I. A. (2014). Protein-protein docking: From interaction to interactome. *Biophysical Journal, 107*(8), 1785–1793. https://doi.org/10.1016/j.bpj.2014.08.033
152. Mysinger, M. M., Carchia, M., Irwin, J. J., & Shoichet, B. K. (2012). Directory of useful decoys, enhanced (DUD-E): Better ligands and decoys for better benchmarking. *Journal of Medicinal Chemistry, 55*(14), 6582–6594. https://doi.org/10.1021/jm300687e
153. DeLuca, S., Khar, K., & Meiler, J. (2015). Fully flexible docking of medium sized ligand libraries with RosettaLigand. *PLoS ONE, 10*(7), e0132508. https://doi.org/10.1371/journal.pone.0132508
154. McNutt, A. T., Francoeur, P., Aggarwal, R., et al. (2021). GNINA 1.0: molecular docking with deep learning. *Journal of Cheminformatics, 13*(1), 43. https://doi.org/10.1186/s13321-021-00522-2.
155. Grosdidier, A., Zoete, V., & Michielin, O. (2011). SwissDock, a protein-small molecule docking web service based on EADock DSS. *Nucleic Acids Research, 39*(Web Server issue), W270–W277. https://doi.org/10.1093/nar/gkr366.
156. Pierce, B. G., Hourai, Y., & Weng, Z. (2011). Accelerating protein docking in ZDOCK using an advanced 3D convolution library. *PLoS ONE, 6*(9), e24657. https://doi.org/10.1371/journal.pone.0024657
157. Cheng, T.M.-K., Blundell, T. L., & Fernandez-Recio, J. (2007). PyDock: Electrostatics and desolvation for effective scoring of rigid-body protein-protein docking. *Proteins, 68*(2), 503–515. https://doi.org/10.1002/prot.21419
158. Kozakov, D., Hall, D. R., Xia, B., et al. (2017). The ClusPro web server for protein-protein docking. *Nature Protocols, 12*(2), 255–278. https://doi.org/10.1038/nprot.2016.169

159. Ghoorah, A. W., Devignes, M.-D., Smaïl-Tabbone, M., & Ritchie, D. W. (2013). Protein docking using case-based reasoning. *Proteins, 81*(12), 2150–2158. https://doi.org/10.1002/prot.24433
160. Trott, O., & Olson, A. J. (2010). AutoDock Vina: Improving the speed and accuracy of docking with a new scoring function, efficient optimization, and multithreading. *Journal of Computational Chemistry, 31*(2), 455–461. https://doi.org/10.1002/jcc.21334
161. Dallakyan, S., & Olson, A. J. (2015). Small-molecule library screening by docking with PyRx. *Methods in Molecular Biology, 1263*, 243–250. https://doi.org/10.1007/978-1-4939-2269-7_19
162. Hollingsworth, S. A., & Dror, R. O. (2018). Molecular dynamics simulation for all. *Neuron, 99*(6), 1129–1143. https://doi.org/10.1016/j.neuron.2018.08.011
163. Hospital, A., Goñi, J. R., Orozco, M., & Gelpí, J. L. (2015). Molecular dynamics simulations: Advances and applications. *Advances and Applications in Bioinformatics and Chemistry (AABC), 8*, 37–47. https://doi.org/10.2147/AABC.S70333
164. Guvench, O., & MacKerell, A. D., Jr. (2008). Comparison of protein force fields for molecular dynamics simulations. *Methods in Molecular Biology, 443*, 63–88. https://doi.org/10.1007/978-1-59745-177-2_4
165. Tzeliou, C. E., Mermigki, M. A., & Tzeli, D. (2022). Review on the QM/MM methodologies and their application to metalloproteins. *Molecules, 27*(9), 2660. https://doi.org/10.3390/molecules27092660
166. Vieira, I. H. P., Botelho, E. B., de Souza Gomes, T. J., Kist, R., Caceres, R. A., & Zanchi, F. B. (2023). Visual dynamics: A WEB application for molecular dynamics simulation using GROMACS. *BMC Bioinformatics, 24*(1), 107. https://doi.org/10.1186/s12859-023-05234-y
167. Hospital, A., Andrio, P., Fenollosa, C., Cicin-Sain, D., Orozco, M., & Gelpí, J. L. (2012). MDWeb and MDMoby: An integrated web-based platform for molecular dynamics simulations. *Bioinformatics, 28*(9), 1278–1279. https://doi.org/10.1093/bioinformatics/bts139
168. Páll, S., Zhmurov, A., Bauer, P., et al. (2020). Heterogeneous parallelization and acceleration of molecular dynamics simulations in GROMACS. *The Journal of Chemical Physics, 153*(13), 134110. https://doi.org/10.1063/5.0018516
169. Case, D. A., Aktulga, H. M., Belfon, K., et al. (2023). AmberTools. *Journal of Chemical Information and Modeling, 63*(20), 6183–6191. https://doi.org/10.1021/acs.jcim.3c01153
170. Phillips, J. C., Braun, R., Wang, W., et al. (2005). Scalable molecular dynamics with NAMD. *Journal of Computational Chemistry, 26*(16), 1781–1802. https://doi.org/10.1002/jcc.20289
171. Arantes, P. R., Polêto, M. D., Pedebos, C., & Ligabue-Braun, R. (2021). Making it rain: Cloud-based Molecular Simulations for everyone. *Journal of Chemical Information and Modeling, 61*(10), 4852–4856. https://doi.org/10.1021/acs.jcim.1c00998
172. Doerr, S., Harvey, M. J., Noé, F., & De Fabritiis, G. (2016). HTMD: High-throughput Molecular Dynamics for molecular discovery. *Journal of Chemical Theory and Computation, 12*(4), 1845–1852. https://doi.org/10.1021/acs.jctc.6b00049
173. Thompson, A. P., Aktulga, H. M., Berger, R. A., et al. (2022). LAMMPS–a flexible simulation tool for particle-based materials modeling at the atomic, meso, and continuum scales. *Computer Physics Communications, 271*, 108171. https://doi.org/10.1016/j.cpc.2021.108171
174. Field, M. J. (2022). PDynamo3 molecular modeling and simulation program. *Journal of Chemical Information and Modeling, 62*(23), 5849–5854. https://doi.org/10.1021/acs.jcim.2c01239
175. Metz, S., Kästner, J., Sokol, A. A., Keal, T. W., & Sherwood, P. (2013). C hem S hell—a modular software package for QM/MM simulations. *Wiley Interdisciplinary Reviews: Computational Molecular Science, 4*(2), 101–110. https://doi.org/10.1002/wcms.1163
176. Brooks, B. R., Brooks, C. L., III., Mackerell, A. D., Jr., Nilsson, L., Petrella, R. J., Roux, B., et al. (2009). CHARMM: The biomolecular simulation program. *Journal of Computational Chemistry, 30*(10), 1545–1614. https://doi.org/10.1002/jcc.21287

How Do Big Data and Generative AI Dawn on Computational Biology?

Shaurya Jauhari

Abstract The theme of Generative AI has opened a nasty *Pandora's* Box in the recent epoch, where there is an open debate on the benefits and caveats that it brings to the table. From current and prospective employees fearing their professional spaces to the grains of the education system, from the ambit of entertainment to the purported dent on humanity and moral values, multitude of facets deem eclipsed. The progress of Science and Technology is robustly tethered to discoveries, innovations, and inventions. This innovation encapsulates the vast expanse of scholarly disciplines including the very nascent, *Bioinformatics*. This essay bears my perspective on how the regeneration of biological data (and hence insights from restructured techniques) can either fortify or challenge our existing wisdom in the domain.

Keywords Big data · Deep learning models · Generative AI · Genomics · Sequencing data

1 What is Generative AI?

We have witnessed disruptive technologies intervening our lifestyle and forcibly altering our perceptions- the industrial revolution, advent of the Internet, mobile communications, and now the *hyped* Artificial Intelligence (AI); perhaps most for the latter. In lay-terms, literally, artificial intelligence would mean synthetic wisdom acquired by exoliving entities. This is an evolving space and robustly hinged over data. The data driven techniques are omnipresent and their capabilities benefit modern businesses, research and development, environment, mobility, and even administration and governance. Speaking economically though, for any technology to spread its wings, it takes some turnaround time; and arguably, the Generative AI has been one with the least. While traditional machine learning is attributed for predictive analysis, the algorithms of Generative AI facilitate creation of synthetic data instances to power the predictions.

S. Jauhari (✉)
Infosys Limited, Mysore, India
e-mail: shauryajauhari@hotmail.com

© The Author(s), under exclusive license to Springer Nature Singapore Pte Ltd. 2024
A. Chaudhary et al. (eds.), *Unraveling New Frontiers and Advances in Bioinformatics*,
https://doi.org/10.1007/978-981-97-7123-3_10

The generalization of AI is a culmination of the AI advances in the first place; I wrote a piece sometime back to highlight the same. We have a healthy suite of algorithms that qualify a cross-section of Mathematics and Statistics. Computer Science too is a *blue-collared* discipline emanating from mathematical paradigms, fundamentally. With the capacity to model data and elucidate patterns implicit to it, we can structure software that embrace some form of intelligence that, indeed, is ersatz. There are some formally identified subdomains under AI, viz. Machine Learning, Neural Networks and Deep Learning, Robotics, Computer Vision and Image Processing, Natural Language Processing; the latter here feeds primarily into the working dogma of Generative AI. We will have a chance to expound on all of these (vide 1.2), notwithstanding, let us take a step back and felicitate the key protagonist driving AI.

1.1 The **Ikigai** *of Big Data*

This is perhaps the most suited commencement to the Generative AI lore. We have been basking in the glory of AI, only if it is for the Big Data [1]. The success of a general, machine learning algorithm is contingent to an overwhelming amount of *clean* data, sedentary- still awaiting analysis. In the early decade of 1990–2000, mostly, we were dealing with a very modest data representation. The relational files had a measurable dimension of rows and columns, and the processing and storage was not too taxing for a humbly configured computer system. With the advent of ameliorated networking and Internet era, we got deluged with the influx of semi-structured and unstructured data as well [2]. The management of this escalated complexity and volume of data was overwhelming for a single machine, and we had to resort to a cluster of computers. Each *node* in the cluster, had its individuality as a computational resource and contributed to the broader objective of data analysis (Fig. 1).

The Princeton Model or the Von-Neumann architecture [3] is the linchpin of traditional, as well as modern-day computing. While it still holds true for distributed processing, the template is now more enhanced. As evidenced from Fig. 2, we still

Fig. 1 Von Neumann architecture: the classic abstraction of computation

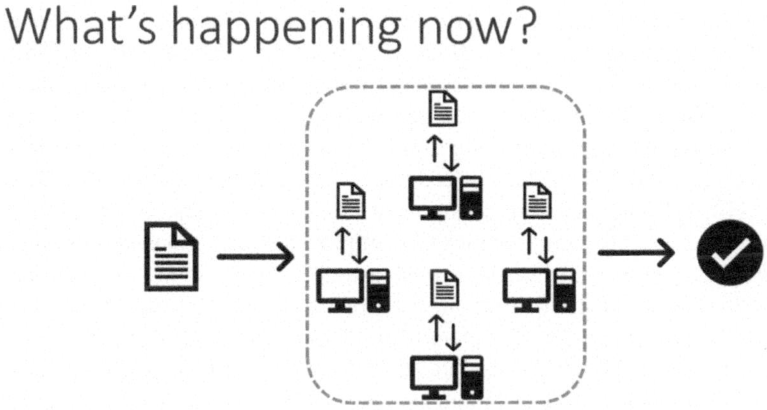

Fig. 2 How distributed computing garnered pertinence as the data evolved

acknowledge the aforementioned architecture, yet is has been scaled to a distributed form. Each node in the cluster is engendering its respective out from a *slice* of input data, and eventually the aggregation of these intermediary results is presented back to the user.

1.1.1 Hadoop Framework

Hadoop was the first to 'launch a counter' on Big Data. It epitomized the paradigm of distributed processing, with a master–slave layout of nodes in a cluster. The Data Nodes (*aka* slave nodes) were encompassing the data and computation, and would report to the Name Node (*aka* master node) the results. To provide a safety-net of fault tolerance in a commodity-oriented, distributed infrastructure, replication of nodes and cross-talks are provisioned so that the integrity of network is maintained. The data residing in *blocks* in one data node, is replicated across blocks in adjacent nodes. Furthermore, a "heartbeat" is relayed by the Name Node to each Data Node provisioned in a cluster to confirm their active state, provided its positive acknowledgment from the respective Data Node. Yet, what if the Name Node becomes inaccessible, for any conceivable reason- power outage, network failure, etc.? It being the "entry-point" to the cluster, the entire infrastructure is rendered dysfunctional. To remedy this scenario, a *secondary Name Node* is instantiated that can fill the void. For the same, it has to stay well synchronized with the primary Name Node, vis-a'-vis the cluster idiosyncrasies [4]. Bioinformatics embraced a substantive followup on Hadoop due to the ability to break down complex jobs into smaller ones, and then aggregate results eventually [4–7]. With dedicated frameworks like Hadoop-BAM [8], BlastReduce [9] and Biodoop [10], that were derived from the direct design tenets of Hadoop Framework, were mandatory improvizations to drive Bioinformatics research.

The core components of a Hadoop deployment include: *MapReduce* (the distributed processing engine), *Hadoop Distributed File System* (HDFS; the distributed storage), and *Yet Another Resource Negotiator* (YARN; the cluster manager). While the former two engage with the intricacies of data management and application of the programming logic, the latter ensures that the cohesion in the system is robust, with the synchronized track of meta-data associated with machines and tasks [4, 11]. The secondary components, on the other hand, include Hive (originally from Facebook; a SQL like interface employed to perform analytics on Hadoop Framework that transforms queries to MapReduce Java programs), Pig (originally from Yahoo; an interface for executing scripts to exploit distributed environment), Sqoop, now moved to Attic (for channeling data to and fro HDFS, from relational database systems), Flume (a step ahead of Sqoop, able to ingest data from non-relational data sources too), Mahout (a library for machine learning algorithms), and not the least, Spark, among others. While I will refer the reader to the official documentation for other modules, let us delve a tad deeper into Spark Framework.

1.1.2 Spark Framework

As aforementioned, it started out as a secondary component of the Hadoop Framework, albeit now, has carved out a niche spot for itself, and is a one-stop destination for largely all analysis pertaining to Big Data. Yet, why did we chose to resort to Spark Framework, with Hadoop Framework already available [12]? The following tenets apply [13].

- As opposed to Hadoop Framework, Spark Framework supports both applications of batch and streaming categories of data.
- The libraries for machine learning and graph data processing pipelines are available implicitly; they come bundled up with the package and upon installation the algorithms are accessible. In Hadoop framework, the same had to be installed explicitly.
- Spark is famed as an "in-memory" computational framework, and works off of the Random Access Memory (RAM). With contingency to the memory-hierarchy as evidenced in the domain of computer hardware, RAM is featured over secondary storage for efficacy of data access. Although paradoxical, Spark Framework does not generally employ RAM in an absolute sense, albeit in a hybrid form with the secondary storage; at the end of the day, we observe that Spark effectively outperforms Hadoop in efficiency.

The operations in Spark are premised over Resilient Distributed Datasets (RDDs) that are rudimentary data objects, having data distributed across specified number of partitions. The granularity of parallel operations over these data slices residing in partitions aids the efficiency in performance. For batch data, the input files are considered in bulk or isolation, while streaming (or near real-time) data, sourced from sensors, CCTV cameras, satellites, and IoT devices, etc. are pulverized into batches and then progressed further with either library methods or SQL queries (actioned

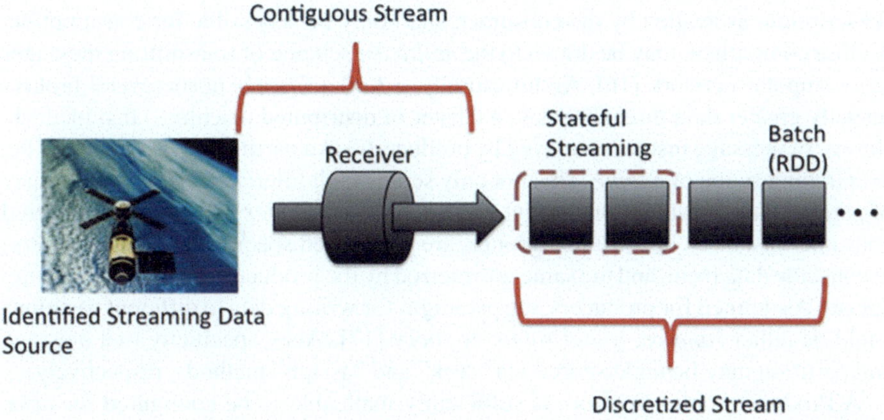

Fig. 3 Spark Framework handles the streaming data differently, than Apache Kafka. Mechanistically, it pulverizes an incessant, influx of data into disjunct batches, which may be considered in aggregation for a collective analysis

on the data frames, rendered from RDDs or even directly from input) Fig. 3. These batches (instances of a **Discretized Stream**) are implicitly RDDs, and are addressed in the technical ambit of distributed processing. They could either be consolidated to maintain collateral relationship with other batches (*stateful streaming*) or otherwise (*stateless streaming*) As a side note, out of the known sets of pros and cons, one of the potential drawbacks of Spark Framework is the inability to offer the developer manual deletion of the RDDs. When not in use frequently or created much earlier during a workflow, an RDD is eliminated from the memory stack following the Least Recently Used (LRU) or First In First Out (FIFO) paradigm, as specified in the Spark configuration. We may, however, assign "persistence" to an RDD engendering it a special privilege of escaping deletion. Spark Framework does not have its own dedicated distributed storage and also carries memory-intensive operations following design. Contemporaneously though, this distributed framework of cluster computing is the state-of-the-art and finds efficient usage in genome informatics [14, 15].

Spark Streaming module can handle the data streams well, but if what is warranted is a pilot analysis that can be carried out in a relatively *lightweight* manner, Kafka may be referred. So, what is Kafka?

1.1.3 Kafka

Apache Kafka hails from the stable of LinkedIn. It is premised over the revered "*Pub-Sub*" or a Publisher-Subscriber model, as illustrated in Fig. 4. This model underlines the communication between two stakeholders: the broadcaster of data, and there could be several of those, and the one(s) ingesting it. An salient point to note here is that only specific communication (data/messages) contingent to a subscription, as availed by the consumer, shall only be accessible for consumption. A close comparison may be drawn to the *multicast* scheme of transmitting messages in a computer network [16]. Architecturally, a *Kafka Cluster* hosts several brokers (usually greater than three; nodes in a cluster of distributed machines) that incubate storage of message instances relayed by producer(s) in a partitioned layout, under the overarching scope of a *topic*. This not only serves fault tolerance, but also efficiency of parallel data consumption through well-defined *consumer groups*. The registered consumer instances in consumer groups, are designated specific partitions (*leaders*) to consume data from, and the same are referred by the producer applications to dump data to. As defined for producers, the paradigm for writing data to different partitions could be either *random*, *round-robin*, or *sticky* [17]. Also, specificity to a message and partition may be underscored via "seek" and "assign" methods, respectively.

A Producer application too, is sufficiently malleable to be configured for *acks*, *retries*, *batch size*, and *buffer memory*, amidst other settings. "Acks" or Acknowledgment is sought from the Kafka Cluster (technically, the 'Leader' and 'Follower(s)' partitions for message logs under the topic) as a confirmation of a message receipt. If '0', no acknowledgment is sought; if set to '1', an acknowledgment is solicited from the "Leader"; and if 'All', both "Leader" and "Follower(s)" are required to respond back to the producer application. Not to mention, while the latter is most welcoming, it adds to the latency of the operation. *Retries* aids to set a threshold for the number of attempts to be made for establishing connection with the cluster. *Buffer Memory* and *Batch Size* emphasize the amount of data held in the memory pool of the producer and the amount to be sent out in one go, respectively.

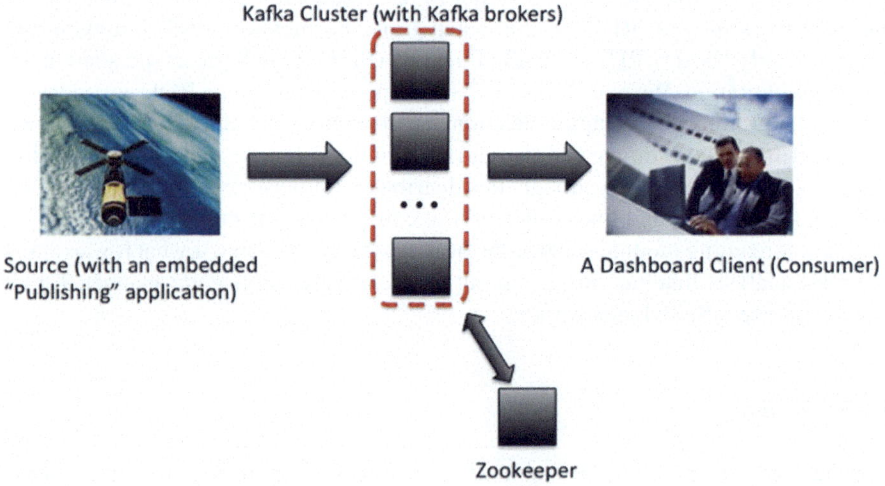

Fig. 4 Kafka is based on a Pub-Sub model, which ascribes a producer entity that relays data and a consumer ingesting data

Finally, *Zookeeper is to Kafka, what YARN is to Hadoop.* It is a cluster manager that traces the metadata of the current Kafka configuration. Besides, Kafka is available in open-source (Apache) and proprietary (Confluent) flavors.

Hitherto, we were able to address batch and streaming data in a distributed manner. It sounds great until we realize that data inflow is literally *perpetual*, so much so that we experience *sedentary* data awaiting acknowledgment for analysis. In all practicality of finance, architectural infrastructure, and manpower, we do *hit a wall* at some stage when capacity threshold exhausts. We have no further ability to store data, manage networks, keep up with the computational prowess- adding more hardware, and the like. To remedy this daunting scenario, Cloud came to the rescue. Let us follow next on this pervasive instrument.

1.1.4 Cloud-Based Platforms

There is a burgeoning body of literature (journal articles, book chapters, conference proceedings, inter alia) on Cloud infrastructure and why it has evolved as a *serf* to the contemporary data analytics tasks. Back in 1961 at the MIT, John McCarthy foresaw the public utility of computing. This was the closest analogy to what we now know as Cloud Computing. It is a technology that enables virtual services on-demand, which are facilitated via Cloud Service Providers (CSPs) [18]. Throughout use, the end-user may never have the *sine qua non* of the idiosyncrasies of services. My attempt here will be to aggregate sufficient ground on the topic, while ensuring it does not go overboard.

If you can imagine yourself in a business-worthy capacity with a generous amount of commodity infrastructure to spare, it would certainly strike you to roll out the excess on lease, and make money. This is also a sensible decision contingent to the fact that technology is very fragile, and advancements in hardware and software materialize before we know it [19]. This is a lucrative exercise as it brings optimum return of investments. There exist offerings from myriad technology enterprises, markedly the high-end platforms like Microsoft Azure, Google Cloud Platform, and Amazon Web Services, have been able to substantively populate the cloud services space [20], with the latter holding the maximum share [21] (Fig. 7). According to the market IT-research firm Canalys, despite the plummet in the growth of Cloud-based services in the second financial quarter of the year 2023 (from 19% in Q1 to 16% in Q2) attributed to economic slowdown pan-globe (vide Fig. 5), the technology market is still replete with Cloud-services.

The hosting of resources relative to computing infrastructure, virtually (over the Intranet or Internet), enable the *anytime-anywhere* ability. In juxtaposition, not all biomedical scientists are computer programmers or would like to delve into it, necessitating either collaboration with the former or adapting to a Graphical User Interface (GUI) platform for data analysis. The availability of data and services can be negotiated via a service-level agreement (SLA), which is a legal contract between the Cloud-service provider and the lessee; more services will surely solicit more

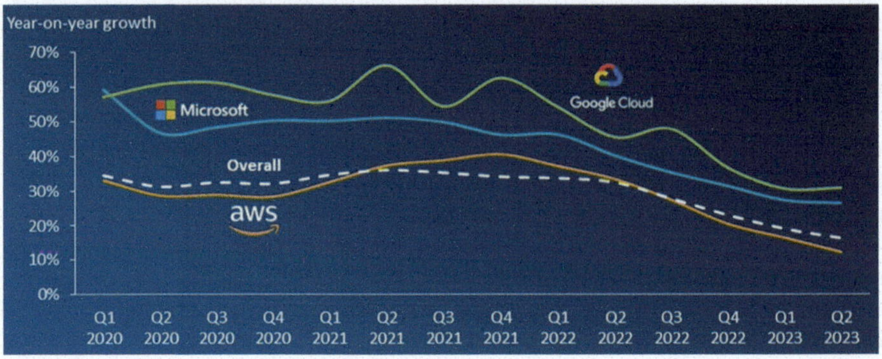

Fig. 5 From 2020–2021, we still observe no general improvement, and hence the growth stays plateaued. *Credit* Canalys; estimates from August 2023 study

monetary liability. Some of the other dominant benefits of Cloud-computing can be recorded as follows:

- As opposed to an *on-premise* computing environment, a rented service on the Cloud is faster to claim and does not involve shedding an initial capital expenditure for erecting infrastructure (hardware, software, and network), upfront. This relieves the end-user from conspicuously contemplating the proportionate spending of the capital, which would have them left anxious otherwise.
- The security is relatively better due to a virtualization, as only defined *principals* can access data and services, which are largely role-based. The Cloud environment follows the "least-privilege" policy, i.e. stakeholders have only *bare-necessary* access, based on their credentials. This is critical, as less would argue that benefits of open and reproducible research benefits Science at large. Albeit, the perseverance of data integrity must be diligently followed [22].
- The elastic nature of deployment makes the securing and release of the services (compute, storage, amongst others), automatic. This has a direct effect on the billing (*pay-per-use*), and if orchestrated finely, may save substantially on costs incurred. This agility of dynamic allocation is tethered to the buffer of resources, available with the connected CSP and the corresponding subscription [23].
- Speaking of costs, the end-user is further alleviated from the maintenance of infrastructure, and can thence focus on the business problems at hand. Again, the fault-tolerance rendered by the distributed nature of data and resources over Cloud ensures worthy customer satisfaction.
- One of the aversions for business organizations to implementing open-source software, was that of implementation and management; software updates, installation of patches and dependencies, and milieu occupy a healthy yet unwanted intervention. Now, with the CSP taking load off the shoulders, the organizations can curtail expenses aligned to software licenses and focus more on their business objectives [24]. This has in fact been the reason why Cloud became reasonably venerable.

The end-users have several Cloud deployment models to choose from, viz. *private*, *public*, *hybrid*, and *community*, according to the grant by the National Institute of Standards and Technology (NIST). As the names suggest, the scope of these cloud platforms is tethered to their scale and reachability [25]. A *private* Cloud is organization-specific and has strict accessibility only to internally authorized stakeholders. A *public* Cloud is available on a paid-subscription basis to anyone soliciting the services; the generic CSPs fall in this category. Furthermore, while a *hybrid* Cloud is a combinatorial deployment of a *private* Cloud, aiming to expand its data and applications over the scaffold of a *public* Cloud, a *community* Cloud is erected to serve similar business-holdings like academia, clinical establishments, etc. You may consider a chain of hospitals or universities, sharing research data or results amongst themselves. One may also spare a thought on the strict pursuance of deployment and adaptation of standards, especially security. Programs like the Federal Risk and Authorization Management Program (FedRAMP) from government of the United States lays a blueprint for approaching risk assessments, authorization, and supervision of information technology (IT) infrastructure, in a generalized manner.

While suitability and choice of an appropriate Cloud deployment model cannot be glossed over, a prospective subscriber will be obliged to consider the service tier to be tap into. Exclusive to a single connection, it could primarily, either be *Software as a Service (SaaS)*, *Platform as a Service (PaaS)*, or *Infrastructure as a Service (IaaS)*. Notwithstanding, services have further granularized in the contemporary epoch as *Data as a Service (DaaS)*; AWS hosts Human Microbiome Project (HMP) data, as an example, *Database as a Service (DBaaS)*, *Desktop as a service (DaaS)*, *Framework as a service (FaaS)*, *Network as a service (NaaS)* [26]. Pointedly, multiple databases usually aggregate their data to a centralized repository called a *datawarehouse*, that facilitates analytics of structured and semi-structured data; biological data streamed from Internet of Medical Things (IoMT) or transactional databases hosting expression data are a few examples. In their cloud "avatar", datawarehouses like Snowflake provide excellent scaling, rectifying any delimitation on data thresholds. A multiuser environment with sundry workloads in parallel pose no real challenges like performance lags, hence supportive of data exchange and collaboration in real-time. Yet, for the risk of digression, let us curtail our discussion on these rudimentary types for now.

The choice of data and tools, is in a true sense, "mission critical". The data that may encapsulate gene expression, copy number variation (CNV), single nucleotide polymorphism (SNP), and DNA methylation artifacts, that may be available under a "single-roof" to aid ease of search, access, and execution [27]. Additionally, the data and the analytics workflows may be shared across in a highly secured environment via platforms like FireCloud linking datasets from National Cancer Institute (NCI), TARGET, and TCGA, inter alia, in addition to projects associated with the Terra ecosystem. Galaxy too is a platform that crosses path with Terra, giving researchers an interface to interact with data and reproduce workflows on variable data. The data in computational space of biology is majorly derivable from transcriptomics, genomics, proteomics, epigenomics, and metabolomics experiments. Manier times, we are seeking analysis of an ensemble of data as opposed to the individual types. We need efficient data management and organization strategies that corroborate workflows that may be user-designed or pulled off-the-shelf (library methods).

1. **Software as a Service**: With this rendering, a lessee may avail services of an application without requiring to bother about the installation over a specific operating system, setting up updates contingent to the business use, amongst others. This is a quintessential feature that is widely adopted by users given the scope. Examples include Microsoft Office 365, DropBox, Google Docs, etc. In terms of expenditure, this amenity happens to be least for an end-user as compared to others, probably corroborating its wide usage. From the vantage point of Computational Biology, BIG Search and BioGPT [28] are contemporary applications premised on rendering correlations, premised over parsing cross-references in the biomedical data.
2. **Platform as a Service**: A *platform*, in crude terms, refers to the operating system. If the work sought so requires that a system software is pre-installed over a desired hardware configuration, we can choose installation of a suite of application software that would facilitate our workflows. This is the crux of subscribing to a *PaaS* [29]. In addition to Galaxy, as alluded to earlier, Le Crom et al. [30] had developed **Eoulsan** that was predicated on MapReduce algorithm (Hadoop Framework) to address high-throughput sequencing data analysis schemes, and as it was hosted on AWS featuring elastic compute (EC2) and storage buckets (S3), the scalability of workflows could be ensured. Furthermore, we also have erstwhile instances like OSDDLinux, that has contributed immensely in its capacity of an open-source platform for drug discovery workflows; it is not available on a cloud-platform, albeit surely puts forth a promise of greater receptivity if scaled thereof.
3. **Infrastructure as a Service**: Engendering the greatest of flexibility, as compared to the aforementioned platforms, *IaaS* gives the subscriber the ability to choose the "whole nine yards"- hardware, network, software. At this level of discretion, the infrastructure may be tuned holistically and propounds granularity of orchestration. BioLinux [31] is one such platform for multi-omics data analysis, offering virtual machines to avail.

Back in 2012, Aspera Inc. (now with IBM) and Beijing Genomics Institute (BGI) partnered in a collaborative endeavor to amalgamate the services of *fasp*™ (from the former), and *Easygenomics*™ (from the latter). In the current times, Globus advances an orchestrated, fast-and-reliable data transfer service with active interfacing (Fig. 6).

I would like to reiterate, that given the ambit of data (which is "Big") and warrants a distributed processing framework for efficient discretion, it is only viable to hinge our discussion on Artificial Intelligence and Generative AI for Big Data analytics. Now that we have learned some pithy points (without getting into the gory details), let us delve into the core domain (Fig. 7).

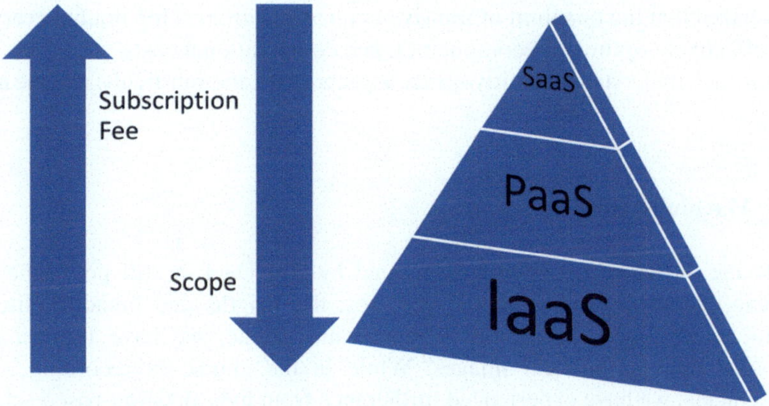

Fig. 6 An illustration summarizing the scope of different cloud-based services. While IaaS has a broader scope and incurs most expense, in contradistinction, SaaS is swift, easy to pursue, and perhaps *less technical* for people seeking a quick deployment

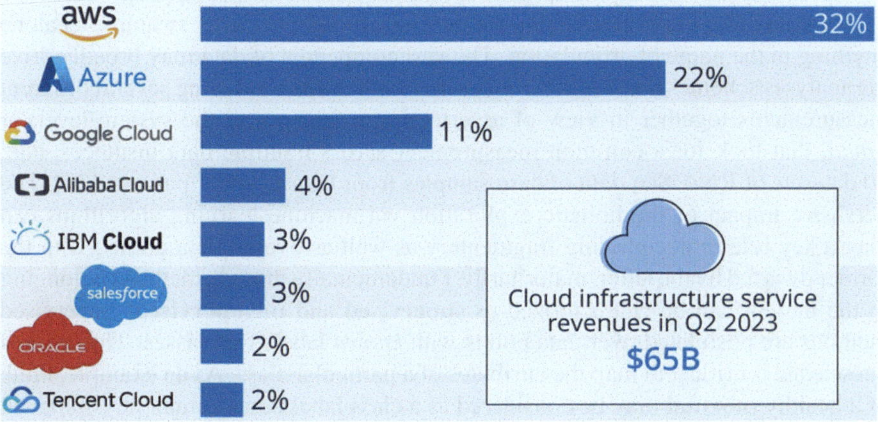

Fig. 7 Inspite of Google and Microsoft exhibiting greater growth potential, Amazon is still able to capture larger market share. *Credit* Statista, Synergy Research Group

1.2 The **Sovereignty** *of Artificial Intelligence*

There exist disparate nuances of intelligence, and for computer systems (machines), they can be as wise as we want them to be. The wisdom that a system exhibits is hypothesized over the *how* and *what* of training- what data and approaches have been adopted for training. Alan Turing in his penning of 1950 alluded to Artificial Intelligence, and explored the idea conceptually [32]. Why is the theme so viable in today's epoch? As aforementioned, the answer is Big Data. Marx [12] argued that the genomic data doubles every year, while Gordon Moore, co-founder on Intel, hypothesized that the quantum of transistors on an electronic chip double every two years, effectively optimizing computation and computational costs.

Let us now understand the nitty–gritty aspects of AI, the sub-fields, and the avant-garde.

1.2.1 Machine Learning

Despite the unrivaled opportunities granted by Big Data, it still poses the most herculean endeavor of mining and analyses; the lengths and breadths, literally, have to be traversed. From the Bioinformatics angle, we have the gamut of sequencing data to medical images. While in the course of recording *omics*-measurements, we have experienced furtherance from hybridization-based schemes to next-generation sequencing (NGS) technologies, even covering more ground now in sequencing approaches. The sequencing data may be engendered by several types of sequencing experiments, including Whole Genome Sequencing (WGS) and Whole Exome Sequencing (WES) for genomics, RNA-Seq for transcriptomics, Methyl-Seq, ChIP-Seq, and ATAC-Seq for epigenomics, Hi-C, and scRNA-Seq [33].

The data deluge is overpowering the analysis, thereby creating swamps, let alone anything in the name of articulation. The vast compendia of data may broadly drive the analyses schemes that could be *integrative*, which are combining several different measurements together in view of ascertaining inferences at the system-level, or *broad*, that look for a common measurement across multiple data instances, e.g., 10 datasets of RNA-Seq data of case samples from breast cancer patients [34]. The pervasive impact of the holistic exploration via machine learning algorithms can play a key role in deciphering fragmentary as well as overall data points, with the former dwarfed by the latter, majoritarily. Fundamentally though, methods belonging to the thereof can be characterized as **supervised** and **unsupervised**. Supervised methods are postulated over data points with *known* labels for classes. This *apriori* knowledge is critical to map the attributes of a particular class. As an example, think of a healthy (**normal** may be considered as a class label here) human physiological state. The variables (dimensions or spaces) attributing to this tag may be seen in the Table 1; some others may include Nutrient Concentration, Acidity (pH), Water Balance, Sodium Levels, Calcium Levels, Oxygen Saturation.

Table 1 Some basic values of health parameters, qualifying as *predictor variables*

Parameter	Value/range
Body temperature	Approximately 37 °C (98.6 °F)
Blood pressure	Systolic BP (90 –120 mmHg), Diastolic BP (60 mmHg and 80 mmHg2)
Heart rate (adults)	Approximately 60–100 beats per minute
Respiratory rate (adults)	12–20 Breaths per minute

In contrast, the unsupervised methods are agnostic (independent) to the class labels, and function on elucidating patterns implicit to the values of the predictor variables. These approaches map organization of the data, that are mainly hinged over derivative dimensions (not apparent in the stated dimensions (with reference to Structured Data) [34, 35]. For example, the **Principal Component Analysis (PCA)** has been substantially applied in a myriad studies to uncover sub-classes, or cohorts that have greater affinity within a cluster, than outside of it; I would refer the reader to a brief tutorial on Clustering approaches in Machine Learning. This is ingrained in the first component that captures the maximum "expanse" of the data, and others in diminishing order. We may also recall **K-means Clustering** that engenders a specified number (k) of clusters that exhibit greater coherence between elements of a particular cluster. Again examples abound in terms of applications, we can relate these approaches in characterizing disease sub-types or biomarker groups that endorse a peculiarity.

1.2.2 Neural Networks and Deep Learning

Neural networks draw inspiration from the nature's impression of interconnected neurons in a brain, and are rudimentarily *discriminative*. These brain cells cross-talk with each other and in the process, generate chemical impulses and electrical, synaptic signals called **action potentials**. In an effort to mimic the same functionality, the artificial neural networks use data and weights channeled through multiple planes of synthetic neuronal layers (reminiscing a complex mesh), subsequently engendering the output. Where Deep Learning takes over, is getting further in complicating the scenario via incorporation of multiple *hidden* layers [36], vide Fig. 8.

There are three major constituents of a neural network- **input layer**, **hidden layers**, and **output layer**. While the first receives raw data (e.g., gene expression levels, protein sequences), the second is tasked to elucidate complex representations embedded in the data, and the third provides the final prediction (e.g., disease classification). For an effective navigation towards a solution, the data and weights are guided with *biases* that characterize errors in prediction and facilitate course correction. Put simply, a bias may be comparable to the *y-intercept (b)* in the following standard linear equation.

$$mx + b = y$$

There are a variety of neural networks, each designed for specific tasks in Computational Biology like prediction of splicing patterns in RNA sequences and their impact on subsequent proteins' functions [37, 38]. Additionally, elucidation of DNA and RNA prospective binding sites to unravel gene regulation mechanisms, parsing

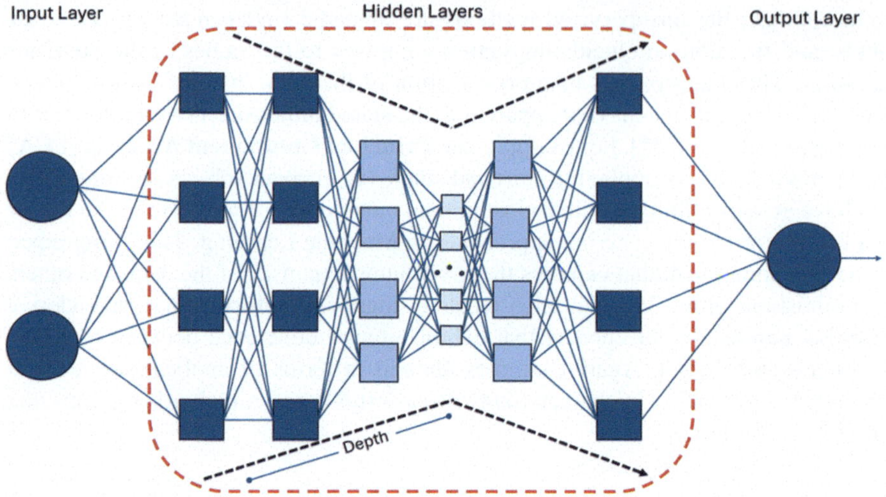

Fig. 8 The depth of a deep learning neural network is determined by the number of hidden layers; denser the hidden layers, more complexity from the inputs may be considered for processing. Note that there could be multiple input and output layers provisioned too

implicit patterns in biomedical images, and prediction of protein structures are salient use-cases. The categories[1] are highlighted below for reference.

1. **Multi-Layer Perceptrons (MLPs)**: An MLP is a class of feedforward artificial neural networks (ANNs). While feedforward networks define the information flow, *backpropagation* enables learning by iteratively updating weights and biases based on the error. It consists of multiple layers of interconnected neurons. The first layer (input layer) receives data, and subsequent layers (hidden layers) process and transform it. The final layer (output layer) produces the network's prediction or classification. Another set of essential components in training and using neural networks are *cost functions* and *activation functions*. The former (mean squared error, cross-entropy loss, etc.) evaluates the overall performance of the network, and the latter (ReLU, Sigmoid, Tanh, etc.) introduces non-linearity and shape the output of individual neurons. These constituents are the building blocks of a robust neural network, and their precise calibrations are warranted. MLPs are commonly used for tasks like image recognition, natural language processing, and regression.
2. **Convolutional Neural Networks (CNNs)**: CNNs are specialized for image processing and computer vision tasks. They use convolutional layers to automatically learn features from input images. In a regular application, convolutional neurons apply filters (kernels) to the input image. Each filter scans across the image, computing a weighted sum of pixel values within its receptive field,

[1] Much of the details have been articulated in the related chapters of this book, and hence the emphasis will be on necessary inklings.

and passes the result to the consecutive (pooling) layers. Convolutional layers learn optimal filters through backpropagation and gradient descent. These layers detect patterns such as edges, textures, and shapes. Pooling layers downsample the feature maps, reducing the spatial dimensions, and the fully connected layers at the extremity make predictions based on the learned features [39]. Understandably, CNNs excel in tasks like image classification, object detection, and facial recognition.

3. **Recurrent Neural Networks (RNNs)**: RNNs are designed for sequential data (e.g., time series, natural language) [40, 41]. They have feedback connections, allowing information to persist across time steps. Each neuron processes an input and its own previous state. Recurrent Neural Networks (RNNs) gain strength through their variants, Long Short-Term Memory (LSTM) and Gated Recurrent Unit (GRU). LSTMs excel at capturing long-range dependencies within sequences, making them particularly adept at tasks like speech recognition, language translation, and even language modeling where understanding context across long stretches of data is crucial. This genre is a star protagonist in LLMs that are the cornerstone of Generative AI, theme that we discuss shortly.
4. **Generative Adversarial Networks (GANs)**: GANs, a prominent framework for generative AI, leverage unsupervised learning to discover patterns and generate new data [42]. They consist of two key components: ***Generator*** and ***Discriminator***. The generator learns to produce novel examples by spawning artificial data that closely resembles the actual data. It attempts to create realistic samples, often starting from random noise. On the other hand, the discriminator's role is to evaluate and distinguish between the genuine and generated instances. It acts as an adversary to the generator, aiming to accurately identify whether an example is real or fake. GANs are like two AI rivals that get better through competition. One, the generator, tries to create new data that is indistinguishable from real data. The other, discriminator, tries to spot the fakes. This back-and-forth training helps the generator learn hidden patterns in the real data, allowing it to create realistic new examples [43]. Analogically, the generator and discriminator networks may reflect the work profile of a software developer and a software tester, respectively. While the former is tasked to structure a piece of code corresponding to a business problem, the latter critiques it iteratively, until we have a potent software ready for deployment. Furthermore, the idea of *zero-sum game*, a notion from game/economic theory also aligns to the same paradigm. Amongst the two stakeholders involved, the gain for one will mean the loss for another. The combined gains of participants are exactly offset by the combined losses, resulting in a net sum of zero. This tenet was propounded by John Von Neumann (we have learnt about his abstraction of computation model, in an erstwhile mention) in 1944, and approximated by John Nash, later in 1950.

1.2.3 Robotics

This field aims to revolutionize healthcare through applications like telesurgery [44, 45], for example, and hence improve patient outcomes. Infact, a major player in the sequencing effort of the Human Genome Project at the Whitehead Institute [46], was a robot called **Sequatron**. This robot was the first, fully-integrated robotic system for DNA sequencing, capable of carrying out DNA purification and setup for feeding gels to the sequencer, without human supervision. In modern sequencers too, miniaturized robotic procedures are tasked with intricate sequencing and automated high-throughput nucleotide screening exercises.

Conventionally, robotics has been deployed in manufacturing, where product assembly lines were driven by programmable, electro-mechanical arms. These certainly provided amelioration in efficiency of the output as compared to human intervention, due to repetitive nature of the job. In a clinical setting, the robot systems embrace precise (optimally invasive procedures) that has broadly transformed surgical practices. Surgeons can now perform intricate procedures with greater conviction and minimal invasiveness. The da Vinci Surgical System [47] is a prominent instance that aids in delicate surgeries such as prostatectomies and cardiac procedures. These robots provide 3D visualization, tremor reduction, and fine motor control, allowing surgeons to achieve remarkable outcomes. Nonetheless, it is noteworthy that the infrastructure supporting the robot ecosystem warrants a robust networking and communication that can withstand serendipitous faults.

Beyond surgery, sensor-based robotic systems cater to personalized treatment plans on rehabilitation and physical therapeutics. The AI algorithms analyze patient movements, adapt exercises, and provide real-time feedback. Robotic exoskeletons and prosthetics restore mobility, enabling patients to regain independence, especially debilitated, aged population.

The enhanced accessibility via telemedicine and remote monitoring facilitates remote consultations [48]. The geographic boundaries have literally shrunk as a team of doctors can be stationed at a distance from one another, yet collaborate through robotic telepresence. Towards stretching the applications further, we have explored *biomimetics*, emulation of natural processes and living creatures [49] (Fig. 9).

1.2.4 Computer Vision and Image Processing

Disparate data modalities like microscopy and histology images, present a compelling scenario of interpreting patterns or nuances that epitomize certain physiological state [50]. To illustrate, ultrasound scans can illuminate abnormalities in tissue structure or relative obstructions that guide a medic on the way forward. More contemporaneously, there are software interfaces now developed that may *see through* these images, and multiples of them concurrently, to draw out pathologically meaningful interpretations. These outcomes can then be confirmed with the alignment to knowledge databanks or human clinicians, thereafter. This automation

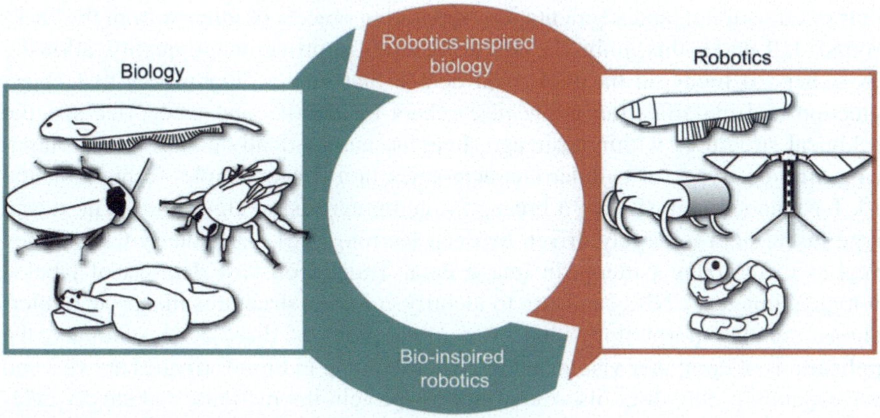

Fig. 9 In line with our deepened interest in mundane processes and life structures, we are building the *humanistic* versions; adapted from Gravish and Lauder [49]

is a great enabler, and a *right hand* assistance to a human doctor bringing in upheaval to their working efficiency.

Bioimage informatics emerges as a specialized field within bioinformatics and computational biology. It focuses on developing and optimizing image data processing and informatics pipelines for large-scale, high-throughput handling, acquisition, scanning, and comparison of biological information from diverse images, particularly those at the cellular and molecular level. The core objective lies in extracting meaningful knowledge from intricate and heterogeneous image data, along with associated metadata. Techniques like image segmentation, feature extraction, and pattern recognition empower researchers to analyze these complex datasets, ultimately aiding diagnoses of pathologies like tumors, fractures, and neurological disorders [51]. Despite its promise, bioimage informatics faces challenges due to the complexity of bio-images. Factors like high cell density, intricate microtubular growth, and multiple color channels pose difficulties for traditional medical imaging techniques. Novel methods are being developed to tackle these issues, which are primarily of the AI descent, including albeit not limited to, accurate segmentation and visualization techniques for complex structures. From the organizational perspective too, efficiently managing large image datasets and annotating relevant information is critical, in addition to the availability of tools, databases, and milieu dependencies for bioimage analysis.

As aforementioned, with the advent of powerful microscopes and other imaging techniques, the amount of visual data—microscopic images of cells, tissues, and organisms—has exploded. This data deluge necessitates sophisticated computational tools to analyze and interpret these images. *Image processing* forms the bedrock upon which computer vision builds. It encompasses techniques for manipulating and enhancing images to prepare them for further analysis. Tasks like noise reduction, contrast adjustment, and segmentation (separating objects of interest from the background) fall under this umbrella. These methods improve image quality, allowing researchers to focus on the biological details that matter. Techniques like object detection and classification allow researchers to identify and categorize specific biological structures within an image. Imagine automatically counting and classifying different types of cellular characteristics in a tissue sample—a monumental task for manual analysis, but a breeze for computer vision algorithms. The recent surge in the field is largely driven by deep learning [52], particularly CNNs; they excel at recognizing patterns in image data. Trained on vast datasets of labeled biological images, CNNs can learn to identify complex structures like protein interactions, cellular morphology, and even subtle signs of disease. In summary, the applications of computer vision and image processing in bioinformatics are vast and ever-expanding, viz. drug discovery, protein–protein interactions, and cancer diagnosis [53], inter alia. However, it stands crucial to remember that these techniques are tools, not replacements for human expertise.

Staying in the realm of image processing and preambling on Generative AI, neural networks could be trained in a supervised manner to resolve classification problems like image recognition. Interestingly though, when used in unsupervised flavor, the neural networks can be tamed to render superfluous images and texts [54]. The recent furore on images that are "DeepFakes" [55], and the ethical considerations surrounding them, is an embodiment of what ***Generative Neural Networks (GNNs)*** can be made to accomplish. The efficacy of the model is contingent to the amount of raw data supplied which is indicative of absolutely no annotation/curation (yet has to be contextual), hence making it relatively easy to collect data and train such networks. This is largely how an infant learns- *by observing*. Technically penning, the problem is of an optimization of weights, so chosen, that the "fake" images are closest to the set of input images.

1.2.5 Natural Language Processing (NLP)

At this stage, after explicating on a plethora of AI approaches and applications, we are closest to "Generative" AI. Before though, let us gain the ultimate ground on the formidable requisite of Large Language Models (LLMs) and their interplay under the aegis of NLP.

NLP is premised on the intersections of **Linguistics** and **Computer Science**. In large **text** repertoires, technologies that corroborate the working of search engines and databases, via indexing and text processing to name a few, are housed in a typical NLP retinue [56]. From the elementary applications of language conversion, a quantum leap was observed about a decade ago, with the introduction of ***Word2Vec*** [57, 58]. It uses a shallow neural network to learn word embeddings based on (input-word, context-word) pairs. Unlike simple n-grams (that we later allude to), the skip-gram model considers the context with a "skipped" word, making it more intricate (Fig. 10).

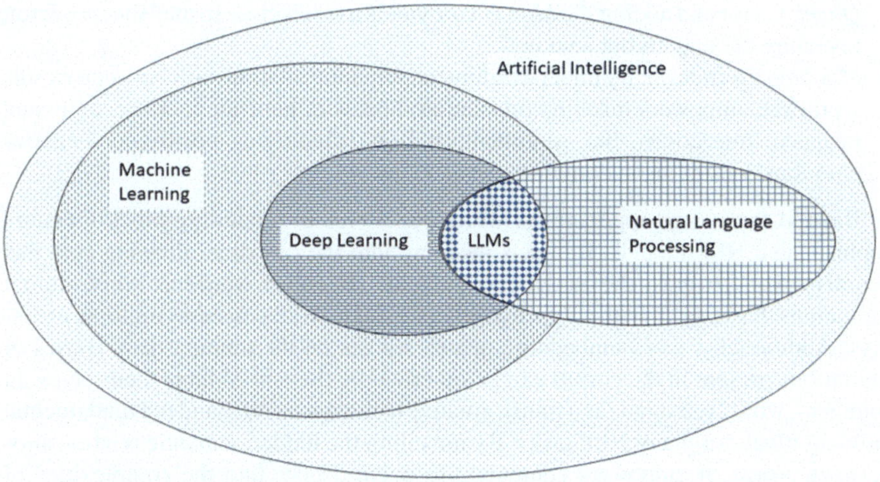

Fig. 10 A Venn Diagram illustrating the intermingling of relevant AI themes

In the realm of Natural Language Processing (NLP), where computers grapple with understanding human language, word2vec stands as a foundational technique and a key protagonist. It delves into the essence of words, transforming them from mere symbols into numerical representations that capture their meaning and relationships; indeed it has a profound impact on various NLP tasks, for example studying protein sequences that are literally strings of amino-acid letters [59]. At the heart of *word2vec* lies the notion of **word embeddings**. These are dense vectors, arrays of numbers, that encapsulate the semantic and syntactic properties of a word. Words with similar meanings or that frequently co-occur tend to have embeddings that reside close together in the vector space. This underlying principle allows the method to unveil the richness hidden within a particular language [60]. The magic behind word2vec unfolds through two primary model architectures: **Continuous Bag-of-Words (CBOW)** and **Skip-gram**. CBOW, as the name suggests, treats a word as the center of attention, flanked by its backdrop. The model is tasked with predicting the surrounding words based on the central word's embedding. Conversely, a Skip-gram takes a single word as input and endeavors to predict other words that appear within a specific window around it. Through this back-and-forth training process, the model refines the word embeddings subsequently, ensuring that words with similar contexts or meanings end up closer in the vector space [61, 62]. The forte of word2vec lies in its ability to unlock a multitude of NLP applications. It paves the way for tasks like:

- *Similarity detection*: Word embeddings enable the identification of words with analogous meanings. Imagine a system searching for synonyms—words close together in the vector space would be prime candidates.
- *Analogy resolution*: Word2vec can crack analogies like "King is to Queen as Man is to…?" by performing vector arithmetic. Subtracting the *King* vector from the

Queen vector and adding the *Man* vector yields a result close to the *Woman* vector, revealing the underlying analogy.
- *Machine translation*: By understanding the semantic relationships between words, word2vec empowers machine translation models to generate more accurate and nuanced translations, like sentiment analysis (identifying positive or negative emotions) or finding synonyms.

Beyond these core applications, word2vec serves as a building block for more sophisticated NLP models. It injects meaning into the raw data, allowing complex neural networks to grasp the subtleties of language. However, word2vec is not without limitations. It struggles with polysemy, where a single word possesses multiple meanings. Additionally, word embedding quality hinges on the training data. Biases or limitations present in the corpus can be inherited by the embeddings themselves. In summary, word2vec stands as a transformative technique in NLP and is a fundamental building block for many NLP tasks. By unveiling the hidden semantic relationships between words, it empowers computers to delve deeper into the complexities of human language as NLP struggles to deal with raw, isolate text. As research in NLP continues to evolve, word2vec will undoubtedly remain a cornerstone, paving the way for advancements in tasks ranging from machine translation to sentiment analysis. The ability to unlock the meaning behind words is a key to understanding human communication, and word2vec provides a powerful tool for achieving just that. Word2vec embeddings are used as input for many advanced NLP techniques like machine translation and question answering, as evidenced in a typical *Chatbot* interface.

While there is a certain eccentricity in the "*quaternary*" *language of the genome* and "*tertiary*" *language of the proteome*, let us familiarize ourselves with the fundamental concepts of language, in general. There are some distinguished levels corresponding to language analysis [63, 64]. These levels interact to create a rich and nuanced communication system, notwithstanding complicated. Linguists study them to unravel the intricacies of languages.

- **Morphological Level**: At this level, we examine the structure of words. It deals with ***morphemes***, which are the smallest meaningful units in a language. The analysis involves examining how words are formed, identification of prefixes, suffixes, and root words. For example, in English, the word "unhappiness" consists of the prefix "un-" (negation), the root "happy," and the suffix "-ness" (forming a noun).
- **Syntactic Level**: Syntax focuses on sentence structure and the rules governing how words combine to form meaningful sentences. It includes understanding word order, sentence constituents (such as subject, verb, and object), and grammatical rules. Again, as an example in English, the sentence "She loves pizza" follows a subject-verb-object (SVO) structure.
- **Semantic Level**: Semantics deals with *meaning*. It explores how words and sentences convey information. It includes understanding word meanings, word senses, and how context influences the interpretation. The word "bank", for

instance, can refer to a financial institution or the side of a river, depending on the context; it has two meanings.
- **Pragmatic Level**: Perhaps the most salient one, pragmatics focuses on the use of a language contingent to a context. It considers how speakers convey meaning beyond literal interpretations. It involves understanding implicatures, speech acts, politeness, and conversational norms. As an illustration, saying "Could you pass the salt?" is not just a request for salt; it's also a polite way of asking.

On the associated terminologies, and in distinction to *skip-gram* architecture introduced by Word2vec, an *n-gram* is a contiguous sequence of *n tokens* (usually words) from a given text or sentence. For instance, in the sentence: "The quick brown fox jumps", a bigram (2-g) would be "quick brown" and the trigram (3-g) would be "brown fox jumps". **Tokenization** implicates breaking down of text into smaller units (words, sentences), for characterizing them exclusively and curating their individual and systemic profiles necessary for information retrieval. Tagging *parts-of-speech* facilitates assignment of a grammatical category to each word (noun, pronoun, verb, adverb, adjective, etc.). Identifying and classifying "named" entities in the input text like people, organizations, locations, amongst others, is referred under Named Entity Recognition (NER). The input text may not always be veracious, and often times when we plug input (inaccurate) expressions in modern-day web-browsers or text-editors, instantaneously the errors are flagged. The resolution to such inconsistencies and correction of spelling errors is termed as **text normalization**. Furthermore, **Stemming** and **Lemmatization** aim to reduce **inflected** words (variations of a word) to their base or root form. While stemming is a good choice for large datasets where speed is precious and some loss of accuracy is acceptable, lemmatization is preferable when dealing with tasks such as sentiment analysis or topic modeling, where context-preserved meaning is utmost desirable (Table 2).

For NLP, both supervised and unsupervised methods of machine learning are applicable (vide Sect. 1.2.1), over and beyond the sole platform that neural networks provide. Naive Bayes (NB) and Support Vector Machines (SVMs) are two popular machine learning algorithms widely used in NLP tasks, particularly in text classification (document characterization). NB is a probabilistic classifier based on Bayes' theorem. It calculates the likelihood of a text belonging to a specific class

Table 2 Stemming and Lemmatization are distinct approaches that feature vividly in realigning text, and work towards a common goal embracing stark differences

Feature	Stemming	Lemmatization
Approach	Trims out prefixes and suffixes; context is disregarded	Uses morphological analysis
Output	A "stem" may not always be an actual word	Always a real word (lemma)
Accuracy	Less accurate	More accurate
Speed	Faster	Relatively slower
Use cases	Large datasets; prioritizing low latency	Preservation of meaning is crucial

based on the presence of individual words, having their respective probabilities within that class. It efficiently handles high-dimensional text data and can work equally well with large datasets, even with limited training data; boasting minimal feature engineering (parameter tuning) and training time, and implementation. A shortcoming may be attributed to the assumption of the fundamental independence of words (Bernoulli NB)/word-counts (multinomial NB)/Term Frequency—Inverse Document Frequency (TF-IDF) (features), which is incongruous, in addition to susceptibility to class imbalance issues.

SVMs find an optimal hyperplane (decision boundary) in the feature space (text categories) that cleanly separates the data points belonging to different classes, ensuring clear distinction even for unseen data. We can visualize documents like tiny dots in a vast landscape. Each dot represents a document, and its position is determined by the words it contains and how often they appear. To categorize these documents, we train an SVM model using labeled examples. These labeled examples are like signposts in the landscape, telling the model which category each document belongs to. Suggestively, it is less interpretable than NB, making it difficult to understand why the model makes certain predictions. Careful thoughts must be spared on data size and complexity, data availability, and task objectives to ensue selecting either of these algorithms. A recent study highlights exploration of the Electronic Health Records (EHRs) data via aforementioned approaches accounting the social needs dimension of the subjects [65].

Applications of NLP range far and wide, from machine translation of language, text summarization, sentiment analysis, to chatbots, automated question answering, and information extraction. Albeit, NLP systems can struggle with understanding complex or ambiguous language, leading to inaccurate or misleading results. This can be a problem in critical applications like machine translation or legal document analysis. NLP systems can inherit biases from the data they are trained on. This could disadvantage certain groups in crucial areas like loan approvals or job applications, potentially leading to discriminatory outcomes. Also, developing and maintaining NLP systems requires expertise in various fields like computer science, linguistics, and artificial intelligence. This complexity can limit accessibility and slow down development [66].

NLP systems need to improve their understanding of context, sarcasm, slang, and other subtleties of human language. Currently, most NLP research focuses on well-resourced languages like English. More work is needed to develop NLP tools for languages with less data available. Combining NLP with other AI advancements like robotics and computer vision could lead to even more capable intelligent systems. There's a growing need for NLP models to be more transparent in their decision-making process, especially in critical applications.

Our discussion on NLP may not materialize sans the following indelible appendages. These algorithms are key drivers of the mechanistic principles in language processing that partakes activities like forecasting the subsequent words ("autofill" mechanism); basically an interplay of encoder-decoder pairing. Without further ado, let us dive in.

1.2.6 Transformer Models

In 2017, Polosukhin et al. [67] from Google Research had an epiphany and proposed an attention (or self-attention which stands as a peculiarity) mechanism based "transformer"; in fact, transformer models power the venerable Bidirectional Encoder Representations from Transformers (BERT) and Generative Pretrained Transformer (GPT) platforms that have gained tremendous traction in recent times, that Liang et al. refer to as **foundation models** [68]. In lay terms, these pre-trained models would bear a compilation of vocabulary, tokens with a standardized vector size (congruent to the number of dimensions). In line with language translation and in juxtaposition to RNN, transformers render greater efficiency with implicit parallelism in processing, especially with larger text inputs. This suitability also extends to training time, which is relatively less than the serialism in RNNs. Unlike a "true" transformer model that preserves the meaning of successive words, an RNN "sequentially, yet seriatim" processes individual words in the source text and furnishes the output accordingly. Understandably, that is not how a language translation must undergo. Tranformers are a leg up in this regard and predicated over ***sequence-to-sequence learning***. As an illustration, consider the following translation. An old saw in Deutsch goes like this: "*wenn schon, denn schon*", which translates in English to "*if a thing is worth doing, it is worth doing well*". This would not necessarily be an output from an RNN, say, which glosses over the context and focuses exclusively on individual words. So, "*wenn*" translates to "*if*", "*schon*" to "*already*", "*denn*" to "*because*", and again "*schon*" to "*already*"; effectively, "*if already then already*", which is a sheer offset. Another example, which is possibly more aligned to our segmented discussion and is a sequential input like a language, could be that of genomic sequence being "translated" to a protein sequence. We now understand that the ordering of base-pairs implicates the construction of specific kinds of proteins; one may refer to the **codon tables**. Interestingly, a single shift in the reading-frame is likely to convert to an unexpected outcome [69–72].

Before proceeding further, let us go to school on *vector algebra*; there is **matrix algebra** too which is undeniably crucial, albeit we will focus on vectors for now. A vector is an entity that has a ***magnitude (length)*** and a *textbfdirection*; in transformers, word embedding vectors hold meaning through both their magnitude and direction. The length of a vector is directly proportional to salience of a word in a context, while the direction aids in ascertaining *proximity* (literally, semantic analogies) of a word with other words; words with similar meanings will have vectors pointing in roughly the same direction, even if their magnitudes differ. Essentially, a projection along the same plane or with some minor directional shifts could possibly underline the cohorts (word pools/clusters). Henceforth, the word embedding vectors to understand the nuances of language and perform various tasks like machine translation, sentiment analysis, and question answering. However, we must be wary of the fact that ***correlation does not necessarily imply causation, yet it is the other way round***. During nascency, neural networks were primarily tasked with (not that they've abandoned it now) image classification before being tuned for language processing. As aforementioned, the *frame-of-reference* is critical for translation. This is also true for

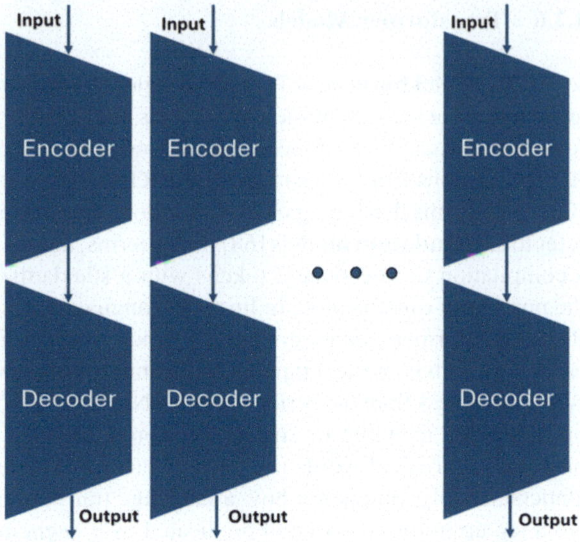

Fig. 11 A bird's eye view of the parallelism, hence efficiency, in transformers. A worm's eye view, however, would allow you to visualize stacks/layers of encoders and decoders within the parent blocks, respectively

sentences in a paragraph, as with the words in a sentence. When descending on the broader meaning, context plays a key role, and that is exactly where the RNN and CNN concede. Furthermore, transformers are ***pre-trained*** models that undergo ***semi-supervised*** training. The *telltale* is that these models are trained on a large corpus of *unlabeled data* (i.e. unsupervised learning), and then fine-tuned on *labeled data* (i.e. supervised learning). For a deeper grasp, let us pay heed to the below cataloged overtures.

Encoder: It works on the **input vectorization**, which is essentially calibration of a *feature tensor*-[2] a numerical representation equivalent to the input word, inclusive of the *surroundings*. It dissects the input, which could be text or some other bioinformatics sequence data, and creates a representation of it that the model can understand. The encoder consists of multiple layers, often stacked on top of each other, with self-attention. Each layer processes the input further attending to different parts (tokens) of the input sequence simultaneously, enriching the understanding. After processing through the layers, the encoder generates a final output which is a refined representation of the entire input sequence, eventually fed into the decoder part of the transformer model (vide Fig. 11).

– ***Input Embedding***: The embedding of input sequences to a vector form; as alluded to earlier. In fact, the Word2Vec (as discussed above) algorithm is one such two-layer neural network setup that outputs such vectors. An instance of such a matrix of sparse vectors for discrete values could be considered via **one-hot encoding**, a simple method that primarily underlines the occurrence (1) or absence (0) of a

[2] It is worth noting the elementary distinction between a vector and a tensor, which is that the former is reflecting data for a direction only, that is a dimension/an axis, while the latter can visit a multidimensional space. Vectors are a special case of *rank*-1 tensors.

word in a binary manner, in the corpus. Example,

	BigData	*GenerativeAI*	*Bioinformatics*
$BigData(token\ id : 1)$	1	0	0
$GenerativeAI(token\ id : 2)$	0	1	0
$Bioinformatics(token\ id : 3)$	0	0	1

Although we do get a vague idea of the availability status of tokens, we will appreciate that the approach is computationally inefficient in a high-dimensionality space, in addition to the delineating the meanings of words' relevance. For continuous values as sequences, we need something better that conveys the relative or absolute position of elements; positional encoding, precisely.

– ***Positional Encoding***: This nicety is sourced from RNN/CNN for safeguarding the positioning of tokens (word/DNA sub-sequence) in a corpus. The relative and absolute mapping of placeholders, again, provides consecution of occurrences which is significant to keep derandomization/shuffling from the final output. This adaptation is imperative and, in a loose way, addresses a potential weakness of transformers of disregarding the order of sub-expressions. Examples like sole- or nested-searching for a word (or a sentence) in biomedical literature, besides parsing a genomic or proteomic sequence for individual nucleotides, the positional embeddings have a pertinent mechanistic role to play. Technically, these are vectors added to the word embeddings (which capture the meaning of a word) in the transformer model, and fed as a consolidated input. Not only is the order of tokens captured through a common approach like **sinusoidal positional encoding**, but the relationships between elements in a sequence too. Why is this important? Because models like transformers do not inherently process sequences in a specific order; yet again, I would refer the inquisitive reader to lucid articulations at Hugging Face and Jay Alammar's Blog on the subject. Again, the idea that windows of sub-sequences are fed as training examples to the model, alongside the correct output labels, we may loosely associate this mechanism to the k-fold cross validation technique in machine learning.

Decoder: The decoder sits at the trailing end of the pipeline and receives two main inputs to process: the *output from corresponding encoder* and *intermediary outputs for maintaining context and coherence when predicting the next element in the output sequence*. While generating the output, one word at a time, the decoder employs *masked self-attention* to consider both these inputs. It then uses feed-forward (Multi-level Perceptron (MLP)) neural network layers to process this information and predict the next most likely element in the output sequence. Pointedly, the quantum of error is calculated for the estimated label and the correct label, and fed back to the model to reorganize its weighted coefficients. After significant iterations of this exercise, the results subsequently tend closer to the actuality. This predicted element is added to the generated sequence, and the process repeats until the desired output length is reached [73]. The "mask" ensures the decoder only attends to the previously

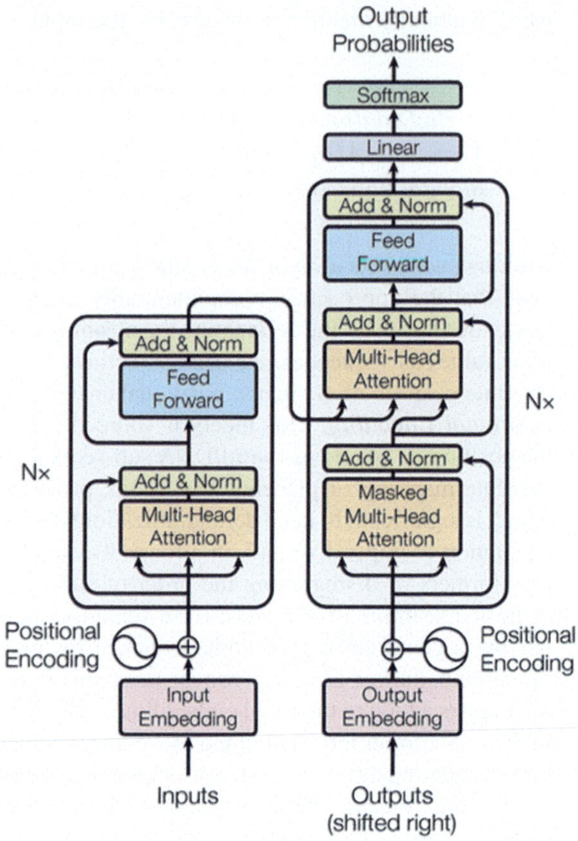

Fig. 12 The central illustration of the transformer-model, as adapted from the original article [67]

generated outputs and refrains from foreseeing advanced words while predicting the current word, thereby preventing information leakage. The scores[3] calculated for all output tokens, post processing, are transformed to probabilities via a *softmax* function (See Fig. 12). Certainly, we can orchestrate a sampling strategy to break the tie, should we have several tokens with equal scores.

P.S. *There is a conceptual similarity between encoders and decoders in electronics and those in transformer models, but they work with different types of information.*

Attention: If I warrant the model to focus on the word "***pen***" and not "***cyan***", when emphasizing on the pronoun "***It***" in the sentence- "***I have a cyan pen. It feels great!***", I would need it to pay "attention" to the context. Extrapolating on the dogma of "attention", and orienting to the following characterizations, we need to get our heads around the ***query***, ***key***, and ***value*** vectors that empower the model to selectively focus on relevant sections of the input sequence during predictions; these vectors are calculated for each embedment vector. The attention layer takes input from the

[3] Discussed later.

encoder and outputs to the feed-forward neural network; this too happens in parallel for each input word. The attention (or **self-attention**, precisely) assists the model to relatively encode a word (token) staying cognizant to other placeholders/positions in the input sequence.

The model projects the input sequence into separate query (Q), key (K), and value (V) vectors using linear transformations, and calculates a score for each element in the sequence by multiplying the query vector (Q) with each key vector (K). This score reflects how relevant each element is to the current focus (Q). By calibrating the attention mechanism using these vectors, transformers can become more adept at focusing on the crucial parts of the input sequence, leading to better performance. As a general principle, the transformer employs a **multi-head attention** (multiple "heads" that run in parallel), where each head learns to attend to different aspects of the relationship between elements in the sequence. Another component called the **projection matrix** perform a linear transformation on the original Q, K, and V vectors, projecting them into a lower dimensional space and reducing the computational cost compared to using the full dimensionality. Finally, the outputs from each head are concatenated to form the final output for the attention layer.

1.2.7 Variational Autoencoders (VAEs)

Autoencoders are **unsupervised neural networks** that record the features or patterns in the input (medical images, for example), and facilitates the ability of improvising on this *mapping*: reproduce it in a higher resolution or a color scheme, crop the main artifact from the background, etc. Now, we submit these as autoencoders, but it is not to say that there are no decoders in the ecosystem. Actually, decoders are taking care of the "generative" aspect when they build upon an entity called **latent vector**. A latent vector is an optimized (from a dimensional perspective) feature vector that harnesses the core of reproducing output [74]. A sub-type of VAEs, called the **disentangled VAEs** attempt to learn a latent space where each dimension (or a small group of dimensions) controls a separate, independent factor of the data. This disentanglement makes the latent space much easier to understand and use.

Rudimentarily, encoders act on the principle of *dimensionality reduction*; they either select cardinal, existing features that best describe the data, or create novel (confounding) features with the same objective. This diminished space is analogously referred to as encoder/latent space, and it does not solely execute the *data compression* objective. The salience of encoding is to elicit key features that are best representatives of the input, sans the background noise. *Sounds familiar*? That is the crux of PCA (1.2.1), too- detemining vectors that best characterize a multidimensional space. Reminds you of something AutoEncoders do not really generative new data, but fall in line of the goal to reduce the reconstruction loss (error between encoder and decoder outputs) via back-propagating the errors in a feedback routine to the encoder to reorganize/tune the latent vector. Where variational autoencoders fill in is to further this tweaking activity by restricting the sampling of data distribution from valid sub-spaces in a multidimensional *universe*, thus improving the specificity of the latent vector and hence closing the gap between original and generated data. In straight terms, the process of training the autoencoder neural network has to be actioned in a manner that curbs overfitting, which is refraining the encoder to take meaningful values from the latent workspace. A close counterpart to a variational autoencoder is GAN. While in GAN, the iterations of output refinements continue until the generator and discriminator find *peace* with each other, there is an explicit preemption warranted in the VAE, however; arguably, the former render enhanced output in comparison. A recent study examined the prudence of six VAE variants to determine the cluster agreement with cancer sub-typing and survival analysis [75]. Another study probed the dexterity of protein–protein interactions are evidenced by VAEs [76]. An undeniable facet of NLP are **Large Language Models (LLMs)** that are built using transformer. A *random walk* over a genomic or protein sequence may unravel some degree of certitude on the hidden liguistic patterns. This is one use-case of LLMs, albeit, it must be noted that not all transformer models are LLMs. Some transformers might be used for tasks like image recognition. LLMs are a subcategory of **Foundation Models**.

Fun Fact: ... Only if Clyde Tombaugh knew about variational autoencoders, while parsing umpteen pictures from the photographic plates, of the night sky (Kuiper belt) to eventually discover the planet (not anymore), Pluto!

1.2.8 Autoregressive Models

Autoregressive models have shone bright in the proteomics domain with wide array of applications [77–79]. Under the hood, an autoregressive model complies with the *stationarity* paradigm of temporal data analysis, especially forecasting. The expression levels of genes and proteins can be temporally modeled given their occurrences (samples) in fixed time- intervals, hence determining their behavioral patterns in specific physiological states from a prospective standpoint. So, how autoregressive models work in predicting the next element in a sequence? The presumption that a set of features will remain the same given the independence of time progression, makes up for a telling proposition that there is some *dependence in freedom*. Irrespective of the snapshots in time, value(s) (lag(s)) can be used to make predictions that bear **constant** statistical properties like *mean* and *variance*; autoregressive models can be perceived as a special case of Markov models where the state space consists of the lagged observations of the time series. The predictions made by the model and the actual observations differ by a random error called, *white noise*. The more number of lags that we incorporate, the "longer" our model is rendered- *long memory models*. All values contribute their bit to the model, with the *farthest making the least*, still doing it. We may consider having permutations (1, 2, 3, ...) of these lags that correspond to the predicted value. Let us spare a thought on the measures of mean and variance. The mean represents the average of all the data points in a given chunk of time. With a smaller number of observations, the mean can be more sensitive to extreme values with a single outlier significantly skewing the average. With escalation in the number of observations, the mean becomes a more stable and reliable estimate of the true population mean. Variance reflects how spread out the data is from the mean.

With fewer observations, the variance might not accurately represent the overall spread of the population. Generally, adding more observations tends to increase the variability captured by the variance. This is because more data points are included, potentially showing a wider range of values.

2 Biology: *Digital Footprints*

The biological data available in the digital format has sprung to explore new frontiers in the domain of advanced clinical medicine. More contemporaneously, the computational biology is *rubbing shoulders* with the experimental version, engendering appreciable results. The diagnostics and prognostics spaces have narrowed paving way for precise therapeutic interventions. This brings us closer to cure than treatment.

The amelioration in laboratory instrumentation have granted access to digital data, which may be harnessed to provide meaningful insights. This information, in tandem with the laboratory techniques, helps elucidate the crucial underlying patterns. As an example, one can think of application of machine learning algorithms to resolve functional and structural implications of proteins. Evolutionary Scale Modeling (ESM) and AlphaFold algorithms have a tremendous reputation in the field of protein structure prediction [80]. Their successive iterations over the years have been progressing in efficiency and scope.

Hitherto, we have learnt that Bioinformatics, the "marriage" of Biology and Computer Science, has become an indispensable tool for unlocking the "secrets" of life. By applying computational methods to analyze vast amounts of biological data, bioinformatics empowers researchers to understand fundamental biological processes, diagnose diseases, and even design new drugs. The study of genes and genomes is probably the most sought scrutiny available in Bioinformatics. The tools for genomics allow researchers to sequence, assemble, and analyze entire genomes, revealing the genetic makeup of variegated organisms. This information is crucial for understanding the genetic basis of diseases, such as cancer, where mutations in specific genes can lead to uncontrolled cell growth. Researchers may also pursue *comparative genomics* to compare the genomes of different species to identify evolutionary relationships and conserved functional elements. Contrastingly, for evaluation of proteins, bioinformatics tools can predict protein function, interactions with other molecules, and potential roles in disease development, by analyzing protein sequences and putative structures. The knowledge offered by proteomic studies is instrumental in the development of targeted therapies that can modulate protein activity to treat various diseases. The field of drug discovery has also been revolutionized by advancements in computational biology. By analyzing large datasets of biological information, researchers can identify potential drug targets, simulate drug interactions, design new drugs with greater efficacy and fewer side effects, as well as expound possibilities in drug repurposing. This approach streamlines (and enriches) the drug development process, leading to the creation of life-saving medications at an accelerated pace.

The comprehensiveness of bioinformatics data is truly staggering. With advancements in high-throughput technologies, researchers are generating an ever-increasing amount of data, encompassing DNA sequences, protein structures, gene expression profiles, and more. This data deluge necessitates sophisticated bioinformatics tools for data storage, retrieval, and analysis. Public databases like GenBank and the Protein Data Bank serve as valuable repositories, providing researchers with open access to a wealth of bioinformatics data. However, the sheer volume of data also presents challenges. Data integration and standardization are crucial aspects of bioinformatics research. Different research groups may generate data using diverse methodologies, leading to inconsistencies and hindering effective data analysis. To efficiently store, share, and analyze this data, specific file formats are crucial. Here's a breakdown of some commonly encountered file formats in bioinformatics [81]:

Sequence Data Formats:

- ***FASTA (FASTA text format)***: FASTA, standing for Fast Alignment Sequences, is a text-based file format widely used in bioinformatics for storing and exchanging nucleotide and protein sequences. It offers a simple yet effective way to organize and share this essential biological data. Each sequence within a FASTA file occupies a distinct entry. This entry starts with a header line, which acts as a label. The label typically functions as a descriptive identifier and often incorporates valuable information such as the organism the sequence originates from and the gene it corresponds to. Following the header line, the actual sequence data itself is presented, typically spanning multiple lines. These lines contain the raw sequence information, represented by a series of letters that correspond to the specific nucleotides (A, C, G, T) for DNA or amino acids for proteins. FASTA is simple and versatile, making it suitable for various sequence analysis tasks.
- ***FASTQ***: This format is an extension of FASTA specifically designed for next-generation sequencing (NGS) data. In addition to the sequence information present in FASTA, FASTQ files also include quality scores for each base in the sequence. These scores indicate the probability that a particular base call is accurate. FASTQ format is essential for analyzing the raw data generated by NGS platforms.

Alignment Data Formats:

- ***SAM (Sequence Alignment/Map)***: This text-based format stores sequence alignments, which show how two or more sequences are related to each other. A SAM file contains multiple columns with information like sequence identifier, alignment position, and mapping quality. While human-readable, the large size of SAM files often necessitates conversion to a compressed format.
- ***BAM (Binary Alignment/Map)***: This is the binary counterpart of SAM, offering significant storage savings by compressing the data. BAM files require specialized software for viewing and analysis but are the preferred format for storing large alignment datasets.

Variant Data Formats:

- ***VCF (Variant Call Format)***: This format is widely used to represent genetic variations like single nucleotide polymorphisms (SNPs) and insertions/deletions (indels). A VCF file stores information about the variant location, reference allele, alternative alleles, and genotype calls for each sample in a study. VCF is a crucial format for analyzing genetic variation data from whole-genome sequencing studies.

Genome Annotation Formats:

- ***GFF (General Feature Format) and GTF (Gene Transfer Format)***: These text-based formats are used to annotate genomic features like genes, exons, introns, and coding regions. They provide information about the location, type, and strand of these features on the genome sequence. GFF and GTF are essential for analyzing gene structure and function.

Other Formats:

- ***BED (Browser Extensible Data)***: This simple format is used to define genomic regions of interest, such as genes or regulatory elements. BED files are often used for visualizing genomic data in genome browsers.
- ***PDB (Protein Data Bank)***: This format stores the 3D structures of biological macromolecules like proteins and nucleic acids. PDB files contain information about the atomic coordinates, bonds, and other structural features of these molecules.

The aforementioned represent just a "tip of the iceberg" when it comes to bioinformatics file formats. The ever-evolving field constantly develops novel formats to accommodate the growing complexity and diversity of biological data. The choice of format remains a crucial step, as it ensures compatibility with analysis tools and accurately reflects the specific data being studied. As the field evolves, new formats are continually being developed to handle increasingly complex data types.

3 Conclusion and Future Directions

Bioinformatics has become an essential pillar of biological research, offering a powerful lens to examine the intricate workings of life. From deciphering genetic diseases to designing personalized therapies, the subject-oriented applications are vast and ever-expanding. As the comprehensiveness of Bioinformatics data continues to grow, researchers will be equipped with even more powerful tools to unlock the mysteries of biology and improve human health. Researchers are using deep learning, specifically transformers inspired by text generation models, to design new molecules for drug discovery [73, 82]. This technique allows them to generate novel drug-like molecules with desired properties by feeding the model with information about existing molecules and the specific features they want in the new ones. The model can even target specific molecular structures (scaffolds) during generation. The future of the domain is undeniably bright, holding the potential for groundbreaking discoveries that will shape the course of medicine and our understanding of life itself.

Generative AI has unboxed a revolution than an evolution in the lineage of Machine Learning, in all perceivable disciplines of research and development. The production of relevant synthetic data fosters increased creativity and innovation, and provides the ability to ascertain nature and application of novel algorithms. With the goal of automating manually driven processes, enhanced efficiency and productivity may be ensued. There are however apprehensions over the explainability and interpretability of models, premised on superfluous data, and efforts like Explainable AI (XAI) are being pursued to make generative models more transparent. The amount of research published in XAI on medical healthcare has observed a "hockey-stick increase" from year 2019 onwards [83–85]. It has been debated that AI methods are largely **domain-agnostic**, which partly explains their obfuscated behavior. The adaptation to Bioinformatics research, say, is critical to examine predictor(s)-response(s) relationships that underline a genomic makeup and corresponding physiological state [86]. This may possibly render a rejoinder to ethical considerations surrounding *deepfake* medical images and the implicit bias.

This chapter attempted to paint a nuanced picture (yet a solid footing) of AI driving bioinformatics research. We looked at the preliminary basis of AI via exploration of big data and corresponding analysis fronts- distributed frameworks. We gently parsed through the AI and its entrenched subdomains, understanding the nature of their applicability and important, specific algorithms. Some fundamental file formats pertaining to biological data that are dealt with on a routine basis in Computational Biology have also been enumerated. En masse, the text may be considered as a "ready-reckoner" for scouting further in the discipline.

Conflict of Interest The views expressed in this text are those of the author and do not necessarily reflect the official policy or position of Infosys Limited.

References

1. Anderson, C. (2008). The end of theory: The data deluge makes the scientific method obsolete. *Wired*. https://www.wired.com/2008/06/pb-theory/.
2. Sagiroglu, S., & Sinanc, D. (2013). *Big data: A review* (pp. 42–47).
3. Wang, D.-S. (2022). A prototype of quantum von Neumann architecture. *Communications in Theoretical Physics, 74*, 095103. https://doi.org/10.1088/1572-9494/ac68d8
4. Taylor, R. C. (2010). An overview of the hadoop/mapreduce/hbase framework and its current applications in bioinformatics. *BMC Bioinformatics, 11*, S1. https://doi.org/10.1186/1471-2105-11-S12-S1
5. Dudley, J. T., & Butte, A. J. (2010). *In silico* research in the era of cloud computing. *Nature Biotechnology, 28*, 1181–1185. https://doi.org/10.1038/nbt1110-1181

6. Stein, L. D. (2010). The case for cloud computing in genome informatics. *Genome Biology, 11*, 207. https://doi.org/10.1186/gb-2010-11-5-207
7. Alnasir, J. J., & Shanahan, H. P. (2018). The application of Hadoop in structural bioinformatics. *Briefings in Bioinformatics, 21*, 96–105. https://doi.org/10.1093/bib/bby106
8. Niemenmaa, M., et al. (2012). Hadoop-BAM: Directly manipulating next generation sequencing data in the cloud. *Bioinformatics, 28*, 876–877. https://doi.org/10.1093/bioinformatics/bts054
9. Schatz, M. C. (2009). CloudBurst: Highly sensitive read mapping with MapReduce. *Bioinformatics, 25*, 1363–1369. https://doi.org/10.1093/bioinformatics/btp236
10. Leo, S., Santoni, F., & Zanetti, G. (2009). *Biodoop: Bioinformatics on hadoop* (pp. 415–422).
11. Tan, Y. S., et al. (2013). Hadoop framework: impact of data organization on performance. *Software: Practice and Experience, 43*, 1241–1260. https://onlinelibrary.wiley.com/doi/abs/, https://doi.org/10.1002/spe.1082.
12. Marx, V. (2013). The big challenges of big data. *Nature, 498*, 255–260. https://doi.org/10.1038/498255a
13. Salloum, S., Dautov, R., Chen, X., Peng, P. X., & Huang, J. Z. (2016). Big data analytics on Apache Spark. *International Journal of Data Science and Analytics, 1*, 145–164. https://doi.org/10.1007/s41060-016-0027-9
14. Ashkouti, F., & Khamforoosh, K. (2023). A distributed computing model for big data anonymization in the networks. *PLoS ONE, 18*, 1–21. https://doi.org/10.1371/journal.pone.0285212
15. Wheeler, N. R., et al. (2020). *Hadoop and PySpark for reproducibility and scalability of genomic sequencing studies* (pp. 523–534). https://www.worldscientific.com/doi/abs/, https://doi.org/10.1142/97898112156360046. https://www.worldscientific.com/doi/pdf/, https://doi.org/10.1142/97898112156360046.
16. Lin, X., & Ni, L. (1993). Multicast communication in multicomputer networks. *IEEE Transactions on Parallel and Distributed Systems, 4*, 1105–1117.
17. Narkhede, N., Shapira, G., & Palino, T. (2017) *Kafka: The definitive guide: Real-time data and stream processing at scale*. O'Reilly Media. https://books.google.co.in/books?id=a3wzDwAAQBAJ.
18. Zhang, Q., Cheng, L., & Boutaba, R. (2010). Cloud computing: State-of-the-art and research challenges. *Journal of Internet Services and Applications, 1*, 7–18. https://doi.org/10.1007/s13174-010-0007-6
19. Wolff, J. (2021). How is technology changing the world, and how should the world change technology? *Global Perspectives, 2*, 27353. https://doi.org/10.1525/gp.2021.27353
20. Alex Smith, Y. Z. (2023) Global cloud services market growth slows to 16Q2 2023. Retrieved May 2, 2024, from https://www.canalys.com/newsroom/global-cloud-services-q2-2023.
21. Richter, F. (2024). Amazon maintains cloud lead as Microsoft edges closer. Retrieved May 2, 2024, from https://www.statista.com/chart/18819/worldwide-market-share-of-leading-cloud-infrastructure-service-providers/.
22. Marx, V. (2012). My data are your data. *Nature Biotechnology, 30*, 509–511. https://doi.org/10.1038/nbt.2243
23. Luo, J., Wu, M., Gopukumar, D., & Zhao, Y. (2016). Big data application in biomedical research and health care: A literature review. *Biomedical Informatics Insights, 8*, BII.S31559. https://doi.org/10.4137/BII.S31559. (PMID: 26843812).
24. Dai, L., Gao, X., Guo, Y., Xiao, J., & Zhang, Z. (2012). Bioinformatics clouds for big data manipulation. *Biology Direct, 7*, 43. https://doi.org/10.1186/1745-6150-7-43
25. Tavbulatova, Z., Zhigalov, K., Kuznetsova, S. Y., & Patrusova, A. (2020). *Types of cloud deployment* (Vol. 1582, p. 012085). IOP Publishing.
26. Odun-Ayo, I., Ananya, M., Agono, F., & Goddy-Worlu, R. (2018) *Cloud computing architecture: A critical analysis* (pp. 1–7). IEEE.
27. Koppad, S. B. A., Gkoutos, G. V., & Acharjee, A. (2021) Cloud computing enabled big multi-omics data analytics. *Bioinformatics and Biology Insights, 15*, 11779322211035921. https://doi.org/10.1177/11779322211035921. (PMID: 34376975).

28. Luo, R., et al. (2022). Biogpt: Generative pre-trained transformer for biomedical text generation and mining. *Briefings in Bioinformatics, 23*. https://www.microsoft.com/en-us/research/publication/biogpt-generative-pre-trained-transformer-for-biomedical-text-generation-and-mining/.
29. Ferrer, A. J., Pérez, D. G., & González, R. S. (2016). Multi-cloud platform-as-a-service model, functionalities and approaches. *Procedia Computer Science, 97*, 63–72. https://www.sciencedirect.com/science/article/pii/S187705091632097X. (2nd International Conference on Cloud Forward: From Distributed to Complete Computing).
30. Jourdren, L., Bernard, M., Dillies, M.-A., & Le Crom, S. (2012). Eoulsan: A cloud computing-based framework facilitating high throughput sequencing analyses. *Bioinformatics, 28*, 1542–1543. https://doi.org/10.1093/bioinformatics/bts165
31. Field, D., et al. (2006). Open software for biologists: From famine to feast. *Nature Biotechnology, 24*, 801–803. https://doi.org/10.1038/nbt0706-801
32. Turing, A. M. (2009). *Computing machinery and intelligence* (pp. 23–65). Dordrecht: Springer Netherlands. https://doi.org/10.1007/978-1-4020-6710-53.
33. Satam, H., et al. (2023). Next-generation sequencing technology: Current trends and advancements. *Biology,* 12. https://www.mdpi.com/2079-7737/12/7/997.
34. Wen Chen, X., & Gao, J. X. (2016). Big data bioinformatics. *Methods, 111*, 1–2 (2016). https://www.sciencedirect.com/science/article/pii/S1046202316304650. (Big Data Bioinformatics).
35. Jauhari, S., & Rizvi, S. (2014). Mining gene expression data focusing cancer therapeutics: A digest. *IEEE/ACM Transactions on Computational Biology and Bioinformatics, 11*, 533–547.
36. Li, H., et al. (2020). Modern deep learning in bioinformatics. *Journal of Molecular Cell Biology, 12*, 823–827.
37. Jurtz, V. I., et al. (2017). An introduction to deep learning on biological sequence data: Examples and solutions. *Bioinformatics, 33*, 3685–3690. https://doi.org/10.1093/bioinformatics/btx531
38. Kang, M. & Oh, J. H. (2022). Editorial of special issue "Deep learning and machine learning in bioinformatics". *International Journal of Molecular Sciences,* 23. https://www.mdpi.com/1422-0067/23/12/6610.
39. Huang, G., Liu, Z., Pleiss, G., Van Der Maaten, L., & Weinberger, K. Q. (2022). Convolutional networks with dense connectivity. *IEEE Transactions on Pattern Analysis and Machine Intelligence, 44*, 8704–8716.
40. Barak, O. (2017). Recurrent neural networks as versatile tools of neuroscience research. *Current Opinion in Neurobiology, 46*, 1–6. https://www.sciencedirect.com/science/article/pii/S0959438817300429. (Computational Neuroscience).
41. Kriegeskorte, N., & Golan, T. (2019). Neural network models and deep learning. *Current Biology, 29*, R231–R236. https://www.sciencedirect.com/science/article/pii/S0960982219302040.
42. Goodfellow, I., et al. (2020). Generative adversarial networks. *Communications of the ACM, 63*, 139–144. https://doi.org/10.1145/3422622
43. Gui, J., Sun, Z., Wen, Y., Tao, D., & Ye, J. (2023). A review on generative adversarial networks: Algorithms, theory, and applications. *IEEE Transactions on Knowledge and Data Engineering, 35*, 3313–3332.
44. Mohan, A., Wara, U. U., Shaikh, M. T. A., Rahman, R. M., & Zaidi, Z. A. (2021). Telesurgery and robotics: An improved and efficient era. *Cureus, 13*, e14124. https://doi.org/10.7759/cureus.14124
45. Lin, Z., Lei, C., & Yang, L. (2023). Modern image-guided surgery: A narrative review of medical image processing and visualization. *Sensors,* 23. https://www.mdpi.com/1424-8220/23/24/9872.
46. Fink, G. R. (1996). The whitehead institute for biomedical research. *Molecular Medicine, 2*, 530–532. https://doi.org/10.1007/BF03401637
47. Douissard, J., Hagen, M. E. & Morel, P. *The da Vinci surgical system* (pp. 13–27). Cham: Springer International Publishing. https://doi.org/10.1007/978-3-030-17223-73.
48. Stasevych, M., & Zvarych, V. (2023). Innovative robotic technologies and artificial intelligence in pharmacy and medicine: Paving the way for the future of health care—a review. *Big Data and Cognitive Computing,* 7. https://www.mdpi.com/2504-2289/7/3/147.

49. Gravish, N., & Lauder, G. V. (2018). Robotics-inspired biology. *Journal of Experimental Biology, 221*, jeb138438. https://doi.org/10.1242/jeb.138438.
50. Stimpel, B., et al. (2020). Multi-modal deep guided filtering for comprehensible medical image processing. *IEEE Transactions on Medical Imaging, 39*, 1703–1711.
51. Karim, M. R., et al. (2020). Deep learning-based clustering approaches for bioinformatics. *Briefings in Bioinformatics, 22*, 393–415. https://doi.org/10.1093/bib/bbz170
52. Maier, A., Syben, C., Lasser, T., & Riess, C. (2019). A gentle introduction to deep learning in medical image processing. *Zeitschrift fu¨r Medizinische Physik, 29*, 86–101. https://www.sciencedirect.com/science/article/pii/S093938891830120X. Special Issue: Deep Learning in Medical Physics.
53. Duggento, A., Conti, A., Mauriello, A., Guerrisi, M., & Toschi, N. (2021). Deep computational pathology in breast cancer. *Seminars in Cancer Biology, 72*, 226–237. https://www.sciencedirect.com/science/article/pii/S1044579X20301784. (Precision Medicine in Breast Cancer).
54. Goodfellow, I. J., et al. (2014). *Generative adversarial nets, NIPS*'14, (pp. 2672–2680). Cambridge, MA, USA: MIT Press.
55. Rana, M. S., Nobi, M. N., Murali, B., & Sung, A. H. (2022). Deepfake detection: A systematic literature review. *IEEE Access, 10*, 25494–25513.
56. Yandell, M. D., & Majoros, W. H. (2002). Genomics and natural language processing. *Nature Reviews Genetics, 3*, 601–610. https://doi.org/10.1038/nrg861
57. Church, K. W. (2017). Word2vec. *Natural Language Engineering, 23*, 155–162.
58. Di Gennaro, G., Buonanno, A., & Palmieri, F. A. N. (2021). Considerations about learning word2vec. *The Journal of Supercomputing, 77*, 12320–12335. https://doi.org/10.1007/s11227-021-03743-2
59. Ofer, D., Brandes, N., & Linial, M. (2021). The language of proteins: NLP, machine learning and protein sequences. *Computational and Structural Biotechnology Journal, 19*, 1750–1758. https://www.sciencedirect.com/science/article/pii/S2001037021000945.
60. Elnaggar, A. Coste, F., Ouardi, F., & Rabusseau, G. (Eds.). (2023). A journey into the generative AI and large language models: From NLP to bioinformatics. In F. Coste, F. Ouardi, & G. Rabusseau (Eds.), *Proceedings of 16th edition of the International Conference on Grammatical Inference, Vol.* 217 *of Proceedings of Machine Learning Research* (pp. 7–7). PMLR. https://proceedings.mlr.press/v217/elnaggar23a.html.
61. Song, B., et al. (2021). Pretraining model for biological sequence data. *Briefings in Functional Genomics, 20*, 181–195. https://doi.org/10.1093/bfgp/elab025
62. Iuchi, H., et al. (2021). Representation learning applications in biological sequence analysis. *Computational and Structural Biotechnology Journal, 19*, 3198–3208. https://doi.org/10.1016/j.csbj.2021.05.039
63. Lovis, C., Michel, P. A., Baud, R., & Scherrer, J. R. (1995). Word segmentation processing: A way to exponentially extend medical dictionaries. *Medinfo, 8*(Pt 1), 28–32.
64. Lee, R. S. T. (2020). *Natural language processing* (pp. 157–192). Singapore: Springer Singapore. https://doi.org/10.1007/978-981-15-7695-96.
65. Bako, A. T., et al. (2021). Using natural language processing to classify social work interventions. *The American Journal of Managed Care, 27*, e24–e31.
66. Kaddari, Z., Mellah, Y., Berrich, J., Belkasmi, M. G., & Bouchentouf, T. Masrour, T., et al. (Eds.). (2021). *Natural language processing: Challenges and future directions.* In T. Masrour, I. El Hassani, & A. Cherrafi (Eds.), *Artificial Intelligence and Industrial Applications* (pp. 236–246). Cham: Springer International Publishing.
67. Vaswani, A., et al. (2017). *Attention is all you need, NIPS*'17 (pp. 6000–6010). Red Hook, NY, USA: Curran Associates Inc.
68. Bommasani, R., et al. (2022). On the opportunities and risks of foundation models. https://doi.org/10.48550/arXiv.2108.07258.
69. Clauwaert, J., Menschaert, G., & Waegeman, W. (2021) Explainability in transformer models for functional genomics. *Briefings in Bioinformatics, 22*, bbab060. https://doi.org/10.1093/bib/bbab060.

70. Choi, S. R., & Lee, M. (2023). Transformer architecture and attention mechanisms in genome data analysis: A comprehensive review. *Biology,* 12. https://www.mdpi.com/2079-7737/12/7/1033.
71. Le, N. Q. K. (2023). Leveraging transformers-based language models in proteome bioinformatics. *PROTEOMICS, 23*, 2300011. https://analyticalsciencejournals.onlinelibrary.wiley.com/doi/abs/, https://doi.org/10.1002/pmic.202300011.
72. Zhang, S., et al. (2023). Applications of transformer-based language models in bioinformatics: a survey. *Bioinformatics Advances, 3,* vbad001. https://doi.org/10.1093/bioadv/vbad001.
73. Bagal, V., Aggarwal, R., Vinod, P. K., & Priyakumar, U. D. (2022). MolGPT: Molecular generation using a transformer-decoder model. *Journal of Chemical Information and Modeling, 62*, 2064–2076. https://doi.org/10.1021/acs.jcim.1c00600. PMID: 34694798.
74. Pinheiro Cinelli, L., Arau´jo Marins, M., Barros da Silva, E. A., & Lima Netto, S. (2021). *Variational Autoencoder* (pp. 111–149). Cham: Springer International Publishing. https://doi.org/10.1007/978-3-030-70679-15.
75. Eltager, M., et al. (2023). Benchmarking variational autoencoders on cancer transcriptomics data. *PLoS ONE, 18*, 1–22. https://doi.org/10.1371/journal.pone.0292126
76. Zhang, Y., Hu, Y., Li, H., & Liu, X. (2022). Drug-protein interaction prediction via variational autoencoders and attention mechanisms. *Frontiers in Genetics,* 13. https://www.frontiersin.org/journals/genetics/articles/, https://doi.org/10.3389/fgene.2022.1032779.
77. Shin, J.-E., et al. (2021). Protein design and variant prediction using autoregressive generative models. *Nature Communications, 12*, 2403. https://doi.org/10.1038/s41467-021-22732-w
78. Trinquier, J., Uguzzoni, G., Pagnani, A., Zamponi, F., & Weigt, M. (2021). Efficient generative modeling of protein sequences using simple autoregressive models. *Nature Communications, 12*, 5800. https://doi.org/10.1038/s41467-021-25756-4
79. Mardikoraem, M., Wang, Z., Pascual, N., & Woldring, D. (2023). Generative models for protein sequence modeling: recent advances and future directions. *Briefings in Bioinformatics, 24*, bbad358. https://doi.org/10.1093/bib/bbad358.
80. Bertoline, L. M. F., Lima, A. N., Krieger, J. E., & Teixeira, S. K. (2023). Before and after alphafold2: An overview of protein structure prediction. *Frontiers in Bioinformatics,* 3. https://www.frontiersin.org/articles/, https://doi.org/10.3389/fbinf.2023.1120370.
81. Mills, L. (2014). Common file formats. *Current Protocols in Bioinformatics, 45*, A.1B.1–A.1B.18. https://currentprotocols.onlinelibrary.wiley.com/doi/abs/, https://doi.org/10.1002/0471250953.bia01bs45.
82. Zeng, X., et al. (2022). Deep generative molecular design reshapes drug discovery. *Cell Reports Medicine, 3*, 100794. https://www.sciencedirect.com/science/article/pii/S2666379122003494.
83. van der Velden, B. H., Kuijf, H. J., Gilhuijs, K. G., & Viergever, M. A. (2022). Explainable artificial intelligence (XAI) in deep learning-based medical image analysis. *Medical Image Analysis, 79*, 102470. https://www.sciencedirect.com/science/article/pii/S1361841522001177.
84. S Band, S., et al. (2023). Application of explainable artificial intelligence in medical health: A systematic review of interpretability methods. *Informatics in Medicine Unlocked, 40*, 101286. https://www.sciencedirect.com/science/article/pii/S2352914823001302.
85. Chaddad, A., Peng, J., Xu, J., & Bouridane, A. (2023). Survey of explainable AI techniques in healthcare. *Sensors (Basel),* 23.
86. Karim, M. R., et al. (2023). Explainable AI for bioinformatics: Methods, tools and applications. *Briefings in Bioinformatics, 24*, bbad236. https://arxiv.org/pdf/2212.13261.

Molecular and Computational Aspect of Microbial Bioprospection in Emerging Era

Himanshu Shekhar, Tamanna, Shamli Chandel, Matruprasad Mohanty, and Dharam Singh

Abstract The emergence of high-throughput technologies like microarray and next-generation sequencing (NGS) has led to the accumulation of large biological data, laying the red carpet for biology in the big data era. The OMICs revolution, which provided global information about different properties of biomolecules at the minimal expense of time and resources ever possible, has successfully postulated itself as the torch-bearer in this grand welcome. With the importance of sustainability being the underlying principle, the *Cradle to Cradle* (C2C) approach has drawn significant attention. This approach focuses on employing microbial biodiversity for industrial processes that are sustainable and considerate to life and future generations. These processes use unique proteins/enzymes to produce finished products that were conventionally produced using chemicals. Bioprospection of extreme environments for such unique enzymes has been a hallmark of such processes, whether the discovery of thermophilic polymerases that revolutionized the PCR-based technologies or the common UV-protectants we apply on our faces today. These extremophiles have a plethora of answers to almost every desired trait with capability of easy scale-up compared to enzymes from higher life forms. This chapter provides an overview of OMIC technologies applied in the bioprospection of extreme niches with a special focus on applying bioinformatics in achieving such goals.

H. Shekhar · Tamanna · S. Chandel · M. Mohanty · D. Singh (✉)
Biotechnology Division, CSIR-Institute of Himalayan Bioresource Technology, Palampur, Himachal Pradesh 176061, India
e-mail: dharamsingh@ihbt.res.in; dharams14@gmail.com

H. Shekhar
e-mail: hishu0811@gmail.com

Tamanna
e-mail: tamannanegi632@gmail.com

S. Chandel
e-mail: shamli.chandel43@gmail.com

M. Mohanty
e-mail: matru3041@gmail.com

Tamanna · S. Chandel · M. Mohanty · D. Singh
Academy of Scientific and Innovative Research (AcSIR), Ghaziabad 201002, India

© The Author(s), under exclusive license to Springer Nature Singapore Pte Ltd. 2024
A. Chaudhary et al. (eds.), *Unraveling New Frontiers and Advances in Bioinformatics*,
https://doi.org/10.1007/978-981-97-7123-3_11

Keywords Bioprospection · OMICs · Extremozymes · Community diversity · Molecular profiling

1 Introduction

Today, the enormous diversity of microorganisms requires dual address: identifying the unexplored diversity and mining this diversity towards novel biotechnological applications. Microbial diversity provides us with an evolving arsenal of tools pre-validated on an evolutionary timescale with the potential for direct industrial applications [1]. As the well-known picture "*Flammarion engraving*" suggests, there is a planet that is metaphorically visible beyond the bright lights. Adapting this in the context of microbial diversity by blending bacterial colonies in place of stars in the original painting (Fig. 1), Dorado-Morales tried to highlight the importance of bioprospection [2]. Unusual habitats are particularly intriguing for bioprospecting investigations for three reasons. The first is their high richness, which increases the likelihood of discovering new species. This is demonstrated by the detection of up to 47 new phyla in Colorado's groundwater and aquifer sediments [3]. Secondly, these microbes are predisposed to conditions that are frequently associated with industrial requirements [2, 4]. Lastly, it will create avenues for novel bio-factories from these sturdy microbes that can withstand a variety of industrial stresses. Whether be the case of carotenoid-producing and scytonemin-producing bacteria from sun-exposed environments with applications in cyborg bugs in next-gen solar panels [5], or the role of bacterial platforms utilizing organisms such as *Deinococcus, Hymenobacter, Erythrobacter*, and *Geobacillus* that are being explored in synthetic biology as potential alternatives to conventional *E. coli* models. In order to fully exploit new species, it will be necessary to develop innovative culturing and screening methods, as well as enhance current OMICs data analysis procedures to effectively identify and utilize them. The development of molecular techniques has revealed that the number of microbe's present is much higher than previously thought. It is now widely accepted that over 99% of microbial diversity is considered to be 'microbial dark matter', which cannot be cultivated [6].

Thanks to high-throughput technologies like NGS, OMICs and microarray; the biology of microbial diversity is entering the big data era. These methods have revolutionized industrial research by providing a more comprehensive insight into utilizable bioresources and associated microbiomes. The influence of these technologies has been monumental, as evidenced by the redefinition of the Tree of Life in recent years [7]. Noteworthy breakthroughs have occurred within the bacterial domain of life [8], unveiling new insights and expanding our understanding. Furthermore, the identification of viral genes has experienced a remarkable 16-fold surge [9]. These developments highlight the profound impact of these technologies, reshaping our understanding of the intricate web of life.

Fig. 1 Previous bioprospecting endeavours have primarily concentrated on nearby habitats. However, outside the familiar territory, there are untapped extraordinary habitats that possess immense potential as a reservoir of biological diversity, which could play a pivotal role in forthcoming biotechnological advancements. Image adapted from the famous wood carving *'Flammarion engraving'* created by blending bacterial colonies in place of stars in the original painting [2]. *Source* Tanner K, Vilanova C, Porcar M. Bioprospecting challenges in unusual environments. Microb Biotechnol 2017 Jul;10(4):671–673. https://doi.org/10.1111/1751-7915.12723

Not delving deep into culture-dependent techniques, we will therefore discuss the culture-independent techniques with special attention to role of OMICs and emerging techniques in bioprospection.

2 Culture-Dependent Bioprospection

Sample collection, keeping in consideration the requirement of microbes as inferred from its source of collection, is followed by the enrichment of microbes on the basis of desired traits. These enriched cultures are then used for strain isolation often yielding pure culture. The integrity of this pure culture is then verified using microscopy, quantitative PCR and/or genomic and transcriptomic analyses. Once the purity of the culture is established the physiological characterization of the pure culture is studied in detail.

3 Culture-Independent Bioprospection

Approximately 99% of all microorganisms are generally uncultivable using standard culturing [6]. Additionally, only half of the already identified prokaryotic phyla have been cultured [10]. Culture-independent techniques, therefore, enable the assessment of microbial diversity of hard-to-maintain cultures. These techniques can be PCR-dependent or PCR-independent.

3.1 PCR-Dependent Techniques

Typically, these approaches rely on PCR and are frequently linked to PCR bias resulting from differential amplification efficiency among the templates. As a result, the evaluation of microbial community structure is impeded. Despite significant experimental endeavours to improve multi-template PCR, which involves strategies like minimizing PCR cycles, refining primers, and optimizing polymerases; PCR bias persists as a complex issue in microbiome investigations. Its full comprehension remains elusive, contributing significantly to errors in studies. In a recent study, McLaren and his team proposed a unique strategy using a mock community to tackle PCR bias by treating it as a compositional perturbation [11]. Through translating the log-ratio of relative abundances between various taxa, they managed to create a model that could rectify PCR bias in sequenced mock communities with predetermined starting composition. Nonetheless, the creation of mock communities involves compiling relevant and comprehensive sets of bacterial taxa for a particular sample type, which might not be feasible for non-culturable and unisolated microbes.

3.1.1 16S rRNA Profiling

The successful application of gene cloning, sequencing, and comparative data analysis has enabled Ribosomal RNA profiling on samples from air [12], lake [13], and soil [14, 15]. A large portion of the molecular research has been dependent on the sequences of SSU rRNA/16S rRNA gene. In recent times, various genes like RuBisCO [16], ammonia monooxygenase (amoA) [16, 17], sulfate thioesterase/thiohydrolase (soxB) [16], and mcrA [18] have also been utilized to investigate specific microbial communities. The basic workflow of 16S library preparation is outlined in Fig. 2.

3.1.2 Denaturing/Temperature Gradient Gel Electrophoresis (DGGE/TGGE)

DGGE and TGGE facilitate the differentiation of DNA fragments that possess equal lengths but vary in their sequences. The separation of a partially melted dsDNA is made possible through the use of polyacrylamide gels that contain a linear gradient of urea and formamide (DNA denaturants) or temperature. This separation occurs due to the reduced electrophoretic mobility of the DNA molecule [19, 20]. DNA fragments melt in discrete melting domains, with specific stretches sharing the same Tm. Once a domain with the lowest melting profile reaches its Tm at a certain position

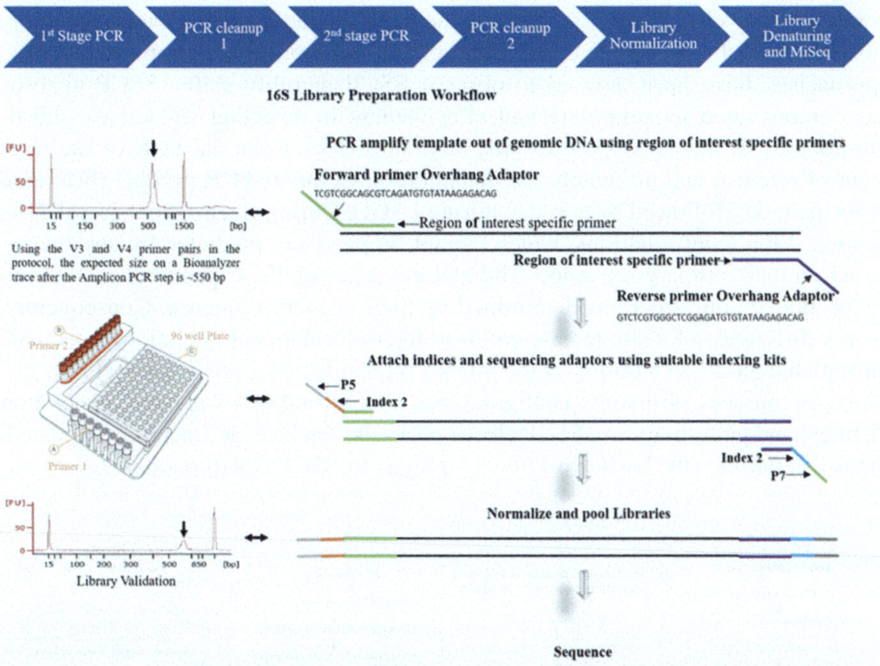

Fig. 2 16S V3-V4 library preparation on Illumina. Amplicon primers targeting V3-V4 region were generated, and an Illumina adapter was added. Amplification of region under focus out of a DNA sample using specific primers with overhang adapters is followed by PCR cleanup to purify the amplified template. Indexing PCR to attach dual indices and sequencing adapters was done. A second round of cleanup of the library, followed by quantification and normalization was done. The size (~630 bp) of the final diluted library was assessed by analysing 1 μl on a DNA analyser chip. For MiSeq sequencing, libraries denatured using NaOH were diluted with buffer, and then subjected to heat denaturation

as visualized by bands on the gel, the molecule undergoes a transition from helical to partially melted state, causing its reptation to almost stop. Thus, molecules with different sequences pause at different locations in the gel owing to variations in melting temperatures as a result of sequence differences. With the use of DGGE or TGGE, it is possible to identify 50% of sequence variants in DNA fragments smaller than 500 bp. By incorporating a GC-clamp, the detection rate can be elevated to nearly 100% [21]. DGGE of amplified 16S rDNA has been successfully employed to assess the community in microbial mats and bacterial biofilms [22].

3.1.3 Single-Strand-Conformation Polymorphism (SSCP)

Under non-denaturing conditions, single-stranded DNA (ssDNA) adopts a folded conformation based on its intramolecular interactions and nucleotide sequence. Even a single base change can cause a mobility shift, leading to distinct band patterns in nondenaturing electrophoresis. Several methods, including radioisotope labelling, silver staining, fluorescent dye-labelled primers, and capillary electrophoresis-based approaches, have been devised to observe SSCP mobility shifts. SSCP analysis has demonstrated its simplicity and effectiveness in detecting single-base substitutions [23]. It involves several steps, including DNA isolation, PCR of the fragment of interest, and ultimately the denaturation of the ds-PCR product (heat and/or formamide), followed by non-denaturing PAGE. During electrophoresis, ssDNAs assume stable conformations, which cannot be precisely predicted by any theory, based on their primary sequence. The method relies on the capacity of ssDNAs to adopt unique conformations determined by their primary sequence. Consequently, even a difference of a single base can lead to conformational alteration, detectable through the modified mobility of the ssDNA molecule in the SSCP method [24, 25]. Thus, the number of distinct configurations for each ssDNA fragment, generating distinct band patterns during SSCP electrophoresis, can be experimentally validated. Figure 3a outlines the basic workflow employed in SSCP electrophoresis.

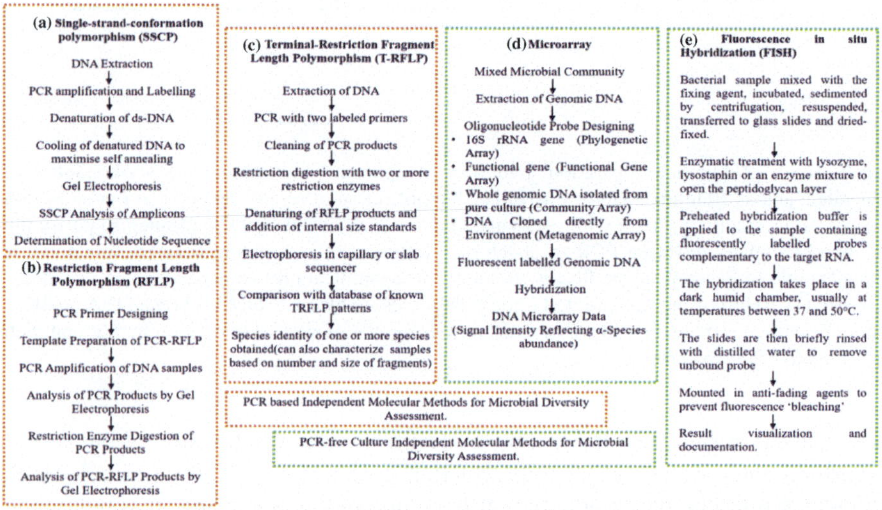

Fig. 3 PCR-based and PCR-free Culture Independent Molecular Methods (CIMMs) for Microbial Diversity Assessment. PCR-based CIMMs- (**a**), (**b**) and (**c**). **a** Single-Strand-Conformation Polymorphism (SSCP) detects mobility shifts in ss-DNA, leading to distinct band patterns in nondenaturing electrophoresis, which can be visualized using radioisotope labelling, silver staining, fluorescent dye-labelled primers, and capillary electrophoresis-based approaches. **b** Restriction Fragment Length Polymorphisms (RFLP) relies on the identification of differences in the lengths of a specific restriction fragment among individuals, which can be ascribed to diverse genotypic variations. These variations may include variances in individual bases that result in the formation or elimination of cleavage sites, as well as the addition or removal of DNA segments within the fragment, ultimately causing alterations in its overall size. **c** Terminal RFLP (T-RFLP) analysis, on the contrary, assesses variation in the size of terminal restriction fragments that originate from a PCR-amplified marker. PCR-free CIMMs- (**d**) and (**e**). **d** Microarray, **e** Fluorescence in-situ Hybridization (FISH)

3.1.4 Restriction Fragment Length Polymorphisms (RFLP) and Terminal RFLP (T-RFLP)

Length variations of a particular restriction digested fragment among individuals can be attributed to different genotypic differences, such as variations in individual bases that lead to the creation or loss of cleavage sites, as well as the INDEL of DNA blocks, ultimately resulting in size changes [26]. These genotypic changes can be identified by examining the differential restriction fragment patterns on agarose gel [27]. Figure 3b outlines the basic workflow employed in RFLP. Additionally, hybridization allows for the detection of sequences of interest within a population of DNA fragments. The DNA isolated from an agarose gel is blotted and then hybridized radioactively on a nitrocellulose membrane. The incorporation of probe sequences with exceptionally high specific radioactivity has allowed for the identification of single base differences in mammalian DNA using this method [28]. RFLP has also been used successfully to estimate the diversity and structure of microcosms [29].

T-RFLP analysis, on the other hand, evaluates variations of terminal restriction fragments (TRF) derived from amplified markers under investigation. This method combines at least three technologies: RFLP, PCR, and electrophoresis [30]. Figure 3c outlines the basic workflow of T-RFLP. The technique's effectiveness lies in its ability to rapidly and precisely determine TRF lengths using sequencing. The use of T-RFLP analysis for microbial communities is becoming more prevalent in the scientific community due to its rapidity and high-resolution [31]. However, it is important to note that this method is susceptible to the same considerations typically associated with molecular approaches, which rely on the efficient community DNA Isolation and amplification of genes of interest. A research tool available at http://www.cme.msu.edu/RDP/html/analyses.html (Ribosomal Database Project website) is designed to assist in microbial community analysis through the use of TRF length polymorphism of 16S ribosomal DNA. This web-based tool, known as TAP T-RFLP, enables users to perform virtual restriction digestions on the entire 16S database. By specifying a primer and a restriction endonuclease target site, users can obtain the sizes of TRFs in base pairs. The resulting output can be studied phylogenetically or by size [32]. T-RFLP analysis involves the restriction endonuclease-mediated digestion of fluorescently end-labelled PCR products, typically with 4 bp cutters. Primers used in the PCR can be labelled in different combinations, single or both labelled and/or both with same probe or different probe. During analysis, only the end-labelled T-RFLP is identified. T-RFLP is a reliable and consistent method for swiftly analysing microbial composition and community dynamics due to alterations in prevailing growth conditions. The specific information of the T-RFs generated, enables gene system comparison from microcosms to *in-silico* predicted T-RFs' databases. This capability also enables immediate fragments for the phylogenetic assignment [33].

3.2 PCR-Independent Techniques

In the case of mixed populations with less than ten different organisms, 5S rRNA was separated using high-resolution gel electrophoresis to isolate species-specific molecules. The individual types of 5S rRNA were subsequently sequenced and compared to databases to determine the phylogenetic relationships of the organisms that produced the analysed 5S rRNAs. In a different method that disregards complexity limitations, the ribosomal genes are "shotgun-cloned". Following their isolation, the diverse types of cloned rRNA genes undergo a meticulous sorting process within the laboratory. These sorted genes are then subjected to precise sequence annotation. This analysis is particularly advantageous for conducting phylogenetic evaluations. By comparing the obtained sequence information with existing reference information, the phylogenetic affinities of each microbe in the community under study are identified.

Similarly, PCR-free molecular techniques, such as fluorescence in situ hybridization (FISH) and microarray, have also been used for studying microbial communities from various environments [34]. Figure 3d and e outline the workflow of microarray and Fluorescence in situ Hybridization (FISH), respectively. Microarrays employ an array of probes (cDNA, oligonucleotides and microbial genomes). The initial report of employing microarray in microbial ecology employed 9 probes, each consisting of 15-20mer oligonucleotides [35]. These probes were utilized to differentiate nitrifying bacteria. Subsequent advancements led to high-density gene-targeting (16S rRNA) microarray, which incorporated approximately 30k 20mer probes [36] and approximately 300k 25mer probes [37]. This sophisticated microarray facilitated the exploration of microbial diversity.

The metagenome microarray (MGA) is an alternative microarray platform designed for the swift analysis of microbial and entire community metagenomes. Unlike traditional probes that are spotted on a glass slide, the MGA format features microarray targets arranged on a glass slide, utilizing a labelled, specific gene probe. This innovative method employed by MGA enables the efficient screening of metagenomic libraries, including cosmid clones from groundwater microcosms [38] and marine fosmid clones [39]. On the other hand, a combination of molecular genetics precision and microscopy visual data in FISH allows for the visualization and identification of individual cells in their natural habitat or diseased tissue. This technique enables the examination of nucleic acid sequences within cells without affecting the cell's morphology or the integrity of its compartments. FISH methods for RNA detection have been integrated into living cells using either fluorophore that can be activated in vivo or probes that fluoresce upon hybridization.

3.2.1 Metagenomics

Metagenomics refers to examining genomic material collected straight from the environment instead of cultured specimens. Since, Handelsman et al. introduced the term 'metagenome' in 1998 [40], the advancement of NGS has allowed for more comprehensive sampling compared to traditional techniques (DGGE, T-RFLP, or rRNA libraries). However, Direct sequencing is suggested to eliminate the inherent error rate of pyrosequencing as it avoids the bias resulting from the amplification of phylogenetic marker genes. Figure 4 outlines the basic metagenomic workflow. Thus far, metagenomic analysis has been extensively utilized in diverse environments, resulting in the sequencing of over 210 distinct metagenomes of various settings. The application of metagenomics has experienced constant expansion in recent years, encompassing activities such as prospection and mining of metagenomes from extreme niches and assessing the diversity of microbial communities at functional and taxonomic levels [41]. Advances in assembly algorithms and associated techniques have resulted in notable enhancements in the precision and effectiveness of sequence assembly [42–48].

Binning is the practice of organizing DNA sequences into groups that may indicate an individual or a genome representative of an entire related microbial community. Binning is performed using machine learning methods, supervised or unsupervised, depending on the use of the reference database. A variety of algorithms to leverage two different layers of information found in a given DNA sequence have been developed. The first type, called compositional binning, capitalizes on the conserved nucleotide composition observed in genomes. This conservation is seen in sequence fragments across the genomes, such as a specific GC content or the distribution of k-mers. The second type of information involves the possibility

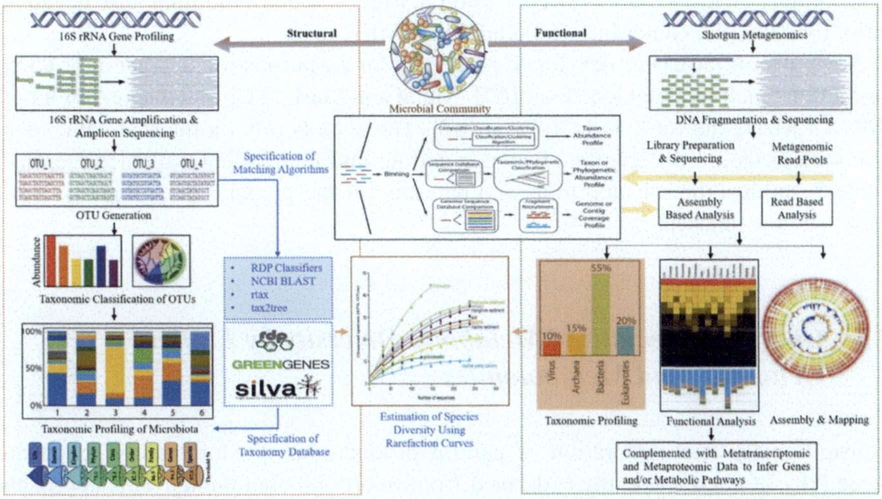

Fig. 4 Structural and functional metagenomic workflow for microbial community. Structural metagenomic strategies rely on rRNA-based targeted amplicon sequencing that basically answers the question of 'who is there (in microbial community)' and functional metagenomic strategies rely on whole metagenome sequencing that basically answers the question of 'what are they doing (in microbial community)'

that DNA fragment under study is encoding a gene. By assessing gene similarity with reference database, it becomes feasible to assign and classify the sequence to a specific category. Metagenome assembly methods for recovering genome bins are of three types based on variations in sampling content, clustering inputs, and utilization of abundance information. These types are NC-based, DA-based, and NCA-based methods signifying nucleotide composition, differential abundance and nucleotide composition and abundance-based approaches. The key difference between these methods is outlined in Table 1. The established superiority of NCA methods has led to the emergence of more tools such as ABAWACA—'a binning algorithm without a cool acronym' (https://github.com/CK7/abawaca), CONCOCT—clustering contigs on coverage and composition [48], MaxBin [49], and GroopM [50].

In the realm of binning algorithms for metagenomic analysis, there are options like Phylopythia [51], S-GSOM [52], PCAHIER [53], and TACAO [54], which focus on composition, as well as software tools like MG-RAST [55], IMG/M [56], CARMA [57], MEGAN [58], SOrt-ITEMS [59], and MetaPhyler [60] that are based on similarity. Furthermore, algorithms such as PhymmBL [61] and MetaCluster [62] consider both composition and similarity. These tools employ various methods like self-organizing maps (SOMs) and hierarchical clustering and can be used in unsupervised or supervised modes with user input for bin definition. The annotation process for metagenomic sequence data generally consists of two steps. Firstly, features of interest, such as genes, are identified through feature prediction. Secondly, the assignment of putative gene functions and taxonomic neighbours through functional annotation. Several algorithms have been created for completed genome sequences to accurately identify CDS with a precision of over 95% and a minimal false negative ratio, for example, GeneMark [63] and GLIMMER [64].

Several tools have been developed especially for the purpose of metagenomic CDS prediction, such as FragGeneScan [65], MetaGeneMark [51], MetaGeneAnnotator (MGA)/ Metagene [66], and Orphelia [67]. These tools rely on internal data, such as codon usage, to differentiate between coding and non-coding sequence regions. Table 2 outlines the selected tools and datasets for the preparation and analysis of metagenomic datasets.

3.3 Charting Specialized Metabolic Diversity by Linking Genomics and Metabolomics

Conventionally, the exploration of natural products has predominantly relied on bioactivity of chemical extracts derived from microbial community. This approach has facilitated the identification and characterization metabolites, thereby forming the foundation of our existing knowledge on the diversity of natural products. However, the limited speed in obtaining essential analytical data, along with the frequent rediscovery of compounds, hampers the efficacy of this method. The advancement of DNA

Table 1 The three main metagenome binning approaches with their key methodological features (*HCL- Hierarchical Clustering Learning, **ESOM- Emerging Self Organizing map)

Method	Starting point	Clustering methods	Negatives	Positives	Computational resources
NC-based	Oligonucleotide frequency and GC screening	HCL*, network graph based on correlation and ESOM**	Inefficient differentiation of closely related genotypes	Individual metagenome assemblies with no temporal variation can be used	**R packages**: qgraph, I graph, pv-clust **Tetramer-Freqs** (https://github.com/tetramerFreqs/Binning) **Databionic ESOM tools** (http://databionic-esom.sourceforge.net) **2 T-binning** (https: hmp.ucalgary.ca/HMP/metagenomes/data/SCADC/454/Binning/2TBinning/)

(continued)

Table 1 (continued)

Method	Starting point	Clustering methods	Negatives	Positives	Computational resources
NCA-based	A composite distance matrix	Gaussian mixture models, K-medioids clustering, and expectation algorithm	Associated with need of large sample size, cost, time, and computational resources	Bettercontig binning than NC method	**MetaBAT** (https://bitbucket.org/berkeleylab/metabat) **CONCOT** (https://github.com/BinPro/CONCOT) **MaxBin** (http://downloads.jbei.org/data/microbial_communities/MaxBin/MaxBin.html) **GroopM** (https://github.com/minillinim/GroupM) **Databionic ESOM tools** (http://databionic-esom.sourceforge.net/)
DA-based	Differential coverage patterns across multiple samples	Correlation cut-off	Associated with large sample size displaying spatio-temporal differential abundance. Hence, high cost, computational time, and resources	Strain level resolution	**Multi-metagenome** (https://github.com/MadsAlbertsen/multi-metagenome) **MGS Canopy algorithm** (https://github.com/fplaza/mgs-canopy-algorithm) **Databionic ESOM tools** (https://databionic-esom.sorceforge.net/)

Table 2 Bioinformatic tools and databases for metagenomic dataset preparation and analysis

Name	Application	Description	Link
Binsanity	Meta-genomics, sequence assembly	Coverage and affinity propagation based unsupervised clustering	https://github.com/edgraham/BinSanity
CAZy	Database	Functional domain annotation of glycosidic enzymes	http://www.cazy.org/
COGs	Database	Protein database of all sequenced microbial genomes	http://www.ncbi.nlm.nih.gov/COG/
FragGeneScan	Sequence analysis, gene prediction	Application for finding (fragmented) genes in short reads	https://sourceforge.net/projects/fraggenescan/
MaxBin	Meta-genomics, sequence assembly	Expectation–maximization algorithm based metagenomic binning software	http://downloads.jbei.org/data/microbial_communities/MaxBin/MaxBin.html
MEGAN	Sequence analysis and taxonomy	Large metagenomic dataset analysis	http://ab.inf.uni-tuebingen.de/software/megan6/
MetaBAT	Meta-genomics, sequence assembly	Adaptive binning genome reconstruction algorithm	https://bitbucket.org/berkeleylab/metabat
MetaGeneMark	Gene prediction	Metagenomic gene prediction program	http://exon.gatech.edu/meta_gmhmmp.cgi
MG-RAST	Meta-genomics	Fully-automated metagenomic analysis tool	http://metagenomics.anl.gov
Mothur	Taxonomy	Expandable software for analyzing 16S rRNA gene sequence data	https://www.mothur.org
NCBI	Database	Database	https://www.ncbi.nlm.nih.gov/
Pfam	Database	Protein family and domain information	http://pfam.xfam.org/
Prodigal	Gene prediction	Bacterial and archaeal gene prediction	https://github.com/hyattpd/Prodigal
QIIME	Microbiome analysis	Open-source microbiome analysis pipeline	http://qiime.org/

sequencing has unveiled that organism producing natural products harbour Biosynthetic Gene Clusters (BGCs) in their genomes, which are accountable for producing metabolites that were previously undiscovered. In fungi and bacteria, most metabolic pathways are contained within BGCs, whereas in plants genomic clustering is also seen. These findings have postulated genome mining as a pinpointing technology for identifying known and unknown metabolites [68]. Progress in technological development and instrumentation, particularly Mass Spectrometry (MS), have enabled the investigation of metabolite extracts with increasing complexity. However, accurately annotating metabolites from MS1 data can often be challenging for various reasons, including facts such as the same mass and formula. Spectra resolution of metabolites provides specific advantages for annotating and identifying products. These MS^2 spectra often considered as metabolic fingerprints, are used to utilize information on network structure with the help of several tools. Figure 5 outlines the various technologies for molecular networking generally employed in metabolic diversity analysis [69].

Over the past ten years, the Global Natural Product Social (GNPS) [70] molecular networking has evolved into a comprehensive mass spectrometry ecosystem. The platform currently operates as a medium for users to upload and store both raw and processed datasets. Additionally, GNPS offers a range of statistical and annotation tools for data analysis. To facilitate public-assess, GNPS provides a virtual Interactive MS Environment, MassIVE. Once MS^2 spectral data of a target compound is acquired, MASST can be employed to query its existence across all the GNPS public datasets. This functionality is comparable to conducting an NCBI BLAST. Furthermore, advancements like ReDU ensure normalization for the associated metadata of the dataset by allowing the comparison of experimental data with public datasets in the repository.

3.4 Metaproteomic Data Analysis Based on MetaProteomeAnalyzer (MPA) and Prophane

The utilization of metagenomics and rRNA-Seq allows for the acquisition of valuable knowledge concerning the functional and taxonomic composition of the microbial community understudy. However, metaproteomics goes a step further by revealing the actual functional profile and providing a deep understanding of the intricate interactions between microbes and the environment [71]. Metaproteomics utilizes complex data analysis methods to secure precise peptide and protein identification, along with aiding in the result interpretation using valuable protein annotations [72]. As a result, researchers face specific hiccups and potential abortions at every stage of the analysis [73] (summarized in Table 3). Since non-model organisms are often not functionally annotated, the functional interpretation of sample composition is hampered most of the time. Protein identification in shotgun proteomics primarily relies on matching sample mass spectra with a customized database. Consequently,

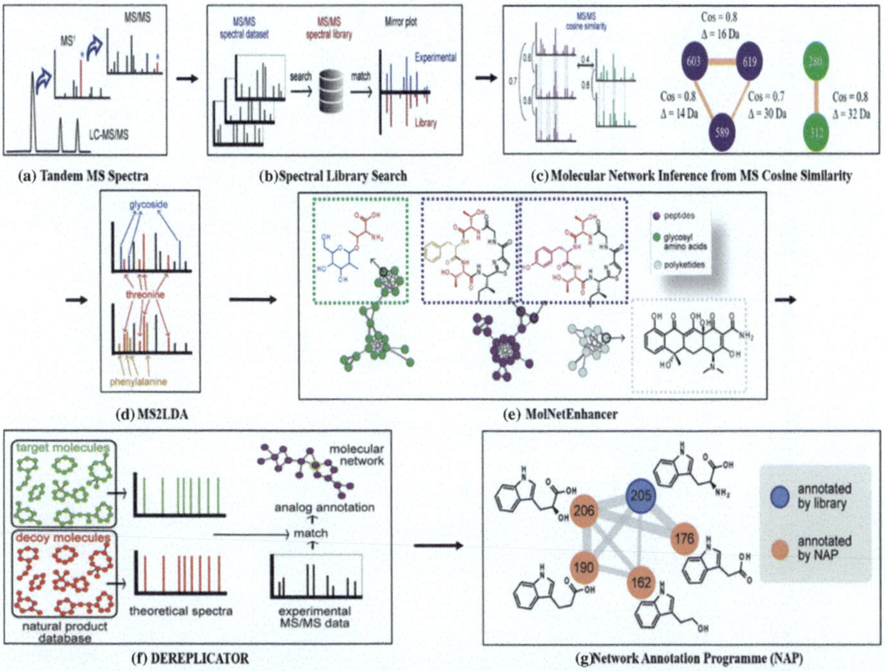

Fig. 5 Technologies for molecular networking towards metabolic diversity analysis. **a** Untargeted LC–MS/MS analysis is employed in metabolomics to examine crude microbial extracts. This involves selecting the most prominent ions in MS1 and fragmenting them to obtain MS/MS spectra. **b** The process of library search (spectral) involves searching each sample MS^2 spectra in a library. The aim is to identify a suitable spectral match that aligns with the given data. **c** The molecular network plays a crucial role in organizing MS/MS datasets based on spectral similarity, measured by the cosine score. High-similarity spectra are grouped together to form networks. **d** By identifying the co-occurrence of ions associated with structural motifs, MS2LDA serves as an effective tool. This substructural information can be further enriched by incorporating annotated data from other *in-silico* tools. **e** MolNetEnhancer merges these outcomes to refine substructures and barcode the chemical classes identified in the dataset being studied. **f** DEREPLICATOR has been developed as an *in-silico* tool with a specific focus on annotating recognized peptic products. **g** NAP, also known as Network Annotation Propagation, utilizes the topology of the network to propagate annotations from the match in the library. Leveraging *in-silico* tools for annotation improves analogue identification by reevaluating the most suitable structure on the basis of structural fingerprint overlaps. *Source* Van Der Hooft et al. [69]

only proteins recorded in the database can be identified. To mitigate such *a-priori* knowledge-based dependency, extensive protein databases, encompassing millions of sequences are employed. Several software solutions, such as MPA [74], MASCOT [75], and Proteome Discoverer, have the capability to handle such databases of large magnitude, even though it may be challenging for certain proteomic database search engines. MPA and Prophane are two free software tools that have been specifically optimized to cater the data analysis needs of this magnitude. MPA, a software solution (Java-based), offers enhanced spectrum correlation by utilizing multiple search

Table 3 Challenges in metaproteomics analysis and possible solution packages

#	Computational task	Major challenge	Solution	mPies	Prophane	MetaGOmics	UniProt	MPA	Unipept	Metalab
1	Protein sequence database generation	Unknown taxonomic composition	Taxon-specific databases	+	–	–	++	–	–	–
2	Peptide-spectrum correlation ("database search")	High computational cost and false discovery rates	Multiple search engines	–	–	–	–	++	–	–
			Multistep search method	–	–	–	–	–	–	+
3	Protein grouping and inference	Distinction in closely related species	Shared peptide-based clustering	–	–	–	–	++	–	+
			Taxonomy based clustering	–	–	–	–	+	–	–
4	Taxonomy of proteins and annotation	Distinct taxonomy fades with increasing distance	LCA based on: NCBI	+	++	–	–	–	–	–
			LCA based on UniProtKB	–	+	+	–	+	+	+
5	Protein functional annotation	Narrow range of comparison	LCA based on UniProtKB	–	+	+	–	+	+	–
			LCA based on eggnog	–	–	–	++	–	–	+
			LCA based on PFAMs	–	–	–	+	–	–	–
			LCA based on TIGRF	–	–	–	+	–	–	–
			LCA based on special databases (CaZy, Foam, Resfams)	–	–	–	+	–	–	–
6	Result visualization	Low quantity and dimension of data	GUI optimized plots	–	–	+	++	+	+	+

engines. It also incorporates specialized protein grouping methods and provides users with a user-friendly graphical user interface. On the other hand, Prophane provides functional and taxonomic annotation for metaproteomes relying on various databases.

It also offers GUI-based result visualization with intuitive web service. MPA can be assessed at http://www.mpa.ovgu.de and Prophane at http://www.prophane.de. MPA is an independent software, available as Galaxy-P module and Conda package as web service and Conda package is available for Prophane. Figure 6 outlines the stage wise metaproteomic data analysis workflow for MPA and Prophane. There are multiple tools designed to assist in the interpretation of metaproteomics data, including UniPept [76] Prophane [73], mPies [77], MetaGOmics [78], MetaLab [79], and iMetaLab [80].

These tools can be categorized as protein-centric (Prophane, mPies), peptide-centric (UniPept), or a combination of both (metaGOmics, MetaLab). Peptide-centric tools utilize identified peptides without determining the associated proteins, thereby sidestepping the challenge of protein inference. Nevertheless, this method fails to obtain information that could be obtained in analysing proteins in totality.

4 Emerging Techniques

4.1 High-Throughput Discovery of Cryptic Microbial Metabolites Based on Elicitor Screening Followed by Induction of Secondary Metabolome

The rise of high-throughput technologies for DNA sequencing has led to a resurgence of interest in microbial natural products (MNPs) for the purpose of drug discovery. Enzymes encoded by co-localized biosynthetic gene clusters (BGCs) are responsible for the production of NPs in general. However, a mismatch exists between the number of identified BGCs that potentially encode the synthesis of NPs and the actual number of annotated metabolites discovered for a particular organism, as revealed by genome sequencing. The consensus is that the majority of microbial GCs are not translated in standard *ex-situ* conditions. These GCs, referred to as 'cryptic GCs', hold a vast reserve of unexplored bioactive compounds that could benefit drug discovery [81, 82]. Cultivation-based approaches used to assess the cryptic metabolic pathways came into the limelight when Bode et al. postulated the "OSMAC" approach (One Strain Many Compounds) in 2002 [83, 84]. Based on this method, every microbial strain can potentially generate various compounds in different conditions. Consequently, alterations in cultivation factors can trigger the synthesis of new NPs. By employing this stochastic method that imitates *in-situ* fluctuations, Bode et al., identified many compounds with more than 25 distinct backbones from just six microorganisms [85]. These chemical cues, along with enzyme probing, Syn-BNPs, untargeted

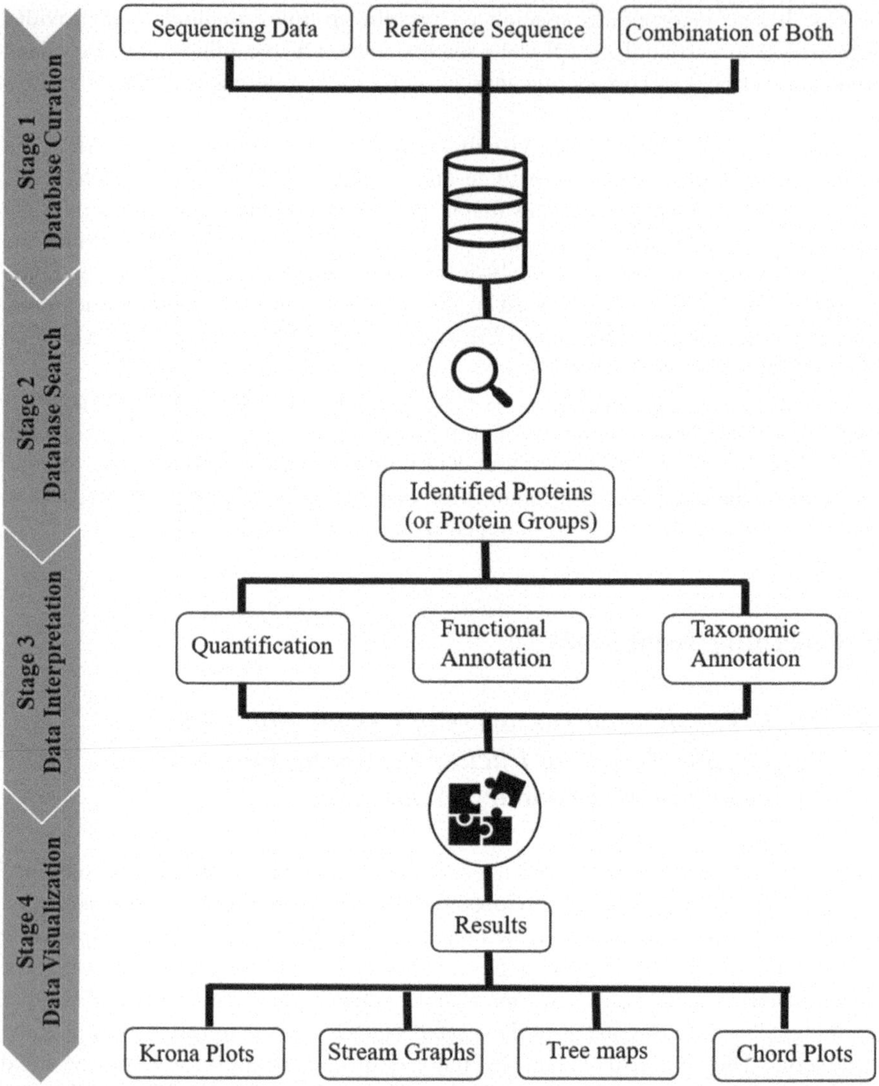

Fig. 6 Stage wise metaproteomic data analysis workflow for MPA and prophane. In **stage-1** creation of protein database is done with the goal of closely representing the sample content. In **stage-2** utilizing X!Tandem* and OMSSA**), the MPA software searches acquired mass spectra against the database, resulting in the identification of proteins. In **stage-3** different databases are utilized to annotate the identified proteins. In instances of multiple protein groups, the annotations are merged to the common ancestor (lowest) using prophane. Moreover, the estimation of protein abundance is conducted by employing normalized spectral abundance factors. In **stage-4** diverse visualization strategies assist in the comprehension of the end results. * X!Tandem is an open source software that can match tandem mass spectra with peptide sequences from a database by employing statistical confidence for all the individual matches, ** OMSA- Open Mass Spectrometry Search Algorithm is a probability-based matching algorithm that calculates a score based on statistical significance of a match between observed peptide fragment and those from sequence library

metabolomics, resistance gene-based mining, Ribosomally synthesized and post-translationally modified peptides (RiPPs) and phylogeny-dependent mining constitute the first-generation methods for assessing BGCs in native hosts. Figure 7 outlines the genomic mining strategies and first-generation methods for assessing cryptic biosynthetic genes (BGCs). Recently the identification of a high-throughput technique based on elicitor screening followed by induction of secondary metabolome called HiTES [69, 86] has revolutionized the activation of BGCs in diverse microorganisms. Through the process of elicitor screening, the secondary metabolome of a particular strain is stimulated, and the resulting metabolomes are subsequently visualized using imaging mass spectroscopy. Importantly, this method circumvents the necessity for genetic manipulation, cloning, or culturing techniques.

HiTES is an endogenous monoculture strategy independent of genetics that elicits and detects the cryptic metabolomes of different bacteria. HiTES is used in conjunction with imaging mass spectroscopy (IMS), known as HiTES–IMS [87], for the early detection of small molecules in an untargeted manner and contains two components: activation of cryptic BGCs by elicitor screening and genetic reporter assays [85, 87]. Since the original work in 2014, several variants of HiTES have evolved with different advantages and applicability *e.g.* Bioactivity HiTES [88], FRET coupled HiTES [89], Assay guided HiTES and Genomisotopic approach followed by elicitor library preparation/screening. Figure 8 outlines the workflow of different variants of HiTES.

4.2 *Vectors for Cloning Whole Metagenomic Libraries into Model Lab Strains*

In large insert metagenomic studies, the preferred vectors, such as *E. coli* F-factor-based vectors are designed for BAC and fosmid cloning. The ability to identify novel activities through functional screening is contingent upon the accurate central dogma of the cloned genes. Although *E. coli* has been employed as a host for expressing novel activities, Gabor et al. [91] demonstrated ~60% cryptic genes from 32 prokaryotic genomes expressed in *E. coli.* A similar report by Martinez et al [92] showed the differential abilities of *E. coli*, *P. putida* and *S. lividans* to express heterologous gene clusters [92] Several BAC vectors have been reported to have the ability to transfer genes of interest to a small number of hosts besides *E. coli.* Nevertheless, these vectors have limitations like inefficient transfer, absence of autonomous replication, or restricted host range. There are several factors that can contribute to this issue in heterologous genes, which are different from the ones used by *E. coli.* These factors hinder the recognition of signals associated with these genes. One such factor is the variation in translation initiation codons. Moreover, differences in codon usage, ribosomal binding sites and transcription promoters can also result in the lack of detectable target expression. Challenges such as protein misfolding, gene toxicity, and low secretion of gene products can impede the identification of new genes

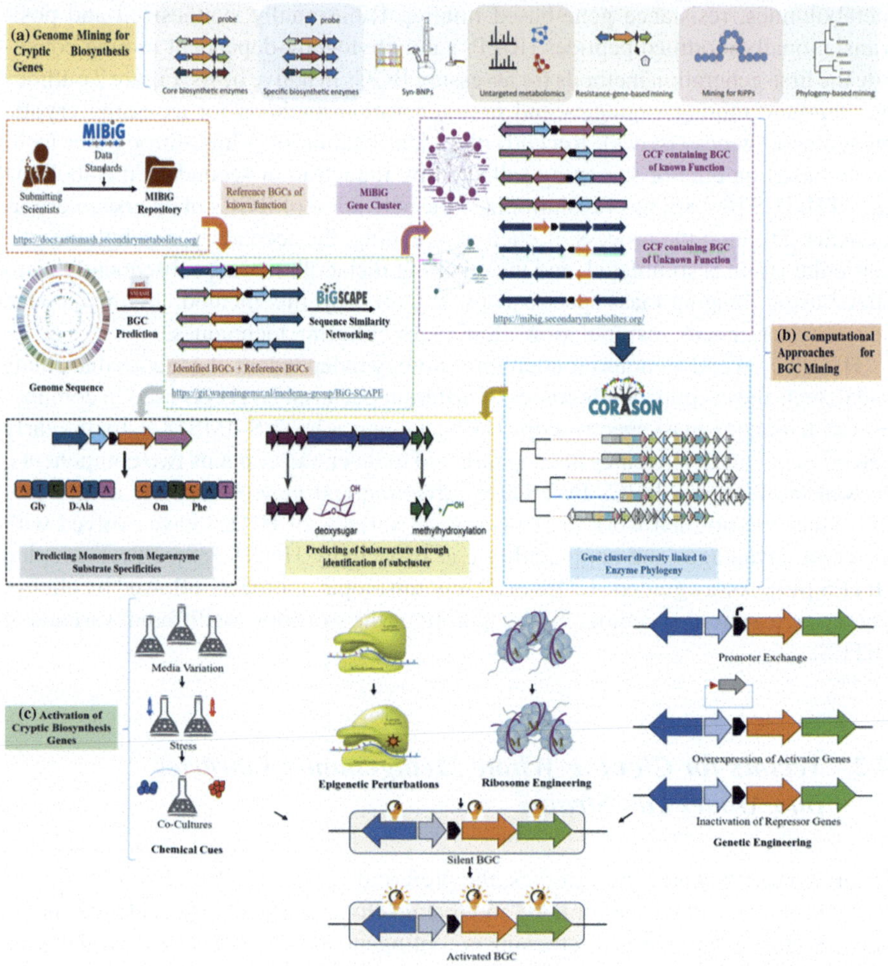

Fig. 7 Genomic mining strategies and first-generation methods for assessing cryptic biosynthetic genes (BGCs). **a** Different methods for genome mining of BGCs viz. probing core biosynthetic genes, probing specific biosynthetic enzymes, synBNPs, Untargeted metabolomics, RiPPs and Phylogeny based mining. **b** Genomic mining for metabolic diversity can be achieved through computational tools like antiSMASH which enables the automatic identification of Biosynthetic Gene Clusters (BGCs) in genome sequences. These BGCs can be compared with databases like MIBiG for dereplication. Additionally, sequencing similarity networking can help in identifying clusters of similar BGCs across extensive datasets, thereby assisting in the annotation of gene cluster families (GCFs). *Source* Van der Hooft et al., with modification. **c** The various first-generation methods for assessing cryptic biosynthetic genes viz. Chemical cues, epigenetic perturbations, ribosome engineering and genetic engineering leading to activation of cryptic biosynthetic genes. *Source* Scherlach et al., with modification [69, 86]

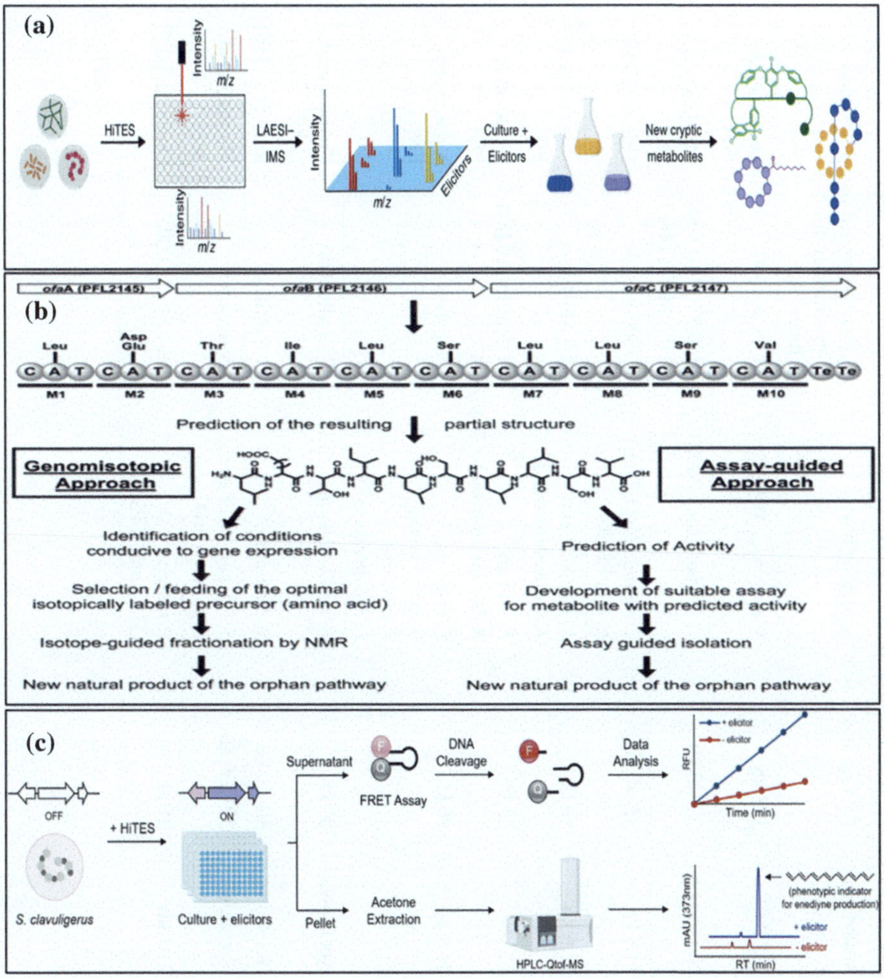

Fig. 8 Different variants of HiTES (**a**) HiTES-IMS workflow. Bacterial culture is assessed in 96-well format by LAESI-IMS. 3D plots are made that link elicitor to metabolites, based on their m/z and intensity values. Isolation and characterization of new cryptic metabolites through large scale cultures along with elicitor. **b** Assay guided and genomisotopic HiTES. The assay-guided technique used by Gross et al. [90] for isolation of orfamides relies on establishing an assay that steers the isolation of natural products. In the genomisotopic procedure, the organism of interest is cultured under conditions conducive to the activation of biosynthetic genes. A labelled amino acid, anticipated to be part of the metabolic product, is included in the culture medium. This label is then employed to direct the purification process of the natural product. *Source* Gross et al. [90] **c** FRET-coupled HiTES workflow. The FRET-coupled HiTES workflow involves the preparation of bacterial cultures in 96-well plates. Cryptic secondary metabolites of these cultures are then activated by subjecting to HiTES. The resulting culture supernatants are analysed with qualitative assay (top path). *Source* Han et al. [89]

Table 4 Host-vector platforms for cloning metagenomic DNA and conducting functional screening. *Source* Adapted from Cheng et al. [94]

Library vector	Vector type	Vector size	Average insert size	Screening hosts	Library size (MB or GB)/Number of clones	References
pLF61	Shuttle/plasmid	5.4 kb	1–3 kb	*S. cerevisiae* W303	1.75×10^6	Damon et al. [95]
pMM436	Shuttle/cosmid	Not shown	30–40 kb	*S. lividans* Δact Δred	367.5 Mb/10,500 2485 Mb/71,000	McMahon et al. [96]
pCT3FK	Shuttle/fosmid	11.6 kb	35–40 kb	*T. thermohilus* BL03, *E. coli* EPI300	≈8000	Leis et al. [97]
pCT3FK	Shuttle/fosmid	11.6 kb	50 kb	*E. coli* K12 and *T. thermohilus* HB27	2.9 Mb/192	Angelov et al. [98]
pLAFR3	Broad-host-range/cosmid	22 kb	25 kb	*E. coli* and *R. leguminosarum* bv. Viciae	2750 Mb/110,000	Wexler et al. [93]
pGNS-BAC-1	Shuttle/BAC	11.9 kb	80 kb	*E. coli* *S. marcescens,* *V. cholerae,* *E. nimipressuralis*	Not shown	Kakirde et al. [99]
pOS700I	Shuttle/cosmid	Not shown	50 kb	*E. coli/S. lividans*	250 Mb/5000	Courtois et al. [100]
pWEB436	Shuttle/cosmid	Not shown	40–50 kb	*S. albus*ara>	1.5 million	Iqbal et al. [101]
pSrpsL14	Shuttle/BAC	Not shown	50–100 kb	*S. lividans* Δact Δred, *P. putida* MBD1	Not shown	Martinez et al. [92]
pKS13S	Broad-host-range/cosmid	21.7 kb	25 kb	*P. putida* G7K2 and KTSK2	294 Mb/24,000	Ono et al. [102]

(continued)

Table 4 (continued)

Library vector	Vector type	Vector size	Average insert size	Screening hosts	Library size (MB or GB)/Number of clones	References
pJWC1	Broad-host-range/cosmid	14 kb	High-molecular weight	*A.tumefaciens* LBA4404, *B. graminis* C4D1M, *C. vibrioides* CB15, *E. coli* Ec 100, *P. putida* KT2440, *R. metallidurans* CH34	750,000	Craig et al. [103]
pJWC1	Broad-host-range/cosmid	14 kb	High-molecular weight	*R. metallidurans* CH34	575,000	Craig et al. [103]
pKS13S	Broad-host-range/cosmid	21.7 kb	25 kb	*P. putida* KTSK2 and *P. putida* G&NAD2	5.2 Gb/208,000	Nagayama et al. [104]
pJC8	Broad-host-range/cosmid	13 kb	33 kb	*P. putida* PpUW2 (PHA-)	363 Gb/9 × 10^6	Cheng and Charles [105]
pRS44	Broad-host-range/fosmid and BAC	10.3 kb	35 kb	*P. fluorescens* and *X. campestris*	20,000	Aakvik et al. [106]

in functional screenings. In contrast, RK2 replicons have the capability to address the construction of metagenomic libraries [93]. Small-insert metagenomic libraries (~10 kb) commonly utilize plasmid vectors, while, cosmid or fosmid vectors are used for large-insert metagenomic libraries (~30–50 kb). Table 4 lists multiple cloning systems for metagenomic heterologous functional screening. Although cosmids can replicate in suitable hosts by virtue of possessing a suitable origin of replication, the absence of a copy number control mechanism renders them unstable. Fosmid vectors, akin to cosmids, employ an *E. coli* oriC (for F-factor) resulting in tightly regulated copy numbers-based replication in *E. coli*. This controlled replication helps mitigates gene toxicity problems in metagenomic analyses [94], with fosmids accommodating DNAs of around 40 to 50 kb. Bacterial artificial chromosomes (BACs) are capable of incorporating ~100–300 kb inserts and employ *E. coli oriC* (for F-factor) with the capability of stably maintaining inserts in *E. coli*. Alternative host organisms have emerged as a viable solution to surpass the metagenomic constraints and enhance the discovery of MNPs. Instead of relying solely on *E. coli*, functional screening tests are now conducted using shuttle vectors and broad-host-range vectors. These vectors, commonly employed in alternative hosts mediated metagenomic screening, possess either a single broad-host origin oriV or multiple oriV, enabling replication in both *E. coli* and other host organisms. Alternatively, metagenomic DNA can be integrated into the chromosome of the screening host [94].

4.3 Checkerboard DNA-DNA Hybridization Technique

Utilizing a non-PCR approach, the Checkerboard DNA-DNA hybridization (CKB) analysis proves to be an immensely effective technique. It simultaneously hybridises 40 digoxigenin (DIG)-labelled whole genome probes (WGPs). These probes are carefully chosen from cultivable ecological indicator microbes within the ecosystem. Developed by Dr. S. S. Socransky and Dr. A. Haffajee at The Forsyth Institute, the CKB analysis allows for the simultaneous quantitative analysis of 40 species in microbial ecosystem samples [107], focusing on dental plaque pathogens [108]. DNA standards and DIG-labelled WGPs are prepared for the target species to facilitate the analysis. These components are then fixed onto a membrane in thin lanes using the specialized "checkerboard" MiniSlot system. Subsequently, the membranes are subjected to simultaneous hybridization with 40 DIG-labelled WGPs in a perpendicular axis, utilizing the specialized checkerboard Miniblotter system. Finally, the resulting hybrids are quantitatively detected.

Although the technique is used for dental plaque microcosms studies, it has the potential to be tweaked for microbial diversity analysis and specially for anaerobic microbes which are often hard to maintain and isolate. Such an advancement could help in profiling unique barophilic microbial diversity which are poorly explore and hold promise for important metabolites with applications in space exploration missions.

4.4 Hybridization-Based Enrichment of Cpn60 Gene Fragments: CaptureSeq

CaptureSeq represents a cutting-edge approach that eliminates the need for amplicons by employing a hybridization-based method with probes derived from the chaperonin-60 gene (*cpn60*) to study intricate microbial communities [109]. The *cpn60* [110] and its archaeal counterpart type II chaperonin [111] have long been acknowledged for their remarkable discriminatory potential across all realms of life [112]. The sequences also fulfil the standard International Barcode of Life criteria [113] and allow for the *de-novo* assembly of OTUs from metagenomics data [114]. Recent sub-OTU identification that varies by a single nucleotide using Illumina sequencing without employing read error correction assembly (also known as amplicon sequence variants, ASV) [115] are demonstrated to be applicable to both 16S [116] and cpn60 [117] amplicon data. Links et al., 2021 developed a custom biotinylated RNA capture bait array for mapping the taxonomy of the entire cpnDB database [118, 119]. This array was evaluated for simultaneous enrichment of complete genomic DNA for respective chaperonin-type sequences. The taxonomic diversity of cpnDB is at par with available 16S reference databases, even with fewer entries [119]. CaptureSeq has exhibited its robustness in delivering dependable results for detecting and quantifying both prokaryotes and eukaryotes and could be a promising technology. The read abundances generated through this technique have consistently demonstrated their quantitative reliability. Furthermore, CaptureSeq has successfully provided comprehensive microbial community profiles across various domains in intricate natural ecosystems.

4.5 Chip-SIP: Stable Isotope Probing Analysed with NanoSIMS

With the underlying objective of linking microbial identity with their function in microbial communities, stable isotope probing (SIP) takes advantage of isotope tracing of biomolecules and even intact cells. While density gradient SIP has been used traditionally, recent development of Chip-SIP technology penetrates beyond simple identification to facilitate qPCR-based calculation of population dynamics of individual taxa [120, 121]. Traditional SIP approaches, already cited in hundreds of publications, can enable activity assays employing a single isotope label at a time instead of multiple isotopes and are limited by the requirement of high levels of isotope (over 8 atom %) [120] and low throughput time-intensive nature. Chip-SIP approaches a similar goal but in a different manner. By inculcating taxonomic identity-based separation of biomolecules followed by measurement of isotopic incorporation of individual rRNAs (Fig. 9), taxa are identified like FISH.

Chip-SIP relies on the utilization of rRNA relative abundance, unamplified isolated rRNA is hybridized to array surface directly (hence avoiding the dilution

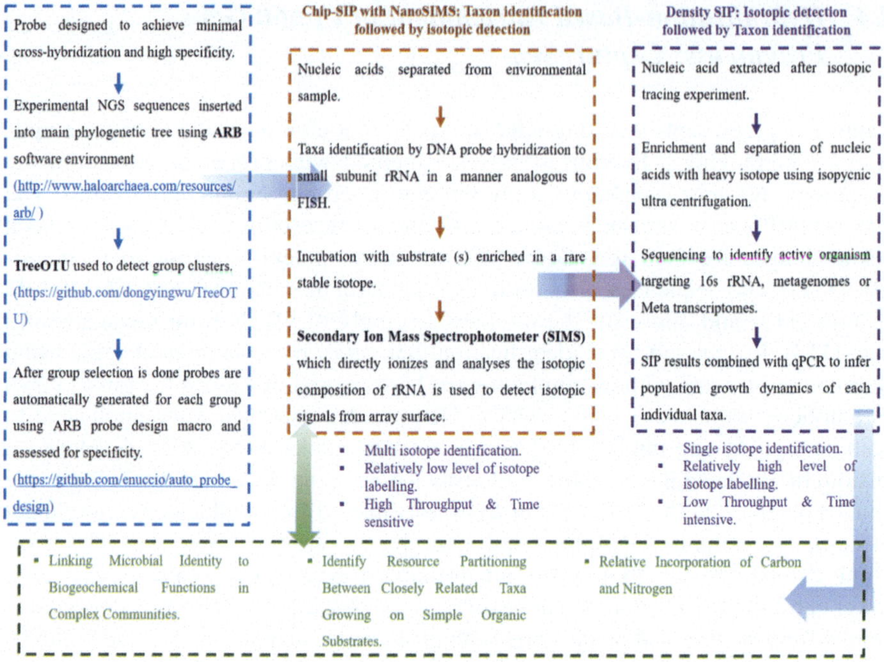

Fig. 9 Workflow of Chip-SIP Technique with NanoSIMS analysis of triple-labelled RNA from purified culture after 16S rRNA targeting microarray

of underrepresented 16S signals below detection limit, a drawback in traditional SIP) followed by Nano-SIMS (Secondary ion mass spectrometer) based detection of isotopic signals. Since its first use in illustration of resource partitioning between closely related microbial communities growing on simple organic substrates [122], the technique has been successfully used to establish the utilization of root and litter detritus by bacteria and eukaryotes [123]. the Chip-SIP technique has been used in several taxonomic differentiation studies [124–126]. Probe designing is the first and most important step in this technique. With the overall goal of minimizing cross-hybridization yet maintaining high specificity, NGS sequences are inserted in the ARB software environment phylogenetic tree (http://www.haloarchaea.com/resources/arb/), and then group clusters are detected using the TreeOTU tool (https://github.com/dongyingwu/TreeOTU) [127]. Once group selection is achieved probes are generated automatically for each group using the ARB Probe Design macro, which is further assessed using ARB Probe Match macro (https://github.com/enuccio/auto_probe_design). Chip-SIP microarrays are made by a photolithographic fabrication method called as MAS (Maskless Array Synthesis) [128, 129], which yields quite small spots (196 μm^2) with a resolution of a single micron. The resultant high-density enables efficient, time-effective isotopic analysis. A high-sensitivity NanoSIMS analysis of these spots after overlaying with a conductive medium, instrument tuning and L'IMAGE software-based stitching of images for each subarray is done to ensure the capture of signals corresponding to even a small amount of rRNA in each probe and resolution of isobaric interferences necessary for reliable measurements. Finally, the isotopic enrichment corresponding to each probe spot matches the probe's identity. Fluorescent intensity is used to quantify the RNA hybridized for each probe spot. The isotopic enrichment is plotted versus probe fluorescence to calculate the linear regression slope (Y-intercept constrained to natural isotope abundance). This slope is used to compare the relative incorporation of a given substrate by different taxa and is called as Hybridization-corrected enrichment (HCE) [130].

5 Limitations of Current Techniques and Path Ahead

Despite all the developments seen in all the microbial diversity assessment techniques, some inherent limitations are associated with them. While PCR bias is the main problem of PCR-dependent approaches, molecular techniques are often associated with problems such as similar melting profiles of different DNA fragments, the quality of DNA isolated, and trade-offs between probe designing time and a cost and applicability. However, there is still scope for the evolution of such technology and ways to mitigate such challenges, which can be tailored as per need by combining the pros of a combination of techniques mentioned in the proceeding sections. For example, although the Checkerboard DNA-DNA hybridization technique is used for dental plaque microcosm studies, it has the potential to be tweaked for microbial diversity analysis and specially for anaerobic microbes, which is often hard to maintain and isolate. Such an advancement could help in profiling unique barophilic microbial diversity, which is poorly explored and holds promise for important metabolites with potential application in space exploration missions. The recent progress in DNA capture, coupled with CaptureSeq technique, has revolutionized the enrichment of samples for specific DNA sequences, even enabling the sequencing of very old pathogenic genomes retrieved from the teeth of over a century-old human carcass [118]. Furthermore, the recent display of interplay between Chip-SIP and Proteomic SIP of marine communities [131] have opened new avenues to link microbial identity with their function in microorganisms. However, there are still many opportunities associated and directions in which such techniques can be directed such as microbial stewardship concept and using CRISPR loci profiling to identify probable hosts for individual microbes in a microbial community. Though, we are not focusing on the cons; still discussing these techniques in brief here.

The concept of microbiome stewardship, which focuses on restoring the microbial diversity among the microbial communities of diverse niches, is a new concept. Loss of global biodiversity and mass extinction of species leading to disturbances of diverse ecosystems is a grave matter of concern that needs immediate address. Microbial stewardship can be an answer to such a problem [132]. Although much of the work in this field is yet focused on host-associated microbial interaction (for example, Gut associated microbiome and oral microbiome in humans, etc.) [133, 134], we believe this concept can play a broader role in restoring microbial diversity at a broader scale also. Restoring the lost microbial diversity in pristine habitats could be a possibility in the future.

CRISPR loci profiling to identify probable hosts for individual microbes in a microbial community, on the other hand, on the other hand, can be useful in identifying a probable host range of newly emerging pathogens that are coming out of rapidly disappearing habitats as a result of humane interventions leading to global warming. Metagenomic analyses of microbes have revealed widespread diversity in the extreme niches, but it remains a challenge to identify specific hosts for a microbial assemblage. The CRISPR system is a putative antiviral immunity mechanism found in both archaea and bacteria [135, 136]. CRISPR loci generally consist of a series of short repeats, each approximately 20–50 bp, interspersed by spacers about 25–75 bp [137]. CRISPR loci are thought to create immunity when short sequences derived from invaders such as viruses or plasmids are incorporated as spacers between the repeat sequences by genes involved in the CRISPR response, known as CRISPR-associated (cas) genes. When introduced genetic elements, such as viruses or plasmids, have a 100% match to a pre-existing CRISPR spacer sequence in the host genome, these elements are recognized as pathogenic invaders [137]. In response, the CRISPR/Cas system cleaves the invading nucleic acid [137]. CRISPR loci effectively act as libraries of previous viral infections; thus, analyses of CRISPR spacers across species have great potential for microbial and viral ecology. Recent studies have found a high degree of variability within CRISPR spacer sequences, implying a rapid rate of host-virus coevolution. These studies have also demonstrated a clear biogeographic distribution in CRISPR spacers. Additionally, Snyder et al. have designed microarrays using CRISPR spacer sequences to detect viruses in environmental samples [137].

To address the problem of identifying specific hosts for a microbial assemblage, Anderson et al., analysed the viral metagenome of a northeast Pacific hydrothermal vent with a comprehensive database of spacers derived from the clustered regularly interspaced short palindromic repeat (CRISPR) putative immune system [138, 139]. Although the method was demonstrated for identifying specific hosts for viral assemblage, it can also be used to extrapolate its application to other microbes (Fig. 10). Such a development will widen the role of bioprospection in the prior-readiness of medical research against 'disease-X' that could erupt as polar caps melt or hot spots of diversity and extreme niches lose their integrity due to future climatic disturbance as a result of humane interventions.

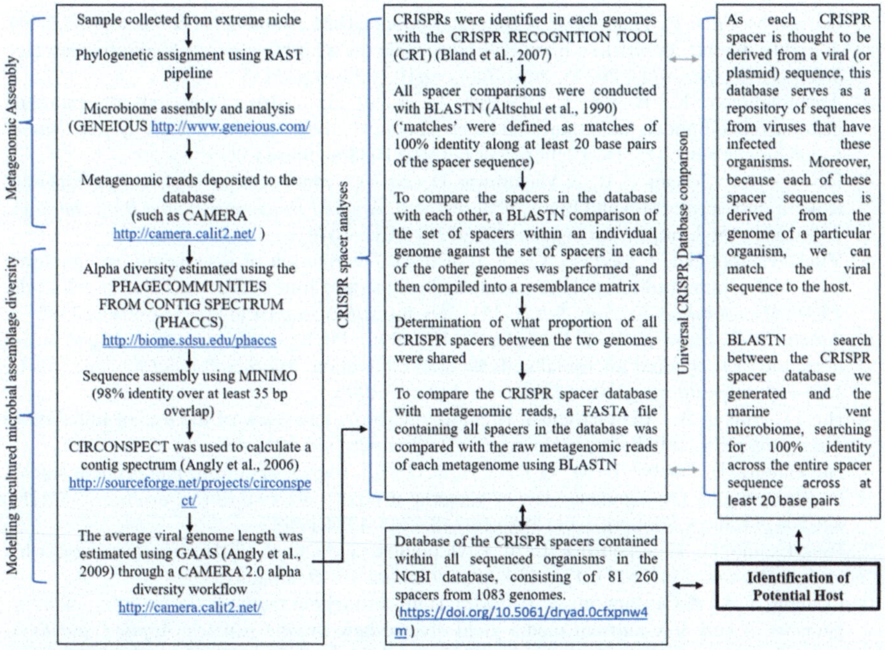

Fig. 10 Using CRISPR loci as library of previous infections and application in bioprospection and finding probable host range of infectious microbes

Declaration All authors declare no conflict of interest.

Acknowledgments DS duly acknowledges the Department of Biotechnology (DBT), New Delhi, for a project GAP0310 under "Himalayan Bioresources Mission". All authors duly acknowledge CSIR, New Delhi, Govt. of India, for the financial support of in-house grant number MLP0201. Tamanna acknowledges financial support from the Council of Scientific and Industrial Research (CSIR) for JRF award no. 31/0054(152257)/2022-EMR-I, SC acknowledges financial support from DBT for JRF award no. DBT/2021- 22/IHBT/1864, and MM acknowledges financial support from the University Grant Commission for JRF award no. 211610215506. The Director, CSIR-IHBT, is duly acknowledged for providing the necessary facilities to carried the work. This manuscript represents CSIR-IHBT communication number 5588.

Authors Contributions HS and DS conceived the project and decided the outline of manuscript. T, SC and MM helped in organizing the manuscript and editing.

References

1. Vilanova, C., & Porcar, M. (2016). Are multi-omics enough? *Nature Microbiology, 1*, 16101. https://doi.org/10.1038/nmicrobiol.2016.101

2. Dorado-Morales, P., Vilanova, C., Peretó, J., Codoñer, F. M., Ramón, D., & Porcar, M. (2016). A highly diverse, desert-like microbial biocenosis on solar panels in a Mediterranean city. *Scientific Reports, 6*(1), 29235. https://doi.org/10.1038/srep29235
3. Anantharaman, K., Brown, C. T., Hug, L. A., et al. (2016). Thousands of microbial genomes shed light on interconnected biogeochemical processes in an aquifer system. *Nature Communications, 7*(1), 13219. https://doi.org/10.1038/ncomms13219
4. Rastogi, R. P., Sonani, R. R., & Madamwar, D. (2015). Cyanobacterial sunscreen scytonemin: Role in photoprotection and biomedical research. *Applied Biochemistry and Biotechnology, 176*, 1551–1563. https://doi.org/10.1007/s12010-015-1676-1
5. Vítek, P., Jehlička, J., Ascaso, C., et al. (2014). Distribution of scytonemin in endolithic microbial communities from halite crusts in the hyperarid zone of the Atacama Desert, Chile. *FEMS Microbiology Ecology, 90*(2), 351–366. https://doi.org/10.1111/1574-6941.12387
6. Amann, R. I., Ludwig, W., & Schleifer, K.-H. (1995). Phylogenetic identification and in situ detection of individual microbial cells without cultivation. *Microbiological Reviews, 59*(1), 143–169. https://doi.org/10.1128/mr.59.1.143-169.1995
7. Hug, L., Baker, B., Anantharaman, K., et al. (2016). A new view of the tree of life. *Nature Microbiology, 1*, 16048. https://doi.org/10.1038/nmicrobiol.2016.48
8. Hugenholtz, P., Goebel, B. M., & Pace, N. R. (1998). Impact of culture-independent studies on the emerging phylogenetic view of bacterial diversity. *Journal of Bacteriology, 180*(18), 4765–4774. https://doi.org/10.1128/jb.180.18.4765-4774.1998
9. Paez-Espino, D., Eloe-Fadrosh, E. A., Pavlopoulos, G. A., et al. (2016). Uncovering Earth's virome. *Nature, 536*(7617), 425–430. https://doi.org/10.1038/nature19094
10. Connon, S. A., & Giovannoni, S. J. (2002). High-throughput methods for culturing microorganisms in very-low-nutrient media yield diverse new marine isolates. *Applied and Environmental Microbiology, 68*(8), 3878–3885. https://doi.org/10.1128/AEM.68.8.3878-3885.2002
11. McLaren, M. R., Willis, A. D., & Callahan, B. J. (2019). Consistent and correctable bias in metagenomic sequencing experiments. *eLife, 8*, e46923. https://doi.org/10.7554/eLife.46923
12. Tringe, S. G., & Hugenholtz, P. (2008). A renaissance for the pioneering 16S rRNA gene. *Current Opinion in Microbiology, 11*(5), 442–446. https://doi.org/10.1016/j.mib.2008.09.011
13. Löhr, A. J., Laverman, A. M., Braster, M., Van Straalen, N. M., & Röling, W. F. (2006). Microbial communities in the world's largest acidic volcanic lake, Kawah Ijen in Indonesia, and in the Banyupahit River originating from it. *Microbial Ecology, 52*, 609–618. https://doi.org/10.1007/s00248-006-9068-2
14. Baik, K. S., Park, S. C., Kim, E. M., et al. (2008). Diversity of bacterial community in freshwater of Woopo wetland. *The Journal of Microbiology, 46*, 647–655. https://doi.org/10.1007/s12275-008-0135-x
15. Bruce, T., Martinez, I. B., Maia Neto, O., Vicente, A. C. P., Kruger, R. H., & Thompson, F. L. (2010). Bacterial community diversity in the Brazilian Atlantic Forest soils. *Microbial Ecology, 60*, 840–849. https://doi.org/10.1007/s00248-010-9750-2
16. Chen, Y., Wu, L., Boden, R., et al. (2009). Life without light: Microbial diversity and evidence of sulfur-and ammonium-based chemolithotrophy in Movile Cave. *The ISME Journal, 3*(9), 1093–1104. https://doi.org/10.1038/ismej.2009.57
17. Shravage, B. V., Dayananda, K. M., Patole, M. S., & Shouche, Y. S. (2007). Molecular microbial diversity of a soil sample and detection of ammonia oxidizers from Cape Evans, Mcmurdo Dry Valley, Antartica. *Microbiological Research, 162*(1), 15–25. https://doi.org/10.1016/j.micres.2006.01.005
18. Vianna, M., Conrads, G., Gomes, B., & Horz, H. (2009). T-RFLP-based mcrA gene analysis of methanogenic archaea in association with oral infections and evidence of a novel Methanobrevibacter phylotype. *Oral Microbiology and Immunology, 24*(5), 417–422. https://doi.org/10.1111/j.1399-302X.2009.00539.x
19. Fischer, S. G., & Lerman, L. S. (1979). Length-independent separation of DNA restriction fragments in two-dimensional gel electrophoresis. *Cell, 16*(1), 191–200. https://doi.org/10.1016/0092-8674(79)90200-9

20. Fischer, S., & Lerman, L. (1983). DNA fragments differing by single base-pair substitutions are separated in denaturing gradient gels: Correspondence with melting theory. *Proceedings of the National Academy of Sciences, 80*(6), 1579–1583. https://doi.org/10.1073/pnas.80.6.1579
21. Myers, R. M., Fischer, S. G., Lerman, L. S., & Maniatis, T. (1985). Nearly all single base substitutions in DNA fragments joined to a GC-clamp can be detected by denaturing gradient gel electrophoresis. *Nucleic Acids Research, 13*(9), 3131–3145. https://doi.org/10.1093/nar/13.9.3131
22. Muyzer, G., Hottentr¨ ager, S., Teske, A., Wawer, C. (1996). Denaturing gradient gel electrophoresis of PCR-amplified 16S rDNA– A new molecular approach to analyse the genetic diversity of mixed microbial communities. In: Akkermans, A.D.L., van Elsas, J.D., de Bruijn, F.J. (Eds) *Molecular Microbial Ecology Manual* (3.4.4: pp. 1–23) Kluwer Academic Publishers, Dordrecht, The Netherlands.
23. Lee, D.-H., Zo, Y.-G., & Kim, S.-J. (1996). Nonradioactive method to study genetic profiles of natural bacterial communities by PCR-single-strand-conformation polymorphism. *Applied and Environmental Microbiology, 62*(9), 3112–3120. https://doi.org/10.1128/aem.62.9.3112-3120.1996
24. Sheffield, V. C., Beck, J. S., Kwitek, A. E., Sandstrom, D. W., & Stone, E. M. (1993). The sensitivity of single-strand conformation polymorphism analysis for the detection of single base substitutions. *Genomics, 16*(2), 325–332. https://doi.org/10.1006/geno.1993.1193
25. Kakavas, K. V. (2021). Sensitivity and applications of the PCR single-strand conformation polymorphism method. *Molecular Biology Reports, 48*(4), 3629–3635. https://doi.org/10.1007/s11033-021-06349-2
26. Jeffreys, A., & Flavell, R. (1977). A physical map of the DNA regions flanking the rabbit β-globin gene. *Cell, 12*(2), 429–439. https://doi.org/10.1016/0092-8674(77)90119-2
27. Laguerre, G., Allard, M.-R., Revoy, F., & Amarger, N. (1994). Rapid identification of rhizobia by restriction fragment length polymorphism analysis of PCR-amplified 16S rRNA genes. *Applied and Environmental Microbiology, 60*(1), 56–63. https://doi.org/10.1128/aem.60.1.56-63.1994
28. Botstein, D., White, R.L., Skolnick, M., Davis, R.W. (1980). Construction of a genetic linkage map in man using restriction fragment length polymorphisms. *American Journal of Human Genetics* 32(3):314. https://pubmed.ncbi.nlm.nih.gov/6247908
29. Moyer, C. L., Dobbs, F. C., & Karl, D. M. (1994). Estimation of diversity and community structure through restriction fragment length polymorphism distribution analysis of bacterial 16S rRNA genes from a microbial mat at an active, hydrothermal vent system, Loihi Seamount, Hawaii. *Applied and Environmental Microbiology, 60*(3), 871–879. https://doi.org/10.1128/aem.60.3.871-879.1994
30. Marsh, T. L. (1999). Terminal restriction fragment length polymorphism (T-RFLP): An emerging method for characterizing diversity among homologous populations of amplification products. *Current Opinion in Microbiology, 2*(3), 323–327. https://doi.org/10.1016/s1369-5274(99)80056-3
31. Morris, T., Robertson, B., & Gallagher, M. (1996). Rapid reverse transcription-PCR detection of hepatitis C virus RNA in serum by using the TaqMan fluorogenic detection system. *Journal of Clinical Microbiology, 34*(12), 2933–2936. https://doi.org/10.1128/jcm.34.12.2933-2936.1996
32. Marsh, T. L., Saxman, P., Cole, J., & Tiedje, J. (2000). Terminal restriction fragment length polymorphism analysis program, a web-based research tool for microbial community analysis. *Applied and Environmental Microbiology, 66*(8), 3616–3620. https://doi.org/10.1128/AEM.66.8.3616-3620.2000
33. Osborn, A. M., Moore, E. R., & Timmis, K. N. (2000). An evaluation of terminal-restriction fragment length polymorphism (T-RFLP) analysis for the study of microbial community structure and dynamics. *Environmental Microbiology, 2*(1), 39–50. https://doi.org/10.1046/j.1462-2920.2000.00081.x

34. Amann, R. I., Krumholz, L., & Stahl, D. A. (1990). Fluorescent-oligonucleotide probing of whole cells for determinative, phylogenetic, and environmental studies in microbiology. *Journal of Bacteriology, 172*(2), 762–770. https://doi.org/10.1128/jb.172.2.762-770.1990
35. Guschin, D. Y., Mobarry, B. K., Proudnikov, D., Stahl, D. A., Rittmann, B. E., & Mirzabekov, A. D. (1997). Oligonucleotide microchips as genosensors for determinative and environmental studies in microbiology. *Applied and Environmental Microbiology, 63*(6), 2397–2402. https://doi.org/10.1128/aem.63.6.2397-2402.1997
36. Wilson, K. H., Wilson, W. J., Radosevich, J. L., et al. (2002). High-density microarray of small-subunit ribosomal DNA probes. *Applied and Environmental Microbiology, 68*(5), 2535–2541. https://doi.org/10.1128/AEM.68.5.2535-2541.2002
37. Brodie, E. L., DeSantis, T. Z., Joyner, D. C., et al. (2006). Application of a high-density oligonucleotide microarray approach to study bacterial population dynamics during uranium reduction and reoxidation. *Applied and Environmental Microbiology, 72*(9), 6288–6298. https://doi.org/10.1128/AEM.00246-06
38. Sebat, J. L., Colwell, F. S., & Crawford, R. L. (2003). Metagenomic profiling: Microarray analysis of an environmental genomic library. *Applied and Environmental Microbiology, 69*(8), 4927–4934. https://doi.org/10.1128/AEM.69.8.4927-4934.2003
39. Park, S.-J., Kang, C.-H., Chae, J.-C., & Rhee, S.-K. (2008). Metagenome microarray for screening of fosmid clones containing specific genes. *FEMS Microbiology Letters, 284*(1), 28–34. https://doi.org/10.1111/j.1574-6968.2008.01180.x
40. Handelsman, J., Rondon, M. R., Brady, S. F., Clardy, J., & Goodman, R. M. (1998). Molecular biological access to the chemistry of unknown soil microbes: A new frontier for natural products. *Chemistry & Biology, 5*(10), R245–R249. https://doi.org/10.1016/s1074-5521(98)90108-9
41. Danko, D., Bezdan, D., Afshin, E.E., et al. (2021). A global metagenomic map of urban microbiomes and antimicrobial resistance. *Cell* 184(13):3376–3393. e17. https://doi.org/10.1016/j.cell.2021.05.002
42. Namiki, T., Hachiya, T., Tanaka, H., Sakakibara, Y. (2011). MetaVelvet: an extension of Velvet assembler to de novo metagenome assembly from short sequence reads. 2011:116–124. https://doi.org/10.1145/2147805.2147818
43. Bankevich, A., Nurk, S., Antipov, D., et al. (2012). SPAdes: A new genome assembly algorithm and its applications to single-cell sequencing. *Journal of Computational Biology, 19*(5), 455–477. https://doi.org/10.1089/cmb.2012.0021
44. Boisvert, S., Raymond, F., Godzaridis, É., Laviolette, F., & Corbeil, J. (2012). Ray Meta: Scalable de novo metagenome assembly and profiling. *Genome Biology, 13*, 1–13. https://doi.org/10.1186/gb-2012-13-12-r122
45. Peng, Y., Leung, H. C., Yiu, S.-M., & Chin, F. Y. (2012). IDBA-UD: A de novo assembler for single-cell and metagenomic sequencing data with highly uneven depth. *Bioinformatics, 28*(11), 1420–1428. https://doi.org/10.1093/bioinformatics/bts174
46. Li, D., Liu, C.-M., Luo, R., Sadakane, K., & Lam, T.-W. (2015). MEGAHIT: An ultra-fast single-node solution for large and complex metagenomics assembly via succinct de Bruijn graph. *Bioinformatics, 31*(10), 1674–1676. https://doi.org/10.1093/bioinformatics/btv033
47. Pell, J., Hintze, A., Canino-Koning, R., Howe, A., Tiedje, J. M., & Brown, C. T. (2012). Scaling metagenome sequence assembly with probabilistic de Bruijn graphs. *Proceedings of the National Academy of Sciences, 109*(33), 13272–13277. https://doi.org/10.1073/pnas.1121464109
48. Alneberg, J., Bjarnason, B. S., De Bruijn, I., et al. (2014). Binning metagenomic contigs by coverage and composition. *Nature Methods, 11*(11), 1144–1146. https://doi.org/10.1038/nmeth.3103
49. Wu Y-W, Simmons BA, Singer SW. (2016). MaxBin 2.0: an automated binning algorithm to recover genomes from multiple metagenomic datasets. *Bioinformatics* 32(4):605–607. https://doi.org/10.1093/bioinformatics/btv638
50. Imelfort, M., Parks, D., Woodcroft, B. J., Dennis, P., Hugenholtz, P., & Tyson, G. W. (2014). GroopM: An automated tool for the recovery of population genomes from related metagenomes. *PeerJ, 2*, e603. https://doi.org/10.7717/peerj.603

51. McHardy, A. C., Martín, H. G., Tsirigos, A., Hugenholtz, P., & Rigoutsos, I. (2007). Accurate phylogenetic classification of variable-length DNA fragments. *Nature Methods, 4*(1), 63–72. https://doi.org/10.1038/nmeth976
52. Chan, C.-K.K., Hsu, A.L., Halgamuge, S.K., Tang, S.-L. (2008). Binning sequences using very sparse labels within a metagenome. *BMC Bioinformatics* 9:1–17. https://doi.org/10.1186/1471-2105-9-215
53. Zheng, H., & Wu, H. (2010). Short prokaryotic DNA fragment binning using a hierarchical classifier based on linear discriminant analysis and principal component analysis. *Journal of Bioinformatics and Computational Biology, 8*(06), 995–1011. https://doi.org/10.1142/S0219720010005051
54. Diaz, N. N., Krause, L., Goesmann, A., Niehaus, K., & Nattkemper, T. W. (2009). TACOA–Taxonomic classification of environmental genomic fragments using a kernelized nearest neighbor approach. *BMC Bioinformatics, 10*, 1–16. https://doi.org/10.1186/1471-2105-10-56
55. Markowitz, V.M., Ivanova, N.N., Szeto, E., et al. (2007). IMG/M: a data management and analysis system for metagenomes. *Nucleic Acids Research* 2007;36(suppl_1):D534-D538. https://doi.org/10.1093/nar/gkm869
56. Glass, E.M., Wilkening, J., Wilke, A., Antonopoulos, D., Meyer, F. (2010). Using the metagenomics RAST server (MG-RAST) for analyzing shotgun metagenomes. *Cold Spring Harbor Protocols* 2010(1):pdb. prot5368. https://doi.org/10.1101/pdb.prot5368
57. Huson, D. H., Auch, A. F., Qi, J., & Schuster, S. C. (2007). MEGAN analysis of metagenomic data. *Genome Research, 17*(3), 377–386. https://doi.org/10.1101/gr.5969107
58. Krause, L., Diaz, N. N., Goesmann, A., et al. (2008). Phylogenetic classification of short environmental DNA fragments. *Nucleic Acids Research, 36*(7), 2230–2239. https://doi.org/10.1093/nar/gkn038
59. Monzoorul Haque, M., Ghosh, T. S., Komanduri, D., & Mande, S. S. (2009). SOrt-ITEMS: Sequence orthology based approach for improved taxonomic estimation of metagenomic sequences. *Bioinformatics, 25*(14), 1722–1730. https://doi.org/10.1093/bioinformatics/btp317
60. Liu, B., Gibbons, T., Ghodsi, M., Treangen, T., & Pop, M. (2011). Accurate and fast estimation of taxonomic profiles from metagenomic shotgun sequences. *Genome Biology, 12*, 1–27. https://doi.org/10.1186/1471-2164-12-S2-S4
61. Brady, A., & Salzberg, S. L. (2009). Phymm and PhymmBL: Metagenomic phylogenetic classification with interpolated Markov models. *Nature Methods, 6*(9), 673–676. https://doi.org/10.1038/nmeth.1358
62. Leung, H. C., Yiu, S.-M., Yang, B., et al. (2011). A robust and accurate binning algorithm for metagenomic sequences with arbitrary species abundance ratio. *Bioinformatics, 27*(11), 1489–1495. https://doi.org/10.1093/bioinformatics/btr186
63. Lukashin, A.V., Borodovsky, M., GeneMark, hmm. (1998). New solutions for gene finding. *Nucleic Acids Research* 26(4):1107–1115. https://doi.org/10.1093/nar/26.4.1107
64. Delcher, A. L., Harmon, D., Kasif, S., White, O., & Salzberg, S. L. (1999). Improved microbial gene identification with GLIMMER. *Nucleic Acids Research, 27*(23), 4636–4641. https://doi.org/10.1093/nar/27.23.4636
65. Rho, M., Tang, H., & Ye, Y. (2010). FragGeneScan: Predicting genes in short and error-prone reads. *Nucleic Acids Research, 38*(20), e191–e191. https://doi.org/10.1093/nar/gkq747
66. Noguchi, H., Taniguchi, T., & Itoh, T. (2008). MetaGeneAnnotator: Detecting species-specific patterns of ribosomal binding site for precise gene prediction in anonymous prokaryotic and phage genomes. *DNA Research, 15*(6), 387–396. https://doi.org/10.1093/dnares/dsn027
67. Hoff, K.J., Lingner, T., Meinicke, P., Orphelia, T.M. (2009). Predicting genes in metagenomic sequencing reads. *Nucleic Acids Research* 37(suppl_2):W101-W105. https://doi.org/10.1093/nar/gkp327
68. Poretsky, R. S., Hewson, I., Sun, S., Allen, A. E., Zehr, J. P., & Moran, M. A. (2009). Comparative day/night metatranscriptomic analysis of microbial communities in the North Pacific subtropical gyre. *Environmental Microbiology, 11*(6), 1358–1375. https://doi.org/10.1111/j.1462-2920.2008.01863.x

69. Van Der Hooft, J. J., Mohimani, H., Bauermeister, A., Dorrestein, P. C., Duncan, K. R., & Medema, M. H. (2020). Linking genomics and metabolomics to chart specialized metabolic diversity. *Chemical Society Reviews, 49*(11), 3297–3314. https://doi.org/10.1039/D0CS00162G
70. Wang, M., Carver, J. J., Phelan, V. V., et al. (2016). Sharing and community curation of mass spectrometry data with Global Natural Products Social Molecular Networking. *Nature Biotechnology, 34*(8), 828–837. https://doi.org/10.1038/nbt.3597
71. Kleiner, M. (2019). Metaproteomics: much more than measuring gene expression in microbial communities. *Msystems* 4(3). https://doi.org/10.1128/msystems.00115-19
72. Hettich, R. L., Pan, C., Chourey, K., & Giannone, R. J. (2013). Metaproteomics: Harnessing the power of high-performance mass spectrometry to identify the suite of proteins that control metabolic activities in microbial communities. *Analytical Chemistry, 85*(9), 4203–4214. https://doi.org/10.1021/ac303053e
73. Schiebenhoefer, H., Schallert, K., Renard, B. Y., et al. (2020). A complete and flexible workflow for metaproteomics data analysis based on MetaProteomeAnalyzer and Prophane. *Nature Protocols, 15*(10), 3212–3239. https://doi.org/10.1038/s41596-020-0368-7
74. Muth, T., Behne, A., Heyer, R., et al. (2015). The MetaProteomeAnalyzer: A powerful open-source software suite for metaproteomics data analysis and interpretation. *Journal of Proteome Research, 14*(3), 1557–1565. https://doi.org/10.1021/pr501246w
75. Perkins, D.N., Pappin, D.J., Creasy, D.M., Cottrell, J.S. (1999). Probability-based protein identification by searching sequence databases using mass spectrometry data. *Electrophoresis: An International Journal* 20(18):3551–3567. https://doi.org/10.1002/(SICI)1522-2683(19991201)20:18%3C3551::AIDELPS3551%3E3.0.CO;2–2
76. Gurdeep Singh, R., Tanca, A., Palomba, A., et al. (2018). Unipept 4.0: functional analysis of metaproteome data. *Journal of Proteome Research* 2018;18(2):606–615. https://doi.org/10.1021/acs.jproteome.8b00716
77. Werner, J., Géron, A., Kerssemakers, J., & Matallana-Surget, S. (2019). MPies: A novel metaproteomics tool for the creation of relevant protein databases and automatized protein annotation. *Biology Direct, 14*, 1–5. https://doi.org/10.1186/s13062-019-0253-x
78. Riffle, M., May, D. H., Timmins-Schiffman, E., et al. (2017). MetaGOmics: A web-based tool for peptide-centric functional and taxonomic analysis of metaproteomics data. *Proteomes, 6*(1), 2. https://doi.org/10.3390/proteomes6010002
79. Cheng, K., Ning, Z., Zhang, X., et al. (2017). MetaLab: An automated pipeline for metaproteomic data analysis. *Microbiome., 5*, 1–10. https://doi.org/10.1186/s40168-017-0375-2
80. Liao, B., Ning, Z., Cheng, K., et al. (2018). iMetaLab 1.0: a web platform for metaproteomics data analysis. *Bioinformatics* 34(22):3954–3956. https://doi.org/10.1093/bioinformatics/bty466
81. Scherlach, K., & Hertweck, C. (2009). Triggering cryptic natural product biosynthesis in microorganisms. *Organic & Biomolecular Chemistry, 7*(9), 1753–1760. https://doi.org/10.1039/B821578B
82. Zarins-Tutt, J. S., Barberi, T. T., Gao, H., et al. (2016). Prospecting for new bacterial metabolites: A glossary of approaches for inducing, activating and upregulating the biosynthesis of bacterial cryptic or silent natural products. *Natural Product Reports, 33*(1), 54–72. https://doi.org/10.1039/C5NP00111K
83. Romano, S., Jackson, S. A., Patry, S., & Dobson, A. D. (2018). Extending the "one strain many compounds" (OSMAC) principle to marine microorganisms. *Marine Drugs, 16*(7), 244. https://doi.org/10.3390/md16070244
84. Bode, H. B., Bethe, B., Höfs, R., & Zeeck, A. (2002). Big effects from small changes: Possible ways to explore nature's chemical diversity. *ChemBioChem, 3*(7), 619–627. https://doi.org/10.1002/1439-7633(20020703)3:7%3C619::AID-CBIC619%3E3.0.CO;2-9
85. Seyedsayamdost, M. R. (2014). High-throughput platform for the discovery of elicitors of silent bacterial gene clusters. *Proceedings of the National Academy of Sciences, 111*(20), 7266–7271. https://doi.org/10.1073/pnas.1400019111

86. Scherlach, K., & Hertweck, C. (2021). Mining and unearthing hidden biosynthetic potential. *Nature Communications, 12*(1), 3864. https://doi.org/10.1038/s41467-021-24133-5
87. Xu, F., Nazari, B., Moon, K., Bushin, L. B., & Seyedsayamdost, M. R. (2017). Discovery of a cryptic antifungal compound from Streptomyces albus J1074 using high-throughput elicitor screens. *Journal of the American Chemical Society, 139*(27), 9203–9212. https://doi.org/10.1021/jacs.7b02716
88. Moon, K., Xu, F., & Seyedsayamdost, M. R. (2019). Cebulantin, a cryptic lanthipeptide antibiotic uncovered using bioactivity-coupled HiTES. *Angewandte Chemie International Edition, 58*(18), 5973–5977. https://doi.org/10.1002/anie.201901342
89. Han, E. J., Lee, S. R., Townsend, C. A., & Seyedsayamdost, M. R. (2023). Targeted discovery of cryptic Enediyne natural products via FRET-coupled high-throughput elicitor screening. *ACS Chemical Biology, 18*(8), 1854–1862. https://doi.org/10.1021/acschembio.3c00281
90. Gross, H., Stockwell, V. O., Henkels, M. D., Nowak-Thompson, B., Loper, J. E., & Gerwick, W. H. (2007). The genomisotopic approach: A systematic method to isolate products of orphan biosynthetic gene clusters. *Chemistry & Biology, 14*(1), 53–63. https://doi.org/10.1016/J.chembiol.2006.11.007
91. Gabor, E. M., Alkema, W. B., & Janssen, D. B. (2004). Quantifying the accessibility of the metagenome by random expression cloning techniques. *Environmental Microbiology, 6*(9), 879–886. https://doi.org/10.1111/j.1462-2920.2004.00640.x
92. Martinez, A., Kolvek, S. J., Yip, C. L. T., et al. (2004). Genetically modified bacterial strains and novel bacterial artificial chromosome shuttle vectors for constructing environmental libraries and detecting heterologous natural products in multiple expression hosts. *Applied and Environmental Microbiology, 70*(4), 2452–2463. https://doi.org/10.1128/AEM.70.4.2452-2463.2004
93. Wexler, M., Bond, P. L., Richardson, D. J., & Johnston, A. W. (2005). A wide host-range metagenomic library from a waste water treatment plant yields a novel alcohol/aldehyde dehydrogenase. *Environmental Microbiology, 7*(12), 1917–1926. https://doi.org/10.1111/j.1462-2920.2005.00854.x
94. Cheng. J., Lam, K.N., Engel, K., Hall, M., Neufeld, J.D., Charles, T.C. (2017). Metagenomic cosmid libraries suitable for functional screening in proteobacteria. *Functional Metagenomics: Tools and Applications.* 2017:1–11. https://doi.org/10.1007/978-3-319-61510-3_1
95. Damon, C., Vallon, L., Zimmermann, S. et al. (2011). A novel fungal family of oligopeptide transporters identified by functional metatranscriptomics of soil eukaryotes. *The ISME Journal, 5*(12), 1871–1880. https://doi.org/10.1038/ismej.2011.67
96. McMahon, M. D., Guan, C., Handelsman, J., & Thomas, M. G. (2012). Metagenomic analysis of *Streptomyces lividans* reveals host-dependent functional expression. *Applied and Environmental Microbiology, 78*(10), 3622–3629. https://doi.org/10.1128/AEM.00044-12
97. Leis, B., Angelov, A., Mientus, M. et al. (2015). Identification of novel esterase-active enzymes from hot environments by use of the host bacterium *Thermus thermophilus. Frontiers in Microbiology, 6*, 275. https://doi.org/10.3389/fmicb.2015.00275
98. Angelov, A., Mientus, M., Liebl, S., Liebl, W. (2009). A two-host fosmid system for functional screening of (meta) genomic libraries from extreme thermophiles. *Systematic and Applied Microbiology, 32*(3), 177–185. https://doi.org/10.1016/J.SYAPM.2008.01.003
99. Kakirde, K. S., Wild, J., Godiska, R. et al. (2011). Gram negative shuttle BAC vector for heterologous expression of metagenomic libraries. *Gene, 475*(2), 57–62. https://doi.org/10.1016/J.GENE.2010.11.004
100. Courtois, S., Cappellano, C. M., Ball, M., et al. (2003). Recombinant environmental libraries provide access to microbial diversity for drug discovery from natural products. *Applied and Environmental Microbiology, 69*(1), 49–55. https://doi.org/10.1128/AEM.69.1.49-55.2003
101. Iqbal, H. A., Low-Beinart, L., Obiajulu, J. U., Brady, S. F. (2016). Natural product discovery through improved functional metagenomics in *Streptomyces. Journal of the American Chemical Society, 138*(30), 9341–9344. https://doi.org/10.1021/jacs.6b02921

102. Ono, A., Miyazaki, R., Sota, M., Ohtsubo, Y., Nagata, Y., Tsuda, M. (2007). Isolation and characterization of naphthalene-catabolic genes and plasmids from oil-contaminated soil by using two cultivation-independent approaches. *Applied Microbiology and Biotechnology, 74*, 501–510. https://doi.org/10.1007/s00253-006-0671-4
103. Craig, J. W., Chang, F.-Y., Kim, J. H., Obiajulu, S. C., Brady, S. F. (2010). Expanding small-molecule functional metagenomics through parallel screening of broad-host-range cosmid environmental DNA libraries in diverse proteobacteria. *Applied and Environmental Microbiology, 76*(5), 1633–1641. https://doi.org/10.1128/AEM.02169-09
104. Nagayama, H., Sugawara, T., Endo, R., et al. (2015). Isolation of oxygenase genes for indigo-forming activity from an artificially polluted soil metagenome by functional screening using *Pseudomonas putida* strains as hosts. *Applied Microbiology and Biotechnology, 99*, 4453–4470. https://doi.org/10.1007/s00253-014-6322-2
105. Cheng, J., Charles, T. C. (2016). Novel polyhydroxyalkanoate copolymers produced in *Pseudomonas putida* by metagenomic polyhydroxyalkanoate synthases. *Applied Microbiology and Biotechnology, 100*(17), 7611–7627. https://doi.org/10.1007/s00253-016-7666-6
106. Aakvik, T., Degnes, K. F., Dahlsrud, R., et al. (2009). A plasmid RK2-based broad-host-range cloning vector useful for transfer of metagenomic libraries to a variety of bacterial species. *FEMS Microbiology Letters, 296*(2), 149–158. https://doi.org/10.1111/j.1574-6968.2009.01639.x
107. Socransky, S., Haffajee, A., Smith, C., et al. (2004). Use of checkerboard DNA–DNA hybridization to study complex microbial ecosystems. *Oral Microbiology and Immunology, 19*(6), 352–362. https://doi.org/10.1111/j.1399-302x.2004.00168.x
108. Socransky SS, Smith C, Martin L, Paster BJ, Dewhirst FE, Levin AE (1994) "Checkerboard" DNA-DNA hybridization. *Biotechniques* 17, 788–792. https://pubmed.ncbi.nlm.nih.gov/7833043
109. Links, M.G., Dumonceaux, T.J., McCarthy, E.L., Hemmingsen, S.M., Topp, E., Town, J.R. (2021) CaptureSeq: hybridization-based enrichment of cpn60 gene fragments reveals the community structures of synthetic and natural microbial ecosystems. *Microorganisms* 9(4):816. https://doi.org/10.3390/microorganisms9040816
110. Hill, J. E., Penny, S. L., Crowell, K. G., Goh, S. H., & Hemmingsen, S. M. (2004). CpnDB: A chaperonin sequence database. *Genome Research, 14*(8), 1669–1675. https://doi.org/10.1101/gr.2649204
111. Chaban, B., & Hill, J. E. (2012). A 'universal'type II chaperonin PCR detection system for the investigation of Archaea in complex microbial communities. *The ISME Journal, 6*(2), 430–439. https://doi.org/10.1038/ismej.2011.96
112. Links, M. G., Demeke, T., Gräfenhan, T., Hill, J. E., Hemmingsen, S. M., & Dumonceaux, T. J. (2014). Simultaneous profiling of seed-associated bacteria and fungi reveals antagonistic interactions between microorganisms within a shared epiphytic microbiome on Triticum and Brassica seeds. *New Phytologist, 202*(2), 542–553. https://doi.org/10.1111/nph.12693
113. Links, M. G., Dumonceaux, T. J., Hemmingsen, S. M., & Hill, J. E. (2012). The chaperonin-60 universal target is a barcode for bacteria that enables de novo assembly of metagenomic sequence data. *PLoS ONE, 7*(11), e49755. https://doi.org/10.1371/journal.pone.0049755
114. Links, M. G., Chaban, B., Hemmingsen, S. M., Muirhead, K., & Hill, J. E. (2013). MPUMA: A computational approach to microbiota analysis by de novo assembly of operational taxonomic units based on protein-coding barcode sequences. *Microbiome, 1*, 1–7. https://doi.org/10.1186/2049-2618-1-23
115. Callahan, B. J., McMurdie, P. J., Rosen, M. J., Han, A. W., Johnson, A. J. A., & Holmes, S. P. (2016). DADA2: High-resolution sample inference from Illumina amplicon data. *Nature Methods, 13*(7), 581–583. https://doi.org/10.1038/nmeth.3869
116. Knight, R., Vrbanac, A., Taylor, B. C., et al. (2018). Best practices for analysing microbiomes. *Nature Reviews Microbiology, 16*(7), 410–422. https://doi.org/10.1038/s41579-018-0029-9
117. Vancuren, S.J., Dos Santos, S.J., Hill, J.E., Team, M.M.L.P. (2020). Evaluation of variant calling for cpn60 barcode sequence-based microbiome profiling. *PLoS One* 15(7):e0235682. https://doi.org/10.1371/journal.pone.0235682

118. Links, M.G., Dumonceaux, T.J., McCarthy, E.L., Hemmingsen, S.M., Topp, E., Town, J.R. (2021). CaptureSeq: hybridization-based enrichment of cpn60 gene fragments reveals the community structures of synthetic and natural microbial ecosystems. *Microorganisms* 2021;9(4):816. https://doi.org/10.3390/microorganisms9040816
119. Hill, J. E., Penny, S. L., Crowell, K. G., Goh, S. H., & Hemmingsen, S. M. (2004). CpnDB: A chaperonin sequence database. *Genome research., 14*(8), 1669–1675. https://doi.org/10.1101/gr.2649204
120. Koch, B. J., McHugh, T. A., Hayer, M., et al. (2018). Estimating taxon-specific population dynamics in diverse microbial communities. *Ecosphere., 9*(1), e02090. https://doi.org/10.1002/ecs22090
121. Hungate, B.A., Mau, R.L., Schwartz, E., et al. (2015). Quantitative microbial ecology through stable isotope probing. *Applied and Environmental Microbiology*. 81(21):7570–7581. https://doi.org/10.1128/AEM.02280-15
122. Mayali, X., Weber, P.K., Brodie, E.L., Mabery, S., Hoeprich, P.D., Pett-Ridge, J. (2012). High-throughput isotopic analysis of RNA microarrays to quantify microbial resource use. *The ISME Journal* 6(6):1210–1221. https://doi.org/10.1038/ismej.2011.175
123. Pett-Ridge, J., & Firestone, M. K. (2017). Using stable isotopes to explore root-microbe-mineral interactions in soil. *Rhizosphere, 3*, 244–253. https://doi.org/10.1016/j.rhisph.2017.04.016
124. Mayali, X., Weber, P. K., & Pett-Ridge, J. (2013). Taxon-specific C/N relative use efficiency for amino acids in an estuarine community. *FEMS Microbiology Ecology, 83*(2), 402–412. https://doi.org/10.1111/j.1574-6941.12000.x
125. Mayali, X., Weber, P. K., Mabery, S., & Pett-Ridge, J. (2014). Phylogenetic patterns in the microbial response to resource availability: Amino acid incorporation in San Francisco Bay. *PLoS ONE, 9*(4), e95842. https://doi.org/10.1371/journal.pone.0095842
126. Mayali, X., Stewart, B., Mabery, S., & Weber, P. K. (2016). Temporal succession in carbon incorporation from macromolecules by particle-attached bacteria in marine microcosms. *Environmental Microbiology Reports, 8*(1), 68–75. https://doi.org/10.1111/1758-2229.12352
127. Wu. D., Doroud, L., Eisen, J.A. (2013). TreeOTU: operational taxonomic unit classification based on phylogenetic trees. *arXiv preprint arXiv:13086333*. https://doi.org/10.48550/arXiv.1308.6333
128. Agbavwe, C., Kim, C., Hong, D., Heinrich, K., Wang, T., & Somoza, M. M. (2011). Efficiency, error and yield in light-directed maskless synthesis of DNA microarrays. *Journal of Nanobiotechnology, 9*, 1–17. https://doi.org/10.1186/1477-3155-9-57
129. Sack, M., Hölz, K., Holik, A.-K., et al. (2016). Express photolithographic DNA microarray synthesis with optimized chemistry and high-efficiency photolabile groups. *Journal of Nanobiotechnology, 14*, 1–13. https://doi.org/10.1186/s12951-016-0166-0
130. Mayali, X., Weber, P.K., Brodie, E.L., Mabery, S., Hoeprich, P.D., Pett-Ridge, J. (2012) High-throughput isotopic analysis of RNA microarrays to quantify microbial resource use. *The ISME Journal* 6(6):1210–1221. https://doi.org/10.1038/ismej.2011.175
131. Bryson S, Li Z, Chavez F, et al. Phylogenetically conserved resource partitioning in the coastal microbial loop. *The ISME journal*. 2017;11(12):2781–2792. https://doi.org/10.1038/ismej.2017.128
132. Enquist, B. J., Abraham, A. J., Harfoot, M. B., Malhi, Y., & Doughty, C. E. (2020). The megabiota are disproportionately important for biosphere functioning. *Nature Communications, 11*(1), 699. https://doi.org/10.1038/s41467-020-14369-y
133. Keller-Costa, T., Lago-Lestón, A., Saraiva, J., et al. (2021). Metagenomic insights into the taxonomy, function, and dysbiosis of prokaryotic communities in octocorals. *Microbiome, 9*, 1–21. https://doi.org/10.1186/s40168-021-01031-y
134. Peixoto, R. S., Voolstra, C. R., Sweet, M., et al. (2022). Harnessing the microbiome to prevent global biodiversity loss. *Nature Microbiology, 7*(11), 1726–1735. https://doi.org/10.1038/s41564-022-01173-1
135. Hale, C.R., Zhao, P., Olson, S., et al. (2009). RNA-guided RNA cleavage by a CRISPR RNA-Cas protein complex. *Cell* 139(5):945–956. https://doi.org/10.1016/j.cell.2009.07.040

136. Garneau, J. E., Dupuis, M. -È., Villion, M., et al. (2010). The CRISPR/Cas bacterial immune system cleaves bacteriophage and plasmid DNA. *Nature, 468*(7320), 67–71. https://doi.org/10.1038/nature09523
137. Snyder, J,C,, Bateson, M.M., Lavin, M., Young, M.J. (2010). Use of cellular CRISPR (clusters of regularly interspaced short palindromic repeats) spacer-based microarrays for detection of viruses in environmental samples. *Applied and Environmental Microbiology*. 76(21):7251–7258. https://doi.org/10.1128/AEM.01109-10
138. Anderson, R. E., Brazelton, W. J., & Baross, J. A. (2011). Using CRISPRs as a metagenomic tool to identify microbial hosts of a diffuse flow hydrothermal vent viral assemblage. *FEMS microbiology ecology., 77*(1), 120–133. https://doi.org/10.1111/j.1574-6941.2011.01090.x
139. Li, H., Wu, Y., Liu, S., Xiao, J., Zhao, W., Chen, J., Alexandrov, G., & Cao, Y. (2022). Decipher soil organic carbon dynamics and driving forces across China using machine learning. *Global Change Biology, 28*(10), 3394–3410. https://doi.org/10.1111/gcb.16154
140. Scholz, M., Lo, C.-C., & Chain, P. S. (2014). Improved assemblies using a source-agnostic pipeline for MetaGenomic Assembly by Merging (MeGAMerge) of contigs. *Scientific Reports, 4*(1), 6480. https://doi.org/10.1038/srep06480

Reinforcement Materials Improve the Characteristics of Metal Matrix Composites—A Review

Ambuj Pathak, Vinay Kumar Singh, Sakshi Chauhan, and Akarsh Verma

Abstract In today's world of engineering materials, every material should have a set of characteristics. It is the driving force behind the creation of several alloys, polymers, composites, and other materials. Metal matrix composites have a wide range of characteristics. The characteristics of metal matrix composites are sometimes determined by the matrix, which is a metallic element. Metal matrix composites' characteristics are occasionally influenced by the reinforcement, which can be metallic or nonmetallic. In this work, we examine the influence of reinforcement on different features of metal matrix composites, such as self-lubrication, self-healing, and self-cleaning. In this review, we also conclude on the impact of reinforcement strategies.

Keywords Self-lubricating · Self-cleaning · Self-healing · Nano-composites · Hybrid composites

1 Introduction

Composite material like a sandwich, a good sandwich consists of various ingredients which give a unique taste to it [1]. A single ingredient cannot provide this distinctive taste. A Composite is a heterogeneous mixture of two or more chemically different components with a special differentiating bond on a macro scale [2]. The availability of a heterogeneous variety of materials (metals, ceramics, polymers, etc.) allows an infinite set of variations in composition. It deals with emerging properties of materials

A. Pathak
Department of Materials Engineering, Indian Institute of Technology Jammu, Jammu, India

V. K. Singh · S. Chauhan
Department of Mechanical Engineering, G. B. Pant University of Agriculture and Technology, Pantnagar, India

A. Verma (✉)
Department of Mechanical Engineering, University of Petroleum and Energy Studies (UPES), Dehradun 248007, India
e-mail: akarshverma007@gmail.com

© The Author(s), under exclusive license to Springer Nature Singapore Pte Ltd. 2024
A. Chaudhary et al. (eds.), *Unraveling New Frontiers and Advances in Bioinformatics*,
https://doi.org/10.1007/978-981-97-7123-3_12

[3, 4] The history of heterogeneous mixing of material is thousand years old [5]. Composite materials consist of matrix and reinforcement [6]. The composite matrix is like the bread of a sandwich, whereas the reinforcement is like the vegetables of a sandwich.

Metal Matrix Composites are a heterogeneous mixture of two or more components, one metal and the other ceramic, metal, and polymer component [7]. The matrix may be a consistent and homogenous material that contains reinforcement. Reinforcement materials are incorporated within the matrix, generally within particles, fibers, whiskers, or wires [8].

2 Al-Based MMCs

Aluminium alloys are the best option as the matrix of MMCs [9]. Aluminium alloy series are categories based on main alloying elements [10, 11] which are given following 1XXX pure aluminium [12], 2XXX copper [13], 3XXX manganese [14], 4XXX silicon [15], 5XXX magnesium [16], 6XXX magnesium and silicon [17], 7XXX zinc [18] and 8XXX other alloying elements (like Aluminium–lithium alloy) [19]. Aluminium-based MMCs (AMCs) are in the limelight due to their high strength-to-weight ratio [20–22], corrosion resistance [23–25], ease of manufacture [26], economical [27, 28] and having low thermal expanssion [29–32].

Advancement in the properties of MMCs can be obtained by changing the various parameters of reinforcing materials, as shown in Table 1 [33–36]. The reinforcement size, like nanoscale, is used to produce nano-MMC [37–41], and the mixture of two or more reinforcements used fabricated hybrid MMCs [37–41]. The advanced properties of MMCs, like self-lubrication, self-healing, and self-cleaning, are obtained with various reinforcing elements, as shown in Fig. 1 [42–44].

In the Hybrid MMC, two or more reinforcement materials are used to fabricate MMC [45, 46]. Mainly, Hybrid MMCs possess a set of characteristics [47]. This subset of hybrid AMCs is essentially focused on performance improvement, with less emphasis on production costs. Certain of the ceramic synthetic particles that are examined include carbide, alumina, carbide boron, carbide (WC), graphite and Silica, as well as carbon nanotubes. Carbide and alumina, on the opposite hand, are more widely employed than other synthetic reinforcing particles. This has necessitated the utilization of two or more synthetic reinforcing particles for property optimization.

In Nano-MMC, the size of reinforcement material on the nano scale [48, 49]. The nano size of reinforcement has improved properties compared to other more extensive reinforcement reinforcement [48, 50].

Self-lubricated MMCs exhibit better tribological properties as compared to matrix material [51, 52]. To be fabricated SMMC reinforced with a solid lubricant like MoS_2, Graphene, Graphite, Diamond like Carbon, etc [53].

Self-healing MMCs exhibit shape memory characteristics [54, 55]. The microcrack of the Self-healing MMCs healed after some duration of time and heat treatment

Table 1 Multi-functional Al-based MMCs

Matrix	Reinforcement and type	Fabrication	Focused properties	Outcome and reference
Al 7075	B_4C (4 wt.%) and Porcelain (4–16 wt.%), Hybrid	Stir Casting	Physico-mechanical and Tribological	Higher density, improved microhardness (71 to 81 HV), maximum tensile strength 223 MPa for 16 wt.% Porcelain improved wear resistance and coefficient of friction for 12 wt.% Porcelain [61]
Al-Si Alloy	SiC (10 wt.%) and Cr (1.5 and 3 wt.%), Hybrid	Stir Casting (Vortex)	Mechanical and Tribological	Improvements in hardness, strength, and wear resistance of 12.44%, 8.62%, and 33.07%, respectively, were made using 10 wt.% SiC. Along with a slight loss of strength and ductility, the chromium particles containing 10 wt.% SiC also boost the hardness and wear resistance also the formation of hard Cr_3C_2 compound [62]
Recycled Al 7075	Pure nano aluminium (5 wt.%), Pure Zn (2 wt.%), Nb_2Al, SiC and ZnO, Hybrid	Powder Metallurgy	Compressive Strength, Tribological and Electrical	Porosity with the sinter + forging process, which produced a solid and hard microstructure, porosity in the microstructure was reduced. Doping and better preparation conditions can increase the ductility and wear resistance of microstructures [63]
Al 6061	SiC and $Ba(NO_3)_2$, Hybrid	Stir Casting	AWJ—Machining Parameters for surface roughness	The optimum process parameters to obtain low surface roughness on the produced composites are the feed rate of 135 mm/min, the distance stand of 0.5, and the abrasive flow rate of 450 gm/min [64]
Al 6351	TiO_2 and Al_2O_3 (2.5–10 wt.%), Hybrid	Stir Casting	Mechanical and Wear Resistance	Due to the impact of reinforcing particle segregation, an increase in Al_2O_3 and TiO_2 particles above 10% results in decreased tensile strength; an increase in TiO_2 also results in a drop in the hardness of the Al-6351 matrix [65]

(continued)

Table 1 (continued)

Matrix	Reinforcement and type	Fabrication	Focused properties	Outcome and reference
Al 6351	Al_2O_3 and C (2.5–10 wt.%), Hybrid	Stir Casting	Mechanical and Microstructure	The composite has an impact strength that is greater than that of the aluminium matrix, with a UTS of 5 wt.% Al2O3 and C. Al matrix and reinforcing materials have strong interfacial bonding, as seen by a SEM micrograph [66]
Al 6061	SiC and Jute Ash, Hybrid	Stir Casting	Mechanical and Tribological	The appropriate distribution of reinforcing particles increased strength, elongation, microhardness, and wear resistance [67]
Al 6103	SiC (2–8 wt.%) and Al_2O_3 (1 wt.%), Hybrid	Stir Casting	Mechanical and Tribological	As reinforcement increases, the density drops, and microstructure confirms that the reinforcement particles are distributed uniformly. With each additional reinforcement, the hardness rises. Due to the lubricating film, the wear resistance is strong at increased reinforcement content [68]
Al 6082	SiC and Fly ash (2.5–7.5 wt.%), Hybrid	Stir Casting	Microstructure and Tribological	In the Al6082, the fly ash particles were distributed relatively unevenly compared to the SiC particles, according to the microstructural analyses. As the weight proportion of reinforcements rises, ultimate tensile strength, hardness, and wear resistance improve [69]
Al 1XXX	SiC (5–30 wt.%) and Mg (1 wt.%), Hybrid	Stir Casting	Wettability and Mechanical	With a rise in reinforcement wt.%, the tensile strength and hardness are enhanced, and the addition of Mg promotes the wettability between the matrix and reinforcement owing to an intermetallic substance (Mg_2Si) [70]

(continued)

Table 1 (continued)

Matrix	Reinforcement and type	Fabrication	Focused properties	Outcome and reference
Al 6061	SiC (5–15 wt.%), B_4C (3 wt.%) and Talc (2 wt.%), Hybrid	Stir Casting	Mechanical, Tribological and Thermal Expansion	Better mechanical characteristics, lower coefficient of friction, increased wear resistance, and lower thermal expansion coefficient are all features of the SiC, B_4C, and talc particle reinforced aluminium hybrid composites [71]
Al 2219	MoS_2 (2 wt.%), K_2TiF_6 (0.4 wt.%) and B_4C (2 wt.%), Hybrid	Stir Casting	High Temperature Wear Behaviour	The wettability is improved by the addition of K_2TiF_6, the wear mechanism of Al2219 exhibits plastic deformation and delamination at low temperatures, while at high temperatures (100 °C), oxide layer fracture and slightly expanded grooves with glazing layer were seen [72]
Al 6061	SiC (10 wt.%) and Gr (3 wt.%), Hybrid	Stir Casting	Tribological	It was shown that the coefficient of friction for h-AMMC/Steel sliding under MWCNT in oil and surfactant functionalized MWCNT-in oil decreased by more than 1.5 and 2 times, respectively, as compared to base oil [73]
Al 7075	MoS_2 (3–15 wt.%) and TiC (2 wt.%), Hybrid	Stir Casting	Thermal expansion	The thermogram of TGA/ DTA showed that the increase in TiC percentage from 3 to 11 percent and moreover the increase in MoS_2 percentage of 2% builds the start time frame [74]
Al 7075	Gr (2–6 wt.% and Fly ash (10 wt.%), Hybrid	Compo Casting	Mechanical and Tribological	In response to the addition of 10% fly ash and 2% Gr, the bulk hardness of the composites increased from 57 to 62 HRB, or by 8.7%. The bulk hardness reduces when Gr wt.% rises progressively. Fly ash boosts the wear resistance of Al7075, and the hybrid MMC with Gr added exhibits even greater resistance [75]

(continued)

Table 1 (continued)

Matrix	Reinforcement and type	Fabrication	Focused properties	Outcome and reference
Al 6061	Carbon Black (5–15 wt.%), Nano	Stir Casting	Corrosion	As the weight percentage of the reinforcement increased, the corrosion caused by weight loss of the composite reduced. The circumstances of the heat treatment have a significant impact on the corrosion rate. The thermal treatment (T6) has increased the corrosion resistance of both the matrix and the composite [76]
Al 6061	Carbon Black (5–15 wt.%), Nano	Stir Casting	Tribological	At the wear transition zone, the wear rates of the reinforced materials were lower than for the unreinforced matrix. In the wear transition area, only the unreinforced alloy maintained elevated wear rates, While the reinforced composites showed significantly reduced wear rates. An enormous plastic distortion indicative of serious wear results from the material's thermal softening [77]
Al 3560	ZrO_2 (2.5 and 5 wt.%), Nano	Stir Casting	Mechanical	In comparison with the unreinforced samples, 5 percent Nanocomposites showed a 2.5 wt.% ZrO_2 improvement in hardness. In comparison with the unreinforced samples, the YS and UTS of the 5% ZrO_2 samples increased by 40 and 39.83%, respectively. The hard and brittle ceramic particles found in the matrix of the aluminium alloy were thought to be the cause of the decreased elongation. Comparing materials with 5% ZrO_2 to samples without reinforcement, the compressive test rose by 23% [78]

(continued)

Table 1 (continued)

Matrix	Reinforcement and type	Fabrication	Focused properties	Outcome and reference
Al 1XXX	MWCNT (2.5–7.5 wt.%) and Al_2O_3 (1 wt.%), Nano and Hybrid	Spark Plasma Sintering	Microstructure and Mechanical	During tensile deformation, a semicoherent contact between the in-situ Al_2O_3 nanoparticles and the Al matrix caused a noticeably higher dislocation density in the matrix surrounding the nanoparticles. These dislocations were precursors to the production of micro voids, which led to the formation of dimples upon additional deformation. CNTs and Al_2O_3 nanoparticles worked together to produce materials with good tensile ductility and a protracted strain-softening stage [79]
Al 1XXX	CuO (0.5–1 wt.%) and SAC (10 wt.%), Nano and Hybrid	Stir Casting	Microstructure and Mechanical	Dual reinforcement has an effect on the aluminium matrix, as seen by the composite's improved mechanical characteristics. Al + 10%, SAC + 1% nano-CuO is the ideal composition, leading to the greatest increases in hardness (77%) and tensile strength (78%) [80]
Al LM9	MWCNT (1–3 wt.%), Nano	Stir Casting	Mechanical and Corrosion	In terms of pitting corrosion, the composite reinforced with % MWCNTs performed best. With the addition of MWCNT reinforcement, the composites' hardness strength has increased, and the mass fraction of MWCNTs is discovered to have a significant impact on this enhancement. The negative E_{pit} decreased, suggesting less nucleation pitting, due to the rise in the weight fraction of MWCNTs used as reinforcements [81]

(continued)

Table 1 (continued)

Matrix	Reinforcement and type	Fabrication	Focused properties	Outcome and reference
Al 7075	B_4C (6 wt.%) and Alloying Element, Nano	Powder Metallurgy	Mechanical	Composites without alloying elements have greater hardness. This may be explained by the fact that the intermetallic that was made can withstand deformations caused by indentations as well as the mechanism of failure, and that compression testing demonstrates the ductility of the composite with alloying element [82]
Al 6063	SiC (2.5–10 wt.%), Al_2O_3 (1 wt.%) and TiO_2 (1 wt.%), Nano and Hybrid	Stir Casting	Mechanical and Surface Characterization	In comparison to pure Al6063 metals, mechanical qualities including hardness, tensile strength, and yield strength are higher, and SiC increases these properties. SEM was used to examine the MMC's surface morphology. The MMC exhibits ploughing, microwear, microcracks, fractures, voids, and pulling out [83]
Al 1XXX	SiC (10 wt.%), Al_2O_3 (2 wt.%) and WS_2 (5 and 9 wt.%), Nano, Hybrid and Self lubrication	Powder Metallurgy	Mechanical and Tribological	The hardness of the composite rose as the proportion of WS_2 solid lubricant increased. The composite with 0 weight percent WS_2 had a hardness of 29.07 HV, while the composite with 9 weight percent WS_2 had a hardness of 31.93 HV. On the composite that included 0 weight percent WS_2, SEM micrographs revealed tiny pits and a worn surface. Composites with an ideal WS_2 concentration (5%) experienced less wear and abrasion. Additionally, their low coefficient of friction [84]

(continued)

Table 1 (continued)

Matrix	Reinforcement and type	Fabrication	Focused properties	Outcome and reference
Al 6061	SiC (8 wt.%), Al_2O_3 (4 wt.%) and Gr (4 wt.%), Hybrid and Self lubrication	Stir Casting	Mechanical and Tribological	Due to the presence and pinning effect of hard SiC and Al_2O_3 particles, microhardness rises. Due to the mechanically mixed layer created between the composite pin and steel disc surfaces, which comprised fractured SiC and Gr, the Al-SiC/Gr surface hybrid composite showed low wear rate. Due to the presence of reinforcing particles that make the matrix brittle, tensile characteristics are reduced in comparison to the base material [85]
Al 6061	SiC (2.5–10 wt.%), and MWCNT (2.5 wt.%), Nano, Hybrid and Self-lubrication	Powder Metallurgy	Mechanical and Tribological	Hybrid composites' wear rate is decreased when SiC content rises. However, the hardness of the composites is lowered and the rate of wear is enhanced when the MWCNTs are introduced individually to the alloy as individual reinforcements. MWCNT agglomeration is to blame for this. At 7.5 wt.% of SiC addition, the addition of SiC nanoparticulate into the matrix results in an increase in the microhardness of the composite and a uniform distribution of hardness throughout the composites [86]

(continued)

Table 1 (continued)

Matrix	Reinforcement and type	Fabrication	Focused properties	Outcome and reference
Al 2024	SiC (3–9 wt.%), Self-lubrication	Stir Casting	Mechanical and Tribological	The prepared composites get harder when SiC particles are added. Results show that AMC was strengthened by 9 weight percent. The hardest material was determined to be SiC, with a hardness of 55.7 HRB. With the inclusion of silicon carbide particles, the produced composites' tensile strength improved. With the addition of SiC particles, the volume loss of Al2024-SiC composites was much less than that of the matrix material [87]
Al 5059	SiC (5–15 wt.%), and MoS_2 (2 wt.%), Hybrid and Self-lubrication	Stir Casting	Tribological	Along with SiC weight %, SiC particle size, increase in sliding velocity, and distance, load has the greatest impact on how quickly the hybrid composites wear out. The weight percentage 15% reinforcement with tiny size particles produces improved wear rate, according to an ANOVA test. The applied load (43.15%) has the largest statistical impact on wear rate, followed by the SiC wt.% (24.89%), SiC particle size (16.12%), sliding speed (13.11%), and sliding distance (2.35%) [88]
Al 7075	SiC (5 wt.%) and Gr (0–10 wt.%), Hybrid and Self lubrication	Powder Metallurgy	Tribological	At constant sliding speeds of 0.5 m/s and constant sliding distance of 1000 m, the combination of Al + 5% SiC + 5% Gr had the lowest wear and coefficient of friction. The research shows that adding an additional 10% Gr in the hybrid composite has no beneficial effect on the tribological characteristic [89]

(continued)

Table 1 (continued)

Matrix	Reinforcement and type	Fabrication	Focused properties	Outcome and reference
Al 7075	SiC (3–9 wt.%), Self-lubrication	Stir Casting (Vortex)	Mechanical and Tribological	Al7075-SiC composites showed an improvement in hardness of around 10%, and microhardness of the composites was shown to rise with increased filler content. The composites' tensile strength characteristics are found to be superior to those of the basic matrix, with Al7075-SiC composites having stronger tensile characteristics than Al7075-Al_2O_3 composites. The composites have a greater wear resistance, and SiC has also greatly improved the wear resistance of Al7075-SiC composites [90]
Al 2219	TiB_2 (6 wt.%) and ZrB_2 (6 wt.%), Hybrid	Flux Assisted Synthesis	Machining	A3 (cutting speed of 150 m/min), B1 (feed rate of 0.05 mm/rev), and C1 (level of factors) produce the best results in machining (Depth of cut of 0.5 mm). According to the response graphs, the feed rate has a stronger impact on the cutting force and surface roughness. The importance of the feed rate has a 50.78% contribution to cutting force and a 76.94% contribution to surface roughness, according to an analysis of variance [91]

(continued)

Table 1 (continued)

Matrix	Reinforcement and type	Fabrication	Focused properties	Outcome and reference
Al 7068	ZnO (5–15 wt.%), Self-lubricating	Powder Metallurgy	Tribological	With the addition of 15% ZnO particles, which were twice as hard as the parent alloy, the composites' hardness was raised to 66 VHN. ZnO particle addition showed increased wear resistance, while AA7068-15% ZnO composite showed the highest level of wear resistance. According to SEM examination, the main wear processes for the matrix material and the AA7068-5 percent ZnO composite were abrasion, delamination, and oxidation, whereas adhesion and oxidation affected the composites supplemented with 10 and 15% ZnO particles, respectively [92]
Al 7075	TiO_2 (5and10 wt.%), Nano	Stir Casting	Mechanical and Tribological	Al7075-10 wt.% nano TiO_2 composite has better and improved mechanical qualities compared to unreinforced material. The fracture surface of the composite material is made up of tiny voids as a result of strain localization. Additionally, when the applied load and speed increased, the volumetric wear loss of matrix and TiO_2 particle enhanced nano composites increased [93]

(continued)

Table 1 (continued)

Matrix	Reinforcement and type	Fabrication	Focused properties	Outcome and reference
Al 6061	60Pb40Sn solder alloy, Self-Healing	Casting, drilling and solder alloy filling	Mechanical	The comparison between self-healing composite with pure matrix material has been developed. And the experimental assessment results show a reduction in tensile force as well as stress by 30%. Reduction by 50% in other properties like braking distance and strain in the fabricated self-healing matrix. Similarly, in the case of the Charpy test, it reduces by 39% for self-healing samples as compared to pure aluminium sample [94]
Al 2014	Sn60Pb40 solder alloy (6 vol%), SMA Wires and Ni55Ti45 (1 vol%), Hybrid and Self-Healing	Casting	Shape Memory	At higher temperatures (i.e., roughly 76% for entirely damaged and 80% for partially damaged samples) at 600 °C healing temperature, the recovery efficiency in the plastic strain (i.e., shape recovery) of deformed sample is noticed greater. To prevent the matrix from melting and to keep the structure's dimensions stable, it is crucial to choose the temperature with consideration for the phase diagram [95]

(continued)

Table 1 (continued)

Matrix	Reinforcement and type	Fabrication	Focused properties	Outcome and reference
Al 3805	SMA Wires, Self-Healing	Casting	Mechanical and Shape Memory	The reinforced component was found to be almost twice as strong and flexible as the unreinforced sample. It was discovered that the sample's dimensions, not the number of wires, were what determined how quickly the break closed. There was no restored strength since there was no bouncing between the damaged surfaces, which might be advantageous in some circumstances [96]
Al 5083	CNT (0–2 wt.%), Ni (0–5 wt.%) and MoS2 (0–4 wt.%), Hybrid and Self-lubrication	Stir Casting	Corrosion, Mechanical and Tribological	With the addition of reinforcements, the wear rate of the specimen is significantly reduced due to the increase in MoS_2 wt.%, but when the weight percentage of nickel rises to 5%, the material tends to show a minor increase in wear rate due to the increase in ductility. With an increase in the wt.% of reinforcement, the specimen's rate of corrosion declines significantly. This is because MWCNT, Ni helps with corrosion resistance. Because nickel helps the compression strength, the specimen's compression strength increases as the wt.% of reinforcements increases [97]

in some case [56]. Solder alloys are used as a reinforcing element in Self-healing MMCs [57].

Surfaces of Self-cleaning MMCs exhibit superhydrophobic properties, also known as lotus surface effect [58]. Self-cleaning characteristics can be developed either with the help of surface modification (LASER etching, plasma etching, etc.) or reinforcing superhydrophobic material in the matrix like SiO_2, TiO_2 etc [59, 60]. Refer Table 2 for the optimum mechanical properties reported in the literature.

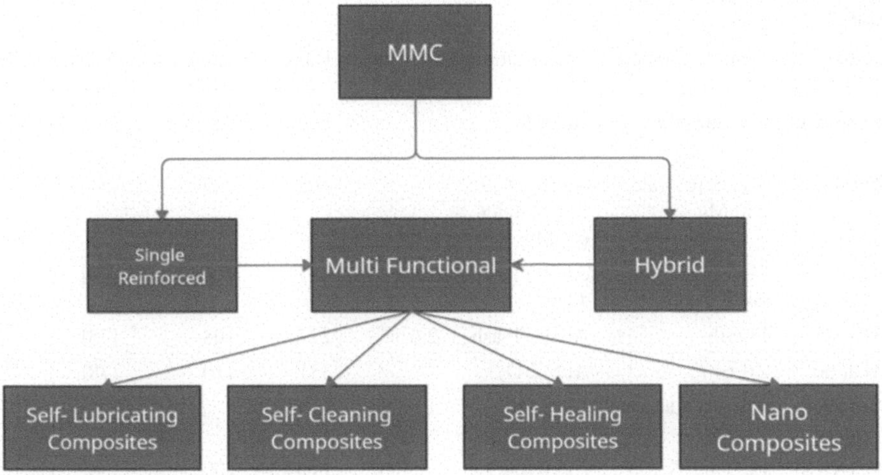

Fig. 1 Multi-functional MMCs

Table 2 Optimum mechanical properties observed in previous research

Matrix	Reinforcements	Composition	Density (g/cm^3)	UTS (MPa)	Hardness (HV)	Reference number
Al 7075	B_4C and Porcelain	Al/4 B_4C/12 Porcelain	2.45	199	77	[61]
Al 7075	B_4C and Porcelain	Al/4 B_4C/16 Porcelain	2.39	223	81	[61]
Al-Si Alloy	SiC and Cr	Al/10 SiC/1.5 Cr	2.765	188	166.67	[62]
Al-Si Alloy	SiC and Cr	Al/10 SiC/3 Cr	2.841	187	123	[62]
Al 7075 Recycled	n-Al, Zn, Nb_2Al, SiC and ZnO	Al/5 n-Al/10 Nb_2Al/10 SiC/30 ZnO	3.5	113	398	[63]
Al 7075 Recycled	n-Al, Zn, Nb_2Al, SiC and ZnO	Al/5 n-Al/10 Nb_2Al/10 SiC/15 ZnO	3.34	186	347	[63]
Al 6351	TiO_2 and Al_2O_3	Al/5 Al_2O_3/5 TiO_2	2.6	61.7	90	[65]
Al 6351	C and Al_2O_3	Al/5 Al_2O_3/5 C	2.44	89.5	90	[66]
Al 6061	SiC	Al/3.4 SiC		123	66	[67]
Al 6082	SiC and Fly Ash	Al/7.5 SiC/7.5 Fly Ash		128	95	[69]

(continued)

Table 2 (continued)

Matrix	Reinforcements	Composition	Density (g/cm^3)	UTS (MPa)	Hardness (HV)	Reference number
Al 1XXX	SiC and Mg	Al/10 SiC/ 1 Mg		105	95	[70]
Al 6061	SiC, B_4C, Talc and Mg	Al/10 SiC/8 B_4C/2 Talc/ 0.5 Mg		168	243	[71]
Al 7075	Flyash and Graphite	Al/10 Flyash/ 4 Gr	2.56	145	116	[75]
Al 7075	Flyash	Al/10 Flyash	2.6	227	108	[75]
Al 3560	ZrO_2	Al/5 ZrO_2		331	110	[78]
Al 1XXX	MWCNT and Al_2O_3	Al/1 Al_2O_3/5 MWCNT *		540		[79]
Al 1XXX	CuO and SAC	Al/10 SAC/ 0.75 nano CuO		111.43	200	[80]
Al 1XXX	CuO and SAC	Al/10 SAC/1 nano CuO		117.21	220	[80]
Al LM9	MWCNT	Al/3 MWCNT			180	[81]
Al 7075	B_4C and Alloying Elements	Al/6 B_4C		210	130	[82]
Al 7075	B_4C and Alloying Elements	Al/6 B_4C/ Alloying Elements		252	100	[82]
Al 6063	SiC, Al_2O_3 and TiO_2	Al/10 SiC/1 Al_2O_3/1 TiO_2		148.62	108	[83]
Al 6063	SiC, Al_2O_3 and TiO_2	Al/7.5 SiC/1 Al_2O_3/1 TiO_2		138.8	86	[83]
Al 1XXX	SiC, Al_2O_3 and WS_2	Al/10 Al2O3/ 9 WS_2	2.757		31.93	[84]
Al 6061	SiC and Gr	Al/8 SiC/4 Gr		219		[85]
Al 6061	SiC and Al_2O_3	Al/8 SiC/4 Al_2O_3		192		[85]
Al 2024	SiC	Al/9 SiC	2.82	247.44		[87]
Al 7075	SiC and Gr	Al/5 SiC/10 Gr	2.84		130	[89]
Al 7075	SiC and Gr	Al/5 SiC/5 Gr	2.86		140	[89]

* vol%

3 Results

See Fig. 2.

4 Conclusions

The study of various type of multifunctional Al based MMCs is conducted successfully. Most of Researchers used Al 6XXX alloy series but ignored Al 4XXX alloy series to fabricate Al based MMCs as shown Fig. 2. Researchers used SiC to reinforce Al based MMCs. It improves the mechanical, tribological, and corrosive behaviors of MMCs. Stir Casting is the most used fabrication method to fabricate Al based MMCs. It gives a proper distribution of reinforcement in the throughout matrix. It is also an economical and easy fabrication method. Mechanical and tribological behavior are

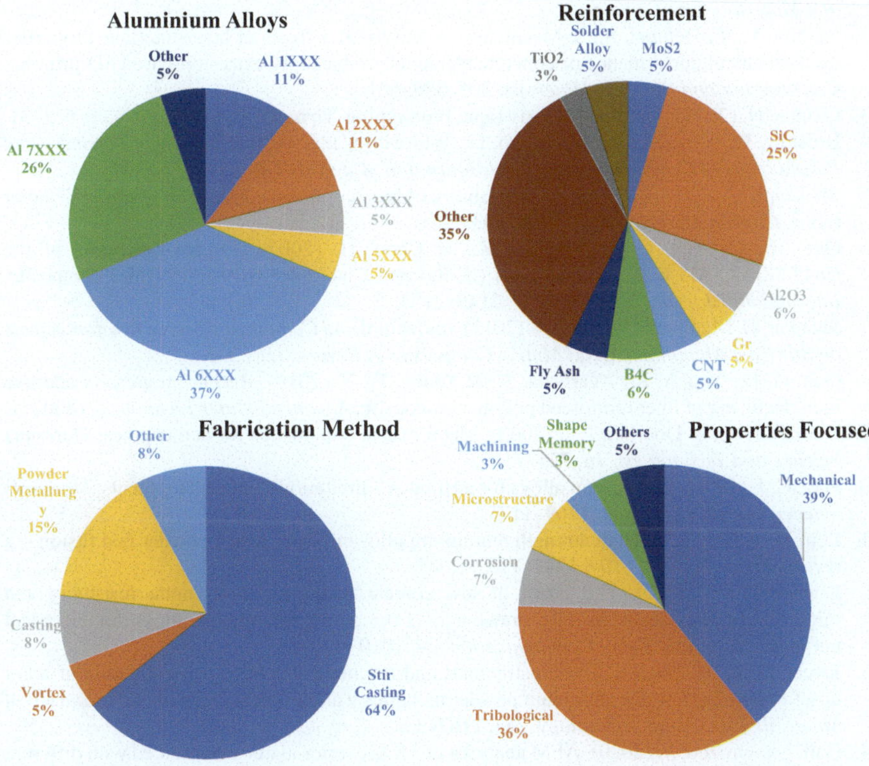

Fig. 2 Pie chart for aluminium alloys, reinforcement, fabrication method and properties focused as per Table 1

the most focused set of behavior in the development of Al based MMCs. The development and characterization of Al based self-cleaning MMC is not in the limelight until now. The development of self-healing MMCs is a very precision-based process. Lastly, the authors have considerable experience in the field of polymer composites and molecular dynamics simulations that proves the worth of writing this chapter [98–152].

Data Availability Statement Data sharing is not applicable to this article as no datasets were generated or analyzed during the current study.

Conflict of Interest There are no conflicts of interest to declare by the authors.

References

1. Haghshenas, M. (2016). Metal–Matrix Composites. In *Reference Module in Materials Science and Materials Engineering*
2. Nerella, V. N., Hempel, S., & Mechtcherine, V. (2019). Effects of layer-interface properties on mechanical performance of concrete elements produced by extrusion-based 3D-printing. *Construction and Building Materials, 205*, 586–601.
3. Oxman, N. (2011). Variable property rapid prototyping. *Virtual Physics Prototyping, 6*, 3–31.
4. Brostow, W., Kovačević, V., Vrsaljko, D., Whitworth, J. (2010). *Tribology of Polymers and Polymer-Based Composites. Whitworth Journal of Materials Education* 32
5. Spokas, K. A. (2010). Review of the stability of biochar in soils: Predictability of O:C molar ratios. *Carbon Management, 1*, 289–303.
6. Qiao, D. C., Fan, G. J., Liaw, P. K., & Choo, H. (2007). Fatigue behaviors of the Cu47.5Zr47.5Al5 bulk-metallic glass (BMG) and Cu47.5Zr38Hf9.5Al5 BMG composite. *International Journal of Fatigue, 29*, 2149–2154.
7. Shankar, G. M,. Shetty, R., Kini, A. (2013). *Individual and Combined Effect of Reinforcements on Stir Cast Aluminium Metal Matrix Composites-A Review. IJECT.* 3
8. Goh, G. D., Yap, Y. L., Agarwala, S., & Yeong, W. Y. (2019). Recent progress in additive manufacturing of fiber reinforced polymer composite. *Advanced Mater Technol., 4*, 1800271.
9. Hooker, J. A., & Doorbar, P. J. (2000). Metal matrix composites for aeroengines. *Materials Science and Technology., 16*, 725–731.
10. Peel, C. J. (1986). Aluminium alloys for airframes – limitations and developments. *Materials Science and Technology, 2*, 1169–1175.
11. Leirmo, J. L. (2021). High strength aluminium alloys in laser-based powder bed fusion – a review. *Procedia CIRP, 104*, 1747–1752.
12. Koprowski, P. (2020). The effect of low content additives on strength, resistivity and microstructural changes in wire drawing of 1xxx series aluminium alloys for electrical purposes. *Materials Today Communication, 24*, 101039.
13. Sweet, G. A. W. (2020). A microstructural and mechanical property investigation of a hot upset forged 2xxx series aluminum powder metallurgy alloy reinforced with AlN. *Journal of Materials Processing Technology, 284*, 116742.
14. Gallegos-Orozco, V. (2020). AFM analyses of 3XXX series al alloy reinforced with different hard nanoparticles produced in liquid state. *Materials, 13*, 272.
15. Best, J. P., Maeder, X., Michler, J., & Spierings, A. B. (2019). Mechanical Anisotropy Investigated in the Complex SLM-Processed Sc- and Zr-Modified Al–Mg Alloy Microstructure. *Advanced Engineering Materials, 21*, 1801113.

16. Liu, W. C., & Morris, J. G. (2004). Quantitative analysis of texture evolution in cold-rolled, continuous-cast AA 5xxx-series aluminum alloys. *Metallurgical and Materials Transactions A, 35*, 265–277.
17. Bhat, K. U., Panemangalore, D. B., Kuruveri, S. B., John, M., & Menezes, P. L. (2022). Surface Modification of 6xxx Series Aluminum Alloys. *Coatings, 12*, 180.
18. Min, K. H., Lee, B.-H., Chang, S.-Y., Kim, Y. do. (2007). Mechanical properties of sintered 7xxx series Al/SiCp composites. *Materials Letters* 61, 2544–2546
19. Engler, O., Wagner, P., Savoie, J., Ponge, D., & Gottstein, G. (1993). Strain rate sensitivity of flow stress and its effect on hot rolling texture development. *Scripta Metallurgica et Materialia, 28*, 1317–1322.
20. Rajesh kumar, L. *et al.* Optimization of wear behaviour for AA2219-MoS2 metal matrix composites in dry and lubricated condition. *Materials Today Proceedings* 27, 2645–2649 (2020).
21. Joel, J., Xavior, A., Aluminium, M. (2018). Alloy composites and its machinability studies: a review. *Materials Today Proceedings* 5, 13556–13562 (2018).
22. Mohapatra, S. K., Ranjan, V., & Tripathy, S. (2022). Study of severe plastic deformations of metallic materials: A move towards Amorphization. *Materials Today: Proceedings, 56*, 735–741.
23. Paciej, R. C., & Agarwala, V. S. (1988). Influence of Processing Variables on the Corrosion Susceptibility of Metal-Matrix Composites. *Corrosion, 44*, 680–684.
24. Hihara, L. H., & Latanision, R. M. (1994). Corrosion of metal matrix composites. *International Materials Reviews, 39*, 245–264.
25. Senthil Kumar, P., Kavimani, V., Soorya Prakash, K., Murali Krishna, V., Shanthos Kumar, G. (2020). Effect of TiB_2 on the Corrosion Resistance Behavior of In Situ Al Composites. *International Journal of Metalcasting* 14, 84–91.
26. Josyula, S. K., & Narala, S. K. R. (2014). A Breif Review on Manufacturing of Al-TiC MMC. *Advanced Mater Research, 980*, 62–68.
27. Huda, D., el Baradie, M. A., & Hashmi, M. S. J. (1993). Metal-matrix composites: Manufacturing aspects, Part I. *Journal of Materials Processing Technology, 37*, 513–528.
28. Naik, S., Das, S. R., Dhupal, D., & Khatua, A. K. (2022). Electrical discharge machining of engineered Al-22%SiC metal matrix composite: surface roughness analysis, optimization, economic analysis, and sustainability assessment. *Process Integration and Optimization for Sustainability, 6*, 223–251.
29. Xu, Z., Yan, J., Zhao, W., Xu, H., & Yang, S. (2005). Thermal expansion behavior and performance of VLP diffusion-bonded joints of SiCp/A356 composites. *Composites Science and Technology, 65*, 1461–1467.
30. Tjong, S. C., & Tam, K. F. (2006). Mechanical and thermal expansion behavior of hipped aluminum–TiB2 composites. *Materials Chemistry and Physics, 97*, 91–97.
31. Cem Okumus, S., Aslan, S., Karslioglu, R., Gultekin, D., Akbulut, H. (2012). Thermal expansion and thermal conductivity behaviors of Al-Si/SiC/graphite hybrid metal matrix composites (MMCs). *Materials Science* 18.
32. Huber, T., Degischer, H. P., Lefranc, G., & Schmitt, T. (2006). Thermal expansion studies on aluminium-matrix composites with different reinforcement architecture of SiC particles. *Composites Science and Technology, 66*, 2206–2217.
33. Khanna, V., Kumar, V., & Bansal, S. A. (2021). Mechanical properties of aluminium-graphene/carbon nanotubes (CNTs) metal matrix composites: Advancement, opportunities and perspective. *Materials Research Bulletin, 138*, 111224.
34. Kumar, A., Singh, R. C., & Chaudhary, R. (2020). Recent progress in production of metal matrix composites by stir casting process: An overview. *Materials Today: Proceedings, 21*, 1453–1457.
35. Prater, T. (2014). Friction stir welding of metal matrix composites for use in aerospace structures. *Acta Astronautica, 93*, 366–373.
36. Samal, P., Vundavilli, P. R., Meher, A., & Mahapatra, M. M. (2020). Recent progress in aluminum metal matrix composites: a review on processing, mechanical and wear properties. *Journal of Manufacturing Processes, 59*, 131–152.

37. Kim, H. H., Babu, J. S. S., & Kang, C. G. (2013). Fabrication of A356 aluminum alloy matrix composite with CNTs/Al2O3 hybrid reinforcements. *Materials Science and Engineering: A, 573*, 92–99.
38. Bodunrin, M. O., Alaneme, K. K., & Chown, L. H. (2015). Aluminium matrix hybrid composites: A review of reinforcement philosophies; mechanical, corrosion and tribological characteristics. *Journal of Materials Research and Technology, 4*, 434–445.
39. Megahed, M., Attia, M. A., Abdelhameed, M., & El-Shafei, A. G. (2017). Tribological characterization of hybrid metal matrix composites processed by powder metallurgy. *Acta Metallurgica Sinica (English Letters), 30*, 781–790.
40. Mahmoud, E. R. I., Takahashi, M., Shibayanagi, T., & Ikeuchi, K. (2009). Fabrication of surface-hybrid-MMCs layer on aluminum plate by friction stir processing and its wear characteristics. *Materials Transactions, 50*, 1824–1831.
41. Mahmoud, E. R. I., Takahashi, M., Shibayanagi, T., & Ikeuchi, K. (2010). Wear characteristics of surface-hybrid-MMCs layer fabricated on aluminum plate by friction stir processing. *Wear, 268*, 1111–1121.
42. Smart, D. S. R., Johns, N. P., & Smart, J. J. (2017). Mechanical characterization of graphite and CNT reinforced aluminium-5083. *Materials Science Forum, 889*, 56–62.
43. Prabhu, T. R., et al. (2019). Tribological behaviour of graphite-reinforced FeNiCrCuMo high-entropy alloy self-lubricating composites for aircraft braking energy applications. *Tribology Letters, 67*, 78.
44. Mertens, A. I. (2021). Metal matrix composites processed by laser additive manufacturing. In: *Additive Manufacturing* 409–425. https://doi.org/10.1016/B978-0-12-818411-0.00005-7.
45. Al-maamari, A. E. A., Iqbal, A. A., & Nuruzzaman, D. M. (2020). Mechanical and tribological characterization of self-lubricating Mg-SiC-Gr hybrid metal matrix composite (MMC) fabricated via mechanical alloying. *Journal of Science: Advanced Materials and Devices, 5*, 535–544.
46. Kumaraswamy, H. S., Bharat, V. and Krishna Rao, T. (2018). Influence of Mechanical and Tribological Behaviour of Al 2024 MMC fabricated by stir casting technique—A Review. *Materials Today: Proceedings* 5, 11962–11970 (2018).
47. Rajmohan, T., Palanikumar, K., & Prakash, S. (2013). Grey-fuzzy algorithm to optimise machining parameters in drilling of hybrid metal matrix composites. *Composites Part B Engineering, 50*, 297–308.
48. Sahraeinejad, S., Izadi, H., Haghshenas, M., & Gerlich, A. P. (2015). Fabrication of metal matrix composites by friction stir processing with different Particles and processing parameters. *Materials Science and Engineering: A, 626*, 505–513.
49. Raj, P., A. K., Kailas, S. v. (2015) A novel in-situ polymer derived nano ceramic MMC by friction stir processing. *Materials and Design* 85, 626–634.
50. Casati, R., & Vedani, M. (2014). Metal matrix composites reinforced by nano-particles—a review. *Metals (Basel), 4*, 65–83.
51. Soorya Prakash, K., Balasundar, P., Nagaraja, S., Gopal, P. M. and Kavimani, V. Mechanical and wear behaviour of Mg–SiC–Gr hybrid composites. *Journal of Magnesium and Alloys* 4, 197–206 (2016).
52. Manjunath, R., Kumar, D., & Kumar, A. (2021). A review on the significance of hybrid particulate reinforcements on the mechanical and tribological properties of stir-casted aluminum metal matrix composites. *Journal of Bio Tribo-corrosion, 7*, 122.
53. Gill, R. S., Samra, P. S., & Kumar, A. (2022). Effect of different types of reinforcement on tribological properties of aluminium metal matrix composites (MMCs)—A review of recent studies. *Materials Today: Proceedings, 56*, 3094–3101.
54. Rohatgi, P. K. (2014). Al-shape memory alloy self-healing metal matrix composite. *Materials Science and Engineering: A, 619*, 73–76.
55. Zhu, P., et al. (2016). Characterization and modeling of three-dimensional self-healing shape memory alloy-reinforced metal-matrix composites. *Mechanics of Materials, 103*, 1–10.
56. Das, R., Melchior, C., Karumbaiah, K. M. (2016) Self-healing composites for aerospace applications. In: *Advanced Composite Materials for Aerospace Engineering* 333–364 (Elsevier, 2016). https://doi.org/10.1016/B978-0-08-100037-3.00011-0.

57. Bellah, M., Nosonovsky, M. and Rohatgi, P. Recent Advances in Self-healing Metal Matrix Composites. in 297–310 (2022). https://doi.org/10.1007/978-3-030-92567-3_18.
58. Hejazi, V., & Nosonovsky, M. (2012). Wetting transitions in two-, three-, and four-phase systems. *Langmuir, 28*, 2173–2180.
59. Yao, L., & He, J. (2014). Recent progress in antireflection and self-cleaning technology – From surface engineering to functional surfaces. *Progress in Materials Science, 61*, 94–143.
60. Meng, L.-Y., & Park, S.-J. (2014). Superhydrophobic carbon-based materials: A review of synthesis, structure, and applications. *Carbon letters, 15*, 89–104.
61. Aherwar, A., Patnaik, A., & Pruncu, C. I. (2020). Effect of B4C and waste porcelain ceramic particulate reinforcements on mechanical and tribological characteristics of high strength AA7075 based hybrid composite. *Journal of Materials Research and Technology, 9*, 9882–9894.
62. Kumar, J., et al. (2020). Comparative study on the mechanical, tribological, morphological and structural properties of vortex casting processed, Al-SiC-Cr hybrid metal matrix composites for high strength wear-resistant applications: Fabrication and characterizations. *Journal of Materials Research and Technology, 9*, 13607–13615.
63. Enginsoy, H. M., Bayraktar, E., Katundi, D., Gatamorta, F. and Miskioglu, I. (2020). Comprehensive analysis and manufacture of recycled aluminum based hybrid metal matrix composites through the combined method; sintering and sintering + forging. *Compos B Eng* 194.
64. Maneiah, D., Shunmugasundaram, M., Reddy, A. R. and Begum, Z. Optimization of machining parameters for surface roughness during abrasive water jet machining of aluminium/magnesium hybrid metal matrix composites. in *Materials Today: Proceedings* vol. 27 1293–1298 (Elsevier Ltd, 2020).
65. Ahamad, N., Mohammad, A., Sadasivuni, K. K., & Gupta, P. (2020). Structural and mechanical characterization of stir cast Al–Al2O3–TiO2 hybrid metal matrix composites. *Journal of Composite Materials, 54*, 2985–2997.
66. Ahamad, N., Mohammad, A., Sadasivuni, K. K., & Gupta, P. (2020). Phase, microstructure and tensile strength of Al–Al2O3–C hybrid metal matrix composites. *Proceedings of the Institution of Mechanical Engineers, Part C: Journal of Mechanical Engineering Science, 234*, 2681–2693.
67. Coyal, A., Yuvaraj, N., Butola, R. and Tyagi, L. An experimental analysis of tensile, hardness and wear properties of aluminium metal matrix composite through stir casting process. *SN Appl Sci* 2, (2020).
68. Hossain, S. *et al.* Fabrication, microstructural and mechanical behavior of Al-Al2O3-SiC hybrid metal matrix composites. in *Materials Today: Proceedings* vol. 21 1458–1461 (Elsevier Ltd, 2020).
69. Reddy, B. R., Srinivas, C. (2018). Fabrication and Characterization of Silicon Carbide and Fly Ash Reinforced Aluminium Metal Matrix Hybrid Composites. in *Materials Today: Proceedings* vol. 5 8374–8381 (Elsevier Ltd).
70. Singh, V. K., Chauhan, S., Gope, P. C., & Chaudhary, A. K. (2015). Enhancement of wettability of aluminum based silicon carbide reinforced particulate metal matrix composite. *High Temperature Materials and Processes, 34*, 163–170.
71. Kumar, C. R., Malarvannan, R. R. R., & JaiGanesh, V. (2020). Role of SiC on mechanical, tribological and thermal expansion characteristics of B4C/talc-reinforced al-6061 hybrid composite. *SILICON, 12*, 1491–1500.
72. Siddesh Kumar, N. G., Suresh, R., Shiva Shankar, G. S. (2020) High temperature wear behavior of Al2219/n-B4C/MoS2 hybrid metal matrix composites. *Composites Communications* 19, 61–73.
73. Singh, H., Bhowmick, H. (2020). Lubrication characteristics and wear mechanism mapping for hybrid aluminium metal matrix composite sliding under surfactant functionalized MWCNT-oil. *Tribol Int* 145.
74. Rajesh Prabha, N., Edwin Raja Dhas, J. (2017). Effect of TiC and MOS2 reinforced aluminium metal matrix composites on microstructure and thermogravimetric analysis. *Rasayan Journal of Chemistry* 10, 729–737.

75. Kumarasamy, S. P., Vijayananth, K., Thankachan, T., Pudhupalayam Muthukutti, G. (2017). Investigations on mechanical and machinability behavior of aluminum/flyash cenosphere/Gr hybrid composites processed through compocasting. *Journal of Applied Research and Technology* 15, 430–441.
76. Manohar, H. S., Reddy Mungara, S., Anand, S. N., Sharan Teja Reddy, K. (2020). Development and study on corrosion properties of Al6061/Nanocarbon black reinforced aluminium metal matrix composites. in *Materials Today: Proceedings* vol. 20 185–190 (Elsevier Ltd).
77. Teja Reddy, S., Manohar, H. S., Anand, S. N. (2020) Effect of carbon black nano-fillers on tribological properties of Al6061-Aluminium metal matrix composites. in *Materials Today: Proceedings* vol. 20 202–207 (Elsevier Ltd).
78. Harsha, R. N., Kulkarni, M. v. and Satish Babu, B. (2019). Study of mechanical properties of aluminium/nano-zirconia metal matrix composites. in *Materials Today: Proceedings* vol. 26 3100–3106 (Elsevier Ltd).
79. Chen, B. et al. (2020) Microstructure, tensile properties and deformation behaviors of aluminium metal matrix composites co-reinforced by ex-situ carbon nanotubes and in-situ alumina nanoparticles. *Materials Science and Engineering A* 795.
80. Gayathri, J., Elansezhian, R. (2020) Influence of dual reinforcement (nano CuO + reused spent alumina catalyst) on microstructure and mechanical properties of aluminium metal matrix composite. *Journal of Alloys Compound* 829.
81. Jayaraj, A., Moorthy, C. V. K. N. S. N., Venkataramana, V. S. N., Jaikumar, S., Srinivas, V. (2020) Corrosion, mechanical and thermal properties of aluminium alloy metal matrix nano composites (AA-MMNCs) with multi-walled carbon nanotubes. *SN Appl Sci* 2.
82. Manohar, G., Pandey, K. M. and Maity, S. R. (2020). Aluminium (AA7075) metal matrix composite reinforced with B4C nano particles and effect of individual alloying elements in al fabricated by powder metallurgy techniques. In *Journal of Physics: Conference Series* vol. 1451 (IOP Publishing Ltd).
83. Shuvho, M. B. A., Chowdhury, M. A., Kchaou, M., Rahman, A., Islam, M. A. (2020) Surface characterization and mechanical behavior of aluminum based metal matrix composite reinforced with nano Al_2O_3, SiC, TiO_2 particles. *Chemical Data Collections* 28.
84. Singh, N., et al. (2018). Synthesis and tribological investigation of Al-SiC based nano hybrid composite. *Alexandria Engineering Journal, 57*, 1323–1330.
85. Aruri, D., Adepu, K., Adepu, K., & Bazavada, K. (2013). Wear and mechanical properties of 6061–T6 aluminum alloy surface hybrid composites [(SiC + Gr) and (SiC + Al2O3)] fabricated by friction stir processing. *Journal of Materials Research and Technology, 2*, 362–369.
86. Sivamaran, V. et al. *Mechanical And Tribological Properties Of Self-Lubricating Al 6061 Hybrid Nano Metal Matrix Composites Reinforced By Nsic And Mwcnts.*
87. Dey, D., Bhowmik, A. and Biswas, A. (2022). Effect of SiC content on mechanical and tribological properties of Al2024-SiC composites. *Silicon* 14.
88. Daniel, A. A., Murugesan, S., Manojkumar, M., & Sukkasamy, S. (2017). Dry sliding wear behaviour of aluminium 5059/SiC/MoS2 hybrid metal matrix composites. *Materials Research, 20*, 1697–1706.
89. Sridhar, A., & Lakshmi, K. P. (2020). Investigation of graphite effect on the mechanical and tribological properties of Al 7075-SiC-graphite hybrid metal matrix composites. *Tribologia, 37*, 26–32.
90. Balaji, V., Sateesh, N. and Hussain, M. M. (2015). Manufacture of aluminium metal matrix composite (Al7075-SiC) by stir casting technique. in *Materials Today: Proceedings* vol. 2 3403–3408 (2015).
91. Mahamani, A. (2014). Influence of process parameters on cutting force and surface roughness during turning of AA2219-TiB 2/ZrB 2 in-situ metal matrix composites. *Procedia Materials Science, 6*, 1178–1186.
92. John Joshua, K., Vijay, S. J., Ramkumar, P., Mohanasundaram, S. (2020). Effect of ZnO particles on microstructure, microhardness and wear behaviour of AA7068 metal matrix composites synthesized by powder metallurgy. in *Materials Science Forum* vol. 979 MSF 68–73 (2020).

93. Nagaral, M., Auradi, V., Kori, S. A., & Shivaprasad, V. (2020). Mechanical characterization and wear behavior of nano TiO2 particulates reinforced Al7075 alloy composites. *Mechanics of Advanced Composite Structures, 7*, 71–78.
94. Gupta, N. K., Kumar, M. and Thakre, G. D. Mechanical Characterization of 60Pb40Sn Reinforced Al6061 Self-healing Composite. in *Lecture Notes in Mechanical Engineering* 49–59 (2020). https://doi.org/10.1007/978-981-15-4059-2_4.
95. Srivastava, V., Gupta, M. (2019) Experimental assessment of self-healing characteristics in AA2014matrix with nitinol wire and solder alloy as healing agents. *Mater Res Express* 6.
96. Rohatgi, P. K. (2014). Al-shape memory alloy self-healing metal matrix composite. *Materials Science and Engineering A, 619*, 73–76.
97. Robinson Smart, D. S., Pradeep Kumar, J. and Sanjit Cyrus, R. (2019). Development and investigations of Al5083/CNT/Ni/MoS2 metal matrix composite for offshore applications. in *Materials Today: Proceedings* vol. 19 682–685 (2019).
98. Rastogi, S., Verma, A., & Singh, V. K. (2020). Experimental response of nonwoven waste cellulose fabric–reinforced epoxy composites for high toughness and coating applications. *Materials Performance and Characterization, 9*(1), 151–172.
99. Verma, A., & Ogata, S. (2023). Magnesium based alloys for reinforcing biopolymer composites and coatings: a critical overview on biomedical materials. *Advanced Industrial and Engineering Polymer Research*. https://doi.org/10.1016/j.aiepr.2023.01.002
100. Kumar, G., Mishra, R.R., Verma, A. (2022). Introduction to Molecular Dynamics Simulations. In *Forcefields for Atomistic-Scale Simulations: Materials and Applications* (pp. 1–19). Springer, Singapore. https://doi.org/10.1007/978-981-19-3092-8_1
101. Arpitha, G.R., Jain, N., Verma, A. and Madhusudhan, M. (2022). Corncob bio-waste and boron nitride particles reinforced epoxy-based composites for lightweight applications: fabrication and characterization. *Biomass Conversion and Biorefinery*, pp.1–8. https://doi.org/10.1007/s13399-022-03717-1
102. Kataria, A., Chaturvedi, V. C., Verma, A., Rangappa Sanjay, N. J. M., Siengchin, S. (2023). Cellulose fiber-reinforced composites—History of evolution, chemistry, and structure. In *Cellulose Fibre Reinforced Composites* (pp. 1–22). Woodhead Publishing. https://doi.org/10.1016/B978-0-323-90125-3.00012-4
103. Chaturvedi, S., Kataria, A., Chaudhary, V., Verma, A., Rangappa Sanjay, N. J. M., Siengchin, S. (20230. Bionanocomposites reinforced with cellulose fibers and agro-industrial wastes. In *Cellulose Fibre Reinforced Composites* (pp. 1–22). Woodhead Publishing. https://doi.org/10.1016/B978-0-323-90125-3.00017-3
104. Arpitha, G. R., Mohit, H., Madhu, P., & Verma, A. (2023). Effect of sugarcane bagasse and alumina reinforcements on physical, mechanical, and thermal characteristics of epoxy composites using artificial neural networks and response surface methodology. *Biomass Conversion and Biorefinery*. https://doi.org/10.1007/s13399-023-03886-7
105. Chaudhary, A., Sharma, S., & Verma, A. (2022). WEDM machining of heat treated ASSAB'88 tool steel: A comprehensive experimental analysis. *Materials Today: Proceedings, 50*, 946–951.
106. Bisht, N., Verma, A., Chauhan, S., & Singh, V. K. (2021). Effect of functionalized silicon carbide nano-particles as additive in cross-linked PVA based composites for vibration damping application. *Journal of Vinyl and Additive Technology, 27*(4), 920–932.
107. Verma, A., Jain, N., Parashar, A., Singh, V.K., Sanjay, M.R., Siengchin, S. (2020). Design and modeling of lightweight polymer composite structures. In *Lightweight Polymer Composite Structures* (pp. 193–224). CRC Press.
108. Dogra, V., Kishore, C., Verma, A., Rana, A. K., & Gaur, A. (2021). Fabrication and experimental testing of hybrid composite material having biodegradable bagasse fiber in a modified epoxy resin: Evaluation of mechanical and morphological behavior. *Applied Science and Engineering Progress, 14*(4), 661–667.
109. Deji, R., Verma, A., Kaur, N., Choudhary, B. C., & Sharma, R. K. (2022). Density functional theory study of carbon monoxide adsorption on transition metal doped armchair graphene nanoribbon. *Materials Today: Proceedings, 54*, 771–776.

110. Verma, A., Jain, N., Parashar, A., Gaur, A., Sanjay, M.R. and Siengchin, S. (2021). Life-cycle assessment of thermoplastic and thermosetting bamboo composites. In *Bamboo Fiber Composites* (pp. 235–246). Springer, Singapore.
111. Deji, R., Verma, A., Choudhary, B. C., & Sharma, R. K. (2022). New insights into NO adsorption on alkali metal and transition metal doped graphene nanoribbon surface: A DFT approach. *Journal of Molecular Graphics and Modelling, 111*, 108109.
112. Verma, A., Jain, N., Parashar, A., Singh, V.K., Sanjay, M.R. and Siengchin, S. (2020). Lightweight Graphene Composite Materials. In *Lightweight Polymer Composite Structures* (pp. 1–20). CRC Press.
113. Verma, A., Parashar, A. (2020). Characterization of 2D nanomaterials for energy storage. In *Recent Advances in Theoretical, Applied, Computational and Experimental Mechanics* (pp. 221–226). Springer, Singapore.
114. Deji, R., Jyoti, R., Verma, A., Choudhary, B. C., & Sharma, R. K. (2022). A theoretical study of HCN adsorption and width effect on co-doped armchair graphene nanoribbon. *Computational and Theoretical Chemistry, 1209*, 113592.
115. Verma, A., & Samant, S. S. (2016). Inspection of hydrodynamic lubrication in infinitely long journal bearing with oscillating journal velocity. *Journal of Applied Mechanical Engineering, 5*(3), 1–7.
116. Verma, A., Parashar, A., & Packirisamy, M. (2019). Effect of grain boundaries on the interfacial behaviour of graphene-polyethylene nanocomposite. *Applied Surface Science, 470*, 1085–1092.
117. Arpitha, G. R., Verma, A., Sanjay, M. R., & Siengchin, S. (2021). Preparation and experimental investigation on mechanical and tribological performance of hemp-glass fiber reinforced laminated composites for lightweight applications. *Adv. Civil Eng. Mater, 10*(1), 427–439.
118. Verma, A., Jain, N., Rangappa, S.M., Siengchin, S., Jawaid, M. (2021). Natural fibers based bio-phenolic composites. In *Phenolic Polymers Based Composite Materials*(pp. 153–168). Springer, Singapore.
119. Verma, A., Parashar, A., Singh, S.K., Jain, N., Sanjay, S.M., Siengchin, S. (2020). Modeling and Simulation in Polymer Coatings. In *Polymer Coatings* (pp. 309–324). CRC Press.
120. Verma, A., Singh, V.K., Experimental characterization of modified epoxy resin assorted with almond shell particles. *ESSENCE-International Journal for Environmental Rehabilitation and Conservation*, 7 (1), pp. 36–44.
121. Deji, R., Verma, A., Kaur, N., Choudhary, B. C., & Sharma, R. K. (2022). Adsorption chemistry of co-doped graphene nanoribbon and its derivatives towards carbon based gases for gas sensing applications: Quantum DFT investigation. *Materials Science in Semiconductor Processing, 146*, 106670.
122. Verma, A. (2022). A perspective on the potential material candidate for railway sector applications: PVA Based functionalized graphene reinforced composite. *Applied Science and Engineering Progress, 15*(2), 5727–5727.
123. Verma, A., Jain, N., Singh, K., Singh, V.K., Rangappa, S.M. and Siengchin, S. (2022). PVA-based blends and composites. In *Biodegradable Polymers, Blends and Composites* (pp. 309–326). Woodhead Publishing.
124. Raja, S., Verma, A., Rangappa, S. M., & Siengchin, S. (2022). Development and experimental analysis of polymer based composite bipolar plate using Aquila Taguchi optimization: Design of experiments. *Polymer Composites, 43*(8), 5522–5533.
125. Prabhakaran, S., Sharma, S., Verma, A., Rangappa, S. M., & Siengchin, S. (2022). Mechanical, thermal, and acoustical studies on natural alternative material for partition walls: A novel experimental investigation. *Polymer Composites, 43*(7), 4711–4720.
126. Sethi, S. K., Gogoi, R., Verma, A., & Manik, G. (2022). How can the geometry of a rough surface affect its wettability?-A coarse-grained simulation analysis. *Progress in Organic Coatings, 172*, 107062.
127. Verma, A., Jain, N., Sethi, S.K. (2022). Modeling and simulation of graphene-based composites. In *Innovations in Graphene-Based Polymer Composites* (pp. 167–198). Woodhead Publishing.

128. Chaturvedi, S., Verma, A., Sethi, S.K., Rangappa, S.M., Siengchin, S. (2022). Stalk fibers (rice, wheat, barley, etc.) composites and applications. In *Plant Fibers, their Composites, and Applications* (pp. 347–362). Woodhead Publishing.
129. Arpitha, G. R., Verma, A., & MR, S., Gorbatyuk, S., Khan, A., Sobahi, T.R., Asiri, A.M. and Siengchin, S. (2022). Bio-composite film from corn starch based vetiver cellulose. *Journal of Natural Fibers, 19*(16), 14634–14644.
130. Thimmaiah, S. H., Narayanappa, K., Thyavihalli Girijappa, Y., Gulihonenahali Rajakumara, A., Hemath, M., Thiagamani, S. M. K., & Verma, A. (2022). An artificial neural network and Taguchi prediction on wear characteristics of Kenaf-Kevlar fabric reinforced hybrid polyester composites. *Polymer Composites, 44*(1), 261–273.
131. Chaturvedi, S., Verma, A., Singh, S.K., Ogata, S. (2022). EAM Inter-Atomic Potential—Its Implication on Nickel, Copper, and Aluminum (and Their Alloys). In *Forcefields for Atomistic-Scale Simulations: Materials and Applications* (pp. 133–156). Springer, Singapore.
132. Verma, A., Sharma, S. (2022). Atomistic Simulations to Study Thermal Effects and Strain Rate on Mechanical and Fracture Properties of Graphene like BC3. In *Forcefields for Atomistic-Scale Simulations: Materials and Applications* (pp. 237–252). Springer, Singapore.
133. Chaturvedi, S., Verma, A., Sethi, S.K. and Ogata, S. (2022). Defect Energy Calculations of Nickel, Copper and Aluminium (and Their Alloys): Molecular Dynamics Approach. In *Forcefields for Atomistic-Scale Simulations: Materials and Applications* (pp. 157–186). Springer, Singapore.
134. Verma, A., Gaur, A., & Singh, V. K. (2017). Mechanical properties and microstructure of starch and sisal fiber biocomposite modified with epoxy resin. *Materials Performance and Characterization, 6*(1), 500–520.
135. Shankar, U., Gogoi, R., Sethi, S.K., Verma, A., (2022). Introduction to Materials Studio Software for the Atomistic-Scale Simulations. In *Forcefields for Atomistic-Scale Simulations: Materials and Applications* (pp. 299–313). Springer, Singapore.
136. Shankar, U., Sethi, S.K., Verma, A. (2022). Forcefields and Modeling of Polymer Coatings and Nanocomposites. In *Forcefields for Atomistic-Scale Simulations: Materials and Applications* (pp. 81–98). Springer, Singapore.
137. Verma, A., Ogata, S. (2022). Computational Modelling of Deformation and Failure of Bone at Molecular Scale. In *Forcefields for Atomistic-Scale Simulations: Materials and Applications* (pp. 253–268). Springer, Singapore.
138. Homer, E.R., Verma, A., Britton, D., Johnson, O.K., Thompson, G.B. (2022) Simulated migration behavior of metastable Σ3 (11 8 5) incoherent twin grain boundaries. In *IOP Conference Series: Materials Science and Engineering*(Vol. 1249, No. 1, p. 012019).
139. Verma, A., Parashar, A., van Duin, A.C. (2022). Graphene-reinforced polymeric membranes for water desalination and gas separation/barrier applications. In *Innovations in Graphene-Based Polymer Composites* (pp. 133–165). Woodhead Publishing.
140. Verma, A., Jain, N., Sanjay, M.R., Siengchin, S., Viscoelastic Properties of Completely Biodegradable Polymer-Based Composites. In *Vibration and Damping Behavior of Biocomposites* (pp. 173–188). CRC Press.
141. Verma, A., Jain, N., Mishra, R.R. (2022). Applications and drawbacks of epoxy/natural fiber composites. In *Handbook of Epoxy/Fiber Composites* (pp. 1–15). Singapore: Springer Singapore.
142. Lila, M.K., Verma, A., Bhurat, S.S. (2022). Impact behaviors of epoxy/synthetic fiber composites. In *Handbook of Epoxy/Fiber Composites* (pp. 1–18). Singapore: Springer Singapore.
143. Verma, A., & Parashar, A. (2017). The effect of STW defects on the mechanical properties and fracture toughness of pristine and hydrogenated graphene. *Physical Chemistry Chemical Physics, 19*(24), 16023–16037.
144. Verma, A., & Parashar, A. (2018). Molecular dynamics based simulations to study failure morphology of hydroxyl and epoxide functionalised graphene. *Computational Materials Science, 143*, 15–26.

145. Verma, A., & Parashar, A. (2018). Molecular dynamics based simulations to study the fracture strength of monolayer graphene oxide. *Nanotechnology, 29*(11), 115706.
146. Verma, A., & Parashar, A. (2018). Structural and chemical insights into thermal transport for strained functionalised graphene: A molecular dynamics study. *Materials Research Express, 5*(11), 115605.
147. Gautam, R., Gupta, H.S., Chauhan, A., Jaiswal, A.K., Verma, A. (2024). Composite Materials for Bio-Energy. In *Hybrid Composite Materials: Experimental and Theoretical Analysis* (pp. 91–104). Singapore: Springer Nature Singapore. https://doi.org/10.1007/978-981-97-2104-7_4
148. Kriti, Kumari, S., Verma, A. (2024). Synthesis and Applications of Different Polymer Composites. In *Hybrid Composite Materials: Experimental and Theoretical Analysis*(pp. 1–44). Singapore: Springer Nature Singapore. https://doi.org/10.1007/978-981-97-2104-7_1
149. Sethi, S.K., Verma, A., Akubueze, E., Maurya, A.K. (2024). Self-Healing of Polymer Composites: Process and Developments. In *Hybrid Composite Materials: Experimental and Theoretical Analysis* (pp. 253–271). Singapore: Springer Nature Singapore. https://doi.org/10.1007/978-981-97-2104-7_10
150. Kafaltiya, S., Agrawal, P.K., Sharma, P., Singh, V.K., Bhaskar, J., Chauhan, S., Verma, A. Multi-response optimization of characteristics for graphite reinforced biodegradable PVA-fumaric acid cross-linked composite: A gray-fuzzy logic-based hybrid approach. Journal of Vinyl and Additive Technology. https://doi.org/10.1002/vnl.22112
151. Santos, T. F., Santos, C. M., Aquino, M. S., Suyambulingam, I., Kamil Hussein, E., Verma, A., Sanjay, M. R., Siengchin, S., & Nascimento, J. H. O. (2024). Towards sustainable and ecofriendly polymer composite materials from bast fibers: A systematic review. *Engineering Research Express, 6*, 012501. https://doi.org/10.1088/2631-8695/ad2640
152. Sharath, B.N., Madhu, P., Verma, A. (2023). Enhancing tribological performance: a review of ceramic reinforced aluminium hybrid composites for high-temperature engineering applications. *Hybrid Advances* 100094.

Bioinformatics in Crop Improvement and Agricultural Genomics

Riya Mishra, Makhan Singh Karada, and Dheer Agnihotri

Abstract The profuse rise in life sciences data has driven bioinformatics, a multidisciplinary field at the network of biology, computer science, statistics and mathematics. Bioinformatics becomes essential in crop development for analyzing genome sequences and decoding plant genomes. Its efficiency in handling massive genomic datasets assists in identifying key genes for desirable traits, advancing targeted breeding programs. It is augmented by omics technologies such as transcriptomics, genomics, proteomics, metabolomics and phenomics. Incorporation of multiomics unveils sophisticated biological mechanisms, expediting molecular breeding for complex crop traits. This convergence of omics and bioinformatics marks a new era in crop improvement, enabling a comprehensive exploration of trait genetics. In the era of climate change, bioinformatics, in collaboration with genomics, is vital for identifying climate-ready crops and understanding traits resilient to environmental challenges. This interdisciplinary association positions bioinformatics as a catalyst for enhanced agricultural productivity and sustainability amid evolving global food demands.

Keywords Bioinformatics · Bioinformatics tools · Agricultural genomics · Crop improvement · Multiomics

R. Mishra (✉)
Department of Genetics and Plant Breeding, Rajmata Vijayaraje Scindia Krishi Vishwa Vidyalaya, Gwalior, Madhya Pradesh 474002, India
e-mail: riyam2710@gmail.com; riyamishra@jnkvv.org

M. S. Karada
Department of Forestry, Jawaharlal Nehru Krishi Vishwa Vidyalaya, Jabalpur, Madhya Pradesh 482004, India

D. Agnihotri
Silviculture, Agroforestry and Forest Management Division, Tropical Forest Research Institute, Jabalpur, Madhya Pradesh 482021, India

© The Author(s), under exclusive license to Springer Nature Singapore Pte Ltd. 2024
A. Chaudhary et al. (eds.), *Unraveling New Frontiers and Advances in Bioinformatics*,
https://doi.org/10.1007/978-981-97-7123-3_13

1 Introduction

The enhancement of crop traits is an ongoing and iterative endeavor, propelled by the escalating global population's demand for food security through the attainment of sustainable agricultural production [1]. There is a progressively acknowledged consensus regarding the alteration of the global climate, attributed primarily to human activities, notably the combustion of fossil fuels leading to an elevation in atmospheric carbon dioxide concentration [2]. What remains under deliberation is the precise nature of the anticipated changes and their ramifications on agriculture. Accurate prediction of meteorological events is a difficult task due to the complicated interactions of numerous elements [3]. To ensure continuous food production, new crops must be cultivated and existing crops must be modified to fit changing environmental conditions. Progress in genomics presents an opportunity to expedite genomics-driven breeding of crop plants. Nevertheless, the integration of data on genomes with climate-responsive agricultural traits for practical application in crop improvement poses a significant problem, demanding the collaboration of varied skills and expertise. The synergistic application of genetics and bioinformatics can safeguard food sustainability amidst climatic crises by hastening the development of crops equipped to thrive in evolving climatic conditions [4]. Bioinformatics is an important component in contemporary plant and crop science, demonstrating indispensable contributions across various applications [5]. Bioinformatics plays a decisive role in the current era marked by remarkable advancements in omics technologies, providing essential assistance in revealing the complicated dimensions of biological functionalities. The increasing understanding of molecules and mechanisms linked to distinct phenotypes and reactions of the plants to biotic and abiotic challenges, as well as the powerful forecasting capabilities of biological informatics, significantly influences agricultural practices. This, in turn, promotes inventive approaches in tracking, surveillance and evaluation, enhancing human benefits with reduced expenses and thereby promoting sustainability [6].

1.1 Overview of Bioinformatics

The surge in biological data, particularly within the genomic era, has witnessed a substantial expansion propelled by advancements in the field of molecular sciences and genome study. This vast influx of genetic data necessitates the development of database systems, sophisticated equipment and programmed methodologies for the systematic organization, storage, indexing, analysis and visualization of the data being produced. With their roots in the nexus of information sciences, life sciences, computer technologies and information sciences, bioinformatics and computer science exhibit a synergistic relationship. Both disciplines closely engage with the life sciences to unlock and comprehend their comprehensive capabilities [7].

According to the Oxford English dictionary, bioinformatics—*(Molecular)* bioinformatics: bioinformatics is conceptualizing biology in terms of molecules (in the sense of physical chemistry) and applying "informatics techniques" (derived from disciplines such as applied math, computer science and statistics) to understand and organize the information associated with these molecules, on a large scale. To put it briefly, bioinformatics is a molecular biology management information system with a wide range of real-world uses [8].

At its fundamental essence, bioinformatics can be delineated as "the organization of biological data with the purpose of facilitating systematic inquiry" [9–12]. Bioinformatics, as defined by the National Centre for Biotechnology Information (NCBI), encompasses research, development and application of computational equipment and methodologies. These tools are designed to enhance the utilization of biological, medicinal and plant databases, facilitating the acquisition, storage, organization, archiving and visualization of generated data [13]. It commonly involves tasks such as the alignment and mapping of diverse protein and DNA sequences, as well as the generation and visualization of three-dimensional models explaining the structures of proteins [14].

The understanding of structural and functional elements included in plant genomes has significantly increased in the last few years due to noteworthy advancements within the domains of genetics and genomics. This progress has unified foundational understanding, which holds great potential for optimizing crop plants for the benefit of humans [15]. Future advances in biotechnology may be accelerated by the utilization of novel omics technologies such as transcriptomics, proteomics, phenomics, metabolomics and genomics. Interestingly, in 1986, Thomas Roderick originally used the term "genomics" during the human genome sequencing project [1]. The complete sequencing of genomes in Arabidopsis and Rice, coupled with an extensive array of plants expressed sequence tags (ESTs), has become readily accessible. Ongoing sequencing initiatives aimed at augmenting our understanding of major crops are underway. The synergy of this emerging genomic knowledge with traditional breeding methodologies is imperative for enhancement of crop improvement. The strides made in genomics offer dual avenues for contributing to crop improvement. Firstly, an enhanced grasp of biological mechanisms can pave the way for the development of novel or refined screening methods, facilitating the more efficient identification of superior genotypes. Secondly, the acquisition of new information may enhance one's ability to make decisions, thereby optimizing crop improvement tactics for greater efficiency [15]. The landscape of crop improvement has undergone a profound transformation due to technological and scientific advancements, encompassing genome sequencing, genotyping for comprehensive genome-wide annotator studies and the formulation of genomic hypothesis [16].

2 Bioinformatics Resources for Crop Improvement and Genomics

Bioinformatics constitutes a critical component in the domain of crop improvement, furnishing researchers with indispensable tools and methodologies tailored for the analysis and interpretation of expansive biological datasets. These computational resources play a decisive role in revealing the genetic underpinnings of agricultural traits, pinpointing important genes and formulating strategic approaches for the enhancement of crop species. The ensuing enumeration encapsulates key bioinformatics resources instrumental in the sphere of crop improvement:

2.1 Databases as Crop Resources

A database is an organized compilation of information structured to facilitate straightforward accessibility, efficient management and seamless updates [13]. Due to the advent of highly efficient sequencing techniques, copious amounts of data are consistently disseminated and frequently deposited into designated databases. These databases can be systematically categorized based on the taxonomic classification of the represented species, distinguishing between general repositories and those specifically tailored to individual species [17].

Repositories that compile data on proteins, genomes and metabolic pathways are included in general databases (Table 1). Genomic database systems serve as centerpiece and publicly accessible repositories of released information, facilitating scientists in obtaining pertinent knowledge related to the proteins and genes they are interested in. These databases support queries based on specific gene or protein names and relevant keywords and enable protein BLAST searches for retrieving sequences of newly identified proteins [18]. Moreover, they aggregate diverse biological pathways, exemplified by the Kyoto Encyclopedia of Genes and Genomes (KEGG), EuKaryotic Orthologous Groups (KOG), Gene Ontology (GO) and metabolic pathways. These resources aid in establishing the affiliation of a candidate protein with recognized pathways [17].

Species-specific databases, tailored to particular organisms, provide valuable resources for swift access to the latest genome information (Table 2). Notably, model organisms like *Arabidopsis thaliana* boast dedicated databases owing to their widespread utilization in research. For example, The Arabidopsis Information Resource (TAIR) database, offers an opportunity for bulk downloading the gene sequences. The SeqViewer tool within TAIR facilitates the visualization of genes, while the database features robust functions for displaying expression maps, detailing gene expression patterns throughout the development of Arabidopsis as well as within the various growth conditions [26].

Table 1 General databases for assessing plant genomes

Database	Description	URL	References
Gramene	A genome viewer and database built on the EnsEMBL platform, housing information relate to several crop species, with a primary focus on the genomes of Maize (*Zea mays*), Arabidopsis and Rice (*Oryza spp.*)	http://www.gramene.org/	[19, 20]
Ensembl Plants	Ensembl Plants, the plant and red algae portal of Ensembl, offers unified interfaces for genomic data, including genetic variation, markers, gene expression, comparative genomics, gene models and reference genome sequences. It undergoes upto four releases annually	http://plants.ensembl.org	[21]
PlantGDB	Innovative prediction algorithms are integrated into software, visualization tools and data access portals through the PlantGDB database. Moreover, it includes an interconnected infrastructure framework featuring tools for development, for protocol sharing, distributed computing and sturdy record analysis	http://www.plantgdb.org/	[17]
AgBioData	Organization of agricultural biological databases whose goal is to bring together guidelines and optimal methods for obtaining, presenting and utilizing genomic, genetic and breeding (GGB) information	https://www.agbiodata.org/	[22]
PlantsDB	The database offers organized and linked genomic and genetic data for twelve key species of plants, spanning from significant crops like wheat to typical plants such as Arabidopsis. With features like an RNAseq Expression Browser and Crows Nest, a comparative map viewer, it provides an infrastructure for holistic and comparable plant genomics research	http://pgsb.helmholtz-muenchen.de/plant/index.jsp	[23]
Plaza	In addition to a wide range of interactive tools for investigating gene function and gene or genome evolution, this resource includes both functional and structural characterization of published plant genomes	http://bioinformatics.psb.ugent.be/plaza/	[24]
Phytozome	The Joint Genome Institute of the DOE's Phytozome serves as a central location for accessing, displaying and evaluating plant genomes that have been sequenced by JGI as well as some sequences and databases from external sources	http://www.phytozome.net/	[25]

Table 2 Species specific databases for different plant genomes

Database	Description	URL	References
GrainGenes	GrainGenes serves as the global database of Rye, oats, barley and wheat genomes. It serves as a central database for data on genes, QTLs, alleles, mapping probes, primers and genetic maps for these particular species of plants	http://wheat.pw.usda.gov/	[27]
Triticeae Repeat Sequence Database (TREP)	A compilation of frequently occurring sequences of DNA representing different species of Triticeae. These sequences serve as valuable resources for the creation of molecular markers	http://wheat.pw.usda.gov/ITMI/Repeats/	[28]
Maize Genetics and Genomic Database (MaizeGDB)	MaizeGDB, utilizes GBrowse, functions as a central repository for maize related information, encompassing maize sequence data, stock details, phenotypic and genotypic variations and chromosomal mapping data	http://www.maizegdb.org/	[29]
Oryzabase	By combining molecular genomic data with biological data, Oryzabase aims to create a comprehensive understanding of *Oryza sativa* as an ideal monocot plant. The database encompasses details on development of *Oryza* and morphology, its mutants and genetic resources, with a particular emphasis on wild rice varieties	http://www.shigen.nig.ac.jp/rice/oryzabase/	[30]
BRAD	Utilizing GBrowse, BRAD functions as an essential community gateway, providing genome data for the species of *Brassica* and allied *Brassicaceae* plants	http://brassicadb.org/	[31]
Legume Information system (LIS)	Cross-species genome and transcriptome evaluations are made easier by LIS, which harmonizes molecular and genetic information from different bean species. Currently, it encompasses information of 17 species, featuring GBrowse databases for soybean, birdsfoot trefoil, barrel medic and pigeon pea	http://www.comparative-legumes.org/	[32]
SOL Genomics Network (SGN)	In its capacity as a clade oriented database (COD), the SGN compiles biological information about species belonging to the Solanaceae family as well as those closely related to them. In addition to phenotypes and accessions, it involves different kinds of data, spanning from genes to chromosomes. It provides entire genomic data for an increasing number of Solanaceae members such as *Lycopersicon esculentum, Solanum tuberosum, Nicotiana tabaccum* and *Nicotiana benthamiana* etc	https://solgenomics.net/	[33, 34]

2.2 Bioinformatics Software/Tools

Tools for bioinformatics are software applications designed with the sole purpose of obtaining valuable data from large datasets, which makes the analysis stage of bioinformatics easier [13]. Extensive repositories of tools of bioinformatics have been established under freely accessible framework, ensuring their unrestricted accessibility, educational utility and capacity for continuous enhancement [35]. A variety of bioinformatics tools are available, ranging from standard to specialized items personalized to match the unique requirements of distinct initiatives. These tools encompass software for data-mining, adept at extracting information from database systems of genome sequences, as well as visualizing tools designed for the analysis and retrieval of data from databases of proteomics. Broadly classified, the tools include protein functional analysis tools, homology and similarity tools, sequence analysis tools and miscellaneous tools, each playing a crucial role in routine bioinformatics tasks [36]. The following provides a succinct overview of tools routinely employed in everyday bioinformatics.

2.2.1 Basic Local Alignment Search Tool (BLAST)

A group of Windows-only search tools called BLAST (Basic Local Alignment Search Tool) is used to quickly perform similarity searches on any type of sequence, including DNA and protein sequences. It facilitates the search of protein databases to find matches with requested protein sequences and makes it easier to compare nucleotide sequences within databases. Q BLAST, the latest queuing system that NCBI has added to BLAST, allows individuals to recover outcomes whenever it's convenient for them and organize results numerous times using different formatting methods [36–40].

BLAST encompasses a suite of specialized algorithms tailored for distinct types of biological sequences in the context of comparative genomics. Specifically, blastp is designed to scrutinize a particular amino acid querying sequence against a comprehensive protein sequence repository, facilitating the identification of homologous proteins. Blastn is skilled at determining how similar a nucleotide query sequence is to a large nucleotide sequence database when it comes to nucleotide sequence comparisons. To bridge the gap between nucleotide queries and protein databases, in all the reading frames, blastx orchestrates a translation of the nucleotide query sequence, enabling the exploration of potential protein homologs. Conversely, tblastn is engineered to provide a flexible method by dynamically translating a protein query sequence in all reading frames and comparing it to a nucleotide sequence database. Lastly, tblastx compares a nucleotide query sequence's six frame translations to the corresponding translations of nucleotide sequence databases, providing a thorough examination of possible nucleotide sequence homologies across several reading frames. This array of BLAST programs thus provides a nuanced and specialized

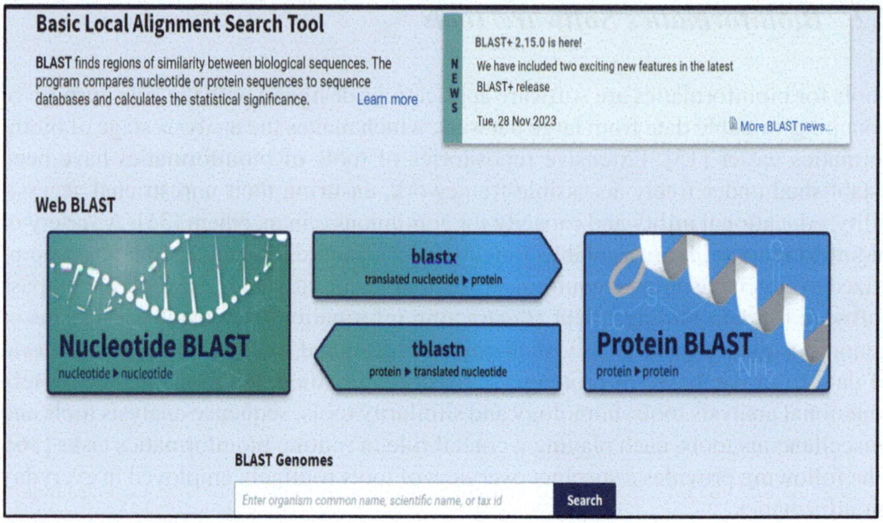

Fig. 1 Homepage of BLAST

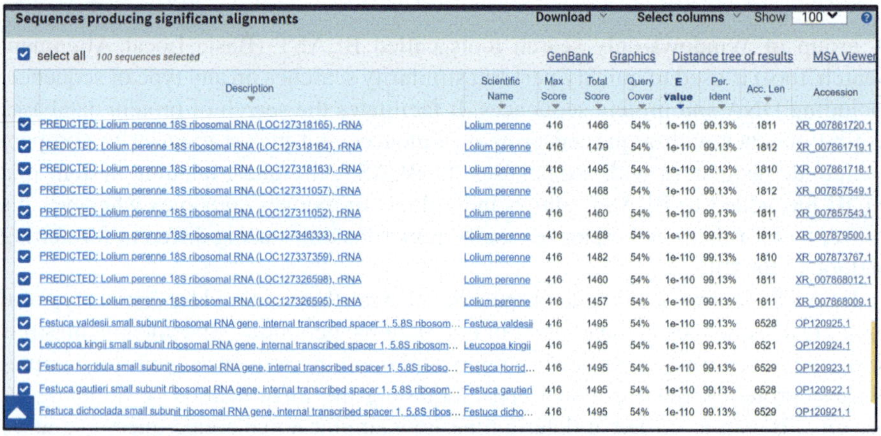
Sequences producing significant alignments — Download — Select columns — Show 100

select all 100 sequences selected — GenBank — Graphics — Distance tree of results — MSA Viewer

Description	Scientific Name	Max Score	Total Score	Query Cover	E value	Per. Ident	Acc. Len	Accession
PREDICTED: Lolium perenne 18S ribosomal RNA (LOC127318165), rRNA	Lolium perenne	416	1466	54%	1e-110	99.13%	1811	XR_007861720.1
PREDICTED: Lolium perenne 18S ribosomal RNA (LOC127318164), rRNA	Lolium perenne	416	1479	54%	1e-110	99.13%	1812	XR_007861719.1
PREDICTED: Lolium perenne 18S ribosomal RNA (LOC127318163), rRNA	Lolium perenne	416	1495	54%	1e-110	99.13%	1810	XR_007861718.1
PREDICTED: Lolium perenne 18S ribosomal RNA (LOC127311057), rRNA	Lolium perenne	416	1468	54%	1e-110	99.13%	1812	XR_007857549.1
PREDICTED: Lolium perenne 18S ribosomal RNA (LOC127311052), rRNA	Lolium perenne	416	1460	54%	1e-110	99.13%	1811	XR_007857543.1
PREDICTED: Lolium perenne 18S ribosomal RNA (LOC127346333), rRNA	Lolium perenne	416	1468	54%	1e-110	99.13%	1811	XR_007879500.1
PREDICTED: Lolium perenne 18S ribosomal RNA (LOC127337359), rRNA	Lolium perenne	416	1482	54%	1e-110	99.13%	1810	XR_007873767.1
PREDICTED: Lolium perenne 18S ribosomal RNA (LOC127326598), rRNA	Lolium perenne	416	1460	54%	1e-110	99.13%	1811	XR_007868012.1
PREDICTED: Lolium perenne 18S ribosomal RNA (LOC127326595), rRNA	Lolium perenne	416	1457	54%	1e-110	99.13%	1811	XR_007868009.1
Festuca valdesii small subunit ribosomal RNA gene, internal transcribed spacer 1, 5.8S ribosom...	Festuca valdesii	416	1495	54%	1e-110	99.13%	6528	OP120925.1
Leucopoa kingii small subunit ribosomal RNA gene, internal transcribed spacer 1, 5.8S ribosom...	Leucopoa kingii	416	1495	54%	1e-110	99.13%	6521	OP120924.1
Festuca horridula small subunit ribosomal RNA gene, internal transcribed spacer 1, 5.8S riboso...	Festuca horrid...	416	1495	54%	1e-110	99.13%	6529	OP120923.1
Festuca gautieri small subunit ribosomal RNA gene, internal transcribed spacer 1, 5.8S ribosom...	Festuca gautieri	416	1495	54%	1e-110	99.13%	6528	OP120922.1
Festuca dichoclada small subunit ribosomal RNA gene, internal transcribed spacer 1, 5.8S ribos...	Festuca dicho...	416	1495	54%	1e-110	99.13%	6529	OP120921.1

Fig. 2 Blastn depicting the nearest neighbor sequences of *Triticum aestivum* 18S rRNA

toolkit for the rigorous examination of diverse biological sequence comparisons [13, 41, 42] (Figs. 1 and 2).

2.2.2 European Molecular Biology Open Software Suite (EMBOSS)

EMBOSS is a highly potent programme that may be used for both local and global alignment of paired sequences. It works especially well with their short DNA

Table 3 Bioinformatics tools and websites for plant research

Name of database	Description	URL	References
STAG-CNS	A tool designed for identification of sequentially conserved non-coding sequences across an arbitrary number of species	https://github.com/srbehera11/stag-cns	[46]
Interactive Tree of Life (Itol)	Online platform for visualizing, manipulation and annotating, managing system development of phylogenetic and other trees	https://itol.embl.de/	[47]
Tbtools	Comprehensive toolkit for interactively analyzing vast amount of biological data	https://bio.tools/tbtools	[48]
Simple Modular Architecture Research Tool (SMART)	Makes it easier to identify genetically mobile domains, annotate them and analyze domain structures	https://smart.embl-heidelberg.de/	[49]
Foreign Element Detector (FED)	Web based tool for identifying foreign components within organisms with altered genomes	http://www.hi-tom.net/FED/	[50]

sequences. The following URL allows users to access EMBOSS-https://www.ebi.ac.uk/Tools/emboss/ [43]. This package includes extensive libraries, encouraging the opportunity for researchers to share their software as freely accessible. EMBOSS is equipped with a suite of sequence analysis programs and offers compatibility with all UNIX platforms [13, 36]. It comprises a confluence of approximately one hundred programs designed for tasks encompassing sequence alignment, exploration of data through sequence patterns, identification of protein motifs, analysis of domains, scrutiny of nucleotide sequence patterns and investigation into codon usage for small genomes, among various other functionalities [44, 45].

Other than these, various tools of bioinformatics and online platforms that perform the key role in predicting protein structures, identifying conserved domains within proteins and annotating genes (Table 3). Additionally, these tools are often specialized for specific tasks, such as forecasting transcription factor binding sites and investigating extensive genomic variation datasets (Table 4).

3 Applications of Bioinformatics in Crop Improvement and Agriculture Genomics

Bioinformatics is crucial in enhancing crop improvement strategies by aiding the thorough analysis of plant genomes, differentiating genetic loci linked to desirable traits and simplifying crop modification processes. The utilization of bioinformatics tools and methodologies expedites the examination of genome sequences, promoting

Table 4 Plant-specific bioinformatics tools for research applications

Tool	Description	URL	References
CRISPR-P	It enables users to identify highly specific Cas9 target sites within designated DNA sequences. The platform additionally offers predictions of potential off-target loci, facilitating further in-depth analysis	http://crispr.hzau.edu.cn/CRISPR2/	[51, 52]
PlantPAN	It serves as a powerful tool for rebuilding transcriptional regulatory networks that are especially suited for plant genes and for anticipating regulatory elements	http://plantpan2.itps.ncku.edu.tw/	[53]
The Bio-Analytic Resource for Plant Biology (BAR)	It serves as a flexible framework for analyzing protein–protein interactions, performing functional categorization enrichment analysis for gene sets, analyzing promoters and examining patterns of gene expression and co-expression	https://www.bar.utoronto.ca/welcome.htm	[54]
Plant MetGenMAP	It is designed to seamlessly integrate and analyze extensive datasets of gene expression and metabolite profiles, complemented by diverse biological information	http://bioinfo.bti.cornell.edu/cgi-bin/MetGenMAP/home.cgi	[55]
SnpHub	Unified web server architecture intended for the investigation of vast amounts of genetic variation data	https://guoweilong.github.io/snpHub/	[56]

the creation of novel algorithms and statistical models for evaluating complex relationships within extensive datasets. Specifically, bioinformatics contributes to crop improvement through tasks such as dissecting sequence information, innovating algorithms and statistical methods, and intricately modeling crop plants.

Moreover, it plays a crucial role in elucidating metabolic networks associated with stress and disease resistance. Metabolomics approaches have proven effective in identifying diverse metabolic networks linked to both biotic and abiotic stress, thereby enhancing crop improvement efforts geared towards achieving superior yield, heightened stress tolerance and increased resistance to diseases [40, 57]. Following are some key applications of bioinformatics in crop improvement and genomics.

3.1 Genomic Data Analysis

Plant biologists have used cutting-edge sequencing technology to unveil the genetic makeup of various plant and microbe species, including characteristics like metabolomics, transcriptomics, proteomics and complex metabolic pathways [58].

Sequence analysis, as a core methodology, serves as the cornerstone for acquiring comprehensive genomic information, including protein, RNA and DNA sequences of the genome of an organism in contemporary scientific endeavors [59]. Whole genome sequencing, in particular, enables a comprehensive understanding of the organizational structure of different species, serving as a foundational resource for unraveling their functionality [60]. The progression of next generation sequencing (NGS) and other sequencing techniques have yielded extensive volumes of biological data, necessitating the establishment of databases for effective information storage. The availability of entire genome sequences in these databases facilitates seamless cross-referencing across genomes concerning gene sequences, putative functions and genetic map positions [59]. Several widely recognized next-generation sequencing (NGS) platforms are commercially available, each employing distinct methodologies. These include the Genome Sequencer by Roche/454 utilizing Pyrosequencing, the Genome Analyzer by Illumina/Solexa employing Sequencing-by-synthesis, SOLiD by Applied Biosystems utilizing Sequencing-by-ligation, Polonator by Dover SystemsP also utilizing Sequencing-by-ligation, Ion Torrent by Life Science, Inc. based on Ion semiconductor technology, Helioscope sequencers by Helicos Bioscience Corporation implementing true single molecule sequencing, PacBio RS sequencers by Pacific Biosciences employing Single molecule, real time sequencing and GrindION and miniaturized MinION sequencers by Oxford Nanopore Technologies utilizing Nanopore sequencing [61, 62].

Leveraging software tools, it becomes feasible to formulate projective hypotheses and integrate phenotypes of interest into plants by scrutinizing genetic markers that exhibit outstanding results and provide enhanced reliability in breeding processes [63, 64]. Beyond the data related to genome sequences, databases containing information on metabolites are essential for exploring the complex connections between proteomics and genomics. These databases are instrumental in investigating how interactions among these molecular components contribute to changes in an organism's phenotype and specific functions [58]. Additionally, genes linked to particular traits can undergo computer-based analysis prior to their introduction into a plant. The outcomes of this analysis guide the selection of specific genetic elements to be further incorporated within the genome of the plant, facilitating accurate phenotypic analysis [17]. The development of lysine-enriched maize kernels [65], vitamin-C-rich *Lactuca sativa* [66] and recently engineered vitamin-D-rich *Lycopersicon esculentum* [67] are some of the famous examples of this approach.

3.2 Comparative Genomics

The comprehensive sequencing of entire genomes necessitates the crucial undertaking of elucidating and characterizing both the coding and non-coding components embedded within these sequences [68]. Consequently, the emergence of plant comparative genomic analysis represents a contemporary biotechnological discipline, primarily aimed at comprehending the functionalities of numerous uncharacterized genes through the examination of noteworthy distinctions and resemblances across diverse species. It is important to note that the identification of these genes relies on their presence within the accessible datasets of orthologs originating from a common ancestral source [69]. Through the application of a comparative genomic approach, an increasing number of genes within species of plants have been systematically cataloged. In this context, the utilization of renowned stress-responsive transcription factors (TFs) from *Arabidopsis thaliana* and *Oryza spp.* has facilitated the accurate prediction of stress-responsive TFs in numerous species of plants, including *Glycine max*, *Zea mays*, *Hordeum vulgare*, *Sorghum bicolor* and *Triticum aestivum* [70–72]. Furthermore, comparative genomics, employing co-expression analysis, proves valuable in discovering previously unidentified biosynthetic genes [73]. Noteworthy tools such as GolmTranscriptome DB [74] and ATTED-II [75] are employed for such analysis in plants. In addition to analytical tools and strategies, robust computational resources are imperative for the storage and management of vast genomic data. To assist comparative genomic studies across distinct species of plants, numerous online platforms have been developed, including Phytozome, PLAZA, GreenPhylDB, PlantsDB and many others [76].

3.3 Disease Resistance

Pathogens constitute formidable constraints to the prosperity and proliferation of crops. Therefore, the primary objective for plant geneticists is the elucidation of the fundamental genetic determinants responsible for conferring resistance to pathogens [77]. In agricultural contexts, the extensive application of bioinformatics proves instrumental in enhancing the efficacy of crop breeding strategies and fortifying plant resistance mechanisms against pathogenic threats [58]. Advancements in sequencing agriculturally important crops have enabled a more comprehensive exploration of interactions between plants and pathogens and mechanisms of resistance in plants. Gene and genome functional aspects may be effectively retrieved, annotated, analyzed and characterized owing to the availability of genetic data from both hosts and pathogens. Several potential resistance (R) genes from various species of plants can be precisely identified and isolated with the use of physical mapping techniques. The identification and characterization of numerous genetic variants, including disease-causing mutations, have been made possible by bioinformatics techniques and have been used to create enhanced disease resistance [78]. In recent years the investigation of plant diseases instigated by pathogens has yielded a substantial body of knowledge, leading to the creation of numerous innovative bioinformatics tools. Notable examples include PhytoPath [79], EffectorP [80], PathoPlant [81] and InterSPPI [82], designed for scrutinizing pathogen pathogenicity mechanisms. Tools such as PRGdb [83] and RRGPredictor [84] are instrumental in identifying resistance (R) proteins, while image-based disease recognition platforms like Plantix

[85] and LeafDoctor [86] serve the purpose of detecting plant diseases. Additionally, resources such as FungiPAD [87], PlantPepDB [88] and APD389 offer agents for the management of plant diseases. These bioinformatics tools have reached a sufficient level of scientific maturity, enhancing the efficiency and speed of decisions related to plant disease management [90].

Moreover, plant pathology has benefitted from the use of bioinformatics, notably in the detection and anticipation of effector proteins that pathogens of plants release to influence their host plants. One of the most important steps in converting sequence data into possible uses in plant pathology is the functional characterization of the pathogen's ability to predict virulence [91].

3.4 *Phylogenetics for Crop Evolution*

Phylogenetics, a branch of molecular phylogenetic analysis, involves the utilization of genomic databases to predict gene functions and explore evolutionary relationships among species [76]. Plant phylogenomics possesses a distinct advantage, allowing for the identification of numerous nuclear genes of low copy numbers. This capability significantly enhances the study of molecular systematics and evolutionary biology of plants [92]. In phylogenomic research, two primary objectives are pursued. The first aims to unveil evolutionary patterns among plant species through the analysis of nuclear genomic information. The second objective involves generating new hypotheses regarding unidentified functions regarding plant genes linked to significant deviation happenings of plant species advancements [93]. Using a set of orthologous genes acquired from genomic sequences in phylogenetic framework, phylogenomics creates biological process and gene-related hypothesis [94]. Orthologs are genes with a shared ancestry due to speciation, while paralogs arise from duplications within a genome, such as polyploidization or tandem duplications [95, 96]. The identification of true orthologs in plants is complicated by their paleopolyploid nature and extensive gene duplication events during evolution [97, 98]. Validated sets of orthologous genes are essential for many aspects of evolutionary biology, systematics and comparative genomics, such as selection analysis, ancestral genome reconstruction, genomic mapping, phylogenetic studies on regions of coding and forecasts for prevalent function of genes within species [99, 100] (Fig. 3).

Wu et al. [101] employed algorithms to identify a substantial number (2869) of single copy orthologs (COSII) shared by most, euasterid plant species and the model species Arabidopsis. This ortholog resource proved valuable in phylogenetic studies and comparative mapping across plant families, exemplified by investigations in Solanaceae (tomato) and Rubiaceae (coffee) families. In addressing the limitations of plant phylogenomics, novel methods and tools have emerged. One such advancement is the entirely new compilation of short read RNA-seq data, which significantly improves gene accessibility by producing transcripts that are non-chimeric and repetitive and are optimized for phylogenomic study later on [102].

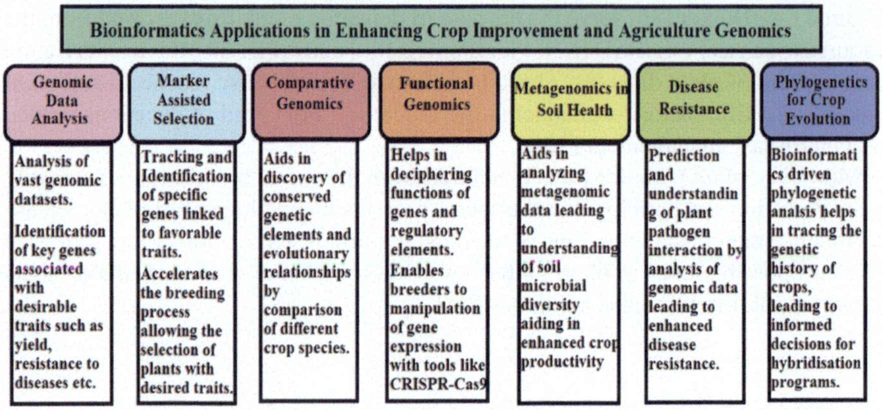

Fig. 3 Bioinformatics applications in enhancing crop improvement and advancing agriculture genomics

A new protocol, termed Hyb-seq, combines genome skimming of high-copy repetitions and organelle genomes with target enrichment of low-copy nuclear exons and surrounding regions. This approach effectively generates genome-scale datasets for phylogenomics of plants [103]. A recent development includes the update of ExamML (Exascale Maximum Likelihood), a code designed for gigantic phylogenetic analysis on the Intel MIC (Many Integrated Core) hardware platform. Version 3 of this program signifies significant progress in the development of advanced phylogenetic analysis algorithms, enabling the analysis of datasets comprising 10–20 genes and up to 55,000 taxa [104].

4 Summary

In summary, the amalgamation of bioinformatics and agricultural genomics presents unprecedented opportunities for improving crop traits in response to global challenges. Bioinformatics, as a key element in this multidisciplinary approach, manages and analyses vast biological datasets, sorting complex molecular functionalities. Combination of advanced omics technologies such as transcriptomics, genomics, phenomics, proteomics, metabolomics, etc., enhances understanding of plant genomes. This, coupled with traditional breeding, enables efficient identification of superior genotypes and optimization of breeding strategies. Critical bioinformatics resources, such as general and species-specific databases, alongside tools like BLAST and EMBOSS, play decisive roles in crop improvement, providing centralized and accessible genomic information. Bioinformatics applications in genomic data analysis, comparative genomics, disease resistance studies and phylogenetics contribute significantly to crop enhancement. Tools enable exploration of plant-pathogen interactions, identification of genetic markers and prediction of gene functions. In plant pathology, bioinformatics strengthens our ability to fortify plant resistance mechanisms by identifying effector proteins produced by pathogens. This synergistic application of bioinformatics and genomics holds tremendous potential for ensuring global food security amid climate change. By accelerating the development of crops suited for evolving climatic conditions, bioinformatics emerges as a crucial ally in enhancing sustainable agricultural production. The collaborative integration of diverse expertise facilitated by bioinformatics fosters innovation in diagnostics, monitoring and traceability, ultimately promoting sustainability in agriculture.

Conflict of Interest The authors do not have any pertinent conflicts of interest to disclose regarding the content of this article.

References

1. Kumar, R., Yadav, M.K., Yadav, S.L., Kumar, M., Sharma, A.K. (2023). Bioinformatics and omics for crop improvement. Octa J Biosci 11(1):24–39. https://www.researchgate.net/profile/Ravi-Kumar572/publication/372746706_Bioinformatics_and_omics_for_crop_improvement/links/64c56e906f28555d86e09a54/Bioinformatics-and-omics-for-crop-improvement.pdf
2. Field, C.B., Barros, V.R., editors. (2014). Climate change 2014–Impacts, adaptation and vulnerability: Regional aspects. Cambridge University Press
3. Asseng, S., Ewert, F., Rosenzweig, C., Jones, J. W., Hatfield, J. L., Ruane, A. C., Boote, K. J., Thorburn, P. J., Rotter, R. P., Cammarano, D., & Brisson, N. (2013). Uncertainty in simulating wheat yields under climate change. *Nature Climate Change, 3*(9), 827–832. https://doi.org/10.1038/nclimate1916
4. Batley, J., & Edwards, D. (2016). The application of genomics and bioinformatics to accelerate crop improvement in a changing climate. *Current Opinion in Plant Biology, 30*, 78–81. https://doi.org/10.1016/j.pbi.2016.02.002
5. Shafi, A., Zahoor, I., Haq, E., Fazili, K.M. (2019). Impact of bioinformatics on plant science research and crop improvement. In: Essentials of Bioinformatics, Volume III: In Silico Life Sciences: Agriculture 29–46. https://doi.org/10.1007/978-3-030-19318-8_2
6. Esposito, A., Colantuono, C., Ruggieri, V., & Chiusano, M. L. (2016). Bioinformatics for agriculture in the next-generation sequencing era. *Chemical and Biological Technologies in Agriculture, 3*, 1–2. https://doi.org/10.1186/s40538-016-0054-8
7. Aslam, Z., Khattak, J.Z., Ahmed, M., Asif, M. (2017). A role of bioinformatics in agriculture. In: Quantification of climate variability, adaptation and mitigation for agricultural sustainability 413–434. https://doi.org/10.1007/978-3-319-32059-5_17
8. Luscombe, N. M., Greenbaum, D., & Gerstein, M. (2001). What is bioinformatics? An introduction and overview. *Yearbook of Medical Informatics, 10*(01), 83–100. https://doi.org/10.1055/s-0038-1638103
9. Edwards, D., & Batley, J. (2004). Plant bioinformatics: From genome to phenome. *Trends in Biotechnology, 22*(5), 232–237. https://doi.org/10.1016/j.tibtech.2004.03.002
10. Lai, K., Lorenc, M. T., & Edwards, D. (2012). Genomic databases for crop improvement. *Agronomy, 2*(1), 62–73. https://doi.org/10.3390/agronomy2010062

11. Edwards, D. (2013). Bioinformatics tools to assist breeding for climate change. In: *Genomics and Breeding for Climate-Resilient Crops: Vol. 1 Concepts and Strategies*. Springer Berlin Heidelberg (pp. 391–414). https://doi.org/10.1007/978-3-642-37045-8_10
12. Prabha, R., Verma, M.K., Singh, D.P. (2017). Bioinformatics in agriculture: translating alphabets for transformation in the field. In: *Plant Bioinformatics: Decoding the Phyta* (pp. 197–214). https://doi.org/10.1007/978-3-319-67156-7_7
13. Gupta, R., Tiwari, S., Tripathi, M.K., Pooniya, S.K. (2020). Bioinformatics and its applications in crop improvement. In: *Recent Trends in Molecular Biology and Biotechnology*, Vol-1. Integrated Publications (pp. 129–151). https://doi.org/10.22271/int.ed.book02
14. Nilges, M., Linge, J.P. (2009). Bio-Informatique Structurale. Institut Pasteur, F-75015 Paris, France.
15. Varshney, R. K., Graner, A., & Sorrells, M. E. (2005). Genomics-assisted breeding for crop improvement. *Trends in Plant Science, 10*(12), 621–630. https://doi.org/10.1016/j.tplants.2005.10.004
16. Liu, D.D., Zhang, L. (2019). Trends in the characteristics of human functional genomic data on the Gene Expression Omnibus, 2001–2017. *Laboratory Investigation, 99(1)*,118–127. https://pubmed.ncbi.nlm.nih.gov/30206311/
17. Mu, H., Wang, B., & Yuan, F. (2022). Bioinformatics in plant breeding and research on disease resistance. *Plants, 11*(22), 3118. https://doi.org/10.3390/plants11223118
18. Pundir, S., Martin, M. J., & O'Donovan, C. (2016). UniProt Consortium UniProt tools. *Current Protocols in Bioinformatics, 53*(1), 1–29. https://doi.org/10.1002/0471250953.bi0129s53
19. ouens-Clark, K., Buckler, E., Casstevens, T., Chen, C., DeClerck, G., Derwent, P., Dharmawardhana, P., Jaiswal, P., Kersey, P., Karthikeyan, A.S., Lu, J. (2010). Gramene database in 2010: updates and extensions. *Nucleic Acids Research 39*(1), 1085–1094. https://doi.org/10.1093/nar/gkq1148
20. Carollo, V., Matthews, D.E., Lazo, G.R., Blake, T.K., Hummel, D.D., Lui, N., Hane, D.L., Anderson, O.D. (2005). Grain Genes 2.0. An improved resource for the small-grains community. *Plant Physiology 139*(2), 643–651. https://doi.org/10.1104/pp.105.064485
21. Contreras-Moreira, B., Naamati, G., Rosello, M., Allen, J.E., Hunt, S.E., Muffato, M., Gall, A., Flicek, P. (2022). Scripting analyses of genomes in ensemble plants. In: *Plant bioinformatics: Methods and protocols*. Springer US (pp. 27–55). https://doi.org/10.1007/978-1-0716-2067-0_2
22. Harper, L., Campbell, J., Cannon, E.K., Jung, S., Poelchau, M., Walls, R., Andorf, C., Arnaud, E., Berardini, T.Z., Birkett, C., Cannon, S., et al. (2018). AgBioData consortium recommendations for sustainable genomics and genetics databases for agriculture. *Database (Oxford) 2018*, bay088. https://doi.org/10.1093/database/bay088
23. Spannagl, M., Nussbaumer, T., Bader, K., Gundlach, H., Mayer, K.F. (2017). PGSB/MIPS PlantsDB database framework for the integration and analysis of plant genome data. In: *Plant genomics databases: Methods and protocols* 2017:33–44. https://doi.org/10.1007/978-1-4939-6658-5_2
24. Proost, S., Van Bel, M., Sterck, L., Billiau, K., Van Parys, T., Van de Peer, Y., & Vandepoele, K. (2009). PLAZA: A comparative genomics resource to study gene and genome evolution in plants. *The Plant Cell, 21*(12), 3718–3731. https://doi.org/10.1105/tpc.109.071506
25. Goodstein, D. M., Shu, S., Howson, R., Neupane, R., Hayes, R. D., Fazo, J., Mitros, T., Dirks, W., Hellsten, U., Putnam, N., & Rokhsar, D. S. (2012). Phytozome: A comparative platform for green plant genomics. *Nucleic Acids Research, 40*(1), 1178–1186. https://doi.org/10.1093/nar/gkr944
26. Reiser, L., Subramaniam, S., Zhang, P., & Berardini, T. (2022). Using the arabidopsis information resource (TAIR) to find information about arabidopsis genes. *Current Protocols, 2*(10), e574. https://doi.org/10.1002/cpz1.574
27. Matthews, D. E., Carollo, V. L., Lazo, G. R., & Anderson, O. D. (2003). GrainGenes, the genome database for small-grain crops. *Nucleic Acids Research, 31*(1), 183–186. https://doi.org/10.1093/nar/gkg058

28. Wicker, T., Buell, C.R. (2009). Gene and repetitive sequence annotation in the Triticeae. In: *Genetics and Genomics of the Triticeae* (pp. 407–425). https://doi.org/10.1007/978-0-387-77489-3_15
29. Lawrence, C. J., Dong, Q., Polacco, M. L., Seigfried, T. E., & Brendel, V. (2004). MaizeGDB, the community database for maize genetics and genomics. *Nucleic Acids Research, 32*(1), 393–397. https://doi.org/10.1093/nar/gkh011
30. Kurata, N., Yamazaki, Y. (2006). Oryzabase. An integrated biological and genome information database for rice. *Plant Physiology 140*(1):12–17. https://doi.org/10.1104/pp.105.063008
31. Chen, H., Wang, T., He, X., Cai, X., Lin, R., Liang, J., Wu, J., King, G., Wang, X. (2022). BRAD V3.0: an upgraded Brassicaceae database. Nucleic Acids Res. 2022;50(1):1432–1441. https://doi.org/10.1093/nar/gkab1057
32. Gonzales, M. D., Archuleta, E., Farmer, A., Gajendran, K., Grant, D., Shoemaker, R., Beavis, W. D., & Waugh, M. E. (2005). The Legume Information system (LIS): An integrated information resource for comparative legume biology. *Nucleic Acids Research, 33*(1), 660–665. https://doi.org/10.1093/nar/gki128
33. Bombarely, A., Menda, N., Tecle, I.Y., Buels, R.M., Strickler, S., Fischer-York, T., Pujar, A., Leto, J., Gosselin, J., Mueller, L.A. (2010). The Sol Genomics Network (solgenomics.net): growing tomatoes using Perl. *Nucleic Acids Research 39*(1):1149–1155. https://doi.org/10.1093/nar/gkq866
34. Fernandez-Pozo, N., Menda, N., Edwards, J. D., Saha, S., Tecle, I. Y., Strickler, S. R., Bombarely, A., Fisher-York, T., Pujar, A., Foerster, H., & Yan, A. (2015). The sol genomics network (SGN)—from genotype to phenotype to breeding. *Nucleic Acids Research, 43*(1), 1036–1041. https://doi.org/10.1093/nar/gku1195
35. Jayashree B, Hoisington D. (2009). Bioinformatics tools for crop research and breeding. In: *Molecular Techniques in Crop Improvement*, 2nd Edition (pp. 97–116). https://doi.org/10.1007/978-90-481-2967-6_4
36. Balakrishnan, M. Application of bioinformatics in agriculture. 105th FoCARS. Digital repository of course materials.
37. Gedil, M. (2009). Tailoring bioinformatics for the genetic improvement of orphan crops. *ATDF Journal 6*(3/4), 34. https://www.academia.edu/download/99313537/ATDF-Journal-2009-Issue3_4-Orphan_Crops-May2010.pdf#page=34
38. Ye, J., McGinnis, S., & Madden, T. L. (2006). BLAST: Improvements for better sequence analysis. *Nucleic Acids Research, 34*(2), 6–9. https://doi.org/10.1093/nar/gkl164
39. Vassilev, D., Leunissen, J., Atanassov, A., Nenov, A., & Dimov, G. (2005). Application of bioinformatics in plant breeding. *Biotechnology and Biotechnological Equipment, 19*(3), 139–152. https://doi.org/10.1080/13102818.2005.10817293
40. Kushwaha, U.K., Deo, I., Jaiswal, J.P., Prasad, B. (2017). Role of bioinformatics in crop improvement. *Global Journal of Science Frontier Research 17*(1), 13–23. https://www.researchgate.net/profile/Ujjawal-Kushwaha/publication/314032951_Role_of_Bioinformatics_in_Crop_Improvement/links/58b186d692851cf7ae8eb1ff/Role-of-Bioinformatics-in-Crop-Improvement.pdf
41. De Filippis, L.F. (2013). Bioinformatic tools in crop improvement. In: *Crop Improvement: New Approaches and Modern Techniques*. Springer US (pp. 49–122). https://doi.org/10.1007/978-1-4614-7028-1_2
42. Singh, B.D., Singh, A.K. (2015). Bioinformatics tools and databases for genomics research. In: *Marker-Assisted Plant Breeding: Principles and Practices*. Springer. https://doi.org/10.1007/978-81-322-2316-0_14
43. Dwivedi, V. D., Bharadwaj, S., Mohanty, P. S., & Gupta, U. D. (2018). Bioinformatics tools for genomic and evolutionary analysis of infectious agents. *Biomed Biotechnol Res J, 2*(3), 163–167. https://doi.org/10.4103/bbrj.bbrj_74_18
44. Dahiya, B.P. (2017) Bioinformatics impacts on medicine, microbial genome and agriculture. *Journal of Pharmacognosy and Phytochemistry 6*(4):1938–1942. https://www.phytojournal.com/archives?year=2017&vol=6&issue=4&ArticleId=1594

45. Rice, P., Longden, I., & Bleasby, A. (2000). EMBOSS: The European molecular biology open software suite. *Trends in Genetics, 16*(6), 276–277. https://doi.org/10.1016/S0168-9525(00)02024-2
46. Lai, X., Behera, S., Liang, Z., Lu, Y., Deogun, J. S., & Schnable, J. C. (2017). STAG-CNS: An order-aware conserved noncoding sequences discovery tool for arbitrary numbers of species. *Molecular Plant, 10*(7), 990–999. https://doi.org/10.1016/j.molp.2017.05.010
47. Letunic, I., & Bork, P. (2021). Interactive Tree of Life (iTOL) v5: An online tool for phylogenetic tree display and annotation. *Nucleic Acids Research, 49*(1), 293–296. https://doi.org/10.1093/nar/gkab301
48. Chen, C., Chen, H., Zhang, Y., Thomas, H. R., Frank, M. H., He, Y., & Xia, R. (2020). TBtools: An integrative toolkit developed for interactive analyses of big biological data. *Molecular Plant, 13*(8), 1194–1202. https://doi.org/10.1016/j.molp.2020.06.009
49. Schultz, J., Copley, R. R., Doerks, T., Ponting, C. P., & Bork, P. (2000). SMART: A web-based tool for the study of genetically mobile domains. *Nucleic Acids Research, 28*(1), 231–234. https://doi.org/10.1093/nar/28.1.231
50. Liu, Q., Jiao, X., Meng, X., Wang, C., Xu, C., Tian, Z., Xie, C., Li, G., Li, J., Yu, H., & Wang, K. (2021). FED: A web tool for foreign element detection of genome-edited organism. *Science China Life Sciences, 64*(1), 167–170. https://doi.org/10.1007/s11427-020-1731-9
51. Lei, Y., Lu, L., Liu, H. Y., Li, S., Xing, F., & Chen, L. L. (2014). CRISPR-P: A web tool for synthetic single-guide RNA design of CRISPR-system in plants. *Molecular Plant, 7*(9), 1494–1496. https://doi.org/10.1093/mp/ssu044
52. Liu, H., Ding, Y., Zhou, Y., Jin, W., Xie, K., Chen, L.L. (2017). CRISPR-P 2.0: an improved CRISPR-Cas9 tool for genome editing in plants. *Molecular Plant 10*(3):530–532. https://doi.org/10.1016/j.molp.2017.01.003
53. Chow, C.N., Lee, T.Y., Hung, Y.C., Li, G.Z., Tseng, K.C., Liu, Y.H., Kuo, P.L., Zheng, H.Q., Chang, W.C. (2019). PlantPAN3.0: a new and updated resource for reconstructing transcriptional regulatory networks from ChIP-seq experiments in plants. *Nucleic Acids Research 47*(1):1155–1163. https://doi.org/10.1093/nar/gky1081
54. Waese, J., Provart, N.J. (2017). The bio-analytic resource for plant biology. In: *Plant genomics databases: Methods and protocols* (pp. 119–148). https://doi.org/10.1007/978-1-4939-6658-5_6
55. Joung, J. G., Corbett, A. M., Fellman, S. M., Tieman, D. M., Klee, H. J., Giovannoni, J. J., & Fei, Z. (2009). Plant MetGenMAP: An integrative analysis system for plant systems biology. *Plant Physiology, 151*(4), 1758–1768. https://doi.org/10.1104/pp.109.145169
56. Wang, W., Wang, Z., Li, X., Ni, Z., Hu, Z., Xin, M., Peng, H., Yao, Y., Sun, Q., Guo, W. (2020). SnpHub: an easy-to-set-up web server framework for exploring large-scale genomic variation data in the post-genomic era with applications in wheat. *Gigascience 9*(6):giaa060. https://doi.org/10.1093/gigascience/giaa060
57. Kumar, A. (2021). Application of bioinformatics in crop improvement. In: *Agricultural biotechnology: Latest research and trends* (pp. 711–726). https://doi.org/10.1007/978-981-16-2339-4_30
58. Gomez-Casati, D.F., Busi, M.V., Barchiesi, J., Peralta, D.A., Hedin, N., Bhadauria, V. (2018). Applications of bioinformatics to plant biotechnology. *Current Issues in Molecular Biology 27*(1):89–104. https://doi.org/10.21775/cimb.027.089
59. Tan, Y. C., Kumar, A. U., Wong, Y. P., & Ling, A. P. (2022). Bioinformatics approaches and applications in plant biotechnology. *Journal, Genetic Engineering & Biotechnology, 20*(1), 1–3. https://doi.org/10.1186/s43141-022-00394-5
60. Normand, E. A., & Van den Veyver, I. B. (2019). Next-generation sequencing for gene panels and clinical exomes. *Human Reproduction Prenatal Genetics, 1*, 553–1575. https://doi.org/10.1016/B978-0-12-813570-9.00025-5
61. Schuster, S. C. (2008). Next-generation sequencing transforms today's biology. *Nature Methods, 5*(1), 16–18. https://doi.org/10.1038/nmeth1156
62. Singh, D., Singh, P. K., Chaudhary, S., Mehla, K., & Kumar, S. (2012). Exome sequencing and advances in crop improvement. *Advances in Genetics, 79*, 87–121. https://doi.org/10.1016/B978-0-12-394395-8.00003-7

63. Maloy, S., Hughes, K., editors. (2013). *Brenner's encyclopedia of genetics*. Academic Press. https://books.google.com/books?hl=en&lr=&id=4cj64BhrnjcC&oi=fnd&pg=PP1&dq=Maloy+S,+Hughes+K,+editors.+Brenner%27s+Encyclopedia+of+Genetics.+Academic+Press%3B+2013.&ots=Brsn2_N1oU&sig=dOnbjxemgjKDpgNFPNLWbFhTi90
64. Platten, J. D., Cobb, J. N., & Zantua, R. E. (2019). Criteria for evaluating molecular markers: Comprehensive quality metrics to improve marker-assisted selection. *PLoS ONE, 14*(1), e0210529. https://doi.org/10.1371/journal.pone.0210529
65. Houmard, N. M., Mainville, J. L., Bonin, C. P., Huang, S., Luethy, M. H., & Malvar, T. M. (2007). High-lysine corn generated by endosperm-specific suppression of lysine catabolism using RNAi. *Plant Biotechnology Journal, 5*(5), 605–614. https://doi.org/10.1111/j.1467-7652.2007.00265.x
66. Guo, X., Liu, R. H., Fu, X., Sun, X., & Tang, K. (2013). Over-expression of l-galactono-γ-lactone dehydrogenase increases vitamin C, total phenolics and antioxidant activity in lettuce through bio-fortification. *Plant Cell, Tissue and Organ Culture, 114*, 225–236. https://doi.org/10.1007/s11240-013-0318-y
67. Li, J., Scarano, A., Gonzalez, N. M., D'Orso, F., Yue, Y., Nemeth, K., Saalbach, G., Hill, L., de Oliveira, M. C., Moran, R., & Santino, A. (2022). Biofortified tomatoes provide a new route to vitamin D sufficiency. *Nature Plants, 8*(6), 611–616. https://doi.org/10.1038/s41477-022-01154-6
68. Collins, F.S., Green, E.D., Guttmacher, A.E., Guyer, M.S. (2003). US National Human Genome Research Institute. A vision for the future of genomics research. *Nature 422*(6934):835–847. https://doi.org/10.1038/nature01626
69. Devos, K. M., & Gale, M. D. (2000). Genome relationships: The grass model in current research. *The Plant Cell, 12*(5), 637–646. https://doi.org/10.1105/tpc.12.5.637
70. Mochida, K. E., Yoshida, T. A., Sakurai, T. E., Yamaguchi-Shinozaki, K. A., Shinozaki, K. A., & Tran, L. S. (2009). In silico analysis of transcription factor repertoire and prediction of stress responsive transcription factors in soybean. *DNA Research, 16*(6), 353–369. https://doi.org/10.1093/dnares/dsp023
71. Tran, L. S., & Mochida, K. (2010). Identification and prediction of abiotic stress responsive transcription factors involved in abiotic stress signaling in soybean. *Plant Signaling & Behavior, 5*(3), 255–257. https://doi.org/10.4161/psb.5.3.10550
72. Mochida, K. E., Yoshida, T. A., Sakurai, T. E., Yamaguchi-Shinozaki, K. A., Shinozaki, K. A., & Tran, L. S. (2011). In silico analysis of transcription factor repertoires and prediction of stress-responsive transcription factors from six major gramineae plants. *DNA Research, 18*(5), 321–332. https://doi.org/10.1093/dnares/dsr019
73. Usadel, B., Obayashi, T., Mutwil, M., Giorgi, F.M., Bassel, G.W., Tanimoto, M., Chow, A., Steinhauser, D., Persson, S., Provart, N.J. (2009). Co-expression tools for plant biology: opportunities for hypothesis generation and caveats. *Plant Cell Environment 32*(12), 1633–1651. https://doi.org/10.1111/j.1365-3040.2009.02040.x
74. Kopka, J., Schauer, N., Krueger, S., Birkemeyer, C., Usadel, B., Bergmüller, E., Dörmann, P., Weckwerth, W., Gibon, Y., Stitt, M., Willmitzer L. (2005) GMD@ CSB. DB: the Golm metabolome database. Bioinformatics 21(8), 1635–1638. https://doi.org/10.1093/bioinformatics/bti236
75. Obayashi, T., Kinoshita, K., Nakai, K., Shibaoka, M., Hayashi, S., Saeki, M., Shibata, D., Saito, K., Ohta, H. (2007). ATTED-II: a database of co-expressed genes and cis elements for identifying co-regulated gene groups in Arabidopsis. *Nucleic Acids Research 35*(suppl_1), D863–D869. https://doi.org/10.1093/nar/gkl783
76. Ong, Q., Nguyen, P., Phuong Thao, N., & Le, L. (2016). Bioinformatics approach in plant genomic research. *Current Genomics, 17*(4), 368–378. https://doi.org/10.2174/1389202917666160331202956
77. Fernandez-Gutierrez, A., & Gutierrez-Gonzalez, J. J. (2021). Bioinformatic-based approaches for disease-resistance gene discovery in plants. *Agronomy, 11*(11), 2259. https://doi.org/10.3390/agronomy11112259

78. Joshi, A., Song, H. G., Yang, S. Y., & Lee, J. H. (2023). Integrated molecular and bioinformatics approaches for disease-related genes in plants. *Plants, 12*(13), 2454. https://doi.org/10.3390/plants12132454
79. Pedro, H., Maheswari, U., Urban, M., Irvine, A. G., Cuzick, A., McDowall, M. D., Staines, D. M., Kulesha, E., Hammond-Kosack, K. E., & Kersey, P. J. (2016). PhytoPath: An integrative resource for plant pathogen genomics. *Nucleic Acids Research, 44*(1), 688–693. https://doi.org/10.1093/nar/gkv1052
80. Sperschneider, J., Dodds, P.N., Gardiner, D.M., Singh, K.B., Taylor, J.M. (2018). Improved prediction of fungal effector proteins from secretomes with EffectorP 2.0. *Molecular Plant Pathology 19*(9):2094–2110. https://doi.org/10.1111/mpp.12682
81. Bolívar, J.C., Machens, F., Brill, Y., Romanov, A., Bülow, L., Hehl, R. (2014) 'In silico expression analysis', a novel PathoPlant web tool to identify abiotic and biotic stress conditions associated with specific cis-regulatory sequences. *Database (Oxford) 2014*:bau030. https://doi.org/10.1093/database/bau030
82. Yang, S., Li, H., He, H., Zhou, Y., & Zhang, Z. (2019). Critical assessment and performance improvement of plant–pathogen protein–protein interaction prediction methods. *Briefings in Bioinformatics, 20*(1), 274–287. https://doi.org/10.1093/bib/bbx123
83. Osuna-Cruz, C.M., Paytuvi-Gallart, A., Di Donato, A., Sundesha, V., Andolfo, G., Aiese Cigliano, R., Sanseverino, W., Ercolano, M.R. (2018). PRGdb 3.0: a comprehensive platform for prediction and analysis of plant disease resistance genes. *Nucleic Acids Research 46*(1):1197–1201. https://doi.org/10.1093/nar/gkx1119
84. Silva, R. J., & Micheli, F. (2020). RRGPredictor, a set-theory-based tool for predicting pathogen-associated molecular pattern receptors (PRRs) and resistance (R) proteins from plants. *Genomics, 112*(3), 2666–2676. https://doi.org/10.1016/j.ygeno.2020.03.001
85. Rupavatharam, S., Kennepohl, A., Kummer, B., Parimi, V. (2018). Automated plant disease diagnosis using innovative android App (Plantix) for farmers in Indian state of Andhra Pradesh. *Phytopathology 108*(10). http://oar.icrisat.org/id/eprint/11014
86. Pethybridge, S. J., & Nelson, S. C. (2015). Leaf Doctor: A new portable application for quantifying plant disease severity. *Plant Disease, 99*(10), 1310–1316. https://doi.org/10.1094/PDIS-03-15-0319-RE
87. Wang, M. Y., Wang, F., Hao, G. F., & Yang, G. F. (2019). FungiPAD: A free web tool for compound property evaluation and fungicide-likeness analysis. *Journal of Agriculture and Food Chemistry, 67*(7), 1823–1830. https://doi.org/10.1021/acs.jafc.8b06596
88. Das, D., Jaiswal, M., Khan, F. N., Ahamad, S., & Kumar, S. (2020). PlantPepDB: A manually curated plant peptide database. *Science and Reports, 10*(1), 2194. https://doi.org/10.1038/s41598-020-59165-2
89. Wang, G., Li, X., & Wang, Z. (2016). APD3: The antimicrobial peptide database as a tool for research and education. *Nucleic Acids Research, 44*(D1), D1087–D1093. https://doi.org/10.1093/nar/gkv1278
90. Dong, A. Y., Wang, Z., Huang, J. J., Song, B. A., & Hao, G. F. (2021). Bioinformatic tools support decision-making in plant disease management. *Trends in Plant Science, 26*(9), 953–967. https://doi.org/10.1016/j.tplants.2021.05.001
91. Cock, P. J., Grüning, B. A., Paszkiewicz, K., & Pritchard, L. (2013). Galaxy tools and workflows for sequence analysis with applications in molecular plant pathology. *PeerJ, 1*, e167. https://doi.org/10.7717/peerj.167
92. Zhang, N., Zeng, L., Shan, H., & Ma, H. (2012). Highly conserved low-copy nuclear genes as effective markers for phylogenetic analyses in angiosperms. *New Phytologist, 195*(4), 923–937. https://doi.org/10.1111/j.1469-8137.2012.04212.x
93. Cibrian-Jaramillo, A., De la Torre-Barcena, J. E., Lee, E. K., Katari, M. S., Little, D. P., Stevenson, D. W., Martienssen, R., Coruzzi, G. M., & DeSalle, R. (2010). Using phylogenomic patterns and gene ontology to identify proteins of importance in plant evolution. *Genome Biology and Evolution, 2*, 225–239. https://doi.org/10.1093/gbe/evq012
94. Conte, M. G., Gaillard, S., Droc, G., & Perin, C. (2008). Phylogenomics of plant genomes: A methodology for genome-wide searches for orthologs in plants. *BMC Genomics, 9*, 1–6. https://doi.org/10.1186/1471-2164-9-183

95. Gogarten, J. P., & Olendzenski, L. (1999). Orthologs, paralogs and genome comparisons. *Current Opinion in Genetics & Development, 9*(6), 630–636. https://doi.org/10.1016/S0959-437X(99)00029-5
96. Sonnhammer, E. L., & Koonin, E. V. (2002). Orthology, paralogy and proposed classification for paralog subtypes. *Trends in Genetics, 18*(12), 619–620. https://doi.org/10.1016/S0168-9525(02)02793-2
97. Fulton, T. M., Van der Hoeven, R., Eannetta, N. T., & Tanksley, S. D. (2002). Identification, analysis, and utilization of conserved ortholog set markers for comparative genomics in higher plants. *The Plant Cell, 14*(7), 1457–1467. https://doi.org/10.1105/tpc.010479
98. O'Brien, K. P., Remm, M., & Sonnhammer, E. L. (2005). Inparanoid: A comprehensive database of eukaryotic orthologs. *Nucleic Acids Research, 33*(1), 476–480. https://doi.org/10.1093/nar/gki107
99. Blanchette, M., Green, E. D., Miller, W., & Haussler, D. (2004). Reconstructing large regions of an ancestral mammalian genome in silico. *Genome Research, 14*(12), 2412–2423. https://doi.org/10.1101/gr.2800104
100. Bustamante, C. D., Fledel-Alon, A., Williamson, S., Nielsen, R., Todd Hubisz, M., Glanowski, S., Tanenbaum, D. M., White, T. J., Sninsky, J. J., Hernandez, R. D., & Civello, D. (2005). Natural selection on protein-coding genes in the human genome. *Nature, 437*(7062), 1153–1157. https://doi.org/10.1038/nature04240
101. Wu, F., Mueller, L. A., Crouzillat, D., Pétiard, V., & Tanksley, S. D. (2006). Combining bioinformatics and phylogenetics to identify large sets of single-copy orthologous genes (COSII) for comparative, evolutionary and systematic studies: A test case in the euasterid plant clade. *Genetics, 174*(3), 1407–1420. https://doi.org/10.1534/genetics.106.062455
102. Yang, Y., & Smith, S. A. (2013). Optimizing de novo assembly of short-read RNA-seq data for phylogenomics. *BMC Genomics, 14*, 1–11. https://doi.org/10.1186/1471-2164-14-328
103. Weitemier, K., Straub, S. C., Cronn, R. C., Fishbein, M., Schmickl, R., McDonnell, A., & Liston, A. (2014). Hyb-Seq: Combining target enrichment and genome skimming for plant phylogenomics. *Applications in Plant Sciences, 2*(9), 1400042. https://doi.org/10.3732/apps.1400042
104. Kozlov, A. M., Aberer, A. J., & Stamatakis, A. (2015). ExaML version 3: A tool for phylogenomic analyses on supercomputers. *Bioinformatics, 31*(15), 2577–2579. https://doi.org/10.1093/bioinformatics/btv184

Investigation on Mechanical Properties of AA6061 Metal Matrix Composite Reinforced with Particulate Titanium Carbide (TiC)

Ajay R. Bhardwaj, A. M. Vaidya, and P. D. Meshram

Abstract Due to advancement in technologies there is always a requirement to develop a new material having better mechanical properties at low cost. Metals and alloys are found to be most suitable for making a component for different applications like automobile, aerospace, marine etc. and their properties can be enhance by adding reinforcement such as aluminium oxide (Al_2O_3), titanium diboride (TiB_2), boron carbide (B_4C), silicon carbide (SiC), titanium carbide (TiC),cetc. This paper is an experimental work to investigate the mechanical properties of AA6061 reinforced with titanium carbide (TiC) particulates having average particle size (APS) 2 micron. In this work, AA6061 act as matrix metal and TiC particulates as the reinforcement to fabricate AMMC by stir casting process. In present investigation the TiC_p are added in varying proportions of 0%, 2%, 4%, 6% and 8% by weight to the AA6061.The results shows that with increased percentage of TiC_p properties UTS, compressive strength and hardness increases but percent elongation and impact energy decreases. Fractography analysis indicates that the failure of the specimen is a combination of dimple ductile and brittle fracture.

Keywords Reinforcement · Metal matrix composite · Titanium carbide · AMMC · Mechanical properties

A. R. Bhardwaj (✉)
SSBT C.O.E.T Bambhori Jalgaon, Jalgaon, India
e-mail: ajaybhardwaj12345@gmail.com

A. M. Vaidya
Mumbai University, Mumbai, India

P. D. Meshram
U.I.C.T KBC NMU Jalgaon, Jalgaon, India

© The Author(s), under exclusive license to Springer Nature Singapore Pte Ltd. 2024
A. Chaudhary et al. (eds.), *Unraveling New Frontiers and Advances in Bioinformatics*,
https://doi.org/10.1007/978-981-97-7123-3_14

1 Introduction

In today's scenario aluminium metal matrix composites (AMMC's) finds wide applications in automobile, aerospace, sports industries etc. and attracted by the industries due to its high strength to weight ratio, better wear resistance, low thermal expansion, excellent elastic modulus etc. The selection of right material is very challenging, and it must be selected carefully to perform required function satisfactory for a longer period without losing its properties. To search a single monolithic material with required properties for a particular application is almost impossible and also the properties of material are greatly affected by the environment in which they are working and also the loading nature, hence the need arises to develop a new material which is a combination of two or more materials as composites to achieve desired properties.

Composite are classified on the basis of matrix phase like metal matrix composite, polymer matrix composite and ceramic matrix composite also there are inter metallic matrix and carbon matrix composite. Aluminium metal matrix composite consist of aluminium or aluminium alloy which act as a continuous phase and the other phase which is termed as reinforcement is embedded in this aluminium/aluminium alloy matrix. The reinforcement is non-metallic and ceramic in nature like silicon carbide (SiC), aluminium oxide (Al_2O_3), boron carbide(B_4 C), titanium diboride (TiB_2) etc. Desired properties of AMMC's can be achieved as per the requirement. As per UK Advisory Council on Science and Technology, aluminium metal matrix composites (AMMC's) replaced conventional materials with excellent properties. Particulate reinforced aluminium matrix composites (PAMCs) consist of equiaxed ceramic reinforcements with an aspect ratio less than 5. Reinforcements are ceramic in nature. This particulate aluminium composite fabricated either by solid state or liquid state process. Particulate aluminium composite is economical when compared to continuous fibre-reinforced AMMCs [1]. AMMCs can be manufactured by various techniques according to their requirement. As per the data available aluminium alloys are the preferred material for the production of airplane parts [2]. Titanium carbide (TiC) as a ceramic reinforcement was not much investigated to aluminium alloy matrix, it comes in picture very late and find attention for its high melting temperature (3160·C), low thermal coefficient of expansion, extraordinary hardness, excellent wear and abrasion resistance [3, 4].

It has been observed that aluminium alloys shows relatively greater strength at room temperature. But when it has to perform at a temperature above 200 °C the mechanical poperies of aluminium alloys get reduced more than 50%. To overcome this issue titanium carbide particles are introduced into the aluminium matrix. Earlier silicon carbide (SiC) was most preferring particles to reinforce aluminium alloys [5]. Reinforcement having smaller size particles are better reinforcement [6]. In today's scenario AMMCs are manufactured by liquid state processing the reason is simple, economical and suitable for mass production [7]. Widely used liquid state processing is stir casting method to produce AMMCs [8].While fabricating AMMC through stir casting mechanical stirring is main element during this process [9, 10]. Composites having 30% volume fractions can easily be fabricated through stir casting technique and production of Al-TiC is highly expensive and tedious. From the available manufacturing methods, stir casting is found to be an attractive and economical method for producing AMMCs [11].

Process parameters stirrer speed, melt temperature, size of reinforcement and design of stirrer blade plays a major role while determining the properties of AMMCs. Hence, process variables should be carefully selected [12]. While comparing with other manufacturing methods the cost of stir casting technique results in as low as 1/3rd to 1/10th for the production of AMMC's [13]. There is a great attention regarding Al-TiC_p due to its thermodynamically stability, increased hardness and lightness that results in the right choice for wear resistance component [14, 15–17]. It has been observed that manufacturing method effect interfacial bond strength in AMMC reinforced with TiC particulates and wear rate is directly proportional to applied load but wear rate decreases with increase in wt. percentage of TiC particulates. Al7075-TiC produced by stir casting route results in increase in hardness and wear properties with increase in % wt. of TiC particulates under heat treated condition (T6) [18, 19]. Titanium carbide (TiC) acts as an effective grain refiner when fabricating AA-TiC by in-situ reaction with K_2TiF_6 salt and particulate graphite with molten aluminium. It has been observed in research that TiC particulates increase the microhardness, ultimate tensile strength but decreases ductility of Al-6061/TiC composites. Available research data reveals that 20% wt. TiC particles have been introduced with LM6 alloy to investigate the properties like tensile strength and hardness [20]. Hardness as well as tensile strength decreases when AA-7075 reinforced with 10%wt TiC particle. The particle size was 2 microns [21].

From available research data it has been observed that maximum 15% wt. of titanium carbide particulates (TiC) having particle size 2 micron were mixed with AA6061 through stir casting to investigate the mechanical properties and found that with increase percentage of TiC particles there is increases in microhardness, ultimate tensile strength and wear resistance but ductility and percentage elongation decreases [22]. Friction and wear characteristics were investigated of AA/TiC composites and it has been observed that there is decrease in wear rate and coefficient of friction with increasing wt.% of TiC [23]. The main disadvantage of aluminium alloys is their poor performance at higher temperatures and wear resistance. To overcome this issue aluminium alloys are reinforced with ceramic particles known as Aluminium Metal Matrix Composite (AMMC's) [24].

1.1 Research Gap

- Limited research data is available regarding investigation of mechanical properties of titanium carbide particulates reinforced with AA-6061 through stir casting route.
- Less work has been done with TiCp having average particle size of 2 micron.

- From available research data limited work has been reported regarding investigation of mechanical properties such as compressive strength and impact energy of AA-6061 reinforced with TiCp.

1.2 Novelty of the Work

- The main objective of this paper is to study and investigate the mechanical properties, microstructure and fractography analysis of AA6061 reinforced with titanium carbide (TiC) particulates having average particle size (APS) 2 micron through stir casting process.
- There are various fabrication techniques available to fabricate Aluminium Metal Matrix Composites (AMMC's) but stir casting method found to be simple and economical.
- From available research data it has been observed that there are less research data available regarding investigation of titanium carbide (TiC) particulates as a ceramic reinforcement for AMMC's through stir casting route.
- SEM image of the titanium carbide (TiC) particles shows sharp edges that provide an abrasive nature to the particles.
- Fractography analysis indicates that the failure of the specimen is a combination of dimple ductile and brittle fracture as various size dimples observed on the surface of the specimen and cleavage facets indicate brittle fracture.
- Titanium carbide (TiC) is very expensive and it comes into picture very late and it receive attention by the researchers and scientists because of its low thermal coefficient of expansion, better hardness, high melting temperature (3100 °C), better wear and abrasion resistance.
- This fabricated AMMC's can be suitable for manufacturing a component where a main criterion is high ultimate strength, high hardness, high compressive strength and low impact energy/strength and low Percent elongation.

2 Experimental Work

The present work is an attempt to manufacture AA-6061 introduced with titanium carbide particulates with varying volume fractions of 2 wt%, 4 wt%, 6 wt% and 8 wt% by stir casting and to assess the mechanical properties and microstructure analysis. The experimental work has been carried out at Nano Composite Research Lab Mechanical Engineering Department MANIT Bhopal (Madhya Pradesh), and VNIT Nagpur (M.S) India.

2.1 Materials

Commercially available matrix material aluminium alloy 6061T6 round bar 10 kg is selected and purchased from "SURESH METALS" 10th Khetwadi Lane, Mangal Murti plaza B601, Mumbai-400004, (M.S) India.

Titanium Carbide (TiC) particulates reinforcement is selected having density—4.93 g/cm^3, Melting point—3200 °C, Boiling point—4820 °C, Average Particle size—2 *micron*, and Quantity—500 g is purchased from Nano Research Elements Ladwa-Babain road City- Ladwa Kurukshetra, Haryana (HR-06) India.

First pre weighted AA6061 rods are kept in a graphite crucible and it is placed in an electrical furnace for melting at a temperature of 800 °C. For evenly dispersion of reinforcement and to increase wettability of AA-6061 magnesium is added at higher temperature. To form a vortex mechanical stirrer was dipped into the molten aluminium and rotated at a constant speed of 300 rpm. TiC particulates 2% wt. of the base metal was then added slowly to the molten aluminium. To remove moisture, dust, unwanted gases etc. titanium carbide was preheated to 300°C. Stirring of the melt and feeding of TiC particulates were continued intermittently for 30 min. Then melted composite was then poured in a metallic die which was preheated to 300 °C. To get the fabricated composites of different weight percentages i.e. 4%, 6%, 8%. the same procedure was followed. Finally fabricated castings were taken with various amounts of 0, 2, 4, 6 and 8 wt.% TiC particulates.

3 Mechanical Properties of AA6061/TiC AMC's

3.1 Tensile Strength Properties

For testing tensile strength the specimen was prepared as per the ASTM B557 standards and evaluated on cast cylindrical rod. The specimen for tensile testing is shown in Fig. 2 with gauge length 60 mm, gauge dia 12 mm and parallel length 66 mm. Analogue universal testing machine (TFUN-1000) was used to conduct tensile test. Figure 3 shows that there is increase in ultimate tensile strength with increase in TiC percentage and it is found maximum 228.9 MPa with 8% TiC particulates.

3.2 Hardness

Hardness test is carried out as per ASTM standards using Rockwell Hardness tester machine (TRSN-B), with steel ball indenter dia 2.5 mm and load of 187.5 kg. The specimen Fig. 5, size was 25 mm diameter and length 20 mm and it was polished with emery papers to remove any scratches or surface defects if any and tested at five locations on each sample to obtain an average value of hardness. Figure 4 shows

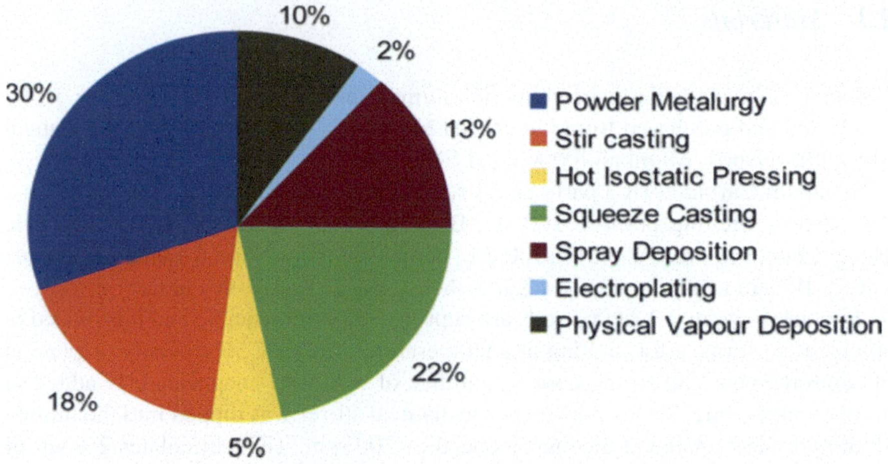

Fig. 1 Research studies conducted on fabrication routes of MMCs for the last 10 years [25]

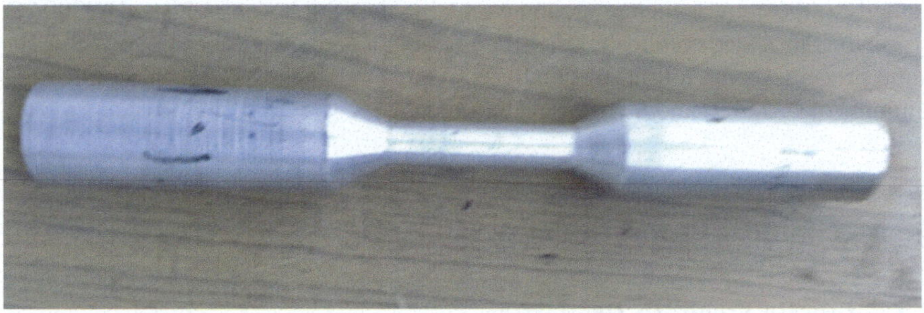

Fig.2 Tensile specimen

there is increase in hardness with increase in TiC percentage and it is found maximum 263.514 HRB with 8% TiC particulates.

3.3 Compressive Strength

For compression test specimen size with length 18 mm, diameter 12 mm and analogue universal testing machine (TFUN-1000) was used to conduct test. Figure 6 shows that there is increase in compressive strength with increase in TiC percentage and it is found maximum 107.6 KN with 8% TiC particulates.

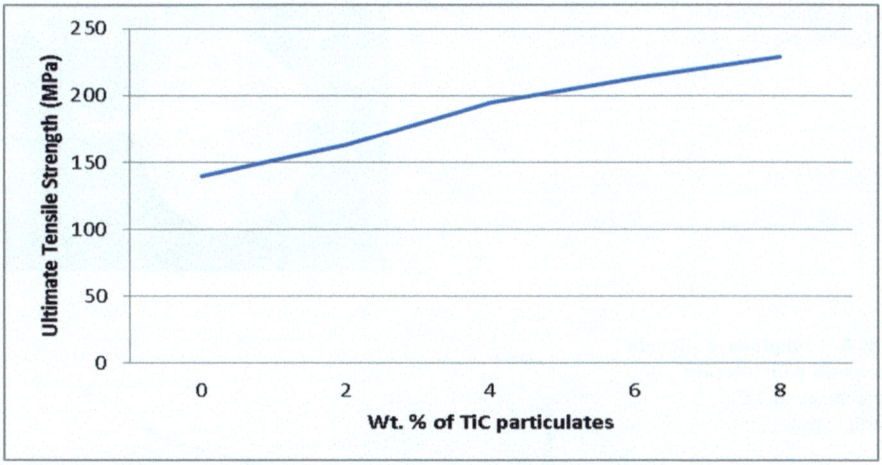

Fig. 3 Ultimate tensile strength increases with increasing percentage of TiC particulates

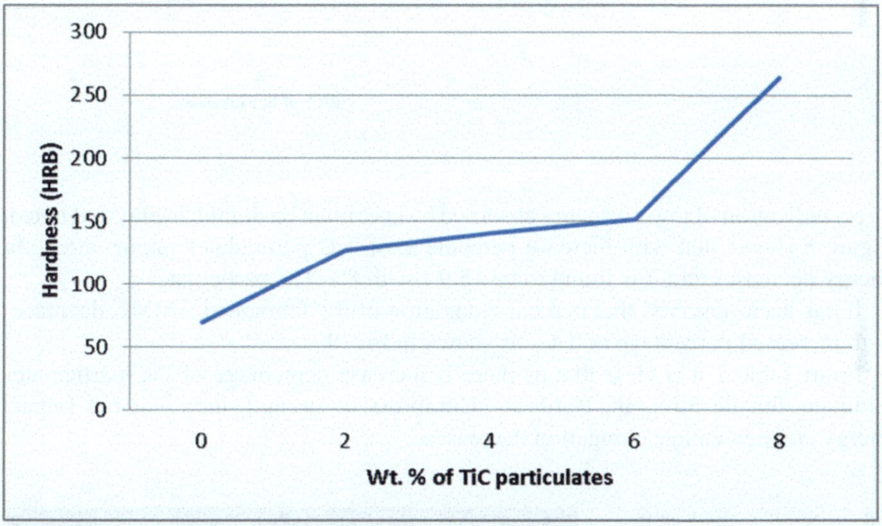

Fig. 4 Hardness increases with increase of TiC particulates

3.4 Charpy Impact Strength/energy Properties

Impact strength of fabricated AMMCs were investigated by conducting charpy impact test of the prepared sample on Charpy Impact Testing Machine FIT 300 D. It was performed on the prepared samples of length 55 mm, height 10 mm and width 10 mm having v notch at the centre with depth 2 mm at room temperature, Fig. 7. The sample was broken by a single blow of freely swinging hammer. There is

Fig. 5 Specimen after hardness test

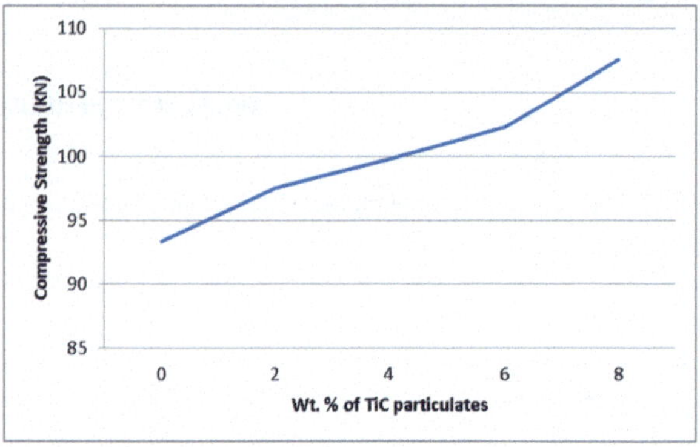

Fig. 6 Compressive strength increases with increase percentage of TiC particulates

direct indication of impact energy absorbed by specimen on digital display and noted. Figure 8 shows that with increase percentage of TiC particulates impact strength/ energy decreases and it is found to be 15.9 J with 8% TiC particulates.

It has been observed that percent elongation of the fabricated AMMC decreases with increased percentage of TiC_p as shown in Fig. 9

From Table 7 it is clear that as there is increase percentage of TiC particulates Ultimate Tensile Strength, Hardness, Compressive strength increases but Impact energy and percentage elongation decreases.

Fig. 7 Specimen for Charpy impact test

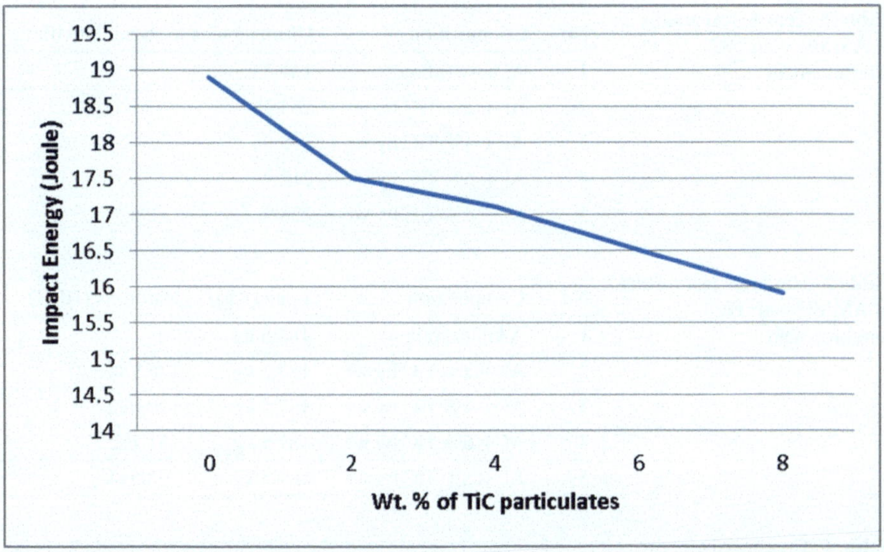

Fig. 8 Impact energy decreases with increasing percentage of TiC particulates

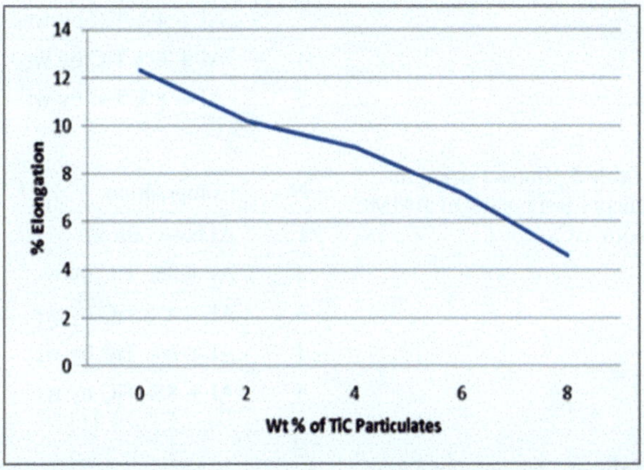

Fig. 9 Percent elongation decreases with increasing percentage of TiC particulates

Table1 Properties of AA 6061-T6

Element	Si	Fe	Cu	Mn	Mg	Cr	Zn	Ti	Aluminium
Wt.%	0.660	0.170	0.260	0.070	1.000	0.145	0.006	0.024	97.66

Table 2 Tensile test results of AMMC with TiC reinforcement

No	Composition	Ultimate tensile strength (MPa)
1	Al base (alloy)	139.33
2	Al + 2% TiC by wt	163.21
3	Al + 4% TiC by wt	195.06
4	Al + 6% TiC by wt	212.9
5	Al + 8% TiC by wt	228.9

Table 3 Hardness test results of AMMC with TiC reinforcement

No	Composition	Load (Kg)	Hardness (HRB)
1	AA base (alloy)	187.5 kg	70
2	Al + 2% TiC by wt	187.5 kg	126.984
3	Al + 4% TiC by wt	187.5 kg	140.082
4	Al + 6% TiC by wt	187.5 kg	151.182
5	Al + 8% TiC by wt	187.5 kg	263.514

Table 4 Compressive strength test results of AMMC with TiC reinforcement

No	Composition	Compressive strength (KN)
1	Al base (alloy)	93.470
2	Al + 2% TiC by wt	97.5
3	Al + 4% TiC by wt	99.8
4	Al + 6% TiC by wt	102.3
5	Al + 8% TiC by wt	107.6

Table 5 Impact strength/energy test results of AMMC with TiC_p

No	Composition	Impact strength/energy (Joule)
1	Al base (alloy)	18.9
2	Al + 2% TiC by wt	17.5
3	Al + 4% TiC by wt	17.1
4	Al + 6% TiC by wt	16.5
5	Al + 8% TiC by wt	15.9

Table.6 Percent elongation test results of AMMC with TiC reinforcement

No	Composition	% Elongation
1	Al base (alloy)	12.3
2	Al + 2% TiC by wt	10.2
3	Al + 4% TiC by wt	9.1
4	Al + 6% TiC by wt	7.2
5	Al + 8% TiC by wt	4.6

Table.7 Mechanical properties of 0 wt.%, 2 wt.%, 4 wt.%, 6 wt.% and 8 wt.% TiC particulates reinforced with A-6061

Material	Ultimate tensile strength (MPa)	Hardness (HRB)	Compressive strength (KN)	Impact strength/ energy (Joule)	% Elongation
Al6061	139.33	70	93.470	18.9	12.3
Al + 2wt% TiC	163.21	126.984	97.5	17.5	10.2
Al + 4wt% TiC	195.06	140.082	99.8	17.1	9.1
Al + 6wt% TiC	212.9	151.182	102.3	16.5	7.2
Al + 8wt% TiC	228.9	263.514	107.6	15.9	4.6

4 Microstructural Analysis of AA6061 Reinforced with TiC Particulates

Scanning electron microscope (SEM) and Energy dispersive X-ray spectroscopy (EDS) test were carried out on JEOL 6380A, Japan made high resolution scanning electron microscope to investigate microstructural of various composites i.e. 0%, 2%, 4%, 6% and 8% TiC particulates. The specimen tested for SEM and EDS were circular having diameter 5 mm and length 5 mm. For microstructural characterization, specimens were polished by standard metallographic polishing then etched with a Keller etchant and SEM images were generated. Figure 10.shows SEM images of TiC particles and it indicates the presence of TiC particles which is uniformly distributed in the matrix. In some places agglomeration and serration of particles are also seen. Figure 11 shows EDS analysis and it indicates presence of Al and TiC particles.

Figure 12 is a SEM image of the titanium carbide (TiC) particles which shows sharp edges that provide an abrasive nature to the particles. Hence there is interfacial bonding within the matrix material during the solidification occur and this results in some significant changes in the mechanical properties when subjected to external loads.

4.1 Fractography Analysis of AA6061 Reinforced with TiC Particulates

Fractured surface of specimen were examined after tensile test by using scanning electron microscope and fractography images are generated as shown in Fig. 13. Fractography analysis indicates that the failure of the specimen is a combination of

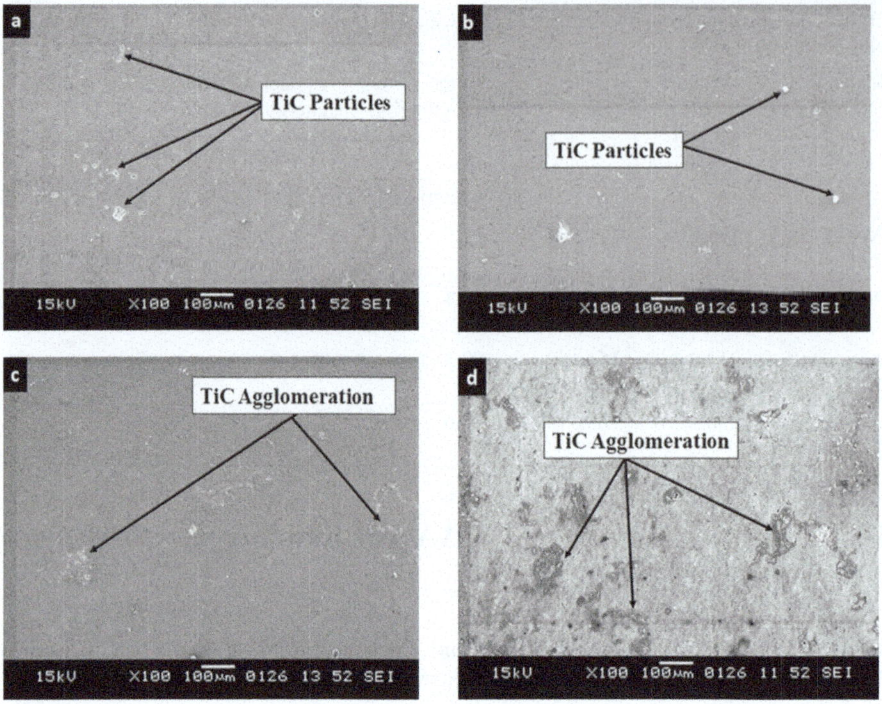

Fig. 10 SEM image for TiC particles **a** AA-6061-2% TiC **b** AA-6061-4% TiC **c** AA-6061-6% TiC **d** AA-6061-8% TiC

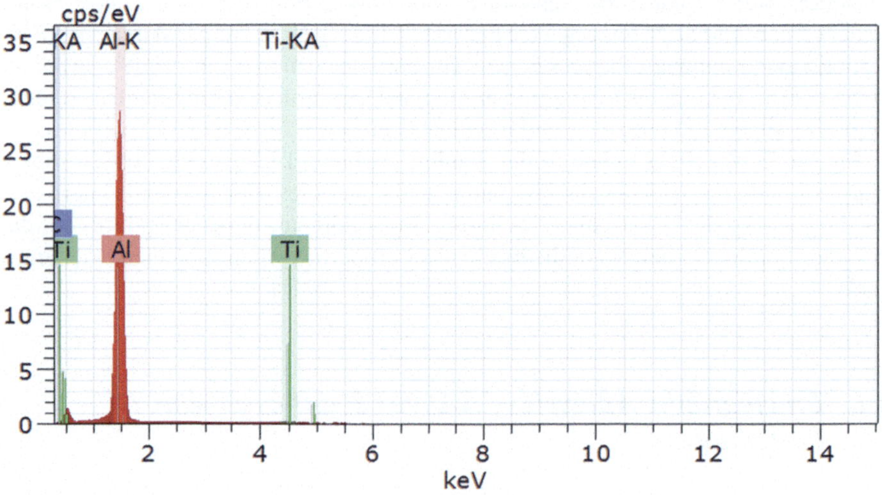

Fig. 11 EDS analysis for AA6061—2% TiC particulates

Fig. 12 SEM image of TiC particles

dimple ductile and brittle fracture as various size dimples observed on the surface of the specimen and cleavage facets indicate brittle fracture as shown in Fig. 13b.

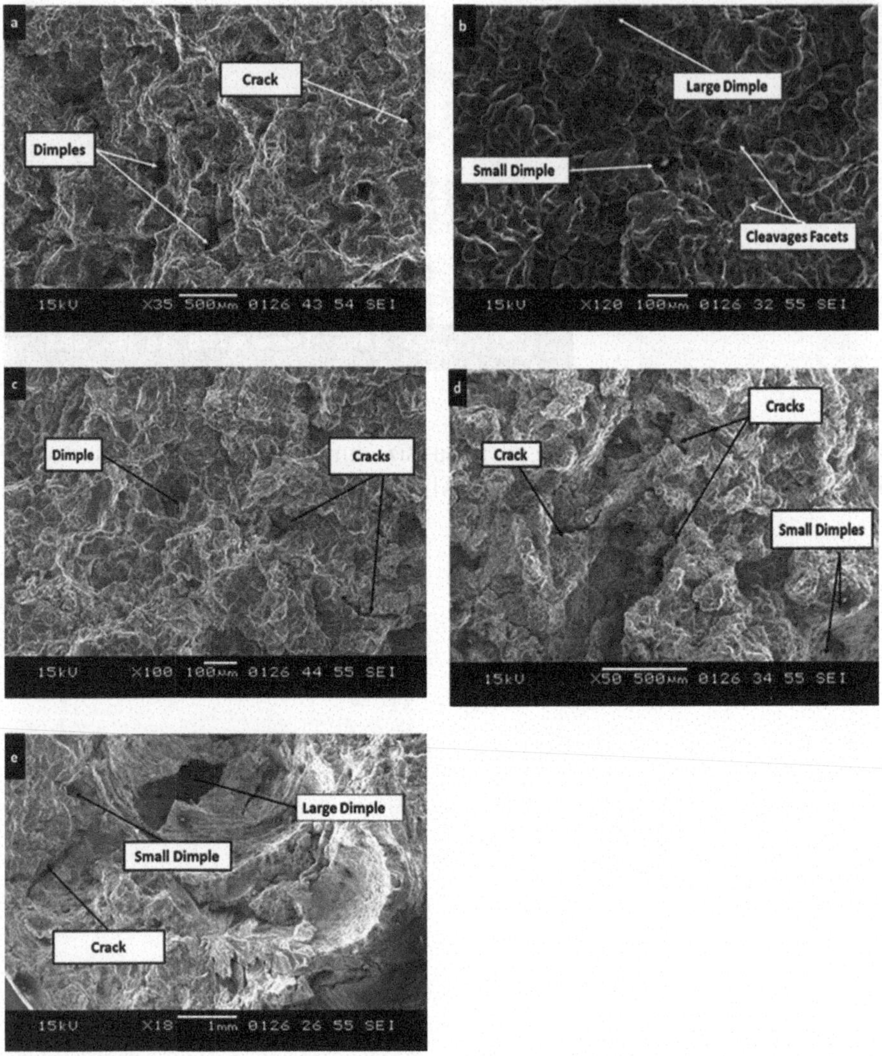

Fig.13 Fractography image of **a** AA-6061 alloy **b** AA-6061-2% TiC **c** AA-6061-4% TiC **d** AA-6061-6% TiC **e** AA-6061–8% TiC

5 Conclusion

In this experimental work AA6061 reinforced with titanium carbide (TiC) particulates was successfully fabricated by stir casting process. The effect of TiC particulates contents on mechanical properties of fabricated sample, fractography analysis and its microstructure was analysed the results are as follows:

- AA6061 was successfully fabricated by stir casting method with 0, 2, 4, 6 and 8 wt.% of TiC particulates reinforcement.
- There is increase in ultimate tensile strength, hardness, compression strength with increase percentage of TiC particulates reinforcement. The extent of increase in said mechanical properties obtained with addition of 2, 4, 6 and 8 wt.% of TiC particulates reinforcement.
- Impact strength/energy and Percent elongation decreases with increase percentage of TiC particulates reinforcement.
- The fabricated AMMC is tested for microstructure and fractography analysis by using Scanning Electron Microscope (SEM) and it has been observed that TiC particles present in the matrix which is uniformly distributed. In some places agglomeration and serration of particles are also seen.
- Regarding fractography analysis it has been observed that the failure of the specimen after tensile testing is a combination of dimple ductile and brittle fracture as various size dimples observed on the surface of the specimen and presence of cleavage facets indicate brittle fracture.
- This fabricated AMMC's can be suitable for manufacturing a component where a main criterion is high ultimate strength, high hardness, high compressive strength and low impact energy/strength and low Percent elongation.
- The process parameters of stir casting method such as the stirrer speed, time of stirring, design of stirrer blade, size of reinforcement and melt temperature majorly effect on the AA-6061TiC_p composite characteristics. To achieve suitable parameters optimization can be done.
- The mechanical properties of AA-6061TiC_p composites depend on the percentage weight of the reinforcement particles. Increased percentage of reinforcement results in the formation of pores and agglomeration, and also affects the properties.

References

1. Surappa, M. K. (2003). Aluminium matrix composites: Challenges and opportunities. *Sadhana, 28*, 319–334.
2. Pandey, U., Purohit, R., Agarwal, P., & Kumar Singh, S. (2018). Study of fabrication, testing and characterization of Al/TiC metal matrix composites through different processing techniques. *Materials. Today Proceedings, 5*(2), 4106–4117.
3. Wang, Y., Rainforth, W. M., Jones, H., & Lieblich, M. (2001). Dry wear behaviour and its relation to microstructure of novel 6092 aluminium alloy–Ni_3Al powder metallurgy composite. *Journal Wear, 251*, 1421–1432.
4. Salem, H. G., El-Eskandarany, S., Kandil, A., Fattah, H. A. (2009). Bulk behavior of ball milled AA2124 nanostructured powders reinforced with TiC. *Journal of Nanomaterials*, (2009), Article ID 479185, 12 pages, 2009.
5. Jia, F., Ma, M. Z., Song, A. J., Zhang, W. G., Zong, H. T., Liang, S. X., Osamu, Y., Liu, R. P. (2009). Creep behavior of in situ TiCp/2618 aluminium matrix composite 506, pp.58–62,
6. Pandey, A. B., Mishra, R. S., & Mahajan, Y. R. (1992). *Steady state creep behaviour of silicon carbide particulate reinforced aluminium composites, 40*, 2045–2052.

7. Taha, M. A. (2001). Practicalization of cast metal matrix composites (MMCCs). *Materials & Design, 22*(6), 431–441.
8. Hashim, J., Looney, L., Hashimi, M. S. J. (1999). Metal matrix composites: production by the stir casting method (92–93), pp.1–7.
9. Himanshu, K., Merb, K.K.S, Sandeep, K. (2014). A review on mechanical and tribological behaviors of stir cast aluminum matrix composites. *Procedia Materials Science* (6), 1951–1960.
10. Saravanan, R.A., Surappa, M.K. (2000). Fabrication and characterisation of pure magnesium-30 vol.% SiCp particle composite. *Materials Science and Engineering A 276*(1–2), pp. 108–116.
11. Thangarasu, A., Murugan, N., Mohankumar, R., Thangapand, P. (2014). Processing and characterization of AA6082/TiC composites by stir casting. *Emerging Materials Research 3*(3), 123.129.
12. Das, B., Roy, S., Rai, R.N., Saha, S.C, Majumder, P. (2016). Effect of in-situ processing parameters on microstructure and mechanical properties of TiC particulate reinforced Al–4.5Cu alloy MMC fabricated by stir-casting technique—optimization using grey based differential. *Evolution Algorithm 93*, pp. 397–408.
13. Surappa, M. K., & Rohatgi, P. K. (1981). Preparation and properties of cast aluminium-ceramic particle composites. *Journal of Materials Science, 16*, 983–993.
14. Mazaheri, Y., Meratian, M., Emadi, R., & Najarian, A. R. (2013). Comparison of microstructural and mechanical properties of Al–TiC, Al–B4C and Al–TiC–B4C composites prepared by casting techniques. *Materials Science and Engineering: A, 560*, 278–287.
15. Jerome, S., Ravisankar, B., Mahato, P. K., Natarajan, S. (2010). Synthesis and evaluation of mechanical and high temperature tribological properties of in-situ Al/TiC composites. *Tribological Internatíonal 43*(11), 2029–2036.
16. Rao, R. V, Ramanaiah, N., Sarcar, M. M. M. (2014). Fabrication and Investigation on properties of TiC reinforced Al7075 metal matrix composites. *Applied Mechanics and Materials* (592–594), 49–353.
17. Rao, R. V., Rramanaiah, N, Sarcar, M. (2015). Dry sliding wear behavior of Al7075 reinforced with titanium carbide (TiC) particulate composites. In: *Proceedings of International Conference on Advances in Materials, Manufacturing and Applications* (pp. 39–44).
18. Sharma, A. (2013). Synthesis of TiC composite by salt route and grain refinement of aluminium alloy. *International Journal of Cast Metals Research, 21*(1–4), 226–230.
19. Moses, J. J., Dinaharan, I., Sekhar, S. J. (2016). Production and characterization of titanium carbide particulate reinforced AA6061aluminium alloy composite using stir casting. *Metallic Materials 54*(4), pp. 257–267.
20. Suraya, S., Sulaiman, S., Nur Najmiyah, J., & Nor Imrah, Y. (2014). Studies on tensile properties of titanium carbide (TiC) particulates composites. *Advanced Materials Research, 903*, 151–156.
21. Rao, R. V., Ramanaiah, N., Sarcar, M. M. M. (2016). Optimization of volumetric wear rate of AA7075-TiC Metal Matrix composite by using Taguchi Technique, Jordan. *Journal of Mechanical and Industrial Engineering 10*(3).
22. Moses, J.J., Dinaharan, I., Sekhar, S.J. (2016). Production and characterization of titanium carbide particulate reinforced AA6061aluminium alloy composite using stir casting. *Metallic Materials 54*(4), 257–267.
23. Tyagi, R. (2005). Synthesis and tribological characterization of in situ cast Al–TiC composites. *Journal Wear, 259*(1–6), 569–576.
24. Bhardwaj, A.R., Vaidya, A.M., Shekhawat, S.P. (2020). Machining of aluminium metal matrix composite: a review. *Materials Today: Proceedings 21*, Part 2, 1396–1402.
25. Bains, P.S., Sidhu, S.S., Payal, H.S. (2015). Fabrication and machining of metal matrix composites: A review. *Materials and Manufacturing Processes 31*(5), 553–573.

Combination of Different Kinds of Materials for Hybrid Composites: A Brief Review

Hariome Sharan Gupta, Deepali Verma, and Akarsh Verma

Abstract In the recent decades, the hybrid composites have been subject of interest to the researchers and engineers from an economic consistency and environmental based. This novel class of materials is made by two or more than two distinct fibers or fillers reinforced in an appropriate polymer matrix to form an inventive material with better properties comparable to those of conventional materials. Therefore, in order to achieve better structural design and properties, a good selection of the hybrid composites ingredients and their properties need to be thoroughly explored. This chapter aims to provide an overview of different kinds of materials used for hybrid composites and how their properties and applications in various fields.

Keywords Polymer matrices · Reinforcing fillers · Hybridizations · Hybrid composites

1 Introduction

The polymer composites materials are made of a fiber-reinforced in a polymer matrix. The composites materials, i.e., base matrix and fillers/fibers materials, are graded according to their content; in the composites, the base matrix binds or holds the fillers/fibers. In this structure, the polymer matrix referred to be as binder materials and fillers/fibers as load-wearing materials. The most promising and discerning materials available in this century is a polymer composites, which has replaced the

---

H. S. Gupta
Department of Polymer and Process Engineering, Indian Institute of Technology Roorkee, Saharanpur 247001, India

D. Verma
Department of Computer Science and Engineering, IIT (BHU), Varanasi 221005, India

A. Verma (✉)
Department of Mechanical Engineering, University of Petroleum and Energy Studies (UPES), Dehradun 248007, India
e-mail: akarshverma007@gmail.com

© The Author(s), under exclusive license to Springer Nature Singapore Pte Ltd. 2024
A. Chaudhary et al. (eds.), *Unraveling New Frontiers and Advances in Bioinformatics*,
https://doi.org/10.1007/978-981-97-7123-3_15

metals, ceramics, alloys, etc., from the daily used materials. The polymer composites reinforced with natural fibers or synthetic fibers are becoming increasingly relevant as the demand for high strength, lightweight, low cost for unique applications is rising in the industry. The polymer composites, used reinforcing materials as a natural fibers/fillers i.e., abaca, alfa, agave, oil palm, hemp, coir, pineapple, pine needle, bagasse, jute, bamboo, flax, ramie, palm, kenaf, banana, sisal, henequen, kapok, cotton, loofah, rice, wheat, corn, silk, wool, hair, etc. Similarly, the polymer composites, used reinforcing materials as a synthetic fibers/fillers i.e., carbon, glass, boron, polyester, aramid, and Kevlar etc. or nanomaterials fillers/fibers i.e., carbon nanotubes, graphene, clays, zeolite, nanocellulose, etc. In addition to the high strength to weight ratio, fibers/fillers reinforced polymer composites often reveal exceptional properties, such as tensile strength, flexural strength, impact strength, stiffness, wear resistance, corrosion resistance, chemical resistance, damping properties, and also fire resistance. In civil constructions, aerospace, automobile, biomedical, marine, and many other manufacturing industries, these large ranges of various features have led polymer composites materials to find applications [1–7].

Much academic and commercial attention has been given to the use of synthetic fibers/fillers such as glass and carbon fibers/fillers in the past. While excellent mechanical, abrasive, wear and thermal properties and also chemical resistance have been achieved in this way, but the limited recyclability and high energy requirements for manufacturing cast and of their end of life processes, life cycle evaluation does not produce favourable results of synthetic fibers/fillers reinforced polymer composites. In many polymer composites, which have used fibers/fillers as reinforcer, that still cause environmental issues, the synthetic fibers/fillers are the main component, recycling and end of life disposal by thermal recycling in both cases created problems [8, 9].

Growing environmental and social issues due to high rate production of petroleum resources such as synthetic fibers/fillers used as reinforcer for polymer composites, required new polymer composites and green materials. For this sort of polymer composites, the special challenges are their eco-friendly characteristics, making them environmentally friendly, fully degradable and renewable. At present trends researcher focused on natural fibers/fillers polymer composites in place of synthetic fibers/fillers polymer composites because of natural fibers/fillers is environmentally friendly, easily available, renewable, less processing or manufacturing cost, sustainable, low cost and can be incinerated. The use of natural fibers/fillers as alternatives to synthetic fibers/fillers is the subjected towards environmentally sustainable polymer composite systems [10, 11].

Innovative ways in material science area and engineering field have been explored by researchers and engineers in their research for materials with improved properties and qualities. In hybrid composites, several materials with different properties are combined, either at the macroscale as layered structures or at the microscale as reinforcement phases. Because of their synergistic relationship, all of the constituent materials can contribute their unique qualities and properties, making a hybrid composite that performs better than any of them alone materials or single fibers/fillers reinforced composites. The processing and compounding of hybrid composites through the blending of two or more reinforcer fillers or fibers and also one or more polymer matrix materials [12, 13]. In the hybrid composites, a revolutionary class of materials known as hybrid composites combines several components, which show special advantages to get beyond individual constraints. This combination of several materials offers up new possibilities for material design and makes it possible to create sophisticated materials that are suited to certain performance requirements in a wide range of applications. The remarkable mechanical properties, chemical resistance, thermal stability, abrasive resistance, electrical conductivity, and other specialised characteristics that these hybrid composites may be made to display make them incredibly flexible and useful for a variety of sectors. Hybrid composites are a novel approach that has emerged to overcome these constraints by offering a technique to attain the best material performance while minimising trade-offs [14–17].

The making of the hybrid composites is not without difficulties, despite the immense promise they provide. Fiber reinforced hybrid composites, particulate reinforced hybrid composites, laminates and functionally graded hybrid composites are just a few of the different methods used to combine materials to made hybrid composites. Particulate reinforced hybrid composites perform well in situations needing improved wear resistance and thermal conductivity, while fiber reinforced hybrid composites are widely used in the automotive industries and also in aerospace area because of their exceptional strength-to-weight ratio [18, 19]. Every variety of the hybrid composite materials has distinct advantages and characteristics, customised to meet specific technical demands. The typical challenges that researchers and engineers must overcome include ensuring materials compatibility and bonding as well as navigating compounding and manufacturing complexity. Furthermore, special consideration must be given to the scalability and cost-effectiveness of the manufacture of the hybrid polymer composites [20–22].

In this review, we examine the different material combinations and related attributes as fascinating into the intriguing world of hybrid composites. We will examine the benefits, drawbacks, and issues surrounding the creation of hybrid composites. We will also go through the many uses for these cutting-edge materials, ranging from sports equipment and infrastructure to aerospace and automotive engineering. We will also examine the most recent studies and developments in the sector, offering an overview of the prospects for the future and the possible effects of hybrid composites on many industries.

2 Types of Polymer Composites

In order to obtain the description of the polymer composites depends, the nature of the polymer matrix reinforcement interface, the geometry of the reinforcement and its distribution, and the nature of the constituents and their attributes. The properties of the polymer composites are mostly determined by the parameters of the polymer matrix, types of reinforcements fibers/fillers, additives and the relative amounts of these components that are incorporated into the formulation, as well as by the implementation compounding or processing. The polymer composites are classified as based on reinforcements, i.e., particulate composites, short fibers/fillers composites, continuous fibers composites and hybrid polymer composites.

2.1 Particulate Polymer Composites

Particulate polymer composites consist of particles as a reinforcement, such as alloys and ceramics, immersed in polymer matrices. As the particulate reinforcer are inserted randomly in the polymer matrices, they are typically showed isotropic behaviour in the polymer composites. The macroscopic behaviour of particulate filled polymer composites is influenced by the condition of the interface between the particles and the polymer matrix, the distribution, scale, and shape of the filler particles. Particulate polymer composites have some benefits such as enhanced mechanical strength, stiffness, physical properties, thermal properties, processing temperature, resistance to oxidation, chemical resistance, reduced shrinkage, etc. [23, 24]. Figure 1 show the particulate fillers, which are used in polymer composites as a reinforcer. Some particulates filler used for polymers composites, such as calcium carbonate, copper oxide, talc, zinc oxide, mica, silica, wollastonite, kaolin, etc. And the average particle size ranges of particulates filler considered from 1 nm to 100 um [25]. Even today, substantial development and research is performed on particulate filled polymeric matrix, considering interest in biomaterials, natural or synthetic fibers/fillers, and nanocomposites. In the early days, to lower costs, fillers were applied to the polymer matrices. Nevertheless, the ever growing technological and aesthetic specifications as well as rising content and manufacturing costs require the use of all possible benefits of fillers or fibers [26].

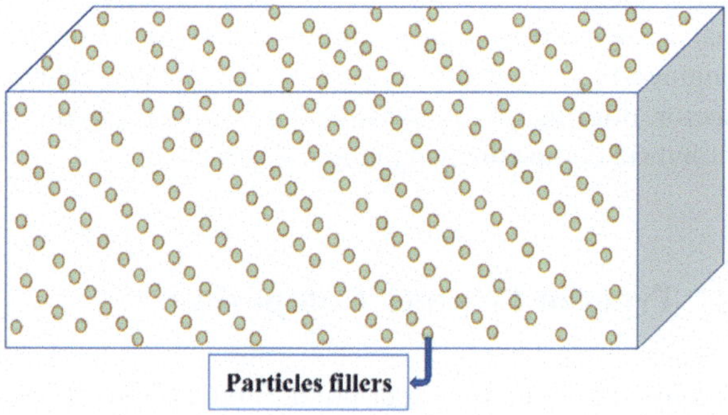

Fig. 1 Particulate fillers for polymer composites

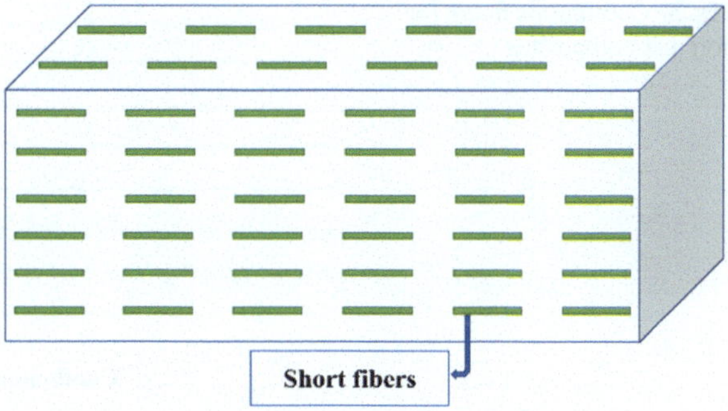

Fig. 2 Short fibers for polymer composites

2.2 Short Fiber Reinforced Polymer Composites

Short fiber reinforced in polymer matrix form short fiber reinforced polymer composites and because of their ease of fabrication, economy and superior mechanical properties, the short fiber reinforced polymer composites are very attractive. To produce short fiber reinforced polymer composites with variety of applications, the extrusion processing or compounding and injection moulding techniques are commonly used [27, 28]. The average length of short fibers considered with or less than 6–8 mm for injection moulded thermoplastic composites and between 8–50 mm for compression moulded thermoset composites [25]. Short glass fibers, short carbon fibers, short natural fibers (i.e., abaca, banana, cotton, bamboo, jute, bagasse, hemp, sisal, flax, pine apple leaf, oil palm, alfa, kenaf, ramie, kapok, coir, corn, henequen, pine fiber, etc.) are extensively used with polymer matrix to improve the performance especially when used for engineering applications. The short fiber polymer composites improve mechanical strength, thermal stability, physical properties, etc. [29–33]. Figure 2 show the short fibers, which are used in polymer composites as a reinforcer.

2.3 Continuous Fiber Reinforced Polymer Composites

Continuous fibre reinforced in polymer matrix form continuous fiber reinforced polymer composites. There are many benefits to the production of continuous fiber reinforced polymer composites, but the cost of processing/manufacturing is high and their process is difficult to control. But mostly prefer pultrusion process and compression moulding process for continuous reinforcement fibers composites are capable to generating such profiles on a large scale. Used recycling/reuse content that used spatial continuous tool route planning strategy to reduce the workload of post manufacturing/processing without cut the fibers for reduced manufacturing costs. The weight/length ratio of profiles made of continuous reinforcement fibers with a given orientation is much better than that of other light weight profiles i.e., aluminium [34, 35]. Developments in continuous fiber reinforcement have led to a

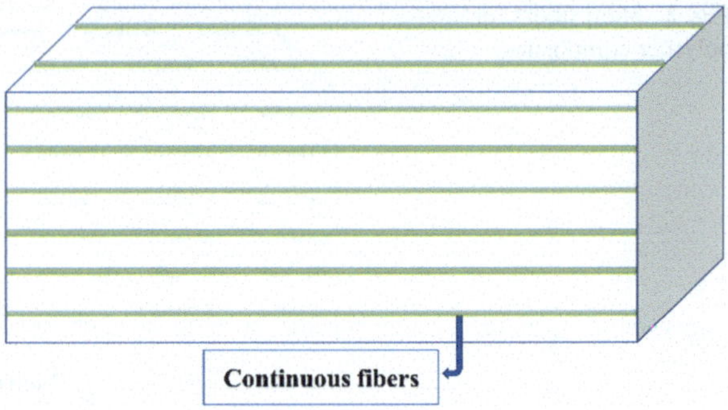

Fig. 3 Continuous fibers for polymer composites

wide range of fibers with a wide range of mechanical characteristics. For example, the high stiffness of carbon fibers and glass fibers allows for extremely efficient crack bridging and thus very low growth rates of crack, contributing to resistance to fatigue [36, 37]. In recent decades, efforts have been focused on the production of fatigue resistant materials that maintain low weight and good mechanical properties [38, 39]. Figure 3 show the continuous fibers, which are used in polymer composites as a reinforcer.

2.4 Hybrid Polymer Composites

Hybrid polymer composites manufactured or naturally occurring materials consist of two or more component materials with substantially different physical or mechanical or chemical properties within the finished structure, which remain separate and distinct. It is easy to describe hybrid composites; with the combinations of hybrid fillers/fibers in composites. Reinforcing materials i.e., fibers, fillers and nanoparticles are combined with the polymer matrix to form hybrid polymer composites. The hybrid polymer composites, consists two or more fillers/fibers are reinforced to polymer matrix to make stronger and different properties [40]. High value hybrid composites can be made by a combination of various types of fibers/fillers in a polymer matrix. They have gained tremendous popularity and can provide new freedom to customize the composites with required properties that can not be accomplished in single fiber reinforced polymer composites. With the help of this combination, the hybrid composites improved physical and mechanical strength, stiffness, electrical and thermal properties while retaining the advantageous and properties of the polymer matrix, such processing simplicity, low density, and also flexibility. It is possible to design hybrid composites with economic advantages for different applications by carefully choosing the fillers/fibers and compounding or processing techniques [41]. Some processing/manufacturing method used for hybrid polymer composites, i.e., injection moulding, compression moulding, extrusion, pultrusion, filament winding, open mould and closed mould process, hand lay-up, resin transfer moulding, vacuum bagging, melt mixing, etc. [17, 42].

At this time, studies on the production of hybrid polymer composites from both polymer matrix thermoplastics (T.P.) and thermosets (T.S.) have been documented using natural fibers, synthetic fibers, nanomaterial fibers/fillers as well as the combination of all fibers/fillers and their elucidated mechanical and thermal properties. The newly designed hybrid composites may be characterised for a wide range of applications by their mechanical, thermal, electrical, freestanding, biomedical and acoustic properties. Mechanical properties are effectively improved by hybridizations of natural, synthetic and nanomaterials fibers/fillers in polymer composites. The explanation for the use of more than one type of fibers/fillers reinforcement polymer composites provides better advantages compare to single used fibers/fillers reinforcement polymer composites advantages, thus enhancing the properties and efficiency of the resulting material of hybrid composites. Compared with untreated composites, hybrid composites with chemical treatments or fibre modifications typically exhibit stronger mechanical properties, resulting from enhanced fibers/fillers polymer matrix bonding in treated composites. The mechanical and thermal properties of hybrid polymer composites are based on several factors, i.e., the dispersion and distribution of the reinforcements in selected polymer matrix, the interfacial adhesion between the polymer and the reinforcements, the aspect ratio of the fibers reinforcer, the mechanical and chemical properties of the reinforcer, the loading effect of the reinforcer, the modification of the surface of the reinforcer, the dimension and orientation of the fibers/fillers. For high performance structural applications, natural, synthetic and nanomaterial fibers/fillers reinforced hybrid composites are therefore promising [22, 43–46].

Through using a wide variety of applications and innovative product technologies, the hybrid polymer composites have been commercialised. There are several hybrid polymer composite applications that are focused on products and components. Even today, the technological potential of thermosets and thermoplastics polymer matrix production is still far from exhausted, since hybrid polymer composites have almost completed their century of developments. Due to the outstanding ability of the hybrid composites, compared to non-hybrid (single fiber reinforced) composites, the hybrid composites have acquired alternative materials to replace metals, ceramics, alloys, etc., especially for transportation sectors aerospace, marine, and automotive fields [47]. In almost every type of advanced engineering applications, the hybrid composites have been used, including the aircrafts, ballistic products, automobiles sector, sporting goods, chemical manufacturing equipment, impact loading, and civil facilities such as buildings and bridges [48]. Today's era is the hybrid composites era, which have replaced heavy metals, ceramics, alloys, etc. The different kinds of matrix have already been used by various researcher, i.e., polymer matrix, metal matrix, and ceramic matrix for the hybrid composites. Figure 4 illustrates the techniques and its implementation for the compounding/processing, properties, characterizations and applications of fibers/fillers reinforced hybrid composites.

The hybrid composites are a novel class of materials that combine the benefits of conventional polymers matrix with various reinforcement materials to provide a

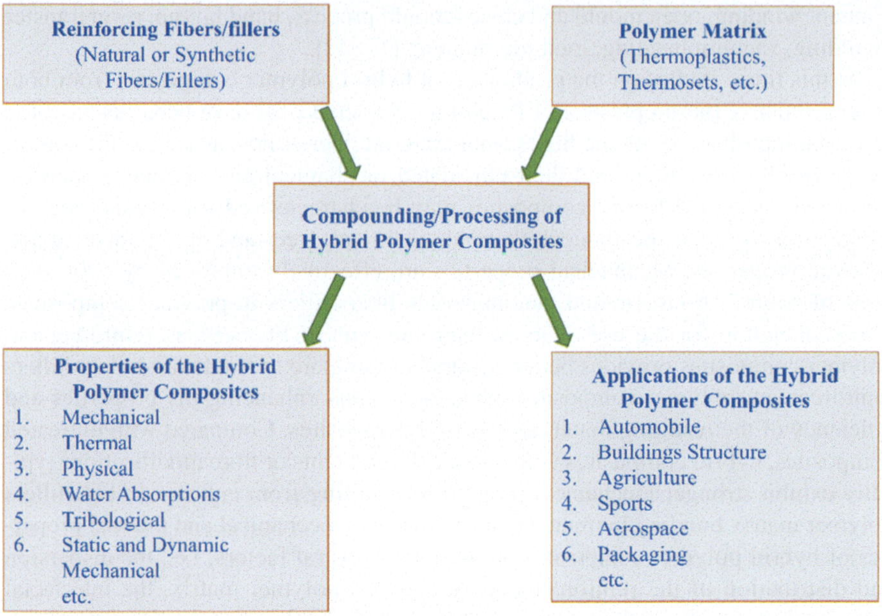

Fig. 4 Fibers/Fillers reinforced hybrid polymer composites; their compounding/processing, properties and applications

synergistic combination of features. Because they may provide customised answers to challenging engineering problems, these cutting-edge materials have attracted a lot of interest recently and have the potential to completely transform a number of industries. The different types of fibers/fillers; used, i.e., natural, synthetic and nanomaterials fibers/fillers for the hybrid composites [49, 50].

3 Reinforcing Fibers/fillers for Hybrid Polymer Composites

The reinforcing fibers/fillers are plays important role in hybrid polymer composites. The reinforcing fibers/fillers are used in hybrid polymer composites as a reinforcement. The hybrid polymer composites properties and characterisations depend on the reinforcing fibers/fillers and also polymer matrix. The different types of fillers/fibers have been classified, as shown in Fig. 5, therefore used for hybrid polymers composites.

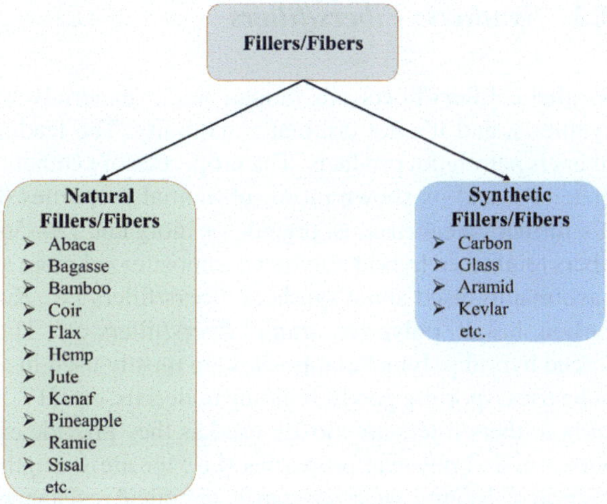

Fig. 5 Classification of fillers/fibers (natural and synthetic fillers/fibers)

3.1 Natural Fibers/fillers

The reinforcing fillers/fibers may be shown fibrous or non-fibrous (particulate) in nature, which has called natural fillers/fibers derived from plants or other living organisms. Natural fillers/fibers are commonly available in nature that is very easy to acquire. They show some excellent materials characteristics such as biodegradability, non-toxic, low cost, unique stiffness and higher mechanical properties. There were several lignocellulosic fillers/fibers in the naturally available. It is possibly classified these fillers/fibers as naturally occurring composites. It was primarily composed of lignocellulosic fillers/fibers embedded in the polymer matrix. Composites made of natural fiber reinforcements tend to have various properties over synthetic fibers, i.e., reduced weight, recyclable, toxicity, contamination of the toxic particles in the atmosphere. Enhancing the economic and environmental advantages of natural fillers/ fibers composites makes them influential for modern applications over synthetic fibers/fillers reinforced composites [51, 52]. Natural fibers/fillers have been found to be excellent materials in recent years for use in hybrid composites that provide an alternative to more costly synthetic fibers/fillers. The capacity of natural fibers/fillers to be hybridized and compare with synthetic fibers/fillers shown lower processing and manufacturing costs and resolving environmental pollutions issues, which are advantageous for economic development and green application implementations [53]. Figure 5 shows the classification of natural fillers/fibers.

3.2 Synthetic Fibers/fillers

Synthetic fillers/fibers are human made materials which produced by chemical synthesis, and it's not originated naturally. The leading source of synthetic fillers/ fibers is petroleum products. The properties of synthetic fillers/fibers depend on their materials, and its shown more substantial properties than natural fibers/fillers, and it's further categorized as organic or inorganic [54–56]. However, synthetic fillers/ fibers reinforced hybrid polymer composites get more attention for researcher, which have mainly used some synthetic fibers/fillers i.e., Kevlar, graphene, glass, boron, carbon, basalt, polyester, aramid fibers/fillers etc. The synthetic fillers/fibers reinforced hybrid polymer composites are mostly used in automobile sectors, aerospace industries, sporting goods, building materials, etc. [17, 56]. Of all synthetic, glass and carbon fibers/fillers are mostly used as they provide excellent thermal, mechanical, abrasive, and physical properties (i.e., tensile strength, impact strength, resilience, friction resistance, wear resistance, chemical resistance, etc.) and also provide better thermal stability. Mostly preferable carbon and glass fibers/fillers reinforced hybrid polymer composites because of machining is relatively sluggish, demanding and shows decreased tool life when operating on traditional machining systems [55]. However, more stiffness is required in some applications, i.e., aerospace industries, automotive sectors, sports goods, etc., the carbon fibers/fillers are used instead of glass fibers/fillers [56]. Figure 5 shows the classification of synthetic fillers/fibers.

4 Different Kinds of Materials Used for Hybrid Polymer Composites

Hybrid polymer composites are extremely desirable for a variety of applications because of the vast range of increased properties and qualities that result from the combination of several types of materials. The characteristics of hybrid polymer composites can be precisely tailored by scientist and engineers by carefully combining reinforcing materials with the polymer matrix. Table 1 shows the different kinds of the reinforcing fibers/fillers and polymer matrices are used for hybrid polymer composites.

Table 1 show the different types of the polymer matrices are hybridizing with different kinds of the fibers and fillers [57–74]. The WF and GF fibers/fillers reinforced with HDPE to form hybrid composites. The WF/GF/HDPE hybrid composites were mixing with 2, 4, 6 and 8% MAPE coupling agents by high speed mixer and mixer dried before compounding by counter rotating twin screw extruder. The extruded samples were injection moulded. The mechanical properties of the hybrid composites enhanced with increasing of GF and MAPE coupling agents due to the interfacial adhesions and effective stress transfer from polymer matrix to reinforcing fibers and fillers [57]. The BNF and COIRF fibers/fillers reinforced with LDPE to form hybrid composites. The BNF/COIRF/LDPE hybrid composites were

Table 1 Different kinds of materials used for hybrid polymer composites

Sr. no	Polymers	Fillers/fibers	Hybridizations	Compounding /processing	Characterisations/ testing	Properties obtained	References
1	Polyethylene (PE)						
a	High density Polyethylene (HDPE), HDPE used fixed 70% for all hybrid composites	Wood flour (WF)	By glass fiber (GF). The ratio of the WF/GF (30/0, 25/5, 20/10, 15/15, 10/20, 5/25, 0/30)	The hybrid composites were mixing by high speed mixer at 3000 rpm (SHR-10a; Huaming Machinery co., China) for 5 min. Then mixer dried samples at 103 °C for 2 h. Mixer are compounded by counter rotating twin screw extruder (HTY-30; Rubber Machinery Factory co., China) at 100–150 °C with 32 rpm screw speed. Extruded samples were injection moulded at 190 °C. Malleated polyethylene (MAPE) 2, 4, 6 and 8% used as coupling agents with 3,000 rpm rotating speed for 5 min	Scanning electron microscope (SEM), Tensile test	The tensile mechanical properties of hybrid polymer composites increases with increasing of GF and coupling agents due to interfacial adhesions and effective stress transfer from polymer matrix to reinforcing fiber and fillers	[57]

(continued)

Table 1 (continued)

Sr. no	Polymers	Fillers/fibers	Hybridizations	Compounding /processing	Characterisations/ testing	Properties obtained	References
b	Low density Polyethylene (LDPE), LDPE used fixed 72% for all hybrid composites	Banana fiber (BNF)	By Coir fiber (COIRF), The ratio of the BNF/COIRF (18.75/6.25, 12.5/12.5, 6.25/18.75)	For homogeneous mixing, the hybrid composites with 3% of malleated polyethylene (MAPE) coupling agents were mixing by mechanical stirrer mixing with rotational speed 2400 rpm for 15 min. The samples were prepared by compression moulding machine at 180 °C with 20 MPa pressure for 20 min	SEM, Thermogravimetric analysis (TGA), Differential thermal analysis (DTA), Tensile test, Water absorption, Flexural test, Impact test	The mechanical (tensile, flexural, and toughness), physical (water absorption resistance) and thermal (thermal stability) properties of the hybrid composites (BNF/COIRF 12.5/12.5) shown better results with compare to other formulated hybrid composites	[58]
c	Linear Low-density polyethylene (LLDPE), LLDPE taken in powder form for this hybrid composites, LLDPE used 60, 70, 80 and 90% for the hybrid composites	Treated and untreated Sugarcane Bagasse fiber (SB)	By treated and untreated Eggshell (ES), The ratio of the SB/ES (6/4, 12/8, 17/13, 20/20, 13/17, 8/12, and 4/6). The size of both fiber taken 53 um. SB fiber were treated by 3% of silane treatment and ES were treated by 1% of silane treatment	The hybrid composites mixing by high speed mixer at 140 °C for 3 min. Mixed hybrid composites held in hot air oven at 80 °C for 24 h. The samples were prepared by compression moulding machine at 140 °C with 250 kg/cm^2 pressure for 1 min	SEM, TGA, Differential scanning calorimetry (DSC), Tensile test, Flexural test, Impact test	The tensile modulus, flexural modulus and hardness of the treated hybrid composites were higher than untreated hybrid composites. The tensile modulus, flexural modulus and hardness of the hybrid composites were increases with increasing of SB/ES content	[59]

(continued)

Table 1 (continued)

Sr. no	Polymers	Fillers/fibers	Hybridizations	Compounding /processing	Characterisations/ testing	Properties obtained	References
2	Polypropylene (PP)						
a	Polypropylene (PP), PP used fixed 70% for all hybrid composites	Oil Palm Empty Fruit Bunch fiber (OPEFBF)	By Coir fiber (COIRF), The ratio of the OPEFBF/COIRF (100/0, 75/25, 50/50, 25/75, 0/100). The (OPEFBF/COIRF) used fixed 30% for all hybrid composite	The hybrid composites were compounded by co-rotating twin screw extruder at 190 °C and their samples prepared by injection moulding machine at 190 °C	SEM, Tensile test, Flexural test, Impact test	The mechanical properties (tensile strength, tensile modulus, flexural strength, flexural modulus and Izod impact strength) of the OPEFBF/COIRF/PP hybrid composites were higher which ratio of the fibers OPEFBF/COIRF have taken 50/50 with comparison of other hybrid composites due to strong fiber/matrix bonding	[60]

(continued)

Table 1 (continued)

Sr. no	Polymers	Fillers/fibers	Hybridizations	Compounding /processing	Characterisations/ testing	Properties obtained	References
b	Isotactic Polypropylene (PP), PP used fixed 70% for all hybrid composites	Sisal fiber (SF)	By glass fiber (GF), The ratio of the SF/GF (25/5, 20/10, 15/15, 10/20, and 5/25)	The hybrid composites were compounded by melt blended counter rotating twin-screw extruder (CTW-100, Haake, Germany) at 155–170 °C with screw speed of 50 rpm. The samples were prepared by Injection moulding machine (80 T, ESS 330/80 HL, Engel, Austria) at 160–180 °C with 1800 bar injection pressure. Malleated polypropylene (MAPP) 1, 2, and 3% used as coupling agents for this hybrid composites	SEM, TGA, DSC, Dynamic mechanical analysis (DMA), Tensile test, Flexural test, Impact test, Water absorption	The mechanical properties (tensile, flexural, and impact strength) of the hybrid composites were shown higher at 15/15 SF/GF ratio in the presence of 2% MAPP, due to interaction between SF/GF and polymer matrix PP. The MAPP improve the fiber/matrix adhesion in the hybrid composites	[61]

(continued)

Table 1 (continued)

Sr. no	Polymers	Fillers/fibers	Hybridizations	Compounding /processing	Characterisations/ testing	Properties obtained	References
c	Polypropylene (PP), PP used 78.8, 87.1, 88.4, 92.3, and 96.3% for the hybrid composites	Long Glass fibers (LGF)	By $CaCO_3$ particulate filler The ratio of the LGF/ $CaCO_3$ (0/ 7.7, 3.7/0, 4.2/7.4, 12.9/0, 14.6/6.6)	The fiber and particulate filler with polymer matrix were compounded by Haake Bucher Rheochord System (40 twin screw extruder) at 180–280 °C with screw speed 40 rpm. The hybrid composites mixer was mechanically mixed. The samples were prepared by injection moulding machine (Arburg Allrounder 320C 600–250) at 225–270 °C	SEM, DSC, Tensile test, Fracture test	The stiffness of the hybrid composites was increase with addition of $CaCO_3$ particulate filler in the hybrid composites but the tensile strength and toughness were decreases because $CaCO_3$ filler did not considerably affect the LGF fibre orientation	[62]

(continued)

Table 1 (continued)

Sr. no	Polymers	Fillers/fibers	Hybridizations	Compounding /processing	Characterisations/ testing	Properties obtained	References
d	Isotactic Polypropylene (PP), PP used fixed 70% for the all hybrid composites	Bamboo fiber (BMBF)	By Glass fiber (GF), The ratio of the BMBF/GF (25/5, 20/10, 15/15, 10/20, and 5/25)	The hybrid composites were compounded by melt blended counter rotating twin-screw extruder (CTW-100, Haake, Germany) at 155–170 °C with screw speed of 50 rpm. The samples were prepared by injection moulding machine (80 T, ESS 330/80 HL, Engel, Austria) at 160–180 °C with 1800 bar injection pressure. The MAPP (1, 2 and 3%) used as coupling agents for this hybrid composites	SEM, TGA, DMA, DSC, Tensile test, Flexural test, Impact test, Water absorption	The mechanical properties (tensile, flexural and impact strength) of the hybrid composites were shown higher which have taken 15/15 BMBF/GF ratio in the presence of the 2% MAPP, due to the interaction between BMBF/GF and polymer matrix PP. The MAPP improve the fiber/matrix adhesion in the hybrid composites	[63]

(continued)

Table 1 (continued)

Sr. no	Polymers	Fillers/fibers	Hybridizations	Compounding /processing	Characterisations/ testing	Properties obtained	References
e	Polypropylene (PP), PP used 77, 82 and 87% for the hybrid composites	Basalt fiber (BF)	By Hazelnut shells filler (HS), the ratio of the BF/HS (5/5, 7.5/7.5 and 10/10)	The hybrid composites were compounded by co-rotating twin-screw extruder at 180–200 °C with screw speed of 63 rpm. The samples were prepared by injection moulding machine (ENGEL GmbH, Schwartzberg, Austria) at 180–200 °C. The 3% MAPP used as coupling agents for this hybrid composites	SEM, Tensile test, Impact test, Water absorption, Creep test	The mechanical, (tensile strength, tensile modulus and impact strength), physical (creep stability, water absorption resistance), thermal (thermal stability) properties of the hybrid composites were shown higher at 10/10 BF/HS ratio due to good interfacial adhesion between fiber/ matrix	[64]
3	Polycaprolactone (PCL)						
a	Polycaprolactone (PCL), PCL used 85.3, 71.2, 57.8, 45, 32.8 and 21.2% for the hybrid composites	Oil Palm Empty Fruit Bunch fiber (OPEFBF)	By nickel oxide filler (NiO), the ratio of the OPEFBF/NiO (12.2/2.5, 23.8/5, 34.7/7.5, 45/10, 54.7/12.5, and 63.8/15)	The hybrid composites have melted blending by Thermo Haake Poly Drive Extruder at 70 °C with 50 rpm for 20 min. The samples were prepared by the melted hybrid composites fill in required shape of mould	SEM, TGA, DTG, Fourier transform infrared spectroscopy (FTIR), X-ray diffraction (XRD), Microwave Absorbing test	The thermal stability of the hybrid composites was increases with increasing of NiO contents in the hybrid composites due to enhancement of the bond between polymer matrix and the reinforcing fiber/ filler	[65]

(continued)

Table 1 (continued)

Sr. no	Polymers	Fillers/fibers	Hybridizations	Compounding /processing	Characterisations/ testing	Properties obtained	References
b	Polycaprolactone (PCL), PCL used 85.3, 71.2, 57.8, 45, 32.8 and 21.2% for the hybrid composites	Oil Palm Empty Fruit Bunch fiber (OPEFBF)	By nickel zinc ferrite (NZF), the ratio of the OPEFBF/NZF taken (12.2/2.5, 23.8/5, 34.7/7.5, 45/10, 54.7/12.5, and 63.8/15)	The hybrid composites have melted blending by Thermo Haake Poly Drive Extruder at 80 °C with 50 rpm. The samples were prepared by the melted hybrid composites fill in required shape of mould	SEM, FTIR, TGA, DTG, Permeability, Permittivity	The thermal stability of hybrid composites was increases with increasing of NZF contents in the hybrid composites due to enhancement of the bond between polymer matrix and the reinforcing fiber/ filler	[66]
4	Polyethylene terephthalate (PET) or polyester						
a	Polyethylene Terephthalate (PET) drinking bottles or Polyester drinking bottles, PET used 90, 80 and 50% for the hybrid composites	Oil Palm Empty Fruit Bunch fiber (OPEFBF)	By IJuk fiber (IJuk), the ratio of the OPEFBF/IJuk (5/5, 10/10 and 25/25)	The hybrid composites have melted blending by twin screw extruder at 140 °C and their samples prepared by hot press machine at 140 °C for 5, 10, 15, 20, and 25 min holding time	SEM, TGA, tensile test, Flexural test	The mechanical properties (tensile and flexural strength) and thermal stability of the 25/25/50 OPEFBF/IJuk/PET hybrid composites at 20 min holding time were shown higher than all hybrid composites because the higher holding time make fiber breaks and lower holding time make not perfect bonding between fiber and matrix	[67]

(continued)

Table 1 (continued)

Sr. no	Polymers	Fillers/fibers	Hybridizations	Compounding /processing	Characterisations/ testing	Properties obtained	References
b	Unsaturated Polyester (UP), UP used 98, 96, 94 and 92% for the hybrid composites	Glass fiber (GF)	By silica sand nanoparticles (SSNp), the ratio of GF/SSNp (1/1, 2/2, 3/3 and 4/4)	The hybrid composites have prepared by hand lay-up method and used cobalt acetate (accelerator) and methyl ethyl ketone peroxide (MEKP) used as hardener. The ratio of accelerator/hardener were taken 1/1 in the hybrid composites	SEM, TGA, Tensile test, Impact test, Hardness test	The mechanical properties (tensile strength, impact strength and hardness) and thermal stability of hybrid composites were increases with increasing of fiber and nanoparticles contents and shown higher value at ratio of 4/4/92 GF/SSNp/UP hybrid composites with the comparisons of the all hybrid composites because the silica nanoparticles filled the cavities in the hybrid composites	[68]
c	Unsaturated Polyester (UP), UP used 90, 85, 80, 85, 80, and 75% for the hybrid composites	Kenaf fiber (KNF) in mat form	By Glass fiber (GF) in mat form, the ratio of KNF/GF (5/5, 10/5, 15/5, 5/10, 10/10 and 15/10)	The hybrid composites have prepared by hand lay-up method and 2% of MEKP used as hardener for the hybrid composites	Tensile test	The mechanical properties (tensile strength and tensile modulus) of the 10/10/80 KNF/GF/UP hybrid composites were shown higher value with compare of the all hybrid composites	[69]
5	Polylactic acid (PLA)						

(continued)

Table 1 (continued)

Sr. no	Polymers	Fillers/fibers	Hybridizations	Compounding /processing	Characterisations/ testing	Properties obtained	References
a	Polylactic acid (PLA), PLA used fixed 40% for all hybrid composites	Borax treated (BT) and untreated (UT) Oil Palm Empty Fruit Bunch fiber (BT-OPEFBF and UT-OPEFBF) and oil palm mesocarp fiber (BT-OPMF and UT-OPMF)	By Borax treated (BT) and untreated (UT) Kenaf core fiber (BT-KCF and UT-KCF), the ratio of the (UT and BT) OPEFBF/ KCF and OPMF/ KCF (55/5)	The hybrid composites have compounded by brabender internal mixer at 170 °C with 50 rpm for 15 min. The samples were prepared by compression moulding machine at 170 °C	SEM, FTIR, Tensile test, Flexural test, Impact test	The tensile strength, tensile modulus, elongation at break, flexural strength, flexural modulus, impact strength and the thermal stability of the 55/5/40 BT-OPEFBF/BT-KCF/PLA hybrid composites shown higher than all BT and UT hybrid composites because borax eliminated the impurities and hemicellulose and enhanced the bond strength between fiber and matrix	[70]
b	Polylactic acid (PLA), PLA used 20, 15 and 10% for the hybrid composites	Oil palm empty fruit bunch fiber (OPEFBF)	By NPK fertilizer, the ratio of the OPEFBF/ NPK fertilizer (5/15, 10/15, 15/15)	The hybrid composites have manually mixing and samples were prepared by compression moulding machine at 170 °C	SEM, TGA, DSC	The glass transition temperature, crystallization temperature and melting temperature of the 15/15/10 OPEFBF/NPK fertilizer/ PLA hybrid composites were higher than others hybrid composites due to hydrolysis of the PLA matrix after adding of the fiber during the thermal processing	[71]

(continued)

Table 1 (continued)

Sr. no	Polymers	Fillers/fibers	Hybridizations	Compounding /processing	Characterisations/ testing	Properties obtained	References
6	Polyvinyl alcohol (PVA)						
a	Polyvinyl alcohol (PVA), PVA used 90, 80 and 70% for the hybrid composites	Nano cellulose Oil Palm Empty Fruit Bunch fiber (OPEFBF)	By chitin, the ratio of the OPEFBF/ chitin (1/10, 1/20 and 1/30)	The hybrid composites have been mixed in distilled water, where magnetic stirrer continuous rotating and hybrid composites slurry found. The slurry poured in non-stick mould and samples were prepared. Before hybrid composites slurry put in non-stick mould, the slurry passes through in ultra-sonication to remove excess contents of the water. The 10, 30 and 50% of the maleic acid (MA) used as crosslinking agents in the hybrid composites	SEM, FTIR, DSC, Tensile test	The mechanical properties (tensile strength, tensile modulus, elongation at break) and thermal properties (glass transition temperature, melting temperature, crystalline temperature and enthalpy) of the hybrid composites were increases with increasing of MA content till 30% for all ratio of hybrid composites then decreases due to excessive of crosslinking and its restrained intermolecular mobility and rigid nature of the chitin backbone chains	[72]
7	Epoxy						

(continued)

Table 1 (continued)

Sr. no	Polymers	Fillers/fibers	Hybridizations	Compounding /processing	Characterisations/ testing	Properties obtained	References
a	Epoxy, Epoxy used fixed 50% for all hybrid composites	Oil Palm Empty Fruit Bunch fiber mat (OPEFBF)	By kenaf fiber mat (KNF), the ratio of the OPEFBF/KNF (4/1, 1/1 and 1/4). The total fixed fiber were taken 50% for the hybrid composites	The hybrid composites have prepared by hand-lay-up method, where the amine 905-3 s used as hardener. The ratio of the epoxy and hardener were 2/1	SEM, FTIR, Tensile test, Flexural test, Impact test	The mechanical properties (tensile strength, flexural strength and impact strength) of the hybrid composites were increasing with increasing of KNF contents due to the hybridisation of fiber	[73]
b	Epoxy, Epoxy used fixed 60% for all hybrid composites	Jute fiber (JF)	By Hemp fiber (HF) and glass fiber (GF), the ratio of the JF/HF (20/20) and the ratio of the JF/HF/GF (13.5/13.12/13.09)	The hybrid composites have prepared by hand-lay-up method, where the epoxy hardener-PAM used as hardener. The ratio of the epoxy and hardener were 10/1	Tensile test, Flexural test, Impact test, Compression test, Shear strength test, Water absorption test	The mechanical properties (tensile strength, flexural strength, impact strength, shear strength and compression strength) and water absorption resistance of JF/HF/GF/Epoxy hybrid composites shown higher than others hybrid composites due to hybridisation of the fibers	[74]

mixing with 3% MAPE coupling agents by mechanically stirrer. The samples were compression moulded. The mechanical, physical and thermal properties of the hybrid composites shown better results which BNF and COIRF fibers ratio were 12.5/12.5 BNF/COIRF [58]. The SB and ES fibers/fillers reinforced with LLDPE to form hybrid composites. The SB/ES/LLDPE hybrid composites were mixing by high speed mixer. The samples were compression moulded. The mechanical properties of the treated (T-SB/T-ES)/LLDPE hybrid composites were higher than the untreated (UT-SB/UT-ES)/LLDPE hybrid composites [59]. The OPEFBF and COIRF fibers/fillers reinforced with PP to form hybrid composites. The OPEFBF/COIRF/PP hybrid composites were compounded by co-rotating twin screw extruder. The samples were injection moulded. The mechanical properties of the hybrid composites shown better results which OPEFBF and COIRF fibers ratio were 50/50 OPEFBF/COIRF [60]. The SF and GF fibers/fillers reinforced with PP to form hybrid composites. The SF/GF/PP hybrid composites with 1, 2, and 3% MAPP coupling agents were compounded by melt blended counter rotating twin-screw extruder. The samples were injection moulded. The mechanical properties of the hybrid composites with 2% MAPP shown better results which SF and GF fibers ratio were 15/15 SF/GF [61]. The LGF and $CaCO_3$ fibers/fillers reinforced with PP to form hybrid composites. The LGF/$CaCO_3$/PP hybrid composites were compounded by Haake Bucher Rheochord. The samples were injection moulded. The stiffness of the LGF/$CaCO_3$/PP hybrid composites were increases with increasing of $CaCO_3$ but the tensile strength and toughness were decreases [62]. The BMBF and GF fibers/fillers reinforced with PP to form hybrid composites. The BMBF/GF/PP hybrid composites with 1, 2 and 3% MAPP coupling agents were compounded by melt blended counter rotating twin screw extruder. The samples were injection moulded. The mechanical properties of the hybrid composites with 2% MAPP shown better results which BMBF and GF fibers ratio were 15/15 BMBF/GF [63]. The BF and HS fibers/fillers reinforced with PP to form hybrid composites. The BF/HS/PP hybrid composites were compounded by co-rotating twin screw extruder. The samples were injection moulded. The mechanical, physical properties and thermal stability of the hybrid composites shown better results which BF and HS fibers ratio were 10/10 BF/HS [64]. The OPEFBF and NiO and NZF fibers/fillers reinforced with PCL to form hybrid composites. The OPEFBF/NiO/PCL and OPEFBF/NZF/PCL hybrid composites were melted blending by Thermo Haake Poly Drive Extruder. The thermal stability of the OPEFBF/NiO/PCL and OPEFBF/NZF/PCL hybrid composites were increases with increasing of NiO and NZF contents [65, 66]. The OPEFBF and IJuk fibers/fillers reinforced with PET to form hybrid composites. The OPEFBF/IJuk/PET hybrid composites were melted blending by twin screw Extruder. The samples were hot press moulded at 5, 10, 15, 20, and 25 min holding time. The mechanical properties and thermal stability of the OPEFBF/IJuk/PET hybrid composites at 20 min holding time were shown higher than all hybrid composites because the higher holding time make fiber breaks and lower holding time make not perfect bonding between fiber and matrix [67]. The GF and SSNp and KNF fibers/fillers reinforced with UP to form hybrid composites. The GF/SSNp/UP and GF/KNF/UP hybrid composites were prepared by hand lay-up method. The mechanical properties and thermal stability of hybrid composites were increases with increasing of fiber and nanoparticles contents [68, 69]. The UT and BT of OPEFBF and OPMF and KCF fibers/fillers reinforced with PLA to form hybrid composites. The OPEFBF/KCF/PLA and OPMF/KCF/PLA hybrid composites were compounded by brabender internal mixer. The mechanical properties and thermal stability of 55/5/40 BT-OPEFBF/BT-KCF/PLA hybrid composites were shown than all BT and UT hybrid composites [70]. The OPEFBF and NPK fibers/fillers reinforced with PLA to form hybrid composites. The OPEFBF/NPK/PLA hybrid composites were manually mixing. The samples were compression moulded. The glass transition temperature, crystallization temperature and melting temperature of the 15/15/10 OPEFBF/NPK fertilizer/PLA hybrid composites were higher than others hybrid composites [71]. The OPEFBF and Chitin fibers/fillers reinforced with PVA to form hybrid composites. The OPEFBF/chitin/PVA hybrid composites with 10, 30 and 50% of MA crosslinking agents were mixing by magnetic stirrer and form slurry. The thermal and mechanical properties of the hybrid composites were increases with increasing of MA content till 30% for all ratio of hybrid composites then decreases due to excessive of crosslinking [72]. The OPEFBF and KNF fibers/fillers reinforced with epoxy to form hybrid composites. The OPEFBF/KNF/epoxy hybrid composites have been prepared by hand-lay-up method. The mechanical properties of the hybrid composites were increasing with increasing of the KNF contents due to the hybridisation of fiber [73]. The JF and HF and GF fibers/fillers reinforced with epoxy to form hybrid composites. The hybrid composites have been prepared by hand-lay-up method. The mechanical properties of the JF/HF/GF/epoxy hybrid composites shown higher than others hybrid composites due to hybridisation of the fibers [74]. Lastly, the authors have considerable experience in the field of polymer composites and atomistic-scale based molecular dynamics simulations which proves the worth of writing this chapter [75–150].

5 Conclusion

Conclusively, the amalgamation of the diverse material types in hybrid composites offers an exceptional prospect for the development of the high-performance materials with customised attributes. The hybrid composites have the potential to transform many technical fields and open the door for a new generation of innovative materials. By adopting this multidisciplinary approach and pushing the frontiers of the material science.

Conflict of Interest "There are no conflicts of interest to declare by the authors."

References

1. Pendhari, S. S., Kant, T., & Desai, Y. M. (2008). Application of polymer composites in civil construction: A general review. *Composite Structures, 84*, 114–124. https://doi.org/10.1016/J.COMPSTRUCT.2007.06.007
2. Muhammad, A., Rahman, M.R., Baini, R., Bin Bakri, M.K. (2021) Applications of sustainable polymer composites in automobile and aerospace industry. Advanced Sustain Polymer Composites 185–207. https://doi.org/10.1016/B978-0-12-820338-5.00008-4
3. Zagho, M.M., Hussein, E.A., Elzatahry, A.A. (2018). Recent overviews in functional polymer composites for biomedical applications. *Polymers 10*, 739. https://doi.org/10.3390/POLYM10070739.
4. Rubino, F., Nisticò, A., Tucci, F., Carlone, P. (2020). Marine application of fiber reinforced composites: A review. *Journal of Marine Science and Engineering 8*, 26. https://doi.org/10.3390/JMSE8010026.
5. Rajak, D.K., Pagar, D.D., Menezes, P.L., Linul, E. (2019). Fiber-reinforced polymer composites: manufacturing, properties, and applications. *Polymers 11*, 1667. https://doi.org/10.3390/POLYM11101667.
6. I.O. Oladele, T.F. Omotosho, A.A. Adediran, Polymer-Based Composites: An Indispensable Material for Present and Future Applications. *International Journal of Polymer Science 2020* (2020). https://doi.org/10.1155/2020/8834518.
7. Hassan, T., Salam, A., Khan, A., Khan, S.U., Khanzada, H., Wasim, M., Khan, M.Q., Kim, I.S. (2021). Functional nanocomposites and their potential applications: A review. *Journal of Polymer Research 282*. 28, 1–22. https://doi.org/10.1007/S10965-021-02408-1.
8. Banik, K., Abraham, T. N., & Karger-Kocsis, J. (2007). Flexural creep behavior of unidirectional and cross-ply all-poly(propylene) (PURE®) composites. *Macromolecular Materials and Engineering, 292*, 1280–1288. https://doi.org/10.1002/MAME.200700180
9. Korabel'nikov, Y.G., Rashkovan, I.A. (2006). Strength and mechanism of fracture of composites randomly reinforced with short carbon fibres. *Fibre Chemistry 38*, 142–146. https://doi.org/10.1007/S10692-006-0059-3.
10. Avella, M., Buzarovska, A., Errico, M.E., Gentile, G., Grozdanov, A. (2009). Eco-challenges of bio-based polymer composites. *Materials 2*, 911–925. https://doi.org/10.3390/MA2030911.
11. Shah, D.U. (2013). Developing plant fibre composites for structural applications by optimising composite parameters: A critical review. *Journal of Material Science 4818*, 6083–6107. https://doi.org/10.1007/S10853-013-7458-7.
12. Swolfs, Y., Verpoest, I., & Gorbatikh, L. (2019). Recent advances in fibre-hybrid composites: Materials selection, opportunities and applications. *International Materials Reviews, 64*, 181–215. https://doi.org/10.1080/09506608.2018.1467365
13. Asim, M., Jawaid, M., Saba, N., Ramengmawii, N., Nasir, M., Sultan, M.T.H. (2017). Processing of hybrid polymer composites—a review, Hybrid Polymer Composites Materials Process 1–22. https://doi.org/10.1016/B978-0-08-100789-1.00001-0.
14. Kistaiah, N., Udaya Kiran, C., Ramachandra Reddy, G., Sreenivasa Rao, M. (2014). Mechanical characterization of hybrid composites: A review. *The Journal of Reinforced Plastics and Composites 33* (2014) 1364–1372. https://doi.org/10.1177/0731684413513050/ASSET/IMAGES/LARGE/10.1177_0731684413513050-FIG5.JPEG.
15. Gupta, H.S., Kumar, P., Sethi, A.K., Verma, A., Sethi, S.K. (2024). Software-based simulations of the creep recovery model of polymer composites. Dynamic Mechanical and Creep Recovery Behavior of Polymer Composites 271–289. https://doi.org/10.1016/B978-0-443-19009-4.00016-3.
16. Gupta, H. S., Shankar, U., Verma, A., Gogoi, R., & Sethi, S. K. (2023). Computational aspects: self-clean coatings, plastics and polymers in coatings. *Materials Horizons From Natural to Nanomaterials Part F, 1086*, 167–184. https://doi.org/10.1007/978-981-99-3549-9_8

17. Sinha, A. K., Narang, H. K., & Bhattacharya, S. (2020). Mechanical properties of hybrid polymer composites: A review. *Journal of Brazilian Social and Mechanical Science Engineering, 42*, 1–13. https://doi.org/10.1007/S40430-020-02517-W/TABLES/3
18. Saba, N., Tahir, P.M., Jawaid, M. (2014) A review on potentiality of nano filler/natural fiber filled polymer hybrid composites. *Polymers 6*, 2247–2273. https://doi.org/10.3390/POLYM6082247.
19. Atmakuri, A., Palevicius, A., Vilkauskas, A., Janusas, G. (2020). Review of hybrid fiber based composites with nano particles—material properties and applications. *Polymers 12*, 2088. https://doi.org/10.3390/POLYM12092088.
20. Gupta, H. S., Sethi, S. K., & Verma, A. (2023). Applications of coating materials: A critical overview. *Materials Horizons From Natural to Nanomaterials Part F, 1086*, 81–109. https://doi.org/10.1007/978-981-99-3549-9_4
21. Suriani, M.J., Ilyas, R.A., Zuhri, M.Y.M., Khalina, A., Sultan, M.T.H., Sapuan, S.M., Ruzaidi, C.M., Wan, F.N., Zulkifli, F., Harussani, M.M., Azman, M.A., Radzi, F.S.M., Sharma, S. (2021). Critical review of natural fiber reinforced hybrid composites: Processing, properties, applications and cost. *Polymers 13*, 3514. https://doi.org/10.3390/POLYM13203514.
22. Alagesan, P.K. (2020). Recent advances of hybrid fiber composites for various applications. *Hybrid Fiber Composites* 381–404. https://doi.org/10.1002/9783527824571.CH17.
23. Ahmed, S., Jones, F.R. (1990). A review of particulate reinforcement theories for polymer composites. *Journal of Material Science 25* 4933–4942. https://doi.org/10.1007/BF00580110.
24. Akca, E., Gursel, A. (2015). A review on the matrix toughness of thermoplastic materials. *Periodicals of Engineering and Natural Sciences 3*. https://doi.org/10.21533/PEN.V3I2.52.
25. Mallick, P.K. (2000). Particulate and short fiber reinforced polymer composites. *Comprehensive Composite Materials* 291–331. https://doi.org/10.1016/B0-08-042993-9/00085-1.
26. Móczó, J., Pukánszky, B. (2015). Particulate fillers in thermoplastics. *Encyclopedia Polymer Composites* 1–35. https://doi.org/10.1007/978-3-642-37179-0_7-1.
27. Denault, J., Vu-Khanh, T., & Foster, B. (1989). Tensile properties of injection molded long fiber thermoplastic composites. *Polymer Composites, 10*, 313–321. https://doi.org/10.1002/PC.750100507
28. Komal, U.K., Lila, M.K., Chaitanya, S., Singh, I. (2019). Fabrication of short fiber reinforced polymer composites. Reinforced Polymer Composites Process Characters. Post Life Cycle Assess 21–38. https://doi.org/10.1002/9783527820979.CH2.
29. Fu, S. Y., Lauke, B., Mäder, E., Yue, C. Y., & Hu, X. (2000). Tensile properties of short-glass-fiber- and short-carbon-fiber-reinforced polypropylene composites. *Composites Part A, Applied Science and Manufacturing, 31*, 1117–1125. https://doi.org/10.1016/S1359-835X(00)00068-3
30. Junaedi, H., Baig, M., Dawood, A., Albahkali, E., Almajid, A. (2020). Mechanical and physical properties of short carbon fiber and nanofiller-reinforced polypropylene hybrid nanocomposites. *Polymers 12*, 2851. https://doi.org/10.3390/POLYM12122851.
31. Lin, J.H., Huang, C.L., Liu, C.F., Chen, C.K., Lin, Z.I., Lou, C.W. (2015). Polypropylene/short glass fibers composites: effects of coupling agents on mechanical properties, thermal behaviors, and morphology. *Materials 8*, 8279–8291. https://doi.org/10.3390/MA8125451.
32. Ku, H., Wang, H., Pattarachaiyakoop, N., & Trada, M. (2011). A review on the tensile properties of natural fiber reinforced polymer composites. *Composites Part B, Engineering, 42*, 856–873. https://doi.org/10.1016/J.COMPOSITESB.2011.01.010
33. Dixit, S., Goel, R., Dubey, A., Shivhare, P.R., Bhalavi, T. (2017). Natural fibre reinforced polymer composite materials - a review 8 (2017) 71–78. https://doi.org/10.1177/204124791700800203.
34. Yao, Y., Li, M., Lackner, M., Herfried, L. (2020). A continuous fiber-reinforced additive manufacturing processing based on PET fiber and PLA. *Materials 13*, 3044. https://doi.org/10.3390/MA13143044.
35. Mitschang, P., Christmann, M. (2012). Continuous fiber reinforced profiles in polymer matrix composites. Manufacturing Techniques for Polymer Matrix Composites 209–242. https://doi.org/10.1533/9780857096258.2.209.

36. St. John, N.A., Brown, J.R. (1998). Flexural and interlaminar shear properties of glass-reinforced phenolic composites. *Composites Part A Appliances and Science Manufacturing 29* 939–946. https://doi.org/10.1016/S1359-835X(98)00019-0.
37. Rikards, R. (2000). Interlaminar fracture behaviour of laminated composites. *Computers & Structures, 76*, 11–18. https://doi.org/10.1016/S0045-7949(99)00148-0
38. Botelho, E. C., Silva, R. A., Pardini, L. C., & Rezende, M. C. (2006). A review on the development and properties of continuous fiber/epoxy/aluminum hybrid composites for aircraft structures. *Materials Research, 9*, 247–256. https://doi.org/10.1590/S1516-14392006000300002
39. Fu, S. Y., Xu, G., & Mai, Y. W. (2002). On the elastic modulus of hybrid particle/short-fiber/polymer composites. *Composites Part B, Engineering, 33*, 291–299. https://doi.org/10.1016/S1359-8368(02)00013-6
40. Panthapulakkal, S., Raghunanan, L., Sain, M., Kc, B., Tjong, J. (2017). Natural fiber and hybrid fiber thermoplastic composites: Advancements in lightweighting applications, Green Compos. *Waste Nature-Based Mater. A Sustainable Future. Second Ed.* 39–72. https://doi.org/10.1016/B978-0-08-100783-9.00003-4.
41. Rakesh, P., Diwakar, V., Venkatesh, K., & Savannananavar, R. N. (2020). A Concise Review on processing of hybrid composites produced by the combination of glass and natural fibers. *Materials Today Proceedings, 22*, 2016–2024. https://doi.org/10.1016/J.MATPR.2020.03.179
42. Safri, S. N. A., Sultan, M. T. H., Jawaid, M., & Jayakrishna, K. (2018). Impact behaviour of hybrid composites for structural applications: A review. *Composites Part B, Engineering, 133*, 112–121. https://doi.org/10.1016/J.COMPOSITESB.2017.09.008
43. Kureemun, U., Ravandi, M., Tran, L. Q. N., Teo, W. S., Tay, T. E., & Lee, H. P. (2018). Effects of hybridization and hybrid fibre dispersion on the mechanical properties of woven flax-carbon epoxy at low carbon fibre volume fractions. *Composites Part B, Engineering, 134*, 28–38. https://doi.org/10.1016/J.COMPOSITESB.2017.09.035
44. Jumaidin, R., Sapuan, S. M., Jawaid, M., Ishak, M. R., & Sahari, J. (2017). Thermal, mechanical, and physical properties of seaweed/sugar palm fibre reinforced thermoplastic sugar palm Starch/Agar hybrid composites. *International Journal of Biological Macromolecules, 97*, 606–615. https://doi.org/10.1016/J.IJBIOMAC.2017.01.079
45. Akil, H. M., Santulli, C., Sarasini, F., Tirillò, J., & Valente, T. (2014). Environmental effects on the mechanical behaviour of pultruded jute/glass fibre-reinforced polyester hybrid composites. *Composites Science and Technology, 94*, 62–70. https://doi.org/10.1016/J.COMPSCITECH.2014.01.017
46. Dhar Malingam, S., Jumaat, F.A., Ng, L.F., Subramaniam, K., Ab Ghani, A.F. (2018). Tensile and impact properties of cost-effective hybrid fiber metal laminate sandwich structures. *Advanced Polymer Technology 37*, 2385–2393. https://doi.org/10.1002/ADV.21913.
47. Sathishkumar, T.P., Naveen, J., Satheeshkumar, S. (2014). Hybrid fiber reinforced polymer composites – a review. 33 (2014) 454–471. https://doi.org/10.1177/0731684413516393.
48. Sathishkumar, T.P., Navaneethakrishnan, P., Shankar, S., Rajasekar, R., Rajini, N. (2013). Characterization of natural fiber and composites – A review. 32 (2013) 1457–1476. https://doi.org/10.1177/0731684413495322.
49. Nair, A.B., Joseph, R. (2014). Eco-friendly bio-composites using natural rubber (NR) matrices and natural fiber reinforcements. Chemistry, Manufacture and Applications of Natural Rubber (2014) 249–283. https://doi.org/10.1533/9780857096913.2.249.
50. Dong, C. (2017). Review of natural fibre-reinforced hybrid composites. *37*, 331–348. https://doi.org/10.1177/0731684417745368.
51. Gangil, B., Ranakoti, L., Verma, S., Singh, T., Kumar, S. (2020). Natural and synthetic fibers for hybrid composites. Hybrid Fiber Composites 1–15. https://doi.org/10.1002/9783527824571.CH1.
52. Rahman, R., Putra, S.Z.F.S. (2019). Tensile properties of natural and synthetic fiber-reinforced polymer composites. *Mechanical and Physical Testing of Biocomposites, Fibre-Reinforced Composites and Hybrid Composites* 81–102. https://doi.org/10.1016/B978-0-08-102292-4.00005-9.

53. Abdellaoui, H., Raji, M., Essabir, H., Bouhfid, R., Qaiss, A.E.K. (2019). Mechanical behavior of carbon/natural fiber-based hybrid composites. *Mechanical and Physical Testing of Biocomposites, Fibre-Reinforced Composites and Hybrid Composites* 103–122. https://doi.org/10.1016/B978-0-08-102292-4.00006-0.
54. Manikandan, A., Rajkumar, R. (2016). Evaluation of mechanical properties of synthetic fiber reinforced polymer composites by mixture design analysis. 24 (2016) 455–462. https://doi.org/10.1177/096739111602400702.
55. Prakash, S. (2019). Experimental investigation of surface defects in low-power CO_2 laser engraving of glass fiber-reinforced polymer composite. *Polymer Composites, 40*, 4704–4715. https://doi.org/10.1002/PC.25339
56. Yi, X.S. (2015). Development of multifunctional composites for aerospace application. *Multi-Functionality of Polymer Composites: Challenges and New Solutions* (2015) 367–418. https://doi.org/10.1016/B978-0-323-26434-1.00011-8.
57. Jiang, J., Mei, C., Pan, M., & Lu, F. (2019). Effects of hybridization and interface modification on mechanical properties of wood flour/polymer composites reinforced by glass fibers. *Polymer Composites, 40*, 3601–3610. https://doi.org/10.1002/PC.25223
58. Prasad, N., Agarwal, V. K., & Sinha, S. (2018). Hybridization effect of coir fiber on physico-mechanical properties of polyethylene-banana/coir fiber hybrid composites. *Science and Engineering of Composite Materials, 25*, 133–141. https://doi.org/10.1515/SECM-2015-0446/MACHINEREADABLECITATION/RIS
59. Sitticharoen, W., Aukaranarakul, S., Kantalue, K. (2018). Study of thermal and mechanical properties of LLDPE/sugarcane bagasse/eggshell hybrid biocomposites. *Walailak Journal of Science and Technology 16*, 739–751. https://doi.org/10.48048/wjst.2019.4261.
60. Zainudin, E. S., Yan, L. H., Haniffah, W. H., Jawaid, M., & Alothman, O. Y. (2014). Effect of coir fiber loading on mechanical and morphological properties of oil palm fibers reinforced polypropylene composites. *Polymer Composites, 35*, 1418–1425. https://doi.org/10.1002/PC.22794
61. Nayak, S.K., Mohanty, S. (2009). Sisal glass fiber reinforced PP hybrid composites: Effect of MAPP on the dynamic mechanical and thermal properties. *29*, 1551–1568. https://doi.org/10.1177/0731684409337632.
62. Hartikainen, J., Hine, P., Szabó, J. S., Lindner, M., Harmia, T., Duckett, R. A., & Friedrich, K. (2005). Polypropylene hybrid composites reinforced with long glass fibres and particulate filler. *Composites Science and Technology, 65*, 257–267. https://doi.org/10.1016/J.COMPSCITECH.2004.07.010
63. Samal, S.K., Mohanty, S., Nayak, S.K. (2008). Polypropylene—bamboo/glass fiber hybrid composites: fabrication and analysis of mechanical, morphological, thermal, and dynamic mechanical behavior. 28 (2008) 2729–2747. https://doi.org/10.1177/0731684408093451.
64. Kufel, A., Kuciel, S. (2019). Hybrid composites based on polypropylene with basalt/hazelnut shell fillers: The influence of temperature, thermal aging, and water absorption on mechanical properties. *Polymers 12*, 18. https://doi.org/10.3390/POLYM12010018.
65. Fahad Ahmad, A., Abbas, Z., Shaari, A.H., Obaiys, S. J., Sa'ad Aliyu, U. (2019). Synthesis, thermal, dielectric, and microwave reflection loss properties of nickel oxide filler with natural fiber-reinforced polymer composite, *Journal of Applied Polymer Science 136* (2019) 46998. https://doi.org/10.1002/APP.46998.
66. Ahmad, A.F., Abbas, Z., Obaiys, S.J., Abdalhadi, D.M. (2017). Improvement of dielectric, magnetic and thermal properties of OPEFB fibre–polycaprolactone composite by adding Ni–Zn ferrite. *Polymers 9*, 12. https://doi.org/10.3390/POLYM9020012.
67. Rihayat, T., Salim, S., Audina, N., Khan, N. S. P., Zaimahwati, Z., Sami, M., Yunus, M., Salisah, Z., Alam, P. N., Saifuddin, S., & Yusuf, I. (2018). Composite material making from empty fruit bunches of palm oil (EFB) and Ijuk (Arengapinnata) using plastic bottle waste as adhesives. *IOP Conference Series Material Science and Engineering, 334*, 012083. https://doi.org/10.1088/1757-899X/334/1/012083
68. Ahmad, T., Raza, S.S., Aleem, E., Kamran, M., Manzoor, U., Makhdoom, A., Ahmad, R., Mukhtar, S. (2017). Improvement in mechanical and thermal properties of unsaturated

polyester- based hybrid composites. *Iranian Polymer Journal (English Ed.) 26*, 305–311. https://doi.org/10.1007/S13726-017-0520-6/FIGURES/8.

69. Zhafer, S. F., Rozyanty, A. R., Shahnaz, S. B. S., Musa, L., & Zuliahani, A. (2016). Kenaf-glass fiber reinforced unsaturated polyester hybrid composites: Tensile properties. *AIP Conference Proceedings, 1756*, 040009. https://doi.org/10.1063/1.4958770
70. Birnin-Yauri, A.U., Ibrahim, N.A., Zainuddin, N., Abdan, K., Then, Y.Y., Chieng, B.W. (2016). Enhancement of the mechanical properties and dimensional stability of oil palm empty fruit bunch-kenaf core and oil palm mesocarp-kenaf core hybrid fiber-reinforced poly(lactic acid) biocomposites by borax decahydrate modification of fibers. *BioResources 11* (2016) 4865–4884. https://doi.org/10.15376/BIORES.11.2.4865-4884.
71. Harmaen, A. S., Khalina, A., Azowa, I., Hassan, M. A., Tarmian, A., & Jawaid, M. (2015). Thermal and biodegradation properties of poly(lactic acid)/fertilizer/oil palm fibers blends biocomposites. *Polymer Composites, 36*, 576–583. https://doi.org/10.1002/PC.22974
72. Mok, C.F., Ching, Y.C., Abu Osman, N.A., Muhamad, F., Mohd Junaidi, M.U., Choo, J.H. (2020). Preparation and characterization study on maleic acid cross-linked poly(vinyl alcohol)/chitin/nanocellulose composites, *Journal of Applied Polymer Science 137* (2020) 49044. https://doi.org/10.1002/APP.49044.
73. Hanan, F., Jawaid, M., Md Tahir, P. (2018). Mechanical performance of oil palm/kenaf fiber-reinforced epoxy-based bilayer hybrid composites. *17* (2018) 155–167.
74. Mechanical Characterization and Water Absorption Studies on Jute/Hemp Reinforced Hybrid Composites, (n.d.). http://article.sapub.org/10.5923.c.materials.201502.27.html (accessed March 30, 2022).
75. Verma, A., Parashar, A., & Packirisamy, M. (2018). Tailoring the failure morphology of 2D bicrystalline graphene oxide. *Journal of Applied Physics, 124*(1), 015102.
76. Verma, A., & Parashar, A. (2018). Reactive force field based atomistic simulations to study fracture toughness of bicrystalline graphene functionalised with oxide groups. *Diamond and Related Materials, 88*, 193–203.
77. Singla, V., Verma, A., & Parashar, A. (2018). A molecular dynamics based study to estimate the point defects formation energies in graphene containing STW defects. *Materials Research Express, 6*(1), 015606.
78. Verma, A., Parashar, A., & Packirisamy, M. (2019). Role of chemical adatoms in fracture mechanics of graphene nanolayer. *Materials Today: Proceedings, 11*, 920–924.
79. Chaudhary, A., Sharma, S., & Verma, A. (2022). Optimization of WEDM process parameters for machining of heat treated ASSAB'88 tool steel using response surface methodology (RSM). *Materials Today: Proceedings, 50*, 917–922.
80. Verma, A., Zhang, W., & Van Duin, A. C. (2021). ReaxFF reactive molecular dynamics simulations to study the interfacial dynamics between defective h-BN nanosheets and water nanodroplets. *Physical Chemistry Chemical Physics, 23*(18), 10822–10834.
81. Verma, A., Parashar, A., & Packirisamy, M. (2018). Atomistic modeling of graphene/hexagonal boron nitride polymer nanocomposites: A review. *Wiley Interdisciplinary Reviews: Computational Molecular Science, 8*(3), e1346.
82. Verma, A., Baurai, K., Sanjay, M. R., & Siengchin, S. (2020). Mechanical, microstructural, and thermal characterization insights of pyrolyzed carbon black from waste tires reinforced epoxy nanocomposites for coating application. *Polymer Composites, 41*(1), 338–349.
83. Verma, A., Budiyal, L., Sanjay, M. R., & Siengchin, S. (2019). Processing and characterization analysis of pyrolyzed oil rubber (from waste tires)-epoxy polymer blend composite for lightweight structures and coatings applications. *Polymer Engineering & Science, 59*(10), 2041–2051.
84. Verma, A., Negi, P., & Singh, V. K. (2019). Experimental analysis on carbon residuum transformed epoxy resin: Chicken feather fiber hybrid composite. *Polymer Composites, 40*(7), 2690–2699.
85. Verma, A., & Singh, V. K. (2018). Mechanical, microstructural and thermal characterization of epoxy-based human hair–reinforced composites. *Journal of Testing and Evaluation, 47*(2), 1193–1215.

86. Jain, N., Verma, A., Ogata, S., Sanjay, M.R. and Siengchin, S., (2022). Application of machine learning in determining the mechanical properties of materials. In *Machine Learning Applied to Composite Materials* (pp. 99–113). Springer.
87. Verma, A., Singh, V. K., Verma, S. K., & Sharma, A. (2016). Human hair: A biodegradable composite fiber–a review. *International Journal of Waste Resources, 6*(206), 2.
88. Kataria, A., Verma, A., Sanjay, M.R., Siengchin, S. and Jawaid, M. (2022). Physical, morphological, structural, thermal, and tensile properties of coir fibers. In *Coir Fiber and its Composites* (pp. 79–107). Woodhead Publishing.
89. Jain, N., Verma, A., & Singh, V. K. (2019). Dynamic mechanical analysis and creep-recovery behaviour of polyvinyl alcohol based cross-linked biocomposite reinforced with basalt fiber. *Materials Research Express, 6*(10), 105373.
90. Chaurasia, A., Verma, A., Parashar, A., & Mulik, R. S. (2019). Experimental and computational studies to analyze the effect of h-BN nanosheets on mechanical behavior of h-BN/polyethylene nanocomposites. *The Journal of Physical Chemistry C, 123*(32), 20059–20070.
91. Verma, A., Kumar, R., & Parashar, A. (2019). Enhanced thermal transport across a bi-crystalline graphene–polymer interface: An atomistic approach. *Physical Chemistry Chemical Physics, 21*(11), 6229–6237.
92. Verma, A., Negi, P., & Singh, V. K. (2018). Physical and thermal characterization of chicken feather fiber and crumb rubber reformed epoxy resin hybrid composite. *Advances in Civil Engineering Materials, 7*(1), 538–557.
93. Bharath, K. N., Madhu, P., Gowda, T. G., Verma, A., Sanjay, M. R., & Siengchin, S. (2020). A novel approach for development of printed circuit board from biofiber based composites. *Polymer Composites, 41*(11), 4550–4558.
94. Verma, A., Joshi, K., Gaur, A. and Singh, V.K. (2018). Starch-jute fiber hybrid biocomposite modified with an epoxy resin coating: fabrication and experimental characterization. *Journal of the Mechanical Behavior of Materials, 27*(5–6).
95. Verma, A., Singh, C., Singh, V. K., & Jain, N. (2019). Fabrication and characterization of chitosan-coated sisal fiber–Phytagel modified soy protein-based green composite. *Journal of Composite Materials, 53*(18), 2481–2504.
96. Rastogi, S., Verma, A., & Singh, V. K. (2020). Experimental response of nonwoven waste cellulose fabric–reinforced epoxy composites for high toughness and coating applications. *Materials Performance and Characterization, 9*(1), 151–172.
97. Verma, A., & Ogata, S. (2023). Magnesium based alloys for reinforcing biopolymer composites and coatings: A critical overview on biomedical materials. *Advanced Industrial and Engineering Polymer Research.* https://doi.org/10.1016/j.aiepr.2023.01.002
98. Kumar, G., Mishra, R.R., Verma, A. (2022). Introduction to Molecular Dynamics Simulations. In *Forcefields for Atomistic-Scale Simulations: Materials and Applications* (pp. 1–19). Springer. https://doi.org/10.1007/978-981-19-3092-8_1
99. Arpitha, G.R., Jain, N., Verma, A. and Madhusudhan, M. (2022). Corncob bio-waste and boron nitride particles reinforced epoxy-based composites for lightweight applications: fabrication and characterization. *Biomass Conversion and Biorefinery*, pp.1–8. https://doi.org/10.1007/s13399-022-03717-1
100. Kataria, A., Chaturvedi, S., Chaudhary, V., Verma, A., Rangappa Sanjay, N. J. M., Siengchin, S. (2023). Cellulose fiber-reinforced composites—History of evolution, chemistry, and structure. In *Cellulose Fibre Reinforced Composites* (pp. 1–22). Woodhead Publishing. https://doi.org/10.1016/B978-0-323-90125-3.00012-4
101. Chaturvedi, S., Kataria, A., Chaudhary, V., Verma, A., Rangappa Sanjay, N. J. M., Siengchin, S. (2023). Bionanocomposites reinforced with cellulose fibers and agro-industrial wastes. In *Cellulose Fibre Reinforced Composites* (pp. 1–22). Woodhead Publishing. https://doi.org/10.1016/B978-0-323-90125-3.00017-3
102. Arpitha, G. R., Mohit, H., Madhu, P., & Verma, A. (2023). Effect of sugarcane bagasse and alumina reinforcements on physical, mechanical, and thermal characteristics of epoxy composites using artificial neural networks and response surface methodology. *Biomass Conversion and Biorefinery.* https://doi.org/10.1007/s13399-023-03886-7

103. Chaudhary, A., Sharma, S., & Verma, A. (2022). WEDM machining of heat treated ASSAB'88 tool steel: A comprehensive experimental analysis. *Materials Today: Proceedings, 50*, 946–951.
104. Bisht, N., Verma, A., Chauhan, S., & Singh, V. K. (2021). Effect of functionalized silicon carbide nano-particles as additive in cross-linked PVA based composites for vibration damping application. *Journal of Vinyl and Additive Technology, 27*(4), 920–932.
105. Verma, A., Jain, N., Parashar, A., Singh, V.K., Sanjay, M.R., Siengchin, S. (2020). Design and modeling of lightweight polymer composite structures. In *Lightweight Polymer Composite Structures* (pp. 193–224). CRC Press.
106. Dogra, V., Kishore, C., Verma, A., Rana, A. K., & Gaur, A. (2021). Fabrication and experimental testing of hybrid composite material having biodegradable bagasse fiber in a modified epoxy resin: Evaluation of mechanical and morphological behavior. *Applied Science and Engineering Progress, 14*(4), 661–667.
107. Deji, R., Verma, A., Kaur, N., Choudhary, B. C., & Sharma, R. K. (2022). Density functional theory study of carbon monoxide adsorption on transition metal doped armchair graphene nanoribbon. *Materials Today: Proceedings, 54*, 771–776.
108. Verma, A., Jain, N., Parashar, A., Gaur, A., Sanjay, M.R., Siengchin, S. (2021). Life-cycle assessment of thermoplastic and thermosetting bamboo composites. In *Bamboo Fiber Composites* (pp. 235–246). Springer.
109. Deji, R., Verma, A., Choudhary, B. C., & Sharma, R. K. (2022). New insights into NO adsorption on alkali metal and transition metal doped graphene nanoribbon surface: A DFT approach. *Journal of Molecular Graphics and Modelling, 111*, 108109.
110. Verma, A., Jain, N., Parashar, A., Singh, V.K., Sanjay, M.R. and Siengchin, S., 2020. Lightweight Graphene Composite Materials. In *Lightweight Polymer Composite Structures* (pp. 1–20). CRC Press.
111. Verma, A., Parashar, A. (2020). Characterization of 2D nanomaterials for energy storage. In *Recent Advances in Theoretical, Applied, Computational and Experimental Mechanics* (pp. 221–226). Springer.
112. Deji, R., Jyoti, R., Verma, A., Choudhary, B. C., & Sharma, R. K. (2022). A theoretical study of HCN adsorption and width effect on co-doped armchair graphene nanoribbon. *Computational and Theoretical Chemistry, 1209*, 113592.
113. Verma, A., & Samant, S. S. (2016). Inspection of hydrodynamic lubrication in infinitely long journal bearing with oscillating journal velocity. *Journal of Applied Mechanical Engineering, 5*(3), 1–7.
114. Verma, A., Parashar, A., & Packirisamy, M. (2019). Effect of grain boundaries on the interfacial behaviour of graphene-polyethylene nanocomposite. *Applied Surface Science, 470*, 1085–1092.
115. Arpitha, G. R., Verma, A., Sanjay, M. R., & Siengchin, S. (2021). Preparation and experimental investigation on mechanical and tribological performance of hemp-glass fiber reinforced laminated composites for lightweight applications. *Advances in Civil Engineering Materials, 10*(1), 427–439.
116. Verma, A., Jain, N., Rangappa, S.M., Siengchin, S., Jawaid, M. (2021). Natural fibers based bio-phenolic composites. In *Phenolic Polymers Based Composite Materials* (pp. 153–168). Springer.
117. Verma, A., Parashar, A., Singh, S.K., Jain, N., Sanjay, S.M., Siengchin, S. (2020). Modeling and simulation in polymer coatings. In *Polymer Coatings* (pp. 309–324). CRC Press.
118. Verma, A., Singh, V.K. Experimental characterization of modified epoxy resin assorted with almond shell particles. *ESSENCE-International Journal for Environmental Rehabilitation and Conservation 7*(1), 36–44.
119. Deji, R., Verma, A., Kaur, N., Choudhary, B. C., & Sharma, R. K. (2022). Adsorption chemistry of co-doped graphene nanoribbon and its derivatives towards carbon based gases for gas sensing applications: Quantum DFT investigation. *Materials Science in Semiconductor Processing, 146*, 106670.

120. Verma, A. (2022). A perspective on the potential material candidate for railway sector applications: PVA based functionalized graphene reinforced composite. *Applied Science and Engineering Progress, 15*(2), 5727–5727.
121. Verma, A., Jain, N., Singh, K., Singh, V.K., Rangappa, S.M., Siengchin, S. (2022). PVA-based blends and composites. In *Biodegradable Polymers, Blends and Composites* (pp. 309–326). Woodhead Publishing.
122. Raja, S., Verma, A., Rangappa, S. M., & Siengchin, S. (2022). Development and experimental analysis of polymer based composite bipolar plate using Aquila Taguchi optimization: Design of experiments. *Polymer Composites, 43*(8), 5522–5533.
123. Prabhakaran, S., Sharma, S., Verma, A., Rangappa, S. M., & Siengchin, S. (2022). Mechanical, thermal, and acoustical studies on natural alternative material for partition walls: A novel experimental investigation. *Polymer Composites, 43*(7), 4711–4720.
124. Sethi, S. K., Gogoi, R., Verma, A., & Manik, G. (2022). How can the geometry of a rough surface affect its wettability? A coarse-grained simulation analysis. *Progress in Organic Coatings, 172*, 107062.
125. Verma, A., Jain, N., Sethi, S.K. (2022). Modeling and simulation of graphene-based composites. In *Innovations in Graphene-Based Polymer Composites* (pp. 167–198). Woodhead Publishing.
126. Chaturvedi, S., Verma, A., Sethi, S.K., Rangappa, S.M., Siengchin, S. (2022). Stalk fibers (rice, wheat, barley, etc.) composites and applications. In *Plant Fibers, their Composites, and Applications* (pp. 347–362). Woodhead Publishing.
127. Arpitha, G. R., Verma, A., & MR, S., Gorbatyuk, S., Khan, A., Sobahi, T.R., Asiri, A.M. and Siengchin, S. (2022). Bio-composite film from corn starch based vetiver cellulose. *Journal of Natural Fibers, 19*(16), 14634–14644.
128. Thimmaiah, S. H., Narayanappa, K., Thyavihalli Girijappa, Y., Gulihonenahali Rajakumara, A., Hemath, M., Thiagamani, S. M. K., & Verma, A. (2022). An artificial neural network and Taguchi prediction on wear characteristics of Kenaf-Kevlar fabric reinforced hybrid polyester composites. *Polymer Composites, 44*(1), 261–273.
129. Chaturvedi, S., Verma, A., Singh, S.K., Ogata, S. (2022). EAM inter-atomic potential—its implication on nickel, copper, and aluminum (and Their Alloys). In *Forcefields for Atomistic-Scale Simulations: Materials and Applications* (pp. 133–156). Springer.
130. Verma, A., Sharma, S. (2022). Atomistic simulations to study thermal effects and strain rate on mechanical and fracture properties of graphene like BC3. In *Forcefields for Atomistic-Scale Simulations: Materials and Applications* (pp. 237–252). Springer.
131. Chaturvedi, S., Verma, A., Sethi, S.K., Ogata, S. (2022). Defect energy calculations of nickel, copper and aluminium (and Their Alloys): Molecular dynamics approach. In *Forcefields for Atomistic-Scale Simulations: Materials and Applications* (pp. 157–186). Springer.
132. Verma, A., Gaur, A., & Singh, V. K. (2017). Mechanical properties and microstructure of starch and sisal fiber biocomposite modified with epoxy resin. *Materials Performance and Characterization, 6*(1), 500–520.
133. Shankar, U., Gogoi, R., Sethi, S.K., Verma, A. (2022). Introduction to materials studio software for the atomistic-scale simulations. In *Forcefields for Atomistic-Scale Simulations: Materials and Applications* (pp. 299–313). Springer.
134. Shankar, U., Sethi, S.K. and Verma, A. (2022). Forcefields and modeling of polymer coatings and nanocomposites. In *Forcefields for Atomistic-Scale Simulations: Materials and Applications* (pp. 81–98). Springer.
135. Verma, A., Ogata, S. (2022). Computational modelling of deformation and failure of bone at molecular scale. In *Forcefields for Atomistic-Scale Simulations: Materials and Applications* (pp. 253–268). Springer.
136. Homer, E.R., Verma, A., Britton, D., Johnson, O.K. and Thompson, G.B., 2022, July. Simulated migration behavior of metastable $\Sigma 3$ (11 8 5) incoherent twin grain boundaries. In *IOP Conference Series: Materials Science and Engineering* (Vol. 1249, No. 1, p. 012019).
137. Verma, A., Parashar, A., van Duin, A.C. (2022). Graphene-reinforced polymeric membranes for water desalination and gas separation/barrier applications. In *Innovations in Graphene-Based Polymer Composites* (pp. 133–165). Woodhead Publishing.

138. Verma, A., Jain, N., Sanjay, M.R., Siengchin, S. Viscoelastic properties of completely biodegradable polymer-based composites. In *Vibration and Damping Behavior of Biocomposites* (pp. 173–188). CRC Press.
139. Verma, A., Jain, N., Mishra, R.R. (2022). Applications and drawbacks of epoxy/natural fiber composites. In *Handbook of Epoxy/Fiber Composites* (pp. 1–15). Springer Singapore.
140. Lila, M.K., Verma, A. and Bhurat, S.S. (2022). Impact behaviors of epoxy/synthetic fiber composites. In *Handbook of Epoxy/Fiber Composites* (pp. 1–18). Springer Singapore.
141. Verma, A., & Parashar, A. (2017). The effect of STW defects on the mechanical properties and fracture toughness of pristine and hydrogenated graphene. *Physical Chemistry Chemical Physics, 19*(24), 16023–16037.
142. Verma, A., & Parashar, A. (2018). Molecular dynamics based simulations to study failure morphology of hydroxyl and epoxide functionalised graphene. *Computational Materials Science, 143*, 15–26.
143. Verma, A., & Parashar, A. (2018). Molecular dynamics based simulations to study the fracture strength of monolayer graphene oxide. *Nanotechnology, 29*(11), 115706.
144. Verma, A., & Parashar, A. (2018). Structural and chemical insights into thermal transport for strained functionalised graphene: A molecular dynamics study. *Materials Research Express, 5*(11), 115605.
145. Gautam, R., Gupta, H.S., Chauhan, A., Jaiswal, A.K., Verma, A. (2024). Composite materials for bio-energy. In *Hybrid Composite Materials: Experimental and Theoretical Analysis* (pp. 91–104). Springer Nature Singapore. https://doi.org/10.1007/978-981-97-2104-7_4
146. Kriti, Kumari, S., Verma, A. (2024). Synthesis and applications of different polymer composites. In *Hybrid Composite Materials: Experimental and Theoretical Analysis* (pp. 1–44). Springer Nature Singapore. https://doi.org/10.1007/978-981-97-2104-7_1
147. Sethi, S.K., Verma, A., Akubueze, E., Maurya, A.K. (2024). Self-healing of polymer composites: process and developments. In *Hybrid Composite Materials: Experimental and Theoretical Analysis* (pp. 253–271). Springer Nature Singapore. https://doi.org/10.1007/978-981-97-2104-7_10
148. Kafaltiya, S., Agrawal, P.K., Sharma, P., Singh, V.K., Bhaskar, J., Chauhan, S., Verma, A., Multi-response optimization of characteristics for graphite reinforced biodegradable PVA-fumaric acid cross-linked composite: A gray-fuzzy logic-based hybrid approach. *Journal of Vinyl and Additive Technology*. https://doi.org/10.1002/vnl.22112
149. Santos, T. F., Santos, C. M., Aquino, M. S., Suyambulingam, I., Kamil Hussein, E., Verma, A., Sanjay, M. R., Siengchin, S., & Nascimento, J. H. O. (2024). Towards sustainable and ecofriendly polymer composite materials from bast fibers: A systematic review. *Engineering Research Express, 6*, 012501. https://doi.org/10.1088/2631-8695/ad2640
150. Sharath, B.N., Madhu, P., Verma, A. (2023). Enhancing tribological performance: A review of ceramic reinforced aluminium hybrid composites for high-temperature engineering applications. Hybrid Advances p.100094.

Evolutionary Techniques in Optimization of Machining Parameters on Hybrid Metal Matrix Composites—A Review

Saurabh Kafaltiya, Pragya Sharma, Neha Belwal, V. K. Singh, Sakshi Chauhan, and Akarsh Verma

Abstract Metal Matrix Composites (MMC) are finding wide application across a range of manufacturing areas such as automotive, defense, and aerospace sectors due to their improved strength-to-weight ratio, superior quality, light weight, high strength, and an appreciable amount of hardness when combined with suitable reinforcement material. Non-Conventional machining has gained popularity in recent years to be the first choice for machining Hybrid MMCs. Faster tool wear, poor machinability, and inefficient production process in traditional machining have led to this shift towards unconventional machining methods. Today in this rapidly growing and aggressive industrial environment, manufacturer objectives are to bring out superior products with minimum cost and in a limited frame of time. Optimization of non-conventional machining process parameters such as pulse off time, peak current, pulse on time, and servo voltage will help us to achieve this ultimate goal. This current paper reviews the research on various hybrid MMCs that took the advantage of evolutionary optimization techniques to optimize non-conventional machining process variables and their impact on machining performance.

Keywords MMC · Non-conventional machining · Evolutionary optimization · Reinforcement · Process parameters

---

S. Kafaltiya (✉) · P. Sharma · V. K. Singh · S. Chauhan
Department of Mechanical Engineering, College of Technology, G. B. Pant, University of Agriculture and Technology, Pantnagar, India
e-mail: skafaltiya@gmail.com

N. Belwal
Department of Electronics & Communication Engineering, Graphic Era (Deemed to Be) University, Dehradun, India

A. Verma
Department of Mechanical Engineering, University of Petroleum and Energy Studies, Dehradun 248007, India

© The Author(s), under exclusive license to Springer Nature Singapore Pte Ltd. 2024

A. Chaudhary et al. (eds.), *Unraveling New Frontiers and Advances in Bioinformatics*,
https://doi.org/10.1007/978-981-97-7123-3_16

1 Introduction

The use of conventional metallic alloys, metals, ceramics, polymers, and monolithic materials is restricted due to their limited properties. There is a need to develop new and advanced materials for modern industrial as well as technical applications having good performance, high-reliability efficiency, and better strength-to-weight ratio to perform under extreme environmental conditions i.e., at high and low pressure and temperature, highly corrosive conditions like marine applications, etc. Many new materials have been developed to meet the need of such extreme conditions. It is possible to create novel material that precisely satisfies the needs of a given application by selecting the right mix of reinforcement and matrix material. Conventional machining has the problem of tool wear and does not provide a good surface finish. Non-conventional machining has come up a long way in generating a quality product. Electric discharge machining (EDM), wire cut electric discharge machining (WEDM), abrasive jet machining (AJM), and ultrasonic machining (UM) has been used successfully by researchers and scientist for machining of MMC.

Real-world situations frequently provide challenging optimization issues that are beyond the scope of current mathematical optimization techniques. Nature-inspired algorithms can be utilized to address these kinds of challenges if the user is not very aware of the precise solution to the issue at hand.

2 Composite Material

Composites are sophisticated materials consisting of two or more constituent materials with distinctly diverse chemical or physical properties that, when mixed, result in a material having unique properties from each component alone. Composites can be broadly classified into natural composite and hybrid composite. Hybrid composites have got ample flexibility in selecting a suitable matrix and reinforcement for achieving desired properties of the end composite.

2.1 Metal Matrix Composites

Composites having a metallic matrix are considered metal matrix composites (MMC). Different matrix materials both ferrous and non-ferrous alloys have been used by researchers and scientists over the years. Typically, MMC include matrix such as titanium alloys, aluminium alloys, and nickel–chromium alloys. These composites can be employed at a higher service temperature than that of their base metals.

2.1.1 Reinforcements

Every MMC's performance is usually determined by the combination of reinforcement materials chosen, which determines the composite's overall performance. The requirement of the industrial application must be taken into consideration while choosing the reinforcing material. Addition of reinforcement material results in the increase of strength, stiffness, and decreases in density simultaneously. Characterization of reinforcement is done based on dimension, volume fraction, property, and spatial distribution in the matrix [1].

3 Wire EDM Process Variables

Wire electric discharge machining (WEDM) is an advanced form of electric discharge machining (EDM). In this method, a metallic wire of a very small diameter is fed to the workpiece, which is submerged into a reservoir of dielectric medium i.e. deionized water or kerosene. Water is normally used as the dielectric fluid. The variables which affect the quality of machining of the WEDM process are:

- Non-electrical parameters like machining time, rate of flushing, wire feed, and work feed rate.
- Electrical parameters like pulse on time, pulse off time, peak current, and servo voltage.
- Electrode parameters like the size of wire and material.
- Dielectric system i.e. flow characteristics and viscosity.

Figure 1 displays the Wire EDM schematic diagram.

4 Evolutionary Algorithms (EAs)

Nature-Inspired Algorithm, or Evolutionary Algorithms are population-based stochastic metaheuristics that mimic natural occurrences to identify the best possible solution to a problem. These algorithms primary concept is that people update themselves by using:

- An understanding of the surroundings and their fitness value.
- The individual's memory—their prior state history.
- The past conditions of the individual's neighborhood.

Figure 2 shows the classification of computation intelligence.

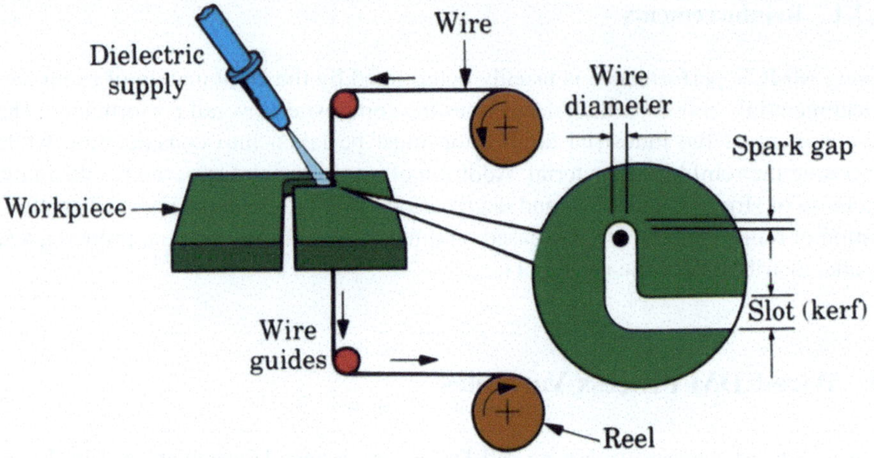

Fig. 1 Schematic diagram of WEDM

Fig. 2 Classification of Computation Intelligence

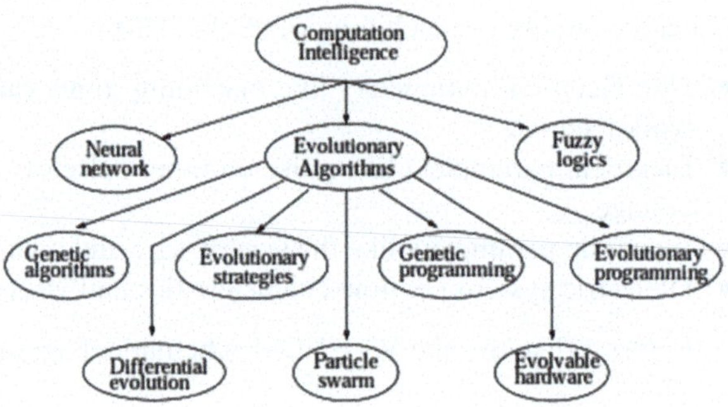

4.1 Artificial Bee Colony Algorithm (ABC)

D. Karaboga developed the Artificial Bee Colony algorithm (ABC) in 2005. It is an optimization algorithm based on swarm intelligence.

Three crucial elements make up the minimum model of food selection that causes honey bee swarms to develop collective intelligence:

(a) Food sources,
(b) Employed foragers
(c) Unemployed foragers
- Scouts
- Onlookers

Food Sources: A food source's worth is determined by its closeness to the nest, richness, and simplicity of nectar extraction.

Employed foragers: They are linked to a certain food source that they either work at or are currently taking advantage of. They share information about this specific source, including its direction and distance from the nest, with a given likelihood.

Unemployed foragers: They are always searching for a food source to take advantage of. Unemployed foragers fall into two categories:

1. **Scouts:** looking for fresh food sources in the area around the nest and
2. **Onlookers:** waiting in the nest and using the knowledge that employed foragers share to create a food source.

4.2 Spider Monkey Optimization Algorithm

Spider monkeys' remarkable foraging techniques served as the model for the development of the Spider Monkey Optimization (SMO) algorithm.

Spider monkeys are classified as animals with a fission–fusion social structure [2]. Based on food scarcity or availability, animals that adhere to fission–fusion social systems divide themselves into larger and smaller groups and vice versa.

4.2.1 Fission—Fusion Social Structure

1. To adjust to the social or environmental conditions, the main parent group in a fission–fusion civilization may split into smaller subdivisions or individuals. For instance, a group may split off from the main group during the day to hunt or gather food, but at night they reunite to share food and engage in other activities. This is known as group fusion.
2. When food is scarce for seasonal reasons, group members compete with one another for food, which is the primary cause of the emergence of the fission–fusion social order.

4.3 Shuffled Frog Leaping Algorithm (SFLA)

Developed in 2003 by Lansey and Eusuff.

Inspired by natural memetics, SFLA is a population-based cooperative search metaphor and memetic meta-heuristic.

The foraging activity of frogs, which mimics infectious information patterns with the social and natural behavior of species, serves as an inspiration for SFLA.

Both global information exchange and local search are incorporated into the algorithm.

Utilizing a technique similar to particle swarm optimization and tailored for discrete situations, the local search is finished locally.

The virtual frogs are regularly shuffled and rearranged into new memplexes in a manner akin to the shuffled complicated evolution algorithm in order to guarantee global exploration.

Furthermore, random virtual frogs are generated and substituted in the population to allow for the random production of improved knowledge.

4.4 Genetic Algorithm (GA)

GA is a nature inspired algorithm (but not copied), although based on natural evolution + genetics. It is a guided search procedure in which offspring are created by crossover, mutation etc. (Natural evolution) and good solutions are retained and bad are deleted (survival of the fittest) and lastly information is being coded (genetics). GA being population based algorithms can be used in optimization, learning and design.

4.4.1 Concept of Dominant and Non-Dominant Solutions

Solution $x^{(1)}$ dominates solution $x^{(2)}$, if.

- $x^{(1)}$ is no worse than $x^{(2)}$ in all objectives.
- $x^{(1)}$ is strictly better than $x^{(2)}$ in at least one objective.

Figure 3 shows the concept of dominant solutions. Example: 3 dominates 2 but 3 does not dominate 5.

Solutions which are not dominated by any member of given set of solutions are non-dominated solutions. Solutions 3, 5 and 6 are non-dominated solutions. Figure 4 shows the concept of non-dominant solutions.

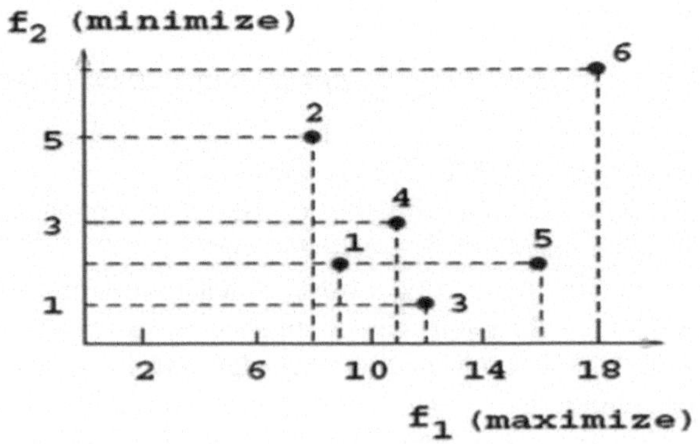

Fig. 3 Concept of dominant solutions

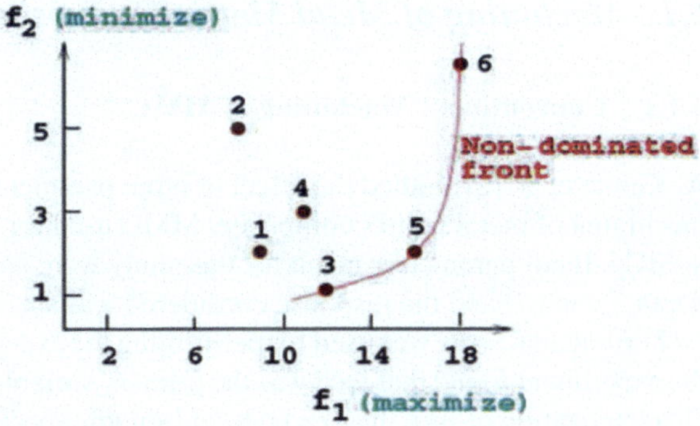

Fig. 4 Concept of non-dominant solutions

4.4.2 Pareto Optimal (PO) Solutions

- P = Non-dominated (P) i.e. solutions which are not dominated by any member of the set P.
- Pareto-Optimal set = Non-dominated(S) ∈ Ω (entire feasible search space).
- A number of solutions are optimal.

Figure 5 shows the obtained feasible region of Pareto optimal solutions.

5 Literature Review

A detailed literature review has been carried out considering the machining investigation of hybrid MMCs, methods used, and their responses, which lead to future research prospects.

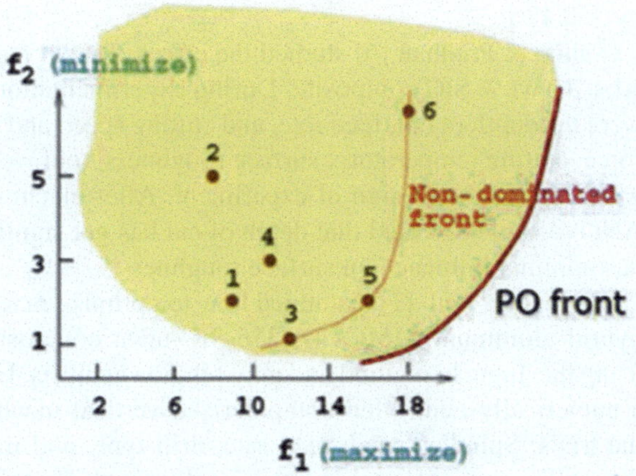

Fig. 5 Feasible Pareto optimal solutions

5.1 *Machining of Metal Matrix Composites (MMC)*

5.1.1 Conventional Machining of MMC

A. Kumar et al. [2] studied the effect of input parameters on output responses during machining of metal matrix composite. MMC used for this investigation was Al2024 + SiCp. Input parameters taken for this study were spindle speed (*S*), feed rate (f), depth of cut (d) and the response considered was surface roughness (*Ra*). Taguchi's L27 orthogonal array was used for performing the experiments. Results obtained after the experiment found that feed was the primary variable affecting the Ra. Mechanical characterization proved that feed rate and spindle speed are the most dominant factor.

Shoba et al. [3] studied parameters affecting the minimization of surface roughness with turning of Al + 6 wt.% SiC + 6 wt.% rice husk ash (RHA) hybrid composite. For this purpose, the Taguchi DoE was used for making the L9 orthogonal array under different experimental conditions. S/N ratio was taken into consideration to study the impact of process variables and the most influencing variable for surface roughness was obtained. Validation of this study was carried out using analysis of variance and the percentage contribution of each parameter was obtained.

Karabulut et al. [4] studied composite based on aluminium and boron carbide i.e. Al6061 + B4C machined using milling operation. The variables taken were speed, feed, depth of cut and corresponding output responses were surface roughness and power consumption. Result obtained found excellent surface quality when high milling speed and lower cutting feed under dry condition was used.

Camposeco-Negrete [5] studied the optimization of cutting parameters like feed rate and depth cut during rough turning of AISI6061 T6 aluminium. Responses considered for the study were energy consumption, surface roughness, and material removal rate. Response surface methodology was used for modeling and optimization. Results obtained found that feed rate and cutting forces are the most influencing parameters for minimizing surface roughness and specific energy. This method of optimization reduced the energy consumption by 14.41% and the surface roughness by 360.47%.

Sahoo & Pradhan [6] studied the effect of input parameters during machining of Al + 10 wt.% SiC composite. During experimentation input parameters considered were the depth of cut, feed rate, and cutting speed and output responses were cutting force, cutting temperature, surface roughness, tool wear, and tool life. Taguchi's L9 was used for the Design of experiment. After machining results obtained from the ANOVA table revealed that depth of cut has got significant effect and cutting speed has minimal influence on surface roughness.

Rajmohan et al. [7] examined how to optimize machining parameters for drilling hybrid aluminum MMCs (Al356-SiC-mica composites). Optimization was done using the Taguchi method and grey relation analysis. Using an L18 orthogonal array, a numerically controlled computerized vertical machining center was utilized for the trials. Spindle speed, feed rate, drill type, and mass percentage of mica were the drilling variables taken into consideration for this investigation depending on the performance properties of burr height, surface roughness, thrust force, and tool wear. The feed rate and drill type were identified to be the most important parameters influencing the drilling process based on the collected results.

Siddesh Kumar et al. [8] investigated aluminium-based composite to find the process parameters like cutting speed and cutting force affecting surface roughness by keeping the depth of cut constant. Composites were fabricated using the stir casting technique. Turning operation on prepared composite i.e. Al2219, Al2219 + 3 wt.% B4C + 3 wt.% MoS2 and Al2219 + 3 wt.% B4C + 5 wt.% MoS2 was performed on an ACE CNC lathe machine. During machining, TiN-coated inserts were used. After machining, surface roughness was measured on a Mitutoyo surface roughness tester, and results obtained were found that with increasing cutting speed and cutting force, surface roughness was decreasing for TiN coated carbide inserts.

Ravindranath et al. [9] studied the effect of machining parameters like speed and feed during the drilling operation of composites i.e. Al2219, A12219 + 8 wt.% B4C and hybrid composite Al2219 + 8 wt.% B4C + 3 wt.% Gr on thrust force and surface roughness. These composites were prepared through a liquid metallurgy process, the prepared composite was machined using an HSS drill bit coated with TiN on varying speeds of 1000, 1500, 2000 rpm and feeds of 0.1, 0.2, 0.3 mm/rev. The study confirmed that surface roughness and thrust force increased with increasing feed rate.

The role of graphene in improving the tribological characteristics of Al 2024-SiC-graphene reinforced hybrid composites was investigated by Kumar & Xavior [10]. 4, 8, and 12 weight % of SiC and 0.5 weight % of graphene were used in the experiment to look at the lubricating characteristic under dry wear circumstances. The composites were subjected to ball milling and ultrasonication. Utilizing a pin-on-disc tribometer, wear testing was done during the study to assess the impact of graphene and SiC percentage in the composite under varied normal load and disc speed circumstances. Following analysis, the findings showed that the addition of graphene and SiC to the nanocomposites significantly improved their tribological characteristics.

Sivasankaran et al. [11] investigated how the inclusion of graphite affected the surface roughness when AA7075-ZrB2 in-situ MMCs were being turned. Employing an in situ wet metallurgical process, the composite of AA7075 alloy, AA7075 + 3 wt.% ZrB2, and AA7075 + 3 wt.% ZrB2 + 1 wt.% Gr were produced. Following the utilization of the TiN-coated carbide-tipped tool, the input variables for the CNC high-speed turning machine were cutting speed (v), feed rate (f), and depth of cut (d). The study focused on investigating the surface roughness, flexural strength, and hardness. Comparing the AA7075 + 3 wt.% ZrB2 and AA7075 alloy to other AA7075 + 3 wt.% ZrB2 + 1 wt.% Gr hybrid in-situ composites, an excellent surface finish was achieved for the latter. This is because Gr has machinability-enhancing, self-lubricating qualities.

By adopting the DoE approach, Razavykia et al. [12] investigated cutting force and surface roughness in the dry turning of Al–Mg_2Si in-situ MMC injected with bismuth. Here, Mg_2Si is in the form of particles and exhibits superior qualities. The machinability characteristics of these composites, particularly those involving the inclusion of bismuth, were examined to assess machining parameters such as feed rate (f), cutting speed (v), and modifier bismuth to maximize cutting force and surface roughness. The statistical observation showed that the cutting force and surface roughness were primarily controlled by the cutting speed, feed rate, and modifier element. Bi, as a modifier reagent, changes the shape of the Mg_2Si reinforcement particulate and produces a Bi compound, which reduces cutting force and improves surface roughness. It has been determined that, when Bi is present, the best cutting parameters for Al + 20 weight % Mg_2Si composite machining are 210 m/min of cutting speed and 0.1 mm/rev of feed rate. The cutting force and surface roughness are impacted by every major effect of the machining parameters, as indicated by the ANOVA study.

Sai Chaitanya Kishore et al. [13] used the Taguchi technique to investigate the optimization of machining variables for enhancing cutting force and surface roughness in the turning of Al6061-TiC in-situ MMC. Using a Taguchi L27 orthogonal array, process variables such as cutting speed (v), feed rate (f), and depth of cut (d) were examined to maximize cutting force and surface roughness. Using the ANOVA technique, the contribution of each variable was examined. It was shown that at greater cutting speeds, lower feed, and a lower depth of cut, the cutting force and surface roughness are low. Additionally, the ANOVA revealed that the depth of cut has a significant impact on the cutting force. When it comes to surface roughness, feed has a significant influence followed by cutting speed and depth of cut.

The surface morphology of Al-SiC particle MMC in hot machining was investigated by Dabade & Jadhav [14]. Stir casting produces Al-SiC MMCs. Experimental trials were conducted on a CNC lathe using PVD-coated carbide inserts under a dry-cutting environment. The L16 orthogonal array is used in Taguchi method experiments on Al-SiC 10 wt.% composites with two separate silicon carbide reinforcement particle sizes: coarse and fine (220 and 600 mesh size, respectively). The experimental investigation demonstrated that cutting force, feed force, thrust force, surface roughness, and micro-hardness fluctuation on the machined surface are significantly impacted by feed, depth of cut, and preheating temperature.

The impact of turning variables on tool wear, surface roughness, and metal removal rate of alumina-reinforced MMC was investigated by Bansal & Upadhyay [15]. Mechanical characteristics were investigated on three sand-cast samples containing 2%, 4%, and 6% of alumina by weight. Mechanical characteristics are improved over pure aluminum, according to experiments. Carbide tools with and without coatings were employed. Input variables considered for the investigation were feed rate, speed, depth of cut, and alumina %. Surface roughness increases with an increase in all input variables except speed. Tool wear increases with the increase of all input variables. MRR increases with the increase of all input variables except the reinforcement ratio. Hardness and Tensile strength increase with the reinforcement ratio.

Nestler & Schubert [16] studied surface properties in Ultrasonic Vibration assisted turning of particle-reinforced Aluminium Matrix Composites. Under powder metallurgy, a high-energy mixing procedure is employed to manufacture this composite. The process of hot isostatic pressing compacts the powder. AA2124 with a 25% volume fraction of SiC particles was used in the experiment. Micro-structured surface in finish cutting was produced by incorporating ultrasonic vibration. A SPINNER precision lathe was used for the external turning trials. The outcomes of the tests indicate that the use of ultrasonic vibration assistance has a great deal of promise for creating microstructures with specific geometries.

Varol & Canakci [17] studied with Al2024 + B4C particulate. The study and optimization of input parameters included B4C weight percentage, milling speed, and milling time. Because of the embedding effect, the fine B4C particle expedited the milling process. The size and composition of the reinforcement were shown to have a significant impact on the composite powder's particle size and microhardness.

Abdollahi et al. [18] studied hardness and wear resistance with input material Al2024 + 5 wt.% B4C and input parameter milling time and milling speed were considered. Al2024's ductility is reduced and its strength, hardness, and wear resistance are increased by the addition of B4C particles.

Response surface methodology (RSM) was applied in a study by Joardar et al. [19] to estimate the cutting force required to turn LM6-SiC MMC. Cutting speed (v), depth of cut (d), and weight percentage of SiC were the input variables. An examination was conducted to determine how the cutting variables affected the tangential, axial, and radial cutting forces of aluminum metal matrix composites made by stir casting. At varying cutting speeds, feed rates, and depths of cut, two distinct K10 cutting tools—uncoated and TiN-coated—were employed. Tool wear was mostly influenced by cutting speed in dry turning conditions, and it increased as cutting speed rose. While for a coated cutting tool, there was less tool wear observed. The outcome demonstrates how the RSM concepts are used to build the functional link between the cutting forces and the cutting variables. For tangential, radial, and axial forces, a quadratic model is fitted. The mathematical models of the cutting forces fit and forecast values of the responses that are near to those readings observed experimentally with a 95% confidence interval, as demonstrated by ANOVA and the performance of confirmation studies. Ultimately, sensitivity analysis reveals that the response variable's most important element was cutting speed.

Bhushan [20] studied the Al7075-SiC particulate. The input parameter was cutting speed, feed rate, depth of cut, and nose radius and output parameters were power consumption and tool life. Optimizing the input parameters reduced power consumption by 30.55% and increase tool life by 22.12%.

Aramesh et al. [21] used the real-time wear value of the worn tool as the input data to study the remaining lifespan of worn cutting inserts under various cutting variables during the turning of Ti-MMCs. The model's covariates include feed, cutting speed rate, and tool wear. This model has the benefit of accounting for both the influence of aging and cutting factors, making it useful for estimating tool life. It would be possible to estimate the tool's remaining lifespan under any intended cutting situation by assessing the extent of tool wear of the worn tool.

The optimization of machining variables on turning of AA6063-T6 aluminum alloy with numerous responses based on the orthogonal array coupled with grey relation analysis was researched by Jayaraman & Mahesh Kumar [22]. Uncoated carbide inserts are employed in dry-cutting environments to perform turning tests. Cutting parameters including depth of cut, feed rate, and speed are optimized while taking into account numerous responses like material removal rate (MRR), roundness (Ø), and surface roughness (Ra). The most significant variable is the feed rate (57.365%), which is followed by the depth of cut (25.11%) and cutting speed (17.35%). The best performance was obtained with a lower cutting speed (119.22 m/min), lower feed rate (0.05 mm/rev), and medium depth of cut of 0.15 mm.

5.1.2 Non-Conventional Machining of MMC

Several non-conventional machining techniques for machining MMC, including electro discharge, laser beam, abrasive water jet, and electrochemical, were examined by Pramanik [23]. For each machining operation, responses such as surface roughness, material removal rate, and machining mechanism have been considered. The study revealed that the primary mechanisms for material removal during non-conventional machining include vaporization, melting, chemical dissolution, and mechanical erosion.

Dey et al. [24] investigated the effect of various machining parameters like (pulse on time, pulse off time, and wire feed) on performance measures such as kerf width, cutting speed, and surface roughness during WEDM machining of composite material i.e. AA6061/cenosphere aluminium matrix composite (AMC). ANOVA was performed to examine the contribution of each process variable. To evaluate performance characteristics, the impact of important machining variables was analyzed using the response surface methodology. The result of the analysis showed pulse on time and percentage of reinforcement were the most significant factors affecting responses.

Gore & Patil [25] studied non-traditional and traditional machining of materials like metal matrix composite (MMC), and ceramics composite which has gotten wider application in the field of aerospace, defense, etc. MMC is advanced material having properties of high strength to weight ratio, low density, low thermal expansion, and good wear resistance. Results obtained found that non-conventional machining (like WEDM) shows a better ability to machine MMC with higher precision and conventional machining results in poor surface finish and higher tool wear.

Mardi et al. [26] studied various non-traditional machining techniques used for the machining of complex shapes which are not possible with traditional machining. The non-traditional technique uses energy in direct form (mechanical, thermo-electrical, etc.) for the removal of material. This investigation includes various non-traditional techniques like laser beam, abrasive water jet, electro-chemical machining (ECM), and electrochemical discharge machining (ECDM). The study found that thermal energy-induced processes have more sub-surface damage than other non-traditional machining processes like ECM and ECDM.

The impact of wire EDM factors on aluminum alloy was investigated by Rao et al. [27]. The Taguchi technique was used to do a parametric analysis of WEDM factors on surface roughness and material removal rate. Using established linear regression models, a hybrid genetic algorithm simultaneously optimized the performance metrics. The results demonstrated a strong agreement with the experimental values. Furthermore, white layer evaluations were also done for the recommended combination of factors.

Mahanta et al. [28] examined maximizing the input machining variables of the EDM technique when cutting metal matrix composites (AMCs) based on aluminum with reinforcement elements SiC and B4C. Here, the key process variables that are used to examine their effects on the surface roughness (Ra) and power consumption (P) of the EDM process machining are servo voltage (SV), pulse current (Ip), pulse duty factor (s), and pulse-on-time (Ton). GA serves as an aid for multi-objective optimization. For varying SR and P values, the GA-optimized Pareto optimal fronts offer distinct optimal cutting settings.

The adaptive neuro-fuzzy inference system (ANFIS) model was examined by Çaydaş et al. [29] for the estimation of average surface roughness and white layer thickness (WLT) as a function of process variables. The model's input characteristics were determined to be pulse duration, open circuit voltage, dielectric flushing pressure, and wire feed rate. By merging the fuzzy inference system with the learning ability of an artificial neural network, a set of rules was directly generated from the experimental data.

Using the artificial bee colony (ABC) method, Das et al. [30] investigated the combination of process variables for optimum response conditions in EDM machining of EN31 tool steel. In order to conduct experiments, the central composite design (CCD) was used to vary the machining parameters, such as pulse on time, pulse off time, discharge current, and voltage. Response surface methodology (RSM) determines second-degree modeling equations for surface roughness and MRR. Single and multi-objective optimization is performed for the output responses. The best combinations of process variables are identified by ABC analysis, and the associated values of output responses are discovered. In the experimental regime, it is observed that MRR and surface roughness were directly proportional to current and pulse on time.

The impact of process parameters on output responses such as material removal rate (MRR) and surface roughness (Ra) were investigated by Kumar et al. [31]. Investigations have been conducted further on the parametric optimization of the MRR and Ra. To validate the relationship between input variables and responses, such as MRR and Ra, a mathematical nonlinear regression analysis have been employed. This system model makes use of the Simulated Annealing (SA) technique to minimize Ra and maximize MRR.

Dewangan et al. [32] looked into how input process variables affected output parameters such as surface roughness (Ra) and material removal rate (MRR). Additionally, parametric optimization of Ra and MRR was performed. Using mathematical nonlinear regression analysis, the relationship between input factors and responses, such as MRR and Ra, has been validated. The Simulated Annealing (SA) method is used in this system model for the conflicting objectives of Ra and MRR.

Vikas et al. [33] used the Grey-Taguchi approach to investigate the impact and optimization of several machining process variables on the surface roughness in EDM for an EN41 material. The process variables taken for the study were voltage across the surface roughness, discharge current, pulse-on time, and pulse-off time. Five distinct output responses—Ra, Rq, Rsk, Rku, and Rsm—that are related to surface roughness are taken into consideration and optimized using the Grey-Taguchi approach. The surface roughness parameter was shown to be more affected by the discharge current. A 95% accuracy level was maintained when comparing the experimental outcomes with the projected results.

Using the robustly constructed Taguchi technique, Goswami & Kumar [34] investigated the optimization of the Nimonic-80A in wire EDM. Surface roughness (SR), material removal rate (MRR), and surface morphology of the machined surface were studied. This work suggests that Nimonic-80A offers fair machinability towards the WEDM Process.

The impact of input variables, specifically pulse on time, pulse off time, peak current, and servo voltage, on cutting rate was examined by Sharma et al. [35]. Here, Al6063-ZrSiO4(p) is the MMC. The experiments are designed using the Box-Behnken approach followed by modeling via response surface methodology (RSM). The statistical significance of input variables is determined by executing the F-test and Analysis of Variance (ANOVA). Experimental research revealed that the cutting rate was directly proportional to the pulse on time and peak current, while inversely proportional to the pulse off time and servo voltage. When pulse on time was increased to a high level and pulse off time was retained at a lesser level, the surface topography of the machined surface revealed that big size craters and cracks were created on the surface.

The temperature variation of the EDM process of Al2618 alloys reinforced with Si3N4, ALN, and Zrb2 in-situ composites was investigated by Kumar et al. [36]. Using the stir casting procedure, he combined Al2618 alloy in different weight percentages (wt.% = 0, 2, 4, 6, 8). It was observed that the mechanical characteristics such as hardness, tensile strength, and compressive strength rose as the weight percentage of reinforcement increased. The input variables such as temperature (ºC), pulse on time (μs), pulse off time (μs), and current (amps) were optimized to minimize the Tool Wear Rate (TWR), Material Removal Rate (MRR), and depth (mm). DoE was done using Taguchi L27 orthogonal array followed by ANOVA method for analysis. The lowest Tool Wear Rate (TWR), Material Removal Rate (MRR), and Depth were discovered to be associated with increasing weight percentages of Al2618 composite. The results showed that the MRR was 0.884 gm/min, the TWR was 0.027 gm/min, and the depth was 3.705 mm.

Sivaprakasam et al. [37] studied Al2024 T351 with input parameter voltage capacitance and feed rate and the main output parameter was surface roughness and metal removal rate. Experiments were carried out using Central Composite Design (CCD). Response Surface Methodology (RSM) and Artificial Neural Network (ANN) models have been developed. ANOVA analysis reveals the importance of all three input process parameters. Capacitance is the most influencing factor for *SR* while capacitance and feed rate are the most significant factor for *MRR*. The fine surface finish of 1.07 m and *MRR* of 0.02154 mm^3/min are achieved using optimum machining conditions of voltage 100 V, capacitance 0.01F, and feed rate of 15 m/sec.

The material removal process in Laser Assisted Machining of Al/SiCp composite was researched by Zhang et al. [38]. A parametric equation that projected the shear zone stress was presented based on the deformation mechanisms related to LAM, and systematic experimental research was used to determine the impacts of operating circumstances on the shear zone stress. When the experimental data were compared with the projections made by the parametric equation, the calculated error of 15% was observed in the stress measurement.

Bhuyan et al. [39] examined Al + 24 wt.% SiC with primary output responses as surface roughness, tool wear rate, and metal removal rate, and input factors including flushing pressure, peak current, and pulse on time. Multi-objective optimization using the TOPSIS method was employed to figure out the ideal machining parameters for Al + 24 wt.% SiC MMC during the EDM. The trials are set up to determine the most favorable and least favorable machining properties based on the closeness coefficient results. Therefore, every machining process can benefit from the optimization of the machining parameters through the use of this multi-response TOPSIS approach.

The impact of wire electrode material type and reinforcement volume percentage on wire EDM performance of A359-SiCp composite was investigated by Patil et al. [40]. As an electrode, they used a copper wire plated with CuZn50 and a plain brass wire. SiC ranges in volume proportion from 10 to 30%. The volume proportion of SiC was the most significant variable for cutting rate followed by the type of electrode material. They discovered that coated wire had a higher cutting rate than plain brass wire with a 58% improvement in cutting rate. It was also discovered that coated wire increased surface roughness.

Using the Taguchi Method, Dixit & Kumar [41] optimized the mechanical characteristics of silica gel-reinforced aluminum MMC. Six distinct proportions of filler were used to fabricate aluminum composites reinforced with silica gel. The tensile and flexural strengths are estimated at three different speeds for the composite. The studies were carried out using Taguchi's L18 orthogonal array, accounting for two design parameters: filler proportion and speed. They concluded that the proportion of filler material has the greatest influence on both tensile and flexural strength. The improvement in the S/N Ratio was found to be 1.97637 dB for tensile strength and 3.94159 dB for flexural strength.

A non-dominated solution set was produced in the Senthilkumar et al. [42] study to optimize the parameters of the ECM process utilizing a non-dominated sorting genetic algorithm (NSGA-II). For optimization, the second-degree mathematical models were generated for MRR and Ra. The electrolyte flow rate (17 gm/l), applied voltage (16 V), and feed rate (0.9 mm/min) are the optimal values. By lowering the machining time, a significant boost was obtained in the production rate. A multi-objective optimization technique was used to optimize the output parameters, such as MRR and Ra.

5.2 Modeling and Optimization of Machining Parameters of MMC

WEDM machining of AISI D3 steel was investigated by Lodhi & Agarwal [43]. They worked on refining the machining conditions to optimize the surface roughness. DoE was done using Taguchi's L9 orthogonal array. Input variables used were peak current, wire feed, pulse-on-time, and pulse-off-time. Taguchi signal-to-noise (S/N) ratio and ANOVA was used for analysis. Discharge current was found to be a significant contributor towards surface roughness.

A mathematical model for surface roughness (Ra) and material removal rate (MRR) during WEDM machining of aluminum oxide-based ceramic material (Al_2O_3 + TiC) was examined by Kung & Chiang [44]. DoE was done using a face-centered central composite design (CCD). The experimental variables were modeled through response surface methodology (RSM). Input variables considered for the study were peak current, pulse on time, duty factor, and wire speed. It was found that the suggested mathematical model had a 95% accuracy.

The prediction and analysis of process variables on squeeze cast A413 + B4C composite was investigated by Soundararajan et al. [45]. The output responses were generated based on feed-forward back propagation and modeled using an artificial neural network. DoE was done taking advantage of full factorial design (FFD). Additionally, statistical and quantitative research was conducted to assess how process parameters affected the composites.

Priyadarshini et al. [46] investigated grey relation optimization combined with the fuzzy methodology to determine optimum process parameters to minimize heat affected zone (*HAZ*) and circularity of hole and material removal rate. The results obtained from experiments (L27 orthogonal array) were used in grey-fuzzy methodology for modeling and further investigation found optimum quality parameters for the responses.

Chalisgaonkar & Kumar [47] investigated the modeling of multi-response optimization with the traditional utility method in conjunction with the weight assignment concept for trim cut using WEDM. Multi-objective optimization is carried out to satisfy multiple users at a time. In contrast to the traditional optimization technique which is realized without considering the priorities of different users. The developed model helped in predicting responses over a range of input parameters.

To predict the machining performance of Al-Si-Cu-Fe cast alloy treated with various additives like bismuth, antimony, and strontium for refinement of surface roughness, Marani Barzani et al. [48] used the fuzzy logic artificial intelligence technique. For determination of optimum machining parameter, an experiment was carried out on CNC turning and variables taken were speed, feed rate, and depth of cut, and response considered was surface roughness value. For confirmation of modeling, an experiment was performed on the developed model. Results obtained showed predicted model has a surface roughness error rate of 5.4%.

Çaydaş et al. [49] investigated the neuro-fuzzy inference system (ANFIS) model, which is used to anticipate average surface roughness and white layer thickness (WLT). Open circuit voltage, pulse duration, wire feed rate, and dielectric flushing were the input variables used in the modeling process. This hybrid model of artificial neural network and fuzzy was created, and a set of rules was based on experiment knowledge. The method was validated by comparing the model's predictions with the outcomes of experiments.

Acilar & Arslan [50] studied the idea proposed for determining membership function and adjusting its shape. Implementation of this idea was developed using the CLO-NALG program for multiple input–output fuzzy systems. Binary particle swarm optimization (PSO) and genetic algorithm (GA) are used for implementing the proposed theory, and they were compared. The result obtained found that the clonel selection algorithm is advantageous in finding out the optimum value of the fuzzy membership function [51].

6 Conclusions

Based on various literature surveys on conventional and non-conventional machining and their parametric optimization following conclusion has been drawn:

- Hybrid MMCs are advanced materials offering improved performance than the individual metal itself.
- The performance of the MMCs can be improved by varying the reinforcement content.
- Many MMCs are investigated using conventional machining techniques, but little research is carried out using the non-conventional technique.
- Non-conventional machining methods seem to be more economical and efficient for machining MMCs.
- Pulse on time, pulse off time, peak current, and servo voltage were considered as the variables while using WEDM for machining of composites.
- Artificial neural network (ANN), response surface methodology (RSM), and fuzzy logic are the widely used modeling techniques by researchers for experimentation.
- Soft computing techniques for optimization are still not employed extensively, therefore it will be an emerging area for multi-objective optimization.

7 Funding Details

The author(s) received no financial support for the research, authorship, and/or publication of this article.

Compliance with Ethical Standards

Conflict of Interest The authors declare that they have no conflict of interest.

References

1. Annigeri, U. K., & Veeresh Kumar, G. B. (2017). Method of stir casting of aluminum metal matrix composites: A review. *Materials Today: Proceedings, 4*(2), 1140–1146. https://doi.org/10.1016/j.matpr.2017.01.130
2. Kumar, A., Kumar, S., Garg, R. (2015). Optimization of machining parameters in end milling of Al 2024-SiC p metal matrix composite using Taguchi method for surface roughness. *WALIA Journal, 31*(1), 6–16. www.Waliaj.com
3. Shoba, C., Ramanaiah, N., & Rao, D. N. (2015). Optimizing the machining parameters for minimum surface roughness in turning Al/6%SiC/6%RHA hybrid composites. *Procedia Materials Science, 10*, 220–229. https://doi.org/10.1016/j.mspro.2015.06.044
4. Karabulut, Ş, Karakoç, H., & Çıtak, R. (2016). Influence of B4C particle reinforcement on mechanical and machining properties of Al6061/B4C composites. *Composites Part B: Engineering, 101*, 87–98. https://doi.org/10.1016/j.compositesb.2016.07.006
5. Camposeco-Negrete, C. (2013). Optimization of cutting parameters for minimizing energy consumption in turning of AISI 6061 T6 using Taguchi methodology and ANOVA. *Journal of Cleaner Production, 53*, 195–203. https://doi.org/10.1016/j.jclepro.2013.03.049
6. Sahoo, A. K., Pradhan, S. (2013). Modeling and optimization of Al/SiCp MMC machining using Taguchi approach. *Measurement: Journal of the International Measurement Confederation, 46*(9), 3064–3072. https://doi.org/10.1016/j.measurement.2013.06.001
7. Rajmohan, T., Palanikumar, K., Madhavan, M., & Harish, G. (2012). Optimizing the machining parameters for minimum burr height in drilling of hybrid composites. *Procedia Engineering, 38*, 56–65. https://doi.org/10.1016/j.proeng.2012.06.009
8. Siddesh Kumar, N. G., Shiva Shankar, G. S., Ganesh, M. N., & Vibudha, L. K. (2017). Experimental investigations to study the cutting force and surface roughness during turning of aluminium metal matrix hybrid composites. *Materials Today: Proceedings, 4*(9), 9371–9374. https://doi.org/10.1016/j.matpr.2017.06.187
9. Ravindranath, V. M., Yerriswamy, M., Vivek, S. V., Shankar, G. S. S., & Siddesh Kumar, N. G. (2017). Drilling of Al2219/B4C/Gr metal matrix hybrid composites. *Materials Today: Proceedings, 4*(9), 9898–9901. https://doi.org/10.1016/j.matpr.2017.06.290
10. Kumar, H. G. P., & Xavior, M. A. (2017). Assessment of mechanical and tribological properties of Al 2024-SiC—graphene hybrid composites. *Procedia Engineering, 174*, 992–999. https://doi.org/10.1016/j.proeng.2017.01.251
11. Sivasankaran, S., Saminathan, E., Sidharth, S., Harisagar, P. T., & Sasikumar, P. (2014). Effect of graphite addition on surface roughness during turning of AA 7075-ZrB2 in-situ metal matrix composites. *Procedia Materials Science, 5*, 2122–2131. https://doi.org/10.1016/j.mspro.2014.07.548
12. Razavykia, A., Farahany, S., Yusof, N. M. (2015). Evaluation of cutting force and surface roughness in the dry turning of Al-Mg2Si in-situ metal matrix composite inoculated with bismuth using DOE approach. *Measurement: Journal of the International Measurement Confederation, 76*, 170–182. https://doi.org/10.1016/j.measurement.2015.08.032
13. Kishore, S. C., Prahlada Rao, D., & Ramesh, A. (2015). Optimization of machining parameters for improving cutting force and surface roughness in turning of Al6061-TiC in-situ metal matrix composites by using Taguchi method. *Materials Today: Proceedings, 2*(4–5), 3075–3083. https://doi.org/10.1016/j.matpr.2015.07.249
14. Dabade, U. A., & Jadhav, M. R. (2016). Experimental study of surface integrity of Al/SiC particulate metal-matrix composites in hot machining. *Procedia CIRP, 41*, 914–919. https://doi.org/10.1016/j.procir.2016.01.024

15. Bansal, P., & Upadhyay, L. (2016). Effect of turning parameters on tool wear, surface roughness and metal removal rate of alumina reinforced aluminum composite. *Procedia Technology, 23*, 304–310. https://doi.org/10.1016/j.protcy.2016.03.031
16. Nestler, A., & Schubert, A. (2014). Surface properties in ultrasonic vibration assisted turning of particle reinforced aluminium matrix composites. *Procedia CIRP, 13*, 125–130. https://doi.org/10.1016/j.procir.2014.04.022
17. Varol, T., & Canakci, A. (2013). Effect of particle size and ratio of B4C reinforcement on properties and morphology of nanocrystalline Al2024-B4C composite powders. *Powder Technology, 246*, 462–472. https://doi.org/10.1016/j.powtec.2013.05.048
18. Abdollahi, A., Alizadeh, A., & Baharvandi, H. R. (2014). Dry sliding tribological behavior and mechanical properties of Al2024-5wt.%B4C nanocomposite produced by mechanical milling and hot extrusion. *Materials and Design, 55*, 471–481. https://doi.org/10.1016/j.matdes.2013.09.024
19. Joardar, H., Das, N. S., Sutradhar, G., Singh, S. (2014). Application of response surface methodology for determining cutting force model in turning of LM6/SiCP metal matrix composite. *Measurement: Journal of the International Measurement Confederation, 47*(1), 452–464. https://doi.org/10.1016/j.measurement.2013.09.023
20. Bhushan, R. K. (2013). Optimization of cutting parameters for minimizing power consumption and maximizing tool life during machining of Al alloy SiC particle composites. *Journal of Cleaner Production, 39*, 242–254. https://doi.org/10.1016/j.jclepro.2012.08.008
21. Aramesh, M., Attia, M. H., Kishawy, H. A., & Balazinski, M. (2016). Estimating the remaining useful tool life of worn tools under different cutting parameters: A survival life analysis during turning of titanium metal matrix composites (Ti-MMCs). *CIRP Journal of Manufacturing Science and Technology, 12*, 35–43. https://doi.org/10.1016/j.cirpj.2015.10.001
22. Jayaraman, P., & Mahesh Kumar, L. (2014). Multi-response optimization of machining parameters of turning AA6063 T6 aluminium alloy using grey relational analysis in Taguchi method. *Procedia Engineering, 97*, 197–204. https://doi.org/10.1016/j.proeng.2014.12.242
23. Pramanik, A. (2014). Developments in the non-traditional machining of particle reinforced metal matrix composites. In *International Journal of Machine Tools and Manufacture* (Vol. 86, pp. 44–61). Elsevier Ltd. https://doi.org/10.1016/j.ijmachtools.2014.07.003
24. Dey, A., Reddy Bandi, V. R., Pandey, K. M. (2018). *Wire electrical discharge machining characteristics of AA6061/cenosphere aluminium matrix composites using RSM*. www.sciencedirect.comwww.materialstoday.com/proceedings2214-7853
25. Gore, A. S., & Patil, N. G. (2018). Wire electro discharge machining of metal matrix composites: A review. *Procedia Manufacturing, 20*, 41–52. https://doi.org/10.1016/j.promfg.2018.02.006
26. Mardi, K. B., Dixit, A. R., & Mallick, A. (2017). Studies on non-traditional machining of metal matrix composites. *Materials Today: Proceedings, 4*(8), 8226–8239. https://doi.org/10.1016/j.matpr.2017.07.165
27. Rao, P. S., Ramji, K., & Satyanarayana, B. (2014). Experimental investigation and optimization of wire EDM parameters for surface roughness, MRR and white layer in machining of aluminium alloy. *Procedia Materials Science, 5*, 2197–2206. https://doi.org/10.1016/j.mspro.2014.07.426
28. Mahanta, S., Chandrasekaran, M., & Samanta, S. (2018). GA based optimization for the production of quality jobs with minimum power consumption in EDM of hybrid MMCs. *Materials Today: Proceedings, 5*(2), 7788–7796. https://doi.org/10.1016/j.matpr.2017.11.457
29. Çaydaş, U., Hasçalik, A., Ekici, S. (2009). An adaptive neuro-fuzzy inference system (ANFIS) model for wire-EDM. *Expert Systems with Applications, 36*(3 PART 2), 6135–6139. https://doi.org/10.1016/j.eswa.2008.07.019
30. Das, M. K., Kumar, K., Barman, T. K., & Sahoo, P. (2014). Application of artificial bee colony algorithm for optimization of MRR and surface roughness in EDM of EN31 tool steel. *Procedia Materials Science, 6*, 741–751. https://doi.org/10.1016/j.mspro.2014.07.090
31. Kumar, A., Mishra, H., Vivekananda, K., & Maity, K. (2017). Multi-objective optimization of wire electrical discharge machining process parameterson Inconel 718. *Materials Today: Proceedings, 4*(2), 2137–2146. https://doi.org/10.1016/j.matpr.2017.02.060

32. Dewangan, S., Gangopadhyay, S., & Biswas, C. K. (2015). Multi-response optimization of surface integrity characteristics of EDM process using grey-fuzzy logic-based hybrid approach. *Engineering Science and Technology, an International Journal, 18*(3), 361–368. https://doi.org/10.1016/j.jestch.2015.01.009
33. Vikas, R., & Kumar, K. (2014). Effect and optimization of various machine process parameters on the surface roughness in EDM for an EN41 material using Grey-Taguchi. *Procedia Materials Science, 6*, 383–390. https://doi.org/10.1016/j.mspro.2014.07.049
34. Goswami, A., & Kumar, J. (2014). Optimization in wire-cut EDM of Nimonic-80A using Taguchi's approach and utility concept. *Engineering Science and Technology, an International Journal, 17*(4), 236–246. https://doi.org/10.1016/j.jestch.2014.07.001
35. Sharma, A., Garg, M. P., & Goyal, K. K. (2014). Prediction of optimal conditions for WEDM of Al 6063/ZrSiO 4(p) MMC. *Procedia Materials Science, 6*, 1024–1033. https://doi.org/10.1016/j.mspro.2014.07.173
36. Kumar, N. M., Senthil Kumaran, S., & Kumaraswamidhas, L. A. (2016). High temperature investigation on EDM process of Al 2618 alloy reinforced with Si3N4, ALN and ZrB2in-situ composites. *Journal of Alloys and Compounds, 663*, 755–768. https://doi.org/10.1016/j.jallcom.2015.12.175
37. Sivaprakasam, P., Hariharan, P., & Gowri, S. (2014). RSM and ANN modeling of micro wire electrical discharge machining of AL 2024 T351. In: *NIRMA UNIVERISTY JOURNAL OF ENGINEERING AND TECHNOLOGY* (Vol. 3, Issue 1).
38. Zhang, H., Kong, X., Yang, L., Wang, Y., & Chi, G. (2015). High temperature deformation mechanisms and constitutive modeling for Al/SiCp/45 metal matrix composites undergoing laser-assisted machining. *Materials Science and Engineering A, 642*, 330–339. https://doi.org/10.1016/j.msea.2015.06.052
39. Bhuyan, R. K., Routara, B. C., & Parida, A. K. (2015). An approach for optimization the process parameter by using TOPSIS Method of Al-24%SiC metal matrix composite during EDM. *Materials Today: Proceedings, 2*(4–5), 3116–3124. https://doi.org/10.1016/j.matpr.2015.07.272
40. Patil, N. G., Brahmankar, P. K., & Thakur, D. G. (2016). On the effects of wire electrode and ceramic volume fraction in wire electrical discharge machining of ceramic particulate reinforced aluminium matrix composites. *Procedia CIRP, 42*, 286–291. https://doi.org/10.1016/j.procir.2016.02.287
41. Dixit, A., & Kumar, K. (2015). Optimization of mechanical properties of silica gel reinforced aluminium MMC by using Taguchi method. *Materials Today: Proceedings, 2*(4–5), 2359–2366. https://doi.org/10.1016/j.matpr.2015.07.301
42. Senthilkumar, C., Ganesan, G., & Karthikeyan, R. (2011). Parametric optimization of electrochemical machining of Al/15 SiC p composites using NSGA-II. *Transactions of Nonferrous Metals Society of China (English Edition), 21*(10), 2294–2300. https://doi.org/10.1016/S1003-6326(11)61010-8
43. Lodhi, B. K., & Agarwal, S. (2014). Optimization of machining parameters in WEDM of AISI D3 steel using taguchi technique. *Procedia CIRP, 14*, 194–199. https://doi.org/10.1016/j.procir.2014.03.080
44. Kung, K. Y., & Chiang, K. T. (2008). Modeling and analysis of machinability evaluation in the wire electrical discharge machining (WEDM) process of aluminum oxide-based ceramic. *Materials and Manufacturing Processes, 23*(3), 241–250. https://doi.org/10.1080/10426910701860616
45. Soundararajan, R., Ramesh, A., Sivasankaran, S., & Vignesh, M. (2017). Modeling and analysis of mechanical properties of aluminium alloy (A413) reinforced with boron carbide (B4C) processed through squeeze casting process using artificial neural network model and statistical technique. *Materials Today: Proceedings, 4*(2), 2008–2030. https://doi.org/10.1016/j.matpr.2017.02.047
46. Priyadarshini, M., Pattnaik, S. K., Mishra, D., Panda, S., & Dhalmahapatra, K. (2015). Multi characteristics optimization of laser drilling process parameter using grey fuzzy reasoning method. *Materials Today: Proceedings, 2*(4–5), 1518–1532. https://doi.org/10.1016/j.matpr.2015.07.078

47. Chalisgaonkar, R., & Kumar, J. (2015). Multi-response optimization and modeling of trim cut WEDM operation of commercially pure titanium (CPTi) considering multiple user's preferences. *Engineering Science and Technology, an International Journal, 18*(2), 125–134. https://doi.org/10.1016/j.jestch.2014.10.006
48. Marani Barzani, M., Zalnezhad, E., Sarhan, A. A. D., Farahany, S., & Ramesh, S. (2015). Fuzzy logic based model for predicting surface roughness of machined Al-Si-Cu-Fe die casting alloy using different additives-turning. *Measurement: Journal of the International Measurement Confederation, 61*, 150–161. https://doi.org/10.1016/j.measurement.2014.10.003
49. Sharma, H., Hazrati, G., & Bansal, J. C. (2019). Spider monkey optimization algorithm. In *Studies in Computational Intelligence* (Vol. 779). Springer International Publishing. https://doi.org/10.1007/978-3-319-91341-4_4
50. Acilar, A. M., & Arslan, A. (2011). Optimization of multiple input-output fuzzy membership functions using clonal selection algorithm. *Expert Systems with Applications, 38*(3), 1374–1381. https://doi.org/10.1016/j.eswa.2010.07.036
51. Kafaltiya, S., Agrawal, P.K., Sharma, P., Singh, V.K., Bhaskar, J., Chauhan, S., Verma, A. Multi-response optimization of characteristics for graphite reinforced biodegradable PVA-fumaric acid cross-linked composite: A gray-fuzzy logic-based hybrid approach. *Journal of Vinyl and Additive Technology*. https://doi.org/10.1002/vnl.22112

Biotechnological Interventions to Mitigate the Influence of Climate Alterations on Ecosystem Viability and Human Well-Being

Madhumita Sarkar and Juhi Gupta

Abstract In today's era of escalating climate unpredictability and frequent extreme weather events, addressing the intersection of environmental change and human health is significant. This chapter describes the multifaceted implications of climate change on mental well-being and offers actionable strategies for mitigation. Climate change, characterized by profound shifts in environmental conditions, poses a significant threat to global health. Beyond ecological impacts like natural disasters and habitat degradation, it profoundly affects mental health. Environmental stressors such as air pollution, chemical exposure, and erratic weather patterns exacerbate anxiety and depression, especially among vulnerable populations. Researchers and policymakers are called to action, seeking to understand the complex dynamics of climate-induced mental health challenges and devise holistic solutions. Initiatives promoting community resilience and sustainable living practices hold promise for addressing eco-anxiety and enhancing mental well-being. Practical measures, from adopting eco-friendly transportation to embracing sustainable diets, can mitigate environmental degradation and safeguard public health. Innovative waste management approaches align with Sustainable Development Goals 2030, underscoring the urgency of sustainable development. A focus on the health and well-being of individuals and communities is key to creating a more sustainable future. By cutting down on greenhouse gas emissions, encouraging responsible environmental practices, and building resilient communities, we can build a world that supports both human health and the vitality of our ecosystems.

Keywords Climate change · Ecosystem viability · Human well-being · Biotechnological solutions · Environmental sustainability

M. Sarkar · J. Gupta (✉)
Amity Institute of Environmental Sciences, Amity University, Noida 201303, India
e-mail: jgupta2@amity.edu

© The Author(s), under exclusive license to Springer Nature Singapore Pte Ltd. 2024

A. Chaudhary et al. (eds.), *Unraveling New Frontiers and Advances in Bioinformatics*,
https://doi.org/10.1007/978-981-97-7123-3_17

1 Introduction

Over the past few years, humans have seen many extreme events and natural disasters like hurricanes, wildfires, and snowstorms. It is because the planet's climate or weather is changing unpredictably [7]. Severe weather changes are happening more often, and rising temperatures are the main culprit. There are multiple underlying causes behind this such as excessive use of pesticides, emerging contaminants in industries, modified micro-organisms and irresponsible consumption by humans [26]. The Intergovernmental Panel on Climate Change (IPCC) defines climate change as long-term alterations in temperature, rainfall patterns, and other atmospheric conditions on Earth, often due to human actions like fossil fuel combustion, deforestation, and industrial activities. These changes can lead to significant disruptions in ecosystems, weather systems, and human societies, creating substantial challenges for adaptation and mitigation efforts. However, climate change isn't the only environmental factor impacting health. Many experts suggest that mental health is currently influenced by various environmental factors [13], including chemical exposure, air pollution, climate change, natural disasters, microbial diseases, inadequate access to healthcare, infrastructure deficiencies, poor water quality, and global environmental issues. All these elements can have both immediate and long-term effects on mental health [13]. For example, individuals residing, working, or attending school in older buildings may face exposure to lead paint, a serious health hazard, particularly for young children [31]. Additionally, climate change is increasingly linked to direct and indirect impacts on mental health and well-being. As global temperatures increase, extreme weather events such as droughts, fires, and floods are becoming more frequent, leading to heightened anxiety and depression in many people [7]. Children, the elderly, the chronically ill, migrants, and refugees are particularly vulnerable to these effects [31].

Research institutions like Imperial College London are dedicated to understanding these problems and finding solutions [51]. Evidence suggests that taking action on climate change and engaging with like-minded communities can reduce eco-anxiety. Climate action can benefit both individuals and society, contributing to a healthier planet and promoting individual well-being [51]. For example, choosing walking or cycling over driving not only cuts greenhouse gas emissions but also lowers the risk of conditions such as obesity, diabetes, and heart disease [39]. Monitoring air quality and taking appropriate measures, such as wearing a mask, can also lead to significant health benefits. Public transportation can play a role in reducing air pollution, while eating more fruits, vegetables, and whole grains, along with minimizing processed and packaged foods, can lead to healthier lifestyles. Composting and recycling can help reduce landfill waste [24, 25]. Driving electric vehicles, cooking with clean energy, and using renewable energy sources for electricity when possible, along with conserving energy by switching off lights and appliances when not in use, can all contribute positively to the environment [20, 51]. Despite ongoing efforts to manage waste, the amount generated often exceeds the capacity for proper treatment, prompting researchers to explore innovative solutions [22, 23].

Sustainable Development goals 2030 highlights the potential action needed for planet Earth. Among the 17 goals, this chapter would like to serve the contribution towards Goal no. 13 (Climate Action) along with others. Acting against climate change is not only good for the environment but also improves health and saves money [8, 39]. Living an active lifestyle has numerous benefits for an individual's health. It contributes to maintaining a healthy weight, regulating cholesterol, and keeping blood pressure within optimal levels. These health benefits not only support personal well-being but also alleviate stress on healthcare systems and reduce environmental impact. Managing serious pollution, particularly air pollution, involves addressing the severity of the symptoms. In extreme cases, a person may require respiratory support for certain periods or even surgical intervention in cases of cancer. Despite these treatment options, the best approach is prevention [8].

2 Impact of Climate Change on Ecosystem's Health

The following section will elaborate on the effects faced by multiple parameters due changing climate.

2.1 Impact of Climate Change on Weather

Human activities, such as burning fossil fuels, are causing the Earth's temperature to rise, which is leading to an increase in extreme weather events like droughts, hurricanes, and floods [7]. These events used to be rare, but now they're happening more often. Not every place will be affected the same way, some areas might face severe droughts, while others could experience more floods. In the past 250 years, the Earth has warmed by 1.1 °C, and if we don't do something about it, scientists say it could get even hotter, reaching up to 4 °C by the year 2100. The main reason of this warming is the burning of fossil fuels like coal, oil, and gas [44]. Table 1 simplifies the weather parameters impacted due to climate change.

2.2 Impact of Climate Change on the Environment

Climate change is deteriorating nature from the North and South Poles to the warm tropics. Even a small temperature change can deteriorate the balance in nature, affecting how animals and plants interact with each other and their surroundings [40]. This can cause big problems in the environment and how food chains work [40] (Table 2).

Table 1 List of weather parameters impacted by climate change

Parameter	Characteristics
Higher average temperature	As the Earth's average temperature goes up, summers are getting hotter, causing more frequent and intense heatwaves. Local weather reporters are noticing longer periods of record-breaking heat, which can be a big issue in places where buildings and infrastructure were not designed to handle such extreme temperatures. Heatwaves are not just uncomfortable; they're the main reason for weather-related deaths in the United States [4]
Long-lasting droughts	When it gets hotter, more water evaporates from the air, causing bad droughts. Right now, in the American West, there's a super serious "megadrought"—the driest it's been in at least 1,200 years for 22 years straight [40]. This is causing problems like less drinking water, crops dying, and making forests more vulnerable to bugs. Droughts can create a loop where dry soil and fewer plants make water evaporate even faster
Intense wildfires	Because the Earth is becoming drier and hotter, it's making wildfires worse. These fires are spreading faster and lasting longer, putting many more lives and homes in danger. The frequency of large wildfires in the western United States has doubled between 1984 and 2015 [6]. In California, the yearly area burned by wildfires has gone up by 500 percent from 1972 to 2018 [57]
Stronger storms	As the Earth gets warmer, the air can hold more moisture, making storms like tropical cyclones stronger and rainier. A recent report from scientists says that for every degree Celsius the world warms up, the amount of rain during heavy storms could go up by about 7 percent, which means a higher risk of flooding. The occurrence of intense hurricanes, specifically Category 4 and 5, is anticipated to rise in frequency. An example of this was Hurricane Harvey in 2017, a very strong storm that caused record-breaking rainfall and serious problems in Houston [55]

2.3 *Impact of Climate Change on Agriculture*

This poses a significant threat, particularly to small family-owned farms worldwide. A severe drought or flood can devastate a whole season's crop yield or a herd of livestock. For example, in June 2022, a very hot period in Kansas caused the death of thousands of cows. Some communities are trying to make their farms more resistant to climate change through regenerative agriculture, but not everyone has equal access to the support needed for these sustainable farming methods (Table 3).

2.4 *Impact of Climate Change on Animals*

See Table 4.

Table 2 List of ecosystem parameters impacted by climate change

Parameter	Characteristics
Melting sea ice	The places getting hit the hardest by climate change are the coldest ones, like the North and South Poles. The Arctic, which is up north, is getting warmer quickly, twice as fast as anywhere else. This is causing the ice on glaciers and polar ice sheets to melt rapidly, and they store a huge amount of water. When the ice on the sea melts, it uncovers darker water that soaks up more sunlight, making it melt even faster. In about 15 years, the Arctic might not have any ice left in the summer
Sea level rise	Scientists say that if we don't reduce emissions, melting ice and warmer water could make sea levels go up by about 3.61 feet by the year 2100. This would be bad for low-lying places, like islands and cities near the coast, such as New York City and Mumbai. Even if sea levels don't go up that much, a foot of rise is expected in the United States by 2050. This can cause regular damage to things like roads, sewage plants, and power plants. Beaches that families enjoy might disappear by the end of the century. It's not just about damage—rising seas can harm the environment too, eroding coastal areas and affecting freshwater that we use for farming and drinking. In places like Bangladesh, where a lot of land is close to sea level, saltwater is already causing problems
Flooding	Climate change not only makes the sea rise and floods coastal areas but also affects flooding in other places. This happens when the snow melts or there's heavy rain. As the Earth gets warmer, floods are expected to happen more often. By the year 2100, the areas prone to flooding might increase by about 45 percent. In 2022, in Pakistan, heavy rains and melting glaciers caused deadly floods that covered a big part of the country [48]. This shows how climate change can make flooding worse in different ways
Marine heat waves & Warmer Ocean waters	Climate change is taking a severe toll on our oceans, which cover more than 70% of Earth's surface. They absorb 93% of the heat retained by greenhouse gases and around 30% of the carbon dioxide produced by burning fossil fuels [58]. Because of this, fish and other sea creatures are changing where they go to find cooler water, messing up the natural balance of ocean life and important fisheries. Heat waves in the ocean are happening more often, causing big numbers of plankton and marine animals to die. On top of that, the ocean is soaking up more carbon dioxide, which makes the water slowly become more acidic. This change in the water's chemistry is a big problem for marine life, especially animals like corals, oysters, and mussels, as it makes it hard for them to build their shells

(continued)

Table 2 (continued)

Parameter	Characteristics
Ecosystem stressors	Climate change is causing problems for land-based areas like forests, savannahs, and rainforests. It's making things tough by increasing pests, invasive species, and diseases in forests. The types of plants that can survive in a place are also changing, affecting the lives of animals and the overall balance of ecosystems. Ecosystems can usually adapt, but many are struggling to handle the stress caused by climate change [46]. As temperatures go up, more issues are expected. These changes are creating a domino effect in nature, and it's hard to predict or completely stop what might happen. Especially, species that play a big role in keeping ecosystems stable are being affected, which can lead to more problems in the long run
Air pollution	Breathing polluted air has similar effects to regularly smoking cigarettes and is considered more harmful than consuming too much salt. Air pollution is one of the most critical issues facing the planet, yet it often receives less attention than it should. This lack of focus is partly because air pollution is not typically recognized as a specific condition, even though it causes numerous diseases. According to data from the World Health Organization (WHO), 9 out of 10 people currently breathe polluted air, and approximately 7 million people die each year due to health complications linked to air pollution [32]. Avoiding something that is invisible is challenging, and in many urban areas, it's nearly impossible to live without exposure to air pollution

Table 3 List of agricultural parameters impacted by climate change

Parameter	Characteristics
Unpredictable growth seasons	As the world gets warmer, growing crops becomes more unpredictable, and raising livestock becomes tougher due to extreme weather. Climate change messes with rainfall, causing unexpected floods and longer droughts. Hurricanes, becoming more frequent and severe, can destroy entire crops in a season [49]. Dealing with pests, diseases, and invasive species also becomes trickier for farmers
Reduced soil health	Healthy soil is characterized by moisture, mineral richness, and a thriving population of beneficial insects, bacteria, fungi, and microbes that support robust crop growth. However, climate change, including extreme heat and altered rainfall patterns, can degrade soil quality and health [42]. This is even worse in places where farming relies heavily on chemicals and grows just one type of crop, as it makes the soil and crops less resilient to environmental changes
Food shortages	The changes in our farming caused by climate change are a big problem for the world's food supply [38]. It means some people might not have enough food, and the prices of food could go up. This won't affect everyone the same way—richer people will still have more choices for getting food, but many others, possibly billions, could face not having enough to eat. This adds to the billions who already struggle to get the food they need

Table 4 List of biodiversity parameters impacted by climate change (in the case of fauna)

Parameter	Characteristics
Biodiversity crisis	It's not just about polar bears; in the most diverse places like the Amazon rainforest and the Galapagos Islands, half of all animal species are in danger of disappearing because of climate change [37]. Moreover, climate change is exacerbating the challenges faced by species already under threat due to our shifting use of land and oceans, such as converting wild habitats into farmland and overexploiting certain species through overfishing and the wildlife trade
Insufficient habitat	Many species are already struggling over 500,000 don't have enough space to survive in the long run. If we don't do something about climate change, it could push millions of species to the brink [16]. Climate change is quickly and drastically changing the places where animals live, places they've adapted to over thousands of years. This is particularly bad for species whose homes are already in trouble for other reasons. Animals like walruses and penguins, which rely on ice, will face a tough time as the ice sheets get smaller
Coral bleaching	Climate change is quickly changing the homes of animals that have lived there for a long time. This is bad, especially for animals whose homes are already in danger for other reasons. Animals like walruses and penguins, who depend on ice, will struggle as the ice gets smaller. Also, the changing temperatures in the oceans are making life difficult for the tiny plants that feed coral reefs, and this often leads to coral reefs becoming unhealthy, which is called coral bleaching [2]
Migrating birds	Wetlands in the Midwest's Prairie Pothole Region are disappearing, taking away important places where many birds go to drink water and have babies. Lots of different kinds of birds are having a tough time because more than 85 percent of these wetlands have been lost since 1700 [60]. Also, rising sea levels will flood or wash away many homes for animals along the coast, including hundreds of types of birds, sea creatures, and other marine animals. The way many animals act—like when they mate, eat, or move—is connected to small changes in the seasons, like temperature and rainfall. But now, the environment is changing faster than some species can keep up with. When plants in an area change or certain species bloom or hatch at different times than before, it affects the food and water available. This, in turn, causes ripples through the food chain

3 Effects of Climate Change on Human Health

3.1 Extreme Weather Events

In recent times, humanity has observed a significant rise in the occurrence of extreme events and natural disasters, including hurricanes, wildfires, and snowstorms, which are mainly linked to climate change. According to recent data, the occurrence of severe weather events has surged globally, accompanied by a significant rise in temperatures across various regions worldwide. Disturbingly, studies reveal that heat waves have been linked to incidents of violence and suicides, posing a heightened risk to vulnerable populations and individuals on psychiatric medications [27]. As global temperatures continue to climb, the prevalence of extreme weather phenomena like droughts, wildfires, and floods has escalated, with Z% more occurrences reported

Fig. 1 Adverse effects of climate change on the ecosystem

annually compared to previous decades. Many individuals are directly impacted, experiencing problems such as post-traumatic stress disorder, while heat waves are associated with heightened mood disorders and more intense symptoms of mental health conditions (Fig. 1).

3.2 *Impacts on Air Quality*

Breathing polluted air is akin to regular smoking and is more hazardous than excessive salt intake. Despite being a critical global issue, air pollution often lacks the attention it deserves. WHO reports indicate that 9 out of 10 people breathe polluted air, causing approximately 7 million deaths annually. Microscopic pollutants infiltrate the body, posing significant risks to the circulatory and respiratory systems, with the lungs being particularly vulnerable. Air pollution is linked to conditions like COPD, heart disease, stroke, and lung cancer. Children, with developing respiratory systems, are especially susceptible. Household air pollution, prevalent in lower-income areas lacking clean fuels, contributes to a majority of pollution-related deaths, primarily in countries with weaker healthcare systems [54].

3.3 Vector-Borne Diseases

Microbes, the most prolific and diverse life forms on Earth, surpass a trillion species, with many yet to be discovered. While non-pathogenic microbes play vital roles in human health, they can become detrimental under specific circumstances. Evidence of human-microbe co-evolution is apparent through shared genetic material, which contributes essential building blocks to the human genome. The human microbiome, inhabiting mucosal surfaces such as the airway, skin, and gut, exerts a profound influence on diseases like diabetes, obesity, heart disease, and neurological disorders. Advanced genome technologies unveil the intricate interplay between human genetics and microbial communities [10]. Lifestyle choices, dietary habits, age, and genetic variations significantly influence the composition of the human microbiota, thereby impacting various physiological functions. Instances of microbiota instability have been associated with conditions like Parkinson's disease and inflammatory bowel disorders, underscoring the pivotal role of the microbiome in averting a spectrum of health ailments.

3.4 Water Quality Impacts

Climate change presents substantial challenges to water quality on a global scale, causing a range of intricate effects on both aquatic ecosystems and human communities. As temperatures increase, precipitation patterns shift, and extreme weather events become more frequent and intense, water quality dynamics undergo significant alterations [27]. Variations in temperature can disturb aquatic ecosystems, influencing the growth and distribution of algae, bacteria, and other microorganisms, which can lead to harmful algal blooms and bacterial contamination. Additionally, shifts in precipitation patterns can exacerbate water scarcity in some regions while intensifying flooding in others, further compromising water quality through soil erosion, sedimentation, and pollutant runoff. Rising sea levels and saltwater intrusion also pose threats to freshwater sources, contaminating groundwater and surface water supplies [27]. Moreover, climate-induced changes in water temperature and chemistry can impact aquatic biodiversity, fisheries, and the overall health of aquatic ecosystems. To tackle the intricate relationship between climate change and water quality, comprehensive approaches are needed. These should include measures to mitigate greenhouse gas emissions, strategies to adapt water resource management, and resilience-building initiatives to protect ecosystems and communities from the effects of climate change [17].

Poor water quality contributes to approximately 485,000 deaths annually due to diarrheal diseases, primarily affecting children under five years old in developing countries. Contaminated water sources are responsible for over 1.8 million deaths each year from waterborne diseases such as cholera, typhoid fever, and dysentery. Unsafe water quality is a leading cause of gastrointestinal illnesses, with an estimated 2 billion cases reported globally each year. Deteriorating water quality has resulted in the loss of biodiversity in freshwater ecosystems, with nearly one-third of freshwater fish species facing extinction threats. Water pollution has severe economic implications, costing billions of dollars annually in healthcare expenses, lost productivity, and water treatment. Waterborne pollutants such as heavy metals, pesticides, and pharmaceuticals pose significant risks to human health, contributing to chronic diseases like cancer and neurological disorders. Addressing water quality issues is critical for achieving Sustainable Development Goal 6, which ensures universal access to clean water and sanitation, as well as for safeguarding human health, ecosystems, and economic prosperity [27].

3.5 Foodborne Diseases

Climate change affects food security and nutrition in significant ways, both directly and indirectly. Changes in temperature and rainfall patterns, along with the increased frequency and intensity of extreme weather events and alterations in growing seasons, can result in variations in crop production. Consequently, there may be a reduction in food production, affecting food availability, and accessibility, and leading to elevated food prices. The quality of food is also at risk, as variations in temperature and precipitation can modify the composition and nutritional value of crops, potentially diminishing food quality and increasing the likelihood of micronutrient deficiencies. Climate change further threatens food access, as variations in temperature and rainfall patterns can obstruct the ability of individuals and communities to cultivate, harvest, transport, and distribute food. This can lead to food shortages, hindering food access and increasing food insecurity [15]. Livestock production also faces these challenges, shifts in temperature and precipitation can affect the availability and quality of animal feed while raising the risk of diseases and parasites in livestock. Consequently, livestock production might decline, leading to reduced meat and dairy supply and worsening food insecurity (Fig. 2).

3.6 Food Security

Climate change constitutes a multifaceted factor influencing food security through various mechanisms. Foremost among these is the inherent dependency of global agriculture on climatic conditions. As climate patterns undergo shifts, agriculture

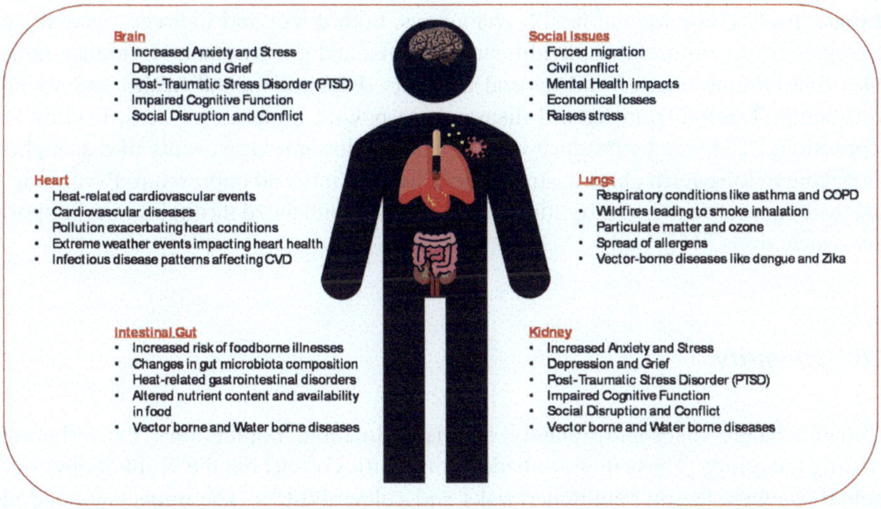

Fig. 2 Negative physiological impacts observed on Humans due climate change

encounters challenges in adapting to these changes, potentially resulting in production shortfalls. Such deficits may lead to heightened food prices and impediments to food accessibility. Extensive research on the impact of climate change on agriculture underscores the pronounced temperature effects on crops. Essentially, climate change disrupts the reliable weather patterns upon which farmers depend, impeding agricultural productivity and diminishing food output. Insufficient food production, in turn, contributes to elevated prices and poses challenges for individuals to obtain essential food resources. Particularly vulnerable are regions already grappling with food scarcity, where rising temperatures and increased aridity exacerbate the existing challenges. Globally, crucial crops like corn, wheat, rice, and soybeans face jeopardy due to their susceptibility to extreme temperatures. Consequently, climate change compounds the complexities of cultivating essential food sources, impacting both caloric and protein availability [56].

3.7 Mental Health

As global temperatures rise, frequent extreme weather disasters like droughts, fires, and floods are on the rise. People directly experience issues like post-traumatic stress disorder from these events, while heat waves are linked to worsened mood and intensified symptoms of mental illnesses. The deleterious impact of human actions on the environment, known as eco-anxiety, is causing heightened emotions and well-being concerns, especially among young people. Witnessing climate change impacts through the news can trigger feelings of loss and distress, termed Solastalgia and climate grief. These mental health challenges, both direct and indirect, necessitate strategies from communities, healthcare systems, and governments. Evidence indicates that climate change and natural disasters damages mental health and social well-being. Trauma from natural disasters starts with immediate stress, leading to depression, PTSD, and substance use over time. Slow-moving events like droughts contribute to long-term chronic stress, including anxiety and depression. Psychological resilience, crucial for mitigating these effects, is enhanced through social support and connections [41].

3.8 Inequity

Climate change disproportionately affects vulnerable populations, exacerbating existing inequities. These marginalized communities often bear the brunt of environmental changes, facing heightened risks and vulnerabilities. The impacts extend to water and food supplies, with variable precipitation patterns and damaging weather events disrupting access to essential resources. As climate change intensifies, vulnerable groups experience increased challenges in securing water and sustaining food production, leading to heightened food insecurity. This further widens social and economic disparities, highlighting the urgent need for targeted efforts to address the disproportionate impact of climate change on vulnerable communities [35].

3.9 Displacement Due to Natural Disasters

Climate change is a significant driver of displacement, particularly due to the increased frequency and severity of natural disasters. Rising global temperatures contribute to extreme weather events such as hurricanes, floods, wildfires, and droughts, leading to the displacement of communities. Natural disasters, exacerbated by climate change, can render areas uninhabitable, forcing people to flee their homes in search of safer environments. Coastal regions are particularly vulnerable to sea-level rise and storm surges, resulting in widespread displacement of populations. The displacement caused by climate-induced natural disasters often leads to complex challenges, including the loss of homes, livelihoods, and community structures. Internally displaced persons (IDPs) and refugees face difficulties in finding suitable shelter, access to resources, and integration into new communities [28]. Addressing the displacement crisis requires not only immediate humanitarian responses but also long-term strategies to mitigate the impacts of climate change, enhance resilience in vulnerable regions, and promote sustainable practices to reduce the risk of future displacements.

3.9.1 Economic Impacts

Climate change poses a formidable threat to the global economy, impacting various sectors with cascading effects. Extreme weather events, altered precipitation patterns, and rising temperatures disrupt agricultural productivity, leading to food shortages and increased prices. Sea-level rise and more frequent storms put coastal regions at economic risk, impacting industries like tourism and fisheries. Damage to infrastructure from extreme weather events incurs significant expenses for governments and businesses [52]. Furthermore, changing climate conditions pose threats to global supply chains, affecting industries across the globe. The financial burden of climate change highlights the need for comprehensive strategies to reduce its impact, shift to sustainable practices, and increase resilience in an evolving climate scenario.

4 Solutions to Overcome the Impacts of Climate Change

Biotechnology presents promising solutions for tackling the effects of climate change on a global level. Through innovations like genetic engineering, bioenergy production, and precision agriculture, biotechnology offers effective tools to reduce greenhouse gas emissions, increase carbon sequestration, and support sustainable resource management. Modifying crops to be more drought-resistant and improve nutrient absorption can boost agricultural output, even as climate conditions change. Bioenergy technologies provide renewable energy sources that can replace fossil fuels, thereby decreasing carbon emissions and combating climate change. Further, advancements in soil microbiology and ecosystem restoration can boost carbon sequestration, helping to restore degraded areas and mitigate the impact of deforestation and land degradation. By leveraging biotechnology, we can devise effective approaches to counter the challenges posed by climate change, promoting environmental sustainability and resilience.

- Climate-Resilient Crop Development:

Climate-resilient crop development is crucial in the face of climate change, as shifting weather patterns and extreme conditions pose significant threats to global food security. Researchers and agricultural scientists are actively engaged in developing crops that can withstand higher temperatures, droughts, and unpredictable rainfall [47]. This involves employing advanced breeding techniques, incorporating stress-tolerant traits, and utilizing precision agriculture technologies. By focusing on climate resilience, we aim to ensure stable agricultural production, safeguarding the livelihoods of farmers and enhancing the sustainability of our food systems [50]. This proactive approach is essential to address the challenges posed by a changing climate and build a more resilient foundation for the future of agriculture.

- Biofortification of Crops:

Biofortification of crops is a promising strategy in the face of climate change, aiming to enhance the nutritional content of food crops. As environmental conditions shift and traditional agricultural practices face challenges, biofortified crops offer a resilient solution [34]. Through selective breeding or genetic modification, these crops are enriched with essential nutrients such as vitamins and minerals. This not only addresses nutritional deficiencies but also contributes to climate resilience by ensuring food security in changing environmental conditions. By cultivating biofortified crops, we not only adapt to the impacts of climate change but also take a proactive step toward building a more robust and sustainable food system for the future [33].

- Genetic Modification for Disease Resistance:

Genetic modification of plants and animals for disease resistance is a promising avenue in the face of climate change. By employing advanced biotechnological techniques, scientists can enhance the genetic traits of organisms to make them more resilient to diseases that thrive in changing environmental conditions [45]. In agriculture, genetically modified crops can withstand pests and diseases, ensuring food security amidst shifting climates. Similarly, animals can be modified to resist diseases that may proliferate due to altered environmental conditions. This approach not only bolsters the resilience of ecosystems but also contributes to sustainable agriculture practices, crucial for adapting to the challenges posed by climate change [59]. However, ethical considerations and careful regulation are imperative to ensure responsible and beneficial applications of genetic modification.

- Microbial Solutions for Soil Health:

Microbial solutions play a crucial role in enhancing soil health, offering a sustainable approach to mitigating climate change impacts. Beneficial microorganisms, such as bacteria and fungi, contribute to soil fertility by promoting nutrient cycling, improving water retention, and aiding plant growth. These microscopic allies form symbiotic relationships with plants, fostering resilience against environmental stressors. Additionally, microbial activities in the soil facilitate carbon sequestration, a key strategy for offsetting greenhouse gas emissions [29]. Harnessing microbial solutions for soil health not only bolsters agricultural productivity but also acts as a natural climate change mitigation strategy, emphasizing the importance of fostering a harmonious balance between microorganisms and the environment [36].

- Biological Carbon Sequestration:

Biological carbon sequestration plays a crucial role in mitigating climate change by harnessing natural processes to capture and store carbon dioxide. This process involves the uptake of carbon by living organisms such as plants and trees through photosynthesis, with the carbon subsequently stored in biomass and soil [18]. Forests, wetlands, and other ecosystems act as vital carbon sinks, absorbing and locking away substantial amounts of atmospheric carbon. Protecting and restoring these ecosystems, as well as adopting sustainable land management practices, are essential strategies for enhancing biological carbon sequestration [21]. By promoting the health of natural habitats, we not only contribute to biodiversity conservation but also address the urgent need to reduce greenhouse gas concentrations, fostering a more resilient and balanced global climate [53].

- Bioremediation of Contaminated Sites:

Bioremediation stands as a promising approach to addressing the environmental challenges posed by climate change, particularly in the context of contaminated sites. This eco-friendly technique utilizes living organisms, such as bacteria, fungi, and plants, to degrade or neutralize pollutants [3]. By harnessing the natural metabolic capabilities of these organisms, bioremediation offers a sustainable solution to mitigate the impacts of climate change on contaminated ecosystems. Through the enhancement of microbial activities, organic contaminants can be broken down into less harmful substances, facilitating the restoration of soil and water quality. Bioremediation not only tackles pollution but also aligns with the broader goal of fostering resilient and sustainable environments in the face of climate-related disruptions [5]. As a cost-effective and environmentally friendly strategy, it exemplifies the potential for synergy between ecological restoration and climate change mitigation efforts.

- Synthetic Biology for Sustainable Materials:

Synthetic biology offers a promising avenue for addressing climate change through the creation of sustainable materials. By leveraging the principles of genetic engineering, researchers can design microorganisms capable of producing eco-friendly materials, such as biodegradable plastics or bio-based textiles. These synthetic organisms can be tailored to efficiently convert renewable resources into valuable materials, reducing reliance on fossil fuels and minimizing environmental impact [19]. The precision and customization afforded by synthetic biology enable the development of materials with specific properties, enhancing their functionality and applicability. This innovative approach holds the potential to revolutionize the production of materials, contributing to a more sustainable and environmentally friendly future in the face of climate change [43].

- Advanced Vaccines for Vector-Borne Diseases:

Advanced vaccines for vector-borne diseases play a crucial role in addressing the evolving challenges posed by climate change. As global temperatures rise and ecosystems shift, vectors like mosquitoes expand their habitats, increasing the prevalence of diseases such as malaria, dengue, and Zika [10]. To combat this, researchers are developing sophisticated vaccines that not only provide immunity but also consider the changing dynamics of vector-borne illnesses. These advanced vaccines often incorporate novel technologies, such as RNA-based platforms, to enhance efficacy and adaptability [11]. By staying ahead of the curve in vaccine development, scientists aim to create robust defences against vector-borne diseases in the face of a changing climate, ultimately safeguarding public health worldwide [14].

- Remote Sensing and Biotechnology Integration:

Remote sensing and biotechnology integration hold significant promise in addressing the challenges posed by climate change. By combining the capabilities of remote sensing technologies, such as satellite imagery and aerial surveys, with advancements in biotechnology, scientists can gain a more comprehensive understanding of ecological systems and climate patterns. This synergy enables precise monitoring of environmental changes, identification of stress factors on ecosystems, and the development of tailored biotechnological solutions to mitigate the impact of climate change [12]. From optimizing agricultural practices to assessing the health of biodiversity hotspots, this interdisciplinary approach facilitates informed decision-making and proactive measures to adapt to the evolving climate scenario. The integration of remote sensing and biotechnology stands as a powerful tool in our collective efforts to foster sustainability and resilience in the face of climate-related challenges [30].

- Precision Agriculture:

Precision Agriculture, in the context of climate change, emerges as a strategic ally in mitigating environmental impact while enhancing agricultural productivity. This advanced farming approach utilizes technology such as sensors, GPS, and data analytics to optimize resource use, minimize waste, and adapt to changing climate conditions [22]. By precisely tailoring irrigation, fertilization, and crop protection measures, Precision Agriculture promotes sustainable practices. It enables farmers to respond dynamically to climate variations, ensuring efficient water usage, reducing greenhouse gas emissions, and fostering resilient crop yields. In essence, Precision Agriculture serves as a crucial tool in the agricultural sector's efforts to adapt and contribute positively to the global challenge of climate change [1].

The entire work has been highly influenced by the following articles [22, 62-68].

5 Conclusion

The current chapter underscores the critical role of biotechnology in addressing the multifaceted challenges posed by climate change. Through innovative biotechnological interventions, the resilience of ecosystems and safeguard human well-being in the face of shifting climatic conditions can be safeguarded. From genetic modification to precision agriculture techniques, biotechnology offers a toolkit for mitigating the adverse impacts of climate alterations on agriculture, biodiversity, and public health. By harnessing the power of biotechnology alongside holistic environmental management strategies, we can pave the way for a more sustainable and resilient future. However, successful implementation of biotechnological solutions requires collaborative efforts between scientists, policymakers, and communities to ensure ethical, equitable, and sustainable outcomes. With proactive engagement and strategic investment in biotechnological research and development, we can navigate the challenges of climate change and foster a thriving, resilient global ecosystem for generations to come.

Acknowledgements The authors are grateful to AIES, Amity University, Noida for their constant support and facilities provided during the work.

Declaration of Interest: There is no conflict of interest.

References

1. Abobatta, W.F. (2021). Precision agriculture to mitigate climate change impacts in horticulture. *Advances in Agricultural Technology and Plant Sciences, 4*(1), p.180054. Available at: https://academicstrive.com/AATPS/AATPS180054.pdf
2. Ainsworth, T.D., Heron, S.F., Ortiz, J.C., Mumby, P.J., Grech, A., Ogawa, D., Eakin, C.M., Leggat, W. (2016). Climate change disables coral bleaching protection on the Great Barrier Reef. *Science, 352*(6283), pp.338–342. Available at: https://www.science.org/doi/abs/https://doi.org/10.1126/science.aac7125
3. Alkorta, I., Epelde, L., Garbisu, C. (2017). Environmental parameters altered by climate change affect the activity of soil microorganisms involved in bioremediation. *FEMS microbiology letters, 364*(19), p.fnx200. Available at: https://academic.oup.com/femsle/article-abstract/364/19/fnx200/4159367
4. Anderson, G.B., Bell, M.L. (2011). Heat waves in the United States: mortality risk during heat waves and effect modification by heat wave characteristics in 43 US communities. *Environmental health perspectives, 119*(2), pp.210–218. Available at: https://ehp.niehs.nih.gov/doi/abs/https://doi.org/10.1289/ehp.1002313
5. Arora, N.K. (2018). Bioremediation: a green approach for restoration of polluted ecosystems. *Environmental Sustainability, 1*(4), pp.305–307. Available at: https://link.springer.com/article/https://doi.org/10.1007/s42398-018-00036-y
6. Barbero, R., Abatzoglou, J.T., Larkin, N.K., Kolden, C.A., Stocks, B. (2015). Climate change presents increased potential for very large fires in the contiguous United States. *International Journal of Wildland Fire, 24*(7), pp.892–899. Available at: https://www.publish.csiro.au/wf/wf15083
7. Bell, J.E., Brown, C.L., Conlon, K., Herring, S., Kunkel, K.E., Lawrimore, J., Luber, G., Schreck, C., Smith, A.. Uejio, C. (2018). Changes in extreme events and the potential impacts on human health. *Journal of the Air & Waste Management Association, 68*(4), pp.265–287. Available at: https://www.tandfonline.com/doi/abs/https://doi.org/10.1080/10962247.2017.1401017
8. Bernard, P., Chevance, G., Kingsbury, C., Baillot, A., Romain, A.J., Molinier, V., Gadais, T., Dancause, K.N. (2021). Climate change, physical activity and sport: a systematic review. *Sports Medicine, 51*, pp.1041–1059. Available at: https://link.springer.com/article/https://doi.org/10.1007/s40279-021-01439-4
9. Caminade, C., McIntyre, K.M. and Jones, A.E., 2019. Impact of recent and future climate change on vector-borne diseases. *Annals of the New York Academy of Sciences, 1436*(1), pp.157–173. Available at: https://nyaspubs.onlinelibrary.wiley.com/doi/abs/https://doi.org/10.1111/nyas.13950?af=R
10. Caminade, C., McIntyre, K.M. and Jones, A.E. (2019). Impact of recent and future climate change on vector-borne diseases. *Annals of the New York Academy of Sciences, 1436*(1), pp.157–173. Available at: https://nyaspubs.onlinelibrary.wiley.com/doi/abs/https://doi.org/10.1111/nyas.13950
11. Campbell-Lendrum, D., Manga, L., Bagayoko, M., Sommerfeld, J. (2015). Climate change and vector-borne diseases: what are the implications for public health research and

policy? *Philosophical Transactions of the Royal Society B: Biological Sciences, 370*(1665), p.20130552. Available at: https://royalsocietypublishing.org/doi/abs/https://doi.org/10.1098/rstb.2013.0552
12. Cavender-Bares, J., Schneider, F.D., Santos, M.J., Armstrong, A., Carnaval, A., Dahlin, K.M., Fatoyinbo, L., Hurtt, G.C., Schimel, D., Townsend, P.A., Ustin, S.L. (2022). Integrating remote sensing with ecology and evolution to advance biodiversity conservation. *Nature Ecology & Evolution, 6*(5), pp.506–519. Available at: https://www.nature.com/articles/s41559-022-01702-5
13. Charlson, F., Ali, S., Benmarhnia, T., Pearl, M., Massazza, A., Augustinavicius, J., Scott, J.G. (2021). Climate change and mental health: a scoping review. *International journal of environmental research and public health, 18*(9), p.4486. Available at: https://www.mdpi.com/1660-4601/18/9/4486
14. Chevalier, V., Courtin, F., Guis, H., Tran, A., Vial, L. (2016). Climate change and vector-borne diseases. *Climate change and agriculture worldwide*, pp.97–108. Available at: https://link.springer.com/chapter/https://doi.org/10.1007/978-94-017-7462-8_8
15. Cissé, G. (2019). Food-borne and water-borne diseases under climate change in low-and middle-income countries: Further efforts needed for reducing environmental health exposure risks. *Acta tropica, 194*, pp.181–188. Available at: https://www.sciencedirect.com/science/article/pii/S0001706X18309525
16. Eigenbrod, F., Gonzalez, P., Dash, J., Steyl, I. (2015). Vulnerability of ecosystems to climate change moderated by habitat intactness. *Global change biology, 21*(1), pp.275–286. Available at: https://onlinelibrary.wiley.com/doi/abs/https://doi.org/10.1111/gcb.12669
17. Fezzi, C., Harwood, A.R., Lovett, A.A., Bateman, I.J. (2017). The environmental impact of climate change adaptation on land use and water quality. In *Building a Climate Resilient Economy and Society* (pp. 27–40). Edward Elgar Publishing. Available at: https://www.elgaronline.com/abstract/edcoll/9781785368448/9781785368448.00013.xml
18. Gayathri, R., Mahboob, S., Govindarajan, M., Al-Ghanim, K.A., Ahmed, Z., Al-Mulhm, N., Vodovnik, M., Vijayalakshmi, S. (2021). A review on biological carbon sequestration: A sustainable solution for a cleaner air environment, less pollution and lower health risks. *Journal of King Saud University-Science, 33*(2), p.101282. Available at: https://www.sciencedirect.com/science/article/pii/S1018364720303955
19. Goodchild-Michelman, I.M., Church, G.M., Schubert, M.G., Tang, T.C. (2023). Light and carbon: Synthetic biology toward new cyanobacteria-based living biomaterials. *Materials Today Bio*, p.100583. Available at: https://www.sciencedirect.com/science/article/pii/S2590006423000431
20. Gupta, J. (2023). E-waste: Policies and legislations for a sustainable green growth. *Waste Management and Resource Recycling in the Developing World*. https://doi.org/10.1016/B978-0-323-90463-6.00029-4.ElsevierInc
21. Gupta, J., Ghosh, P., Kumari, M, Swati, Thakur, I.S. (2022c). Solid waste landfill sites for the mitigation of greenhouse gases. In Biomass, Biofuels, Biochemicals (pp. 315–340). Elsevier.
22. Gupta, J., Kumari, M., Mishra, A., Akram, M., & Thakur, I. S. (2022). Agro-forestry waste management-A review. *Chemosphere, 287*, 132321.
23. Gupta, J., Rathour, R., Dupont, C., Mishra, A., & Thakur, I. S. (2022). Biogeochemical profiling and taxonomic characterization of municipal landfill site by metagenomic sequencing. *Bioresource Technology, 351*, 126936.
24. Gupta, J., Rathour, R., Dupont, C.L., Kaul, D., Thakur, I.S. (2020). Genomic insights into waste valorised extracellular polymeric substances (EPS) produced by Bacillus sp. ISTL8. *Environmental Research 192*, p.110277.
25. Gupta, J., Rathour, R., Kumar, M., & Thakur, I. S. (2017). Metagenomic analysis of microbial diversity in landfill lysimeter soil of Ghazipur Landfill Site, New Delhi, India. *Genome announcement, 5*(42), e01104-e1117.
26. Gupta, J., Rathour, R., Singh, R., & Thakur, I. S. (2019). Production and characterization of extracellular polymeric substances (EPS) generated by a carbofuran degrading strain Cupriavidus sp. ISTL7. *Bioresource Technology, 282*, 417–424.

27. Hashim, J.H., Hashim, Z. (2016). Climate change, extreme weather events, and human health implications in the Asia Pacific region. *Asia Pacific Journal of Public Health*, *28*(2_suppl), pp.8S-14S. Available at: https://journals.sagepub.com/doi/abs/https://doi.org/10.1177/1010539515599030
28. Islam, M.R., Khan, N.A. (2020). Threats, vulnerability, resilience and displacement among the climate change and natural disaster-affected people in South-East Asia: an overview. *Climate Change Mitigation and Sustainable Development*, pp. 111–138. Available at: https://www.taylorfrancis.com/chapters/edit/https://doi.org/10.4324/9780429029035-8/threats-vulnerability-resilience-displacement-among-climate-change-natural-disaster-affected-people-south-east-asia-overview-rezaul-islam-niaz-ahmed-khan
29. Jansson, J.K., Hofmockel, K.S. (2020). Soil microbiomes and climate change. *Nature Reviews Microbiology*, *18*(1), pp. 35–46. Available at: https://www.nature.com/articles/s41579-019-0265-7
30. Jung, J., Maeda, M., Chang, A., Bhandari, M., Ashapure, A., Landivar-Bowles, J. (2021). The potential of remote sensing and artificial intelligence as tools to improve the resilience of agriculture production systems. *Current Opinion in Biotechnology*, *70*, pp. 15–22. Available at: https://www.sciencedirect.com/science/article/pii/S0958166920301257
31. Kim, H.C., Jang, T.W., Chae, H.J., Choi, W.J., Ha, M.N., Ye, B.J., Kim, B.G., Jeon, M.J., Kim, S.Y., Hong, Y.S. (2015). Evaluation and management of lead exposure. *Annals of occupational and environmental medicine*, *27*(1), pp. 1–9. Available at: https://aoemj.biomedcentral.com/articles/https://doi.org/10.1186/s40557-015-0085-9
32. Kinney, P.L. (2018). Interactions of climate change, air pollution, and human health. *Current environmental health reports*, *5*, 179–186. Available at: https://link.springer.com/article/https://doi.org/10.1007/s40572-018-0188-x
33. Mandal, O., Krishnamoorthi, A., Singh, A., Kaur, R., Kalaiselvi, P., Tiwari, U., Mitra, M. (2024). Biofortification: Enhancing nutritional content in crops through biotechnology and fighting climate change. *Journal of Advances in Biology & Biotechnology* *27*(2), 186–210. Available at: http://public.paper4promo.com/id/eprint/1871/
34. Maqbool, A., Abrar, M., Bakhsh, A., Çalışkan, S., Khan, H.Z., Aslam, M. and Aksoy, E. (2020). Biofortification under climate change: the fight between quality and quantity. *Environment, climate, plant and vegetation growth*, pp.173–227. Available at: https://link.springer.com/chapter/https://doi.org/10.1007/978-3-030-49732-3_9
35. Markkanen, S., Anger-Kraavi, A. (2019). Social impacts of climate change mitigation policies and their implications for inequality. *Climate Policy* *19*(7), 827–844. Available at: https://www.tandfonline.com/doi/abs/https://doi.org/10.1080/14693062.2019.1596873
36. Mason, A.R.G., Salomon, M.J., Lowe, A.J., Cavagnaro, T.R. (2023). Microbial solutions to soil carbon sequestration. *Journal of cleaner production*, 137993. Available at: https://www.sciencedirect.com/science/article/pii/S0959652623021510
37. McLaughlin, B.C., Skikne, S.A., Beller, E., Blakey, R.V., Olliff-Yang, R.L., Morueta-Holme, N., Heller, N.E., Brown, B.J., Zavaleta, E.S. (2022). Conservation strategies for the climate crisis: An update on three decades of biodiversity management recommendations from science. *Biological Conservation*, *268*, 109497. Available at: https://www.sciencedirect.com/science/article/pii/S0006320722000507
38. Misra, A.K. (2014). Climate change and challenges of water and food security. *International Journal of Sustainable Built Environment*, *3*(1), 153–165. Available at: https://www.sciencedirect.com/science/article/pii/S221260901400020X
39. Mizdrak, A., Cobiac, L.J., Cleghorn, C.L., Woodward, A., Blakely, T. (2020). Fuelling walking and cycling: human powered locomotion is associated with non-negligible greenhouse gas emissions. *Scientific Reports*, *10*(1), 9196. Available at: https://www.nature.com/articles/s41598-020-66170-y
40. Mukherjee, S., Mishra, A., Trenberth, K.E. (2018). Climate change and drought: a perspective on drought indices. *Current climate change reports*, *4*, 145–163. Available at: https://link.springer.com/article/https://doi.org/10.1007/s40641-018-0098-x

41. Palinkas, L.A. and Wong, M. (2020). Global climate change and mental health. *Current Opinion in Psychology 32*, 12–16. Available at: https://www.sciencedirect.com/science/article/pii/S2352250X19300661
42. Patil, A. and Lamnganbi, M. (2018). Impact of climate change on soil health: A review. *International Journal of Chemical Studies 6*(3), 2399–2404. Available at: https://www.academia.edu/download/89092033/6-3-276-725.pdf
43. Potočnik, V., Gorgieva, S., Trček, J. (2023). From nature to lab: Sustainable bacterial cellulose production and modification with synthetic biology. *Polymers 15*(16), 3466. Available at: https://www.mdpi.com/2073-4360/15/16/3466
44. Raftery, A.E., Zimmer, A., Frierson, D.M., Startz, R., Liu, P. (2017). Less than 2 C warming by 2100 unlikely. *Nature Climate Change 7*(9), 637–641. Available at: https://www.nature.com/articles/nclimate3352.
45. Ramesh, S.V., Mangruthia, S.K., Praveen, S. (2021). Genetic modifications for disease resistance in crops. *Genetically Modified Crops in Asia Pacific*, p. 273. Available at: https://books.google.com/books?hl=en&lr=&id=KDMcEAAAQBAJ&oi=fnd&pg=PA273&dq=climate+change+Genetic+Modification+for+Disease+Resistance&ots=337TzSwVOf&sig=l1DhJ_FQ7HBaVSn6L3f6oZ_O6dg
46. Räsänen, A., Juhola, S., Nygren, A., Käkönen, M., Kallio, M., Monge Monge, A., Kanninen, M. (2016). Climate change, multiple stressors and human vulnerability: a systematic review. *Regional Environmental Change 16*, 2291–2302. Available at: https://link.springer.com/article/https://doi.org/10.1007/s10113-016-0974-7
47. Rivero, R.M., Mittler, R., Blumwald, E., Zandalinas, S.I. (2022). Developing climate-resilient crops: improving plant tolerance to stress combination. *The Plant Journal 109*(2), 373–389. Available at: https://onlinelibrary.wiley.com/doi/abs/https://doi.org/10.1111/tpj.15483
48. Shehzad, K. (2023). Extreme flood in Pakistan: Is Pakistan paying the cost of climate change? A short communication. *Science of The Total Environment*, *880*, 162973. Available at: https://www.sciencedirect.com/science/article/pii/S0048969723015917
49. Sparks, T.H., Menzel, A. (2002). Observed changes in seasons: an overview. *International Journal of Climatology: A Journal of the Royal Meteorological Society*, *22*(14), 1715–1725. Available at: https://rmets.onlinelibrary.wiley.com/doi/abs/https://doi.org/10.1002/joc.821?casa_token=mg1vhXwHEmAAAAAA:UEQHfBIAGO75QzJHUyXJhe_y1JE-c9U6VNGWoEEeV1Hakq2ZALJHt_vx_XPMH7aAt19a4hnURvS0wO8x
50. Srivastav, A.L., Dhyani, R., Ranjan, M., Madhav, S., Sillanpää, M. (2021). Climate-resilient strategies for sustainable management of water resources and agriculture. *Environmental Science and Pollution Research 28*(31), 41576–41595. Available at: https://link.springer.com/article/https://doi.org/10.1007/s11356-021-14332-4
51. Thoma, M.V., Rohleder, N., Rohner, S.L. (2021). Clinical ecopsychology: The mental health impacts and underlying pathways of the climate and environmental crisis. *Frontiers in Psychiatry*, *12*, 675936. Available at: https://www.frontiersin.org/articles/https://doi.org/10.3389/fpsyt.2021.675936/full
52. Tol, R.S. (2018). The economic impacts of climate change. *Review of Environmental Economics and Policy*. Available at: https://www.journals.uchicago.edu/doi/full/https://doi.org/10.1093/reep/rex027
53. Ussiri, D.A., Lal, R. (2017). *Carbon sequestration for climate change mitigation and adaptation* (pp. 287–325). Springer International Publishing. Available at: https://link.springer.com/content/pdf/https://doi.org/10.1007/978-3-319-53845-7.pdf
54. Von Schneidemesser, E., Monks, P.S., Allan, J.D., Bruhwiler, L., Forster, P., Fowler, D., Lauer, A., Morgan, W.T., Paasonen, P., Righi, M., Sindelarova, K. (2015). Chemistry and the linkages between air quality and climate change. *Chemical reviews 115*(10), 3856–3897. Available at: https://pubs.acs.org/doi/full/https://doi.org/10.1021/acs.chemrev.5b00089

55. Walsh, K.J., McBride, J.L., Klotzbach, P.J., Balachandran, S., Camargo, S.J., Holland, G., Knutson, T.R., Kossin, J.P., Lee, T.C., Sobel, A., Sugi, M. (2016). Tropical cyclones and climate change. *Wiley Interdisciplinary Reviews: Climate Change*, *7*(1), 65–89. Available at: https://wires.onlinelibrary.wiley.com/doi/abs/https://doi.org/10.1002/wcc.371?casa_token=KHTR0rKfjyMAAAAA:Hi6t4A3Jhl3Kkks4O3bHgTrzG6H1yyKxWcjfj0ue1oTVpnR0t4kUdwYHZwMckh2hp2COYuX4JrsT4EHs
56. Wheeler, T., Von Braun, J. (2013). Climate change impacts on global food security. *Science*, *341*(6145), 508–513. Available at: https://www.science.org/doi/abs/https://doi.org/10.1126/science.1239402
57. Williams, A.P., Abatzoglou, J.T., Gershunov, A., Guzman-Morales, J., Bishop, D.A., Balch, J.K., Lettenmaier, D.P. (2019). Observed impacts of anthropogenic climate change on wildfire in California. *Earth's Future* *7*(8), 892–910. Available at: https://agupubs.onlinelibrary.wiley.com/doi/abs/https://doi.org/10.1029/2019EF001210
58. Xie, S.P. (2020). Ocean warming pattern effect on global and regional climate change. *AGU Advances* *1*(1), e2019AV000130. Available at: https://agupubs.onlinelibrary.wiley.com/doi/abs/https://doi.org/10.1029/2019av000130
59. Zaidi, S.S.E.A., Mahas, A., Vanderschuren, H., Mahfouz, M.M. (2020). Engineering crops of the future: CRISPR approaches to develop climate-resilient and disease-resistant plants. *Genome Biology* *21*(1), 289. Available at: https://link.springer.com/article/https://doi.org/10.1186/s13059-020-02204-y
60. Zurell, D., Graham, C.H., Gallien, L., Thuiller, W., Zimmermann, N.E. (2018). Long-distance migratory birds threatened by multiple independent risks from global change. *Nature Climate Change* *8*(11), 992–996. Available at: https://www.nature.com/articles/s41558-018-0312-9
61. Gupta, J., Verma, A. (2024a). Microbiology: The Science to the World of Microbes. In: Gupta, J., Verma, A. (eds) Microbiology-2.0 Update for a Sustainable Future. Springer. https://doi.org/10.1007/978-981-99-9617-9_1
62. Sivakumar, H., Gupta, J., Verma, A. (2024). Revolutionizing water quality through microbial interactions in aquatic ecosystems for pollution abatement. In: Gupta, J., Verma, A. (eds). *Microbiology-2.0 Update for a Sustainable Future*. Springer. https://doi.org/10.1007/978-981-99-9617-9_2
63. Eusebius, A., Jacob, C., Nandra, A., Gupta, J. (2024). Microbial communities in glacial environments: key players in cryosphere carbon cycling and the emergence of CECs. In: Gupta, J., Verma, A. (eds) *Microbiology-2.0 update for a sustainable future*. Springer. https://doi.org/10.1007/978-981-99-9617-9_3
64. Bajaj, T., Bhattacharya, S., Gupta, J. (2024). Microbial Genomics—the changing technological landscape of microbiology via NGS. In: Gupta, J., Verma, A. (eds) *Microbiology-2.0 update for a sustainable future*. Springer. https://doi.org/10.1007/978-981-99-9617-9_14
65. Chauhan, K., Kriti, Verma, A. (2024). Biofuel: generations, current status, sustainability impact, feedstock, and production technologies. In: Gupta, J., Verma, A. (eds) *Microbiology-2.0 update for a sustainable future*. Springer. https://doi.org/10.1007/978-981-99-9617-9_16
66. Sharma, R.K., Singh, T.P., Azad, D., Adhikari, H., Verma, A. (2024). Role of biochar as a sustainable sorbent for fipronil removal from aqueous and soil environments. In: Gupta, J., Verma, A. (eds) *Microbiology-2.0 update for a sustainable future*. Springer. https://doi.org/10.1007/978-981-99-9617-9_9
67. Gupta, S., Bhattacharyya, A., Gupta, J. (2024b). Microbial marvels: Transforming the way we diagnose. In: Gupta, J., Verma, A. (eds) *Microbiology-2.0 update for a sustainable future*. Springer. https://doi.org/10.1007/978-981-99-9617-9_8
68. Kriti, Kumari, S., Ahuja, M., Verma, A. (2024). Advances of graphene oxide in the field of microbiology. In: Gupta, J., Verma, A. (eds). *Microbiology-2.0 update for a sustainable future*. Springer. https://doi.org/10.1007/978-f981-99-9617-9_11

Design and Fabrication of Industrial Pick and Place Delta Robot

Piyush Kushwaha, Jatin Kumar, Dhruv Ajit, Abhinav Chauhan, Naman Jain, and Akarsh Verma

Abstract The need for small and medium industries to reduce errors and downtime in their production line has led to a serious consideration for the development of economical and optimized delta robots for various operations required in assembly lines and warehouses. For operations like pick and place, packaging, and sorting, the delta robot shines above all other types of solutions till now in parameters such as high accuracy, high pick and place rate, low power consumption. This study is a brief process of design, development, and fabrication of delta robots featuring high performance and customization options, which is required in a variety of industries. The dimensions of all the links of delta robot are highly optimized by forward and inverse kinematics. Followed by verifications of these driving equations in MATLAB Simulink. Finally obtaining required dimensions of links using workspace analysis. For fabrication of delta robot, new methods are utilized to reduce the overall inertia of moving components and enable high acceleration and deceleration of end effector to increase pick and place rate. These methods include the use of belt and pully arrangements to obtain perfect blend of torque and speed and Fused Deposition Modeling (FDM) method for lightweight components.

Keywords Pick and place robot · Delta robot · Magnetic gripper · Parallel robot · Closed chain kinematics

P. Kushwaha (✉) · J. Kumar · D. Ajit · A. Chauhan · N. Jain (✉)
Department of Mechanical Engineering, ABES Engineering College, Ghaziabad, India
e-mail: piyushkushwaha116@gmail.com

N. Jain
e-mail: namanjainyati@gmail.com

A. Verma
Department of Mechanical Engineering, University of Petroleum and Energy Studies (UPES), Dehradun 248007, India

© The Author(s), under exclusive license to Springer Nature Singapore Pte Ltd. 2024
A. Chaudhary et al. (eds.), *Unraveling New Frontiers and Advances in Bioinformatics*,
https://doi.org/10.1007/978-981-97-7123-3_18

1 Introduction

Industrial automation is a driving force behind the modern manufacturing ecosystem. The need for precise and efficient pick-and-place operations has led to the increasing adoption of new advanced robots such as Cartesian robots [1] which shine in confined spaces with precise movement needs, while articulated robots [2] offer versatility and large workspaces. Delta robots [3] reign supreme in high-speed pick-and-place situations, whereas SCARA [4] robots excel at rapid horizontal tasks within limited space. Cobots (Collaborative Robot) [5] bring safety and ease of use to collaborative workspaces.

Among the various types of industrial robots, the Delta robot stands out for its unique advantages. They belong to the class of parallel robots, meaning their end effector, typically a gripper or tool, is manipulated by multiple independent robotic arms working in concert. Therefore, with their distinctive three-armed architecture connected by parallelogram joints, offers high structural rigidity, speed with high acceleration and deceleration capabilities with only disadvantage is its small working volume compared to other open kinematic serial robots. Emerging in the early 1980s from the mind of Professor Reymond Clavel [6] at EPEL, Switzerland, since then, continuous advancements in materials, motor technology, and control systems helps it to revolutionize industrial pick-and-place tasks. The unique design of closed chain kinematics of Delta robot [7] offers advantages over its conventional counterparts as it eliminates the need for bulky gears and transmissions in each joint, resulting in a compact and lightweight design. This translates to impressive speed and agility, allowing it to perform rapid pick-and-place tasks with remarkable accuracy. Among all kinds of industrial robots which exist presently, delta robots are much simpler and more economical than other solutions. They may require an initial investment, but the long-term benefits are hard to ignore. Delta robots, with their distinctive three-armed architecture and lightning speed, are poised to revolutionize industries across the globe. The future of these robotic marvels is even more exciting like AI-powered robots learning and optimizing their movements, or Cobots seamlessly collaborating with humans in shared workspaces. Industry 4.0 and automation are presently taking shape thanks to technologies such as AI, 5G, advanced automation, advanced sensing, and many others. These are significantly altering how various industries are operating and how their processes are carried out.

Joochim et al. [8] discusses the design, construction, and testing of a prototype of a delta robot. The paper looks into the kinematics of the robot, which is the study of its motion and geometry. It covers both forward and inverse kinematics, which involve the calculation of the end-effector position from the joint angles and vice versa. The mechanical design of the robot, including its base, arms, joints, motors, belts, gears, and end-effector, is also discussed. Additionally, the paper explores the electrical and control design, which involves the design and implementation of the electrical and electronic systems that enable the robot to function. This includes the power supply, sensors, actuators, controllers, communication protocols, and programming languages. The paper presents the results of testing the robot's performance and accuracy, and discusses its potential applications in industry, particularly in food packaging. The paper concludes with a summary of its contributions and limitations, and suggestions for future work.

Hadfield et al. [9] presents the kinematic analysis of a Delta robot, using Conformal Geometric Algebra (CGA). The authors derive the forward and inverse kinematic solutions, as well as the forward and inverse Jacobian matrices, for a Delta robot in CGA. They also implement mathematics in Python and C and simulate a physical robot model in Unity3D. The paper demonstrates the advantages of using CGA for robotic analysis, such as compact and intuitive expressions, geometric interpretation, and easy computation.

Khorasani et al. [10] presents an optimization problem for the design and implementation of a Delta robot, a parallel robot with three degrees of freedom, for pick and place applications. The authors propose to optimize the installation configuration of the robot, which refers to the height and orientation of the robot base, to achieve the highest rate of pick and place operations per minute (PPM) and the lowest energy consumption. The optimization problem is formulated as a function of the actuator torque and speed limitations and solved using a genetic algorithm. The optimal configuration is used to design and implement a prototype robot, called KNTU Delta, and its performance is evaluated in real-time experiments. The paper provides a detailed mathematical analysis of the kinematics, dynamics, and trajectory planning of the Delta robot, and shows the results of the optimization in graphical and tabular forms. The paper also discusses the physical intuition behind the optimal configuration and compares it with a non-optimized one.

Makwana et al. [11] presents a model-based motion simulation of a delta parallel robot using Simulink's SimScape environment. The authors describe the kinematics and dynamics of the robot and use a physical modeling approach to solve the forward kinematics problem. The paper also reports the results of an experimental analysis, where the robot is controlled to follow a circular path. The paper claims that the simulation can provide a natural reference for the design and control of parallel robots, and that the model can handle complex trajectories and dynamics.

Hoai et al. [12] discussed the design and implementation of a low-cost delta robot arm for pick and place applications, utilizing computer vision. The authors have used cost-effective materials and methods such as stepper motors, Arduino Uno, laser cutting, and 3D printing to reduce the fabrication and maintenance costs. They have developed a vision system to determine the 3D coordinates and sizes of the objects in the workspace and control the gripper to open at a distance equal to the object size. The paper also presents an application where the robot arm picks up objects on a moving conveyor belt, validating the design. The experimental results show that the robot system works correctly, the robot arm moves smoothly, and the vision system has a small error.

2 Material and Methods

2.1 Design and Operating Principle

The delta robot, an example of parallel-link manipulator design [13], boasts a unique architecture that allows its end effector's three degrees of freedom: movement along the X, Y, and Z axes. Its core structure comprises a fixed base and a moving platform, the latter housing the end effector. This distinct design categorizes the delta robot as a parallel manipulator, as the moving platform, guided by the interconnected arms, remains consistently parallel to the fixed base throughout its operation, avoids any unwanted rotational or tilting tendencies. This unparalleled characteristic, ensuring precise and predictable movement, stems from the robot's closed-loop kinematic chain mechanism [14].

The delta robot's ability owes much to its three identical kinematic loops, each composed of an upper arm and a lower arm. The upper arms, independently actuated by dedicated actuators, extend from the fixed base, while the lower arms connect to the moving platform via universal joints (or ball joints) and pivot at their connection points with the upper arms. This configuration forms a parallelogram between the upper arm and the base for each loop.

Furthermore, an additional degree of rotation [15] can be mounted in between the moving platform and end-effectors to increase the ability of robot to do distinct works without any additional configuration. Delta robots require more advanced computational methods due to the interconnected nature of their linkages. This complexity necessitates the use of powerful control systems capable of handling the intensive calculations required for rapid and precise robot movements. The foundation of robot control resides in the kinematic Eqs. (16), which establish the fundamental relationship between joint angles (actuator inputs), link lengths (geometric parameters), and the position and orientation (X, Y, Z, and potentially additional degrees of freedom) of the end effector. Robots typically employ two key sets of kinematic equations: forward kinematics and inverse kinematics [17–19]. The vast majority of robot control systems leverage inverse kinematics due to the prevalence of pre-programmed trajectories or real-time position inputs (e.g., from vision systems).

Knowing the precise dimensions of a robot's workspace [20, 21] is crucial when selecting it for any specific task. The workspace defines the reachable area of the robot's end effector, essentially dictating its operational limits. For tasks like pick and place and sorting, delta robots excel due to their unique workspace shape, resembling a truncated cone. The truncated cone shape of a delta robot's workspace offers several advantages for these tasks. Firstly, it provides a large reachable area directly below the robot, ideal for picking up objects from a conveyor belt or placing them into containers. Secondly, the workspace wider towards the top, makes the operations more versatile and improves the agility of robot if it's using some additional rotational axis.

The size of a delta robot's workspace can be customized to meet specific needs as shown in Fig. 1. During the design phase, engineers can adjust the lengths of

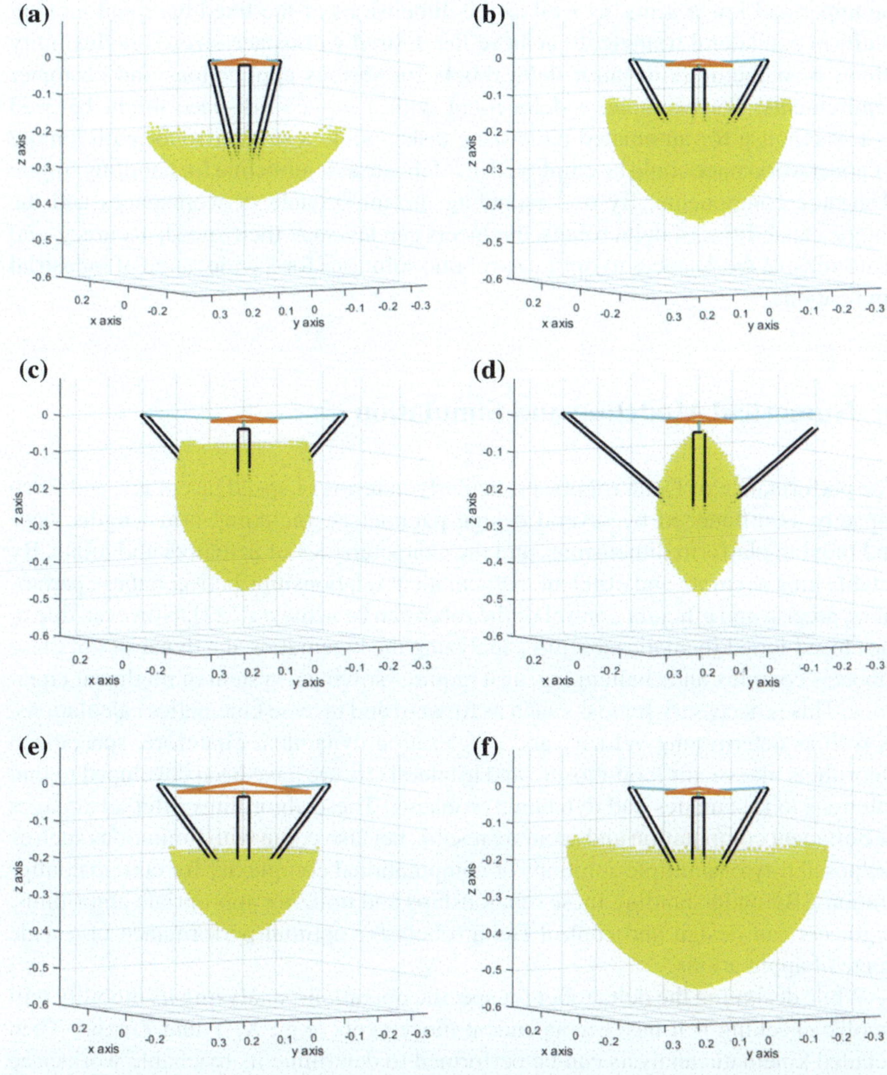

Fig. 1 The working envelope of Delta robot **a** Robot has a working envelope with large horizontal space but smaller in vertical axis when upper arm is much smaller than lower arms. **b** Robot has large working envelope in both axis when ratio of upper arm length to lower arm length is about 0.5. **c** Robot has large working space in vertical axis but smaller shape in horizontal axis when ratio of upper arm to lower arm is 0.7. **d** Robot has large working space only in vertical axis only when ration of upper arm length to lower arm length is unity. **e** Robot work envelope becomes smaller in size when length of arms is same as in case B, but size of base equilateral triangle is increased. **f** Robot has very large work envelope when ratio of upper arm length to lower arm length is about 0.25

the upper and lower arms, as well as the dimensions of the fixed base and moving platform equilateral triangle, to achieve the desired workspace size. This flexibility allows manufacturers to tailor delta robots for various applications and customer requirements. For example, a delta robot with a large workspace might be used in a warehouse for automated palletizing tasks, while a smaller robot with a more compact workspace could be employed in a delicate assembly line for handling fragile electronic components. By understanding the importance of workspaces and the unique capabilities of delta robots, engineers can leverage their speed, accuracy, and customizable workspaces to optimize robotic solutions for a wide range of industrial applications.

3 Numerical Modeling and Simulation

The performance of Delta robots, particularly in terms of speed, accuracy, and force capacity, is influenced by several design parameters, including arm lengths, base and moving platform dimensions, and the characteristics of actuators and joints. By establishing accurate and efficient mathematical relationships between these parameters, precise and efficient control of the robot can be achieved [22]. However, due to the closed-loop kinematic structure, analyzing the kinematics and dynamics of these robots is complex and challenging, as it requires solving a system of nonlinear equations. This is necessary for tasks such as forward and inverse kinematics calculations, as well as determining velocity and acceleration dynamics. Therefore, specialized algorithms like vector loop-closure and trilateral closure have been developed to find solutions to kinematics and dynamics problems. These algorithms offer advantages like efficient computation and accurate results, but also come with limitations such as the possibility of multiple solutions or computational complexity for certain configurations. By understanding these relationships and utilizing appropriate algorithms, engineers can design and control Delta robots for optimal performance in a wide range of applications.

While designing the delta robot for specific operations, analyzing its mobility will involve checking if it has 3 independent movements (e.g., X, Y and Z axis). Then detailed kinematic analysis can be performed to determine its reachable workspace and path planning for picking and placing objects. Finally, dynamics analysis would consider the forces and torques involved in these movements to ensure the robot operates safely and efficiently.

Figure 2a shows the kinematic diagram of Delta Robot where the origin of Cartesian reference frame {B} is the centroid of fixed base equilateral triangle with three upper arm length L and lower arm length l. The origin of reference frame {P} is the centroid of the moving platform equilateral triangle. B_i (for i = 1,2,3) are the revolve joints connection points on fixed-base similarly P_i (for i = 1,2,3) are the universal joint connection point for the moving platform.

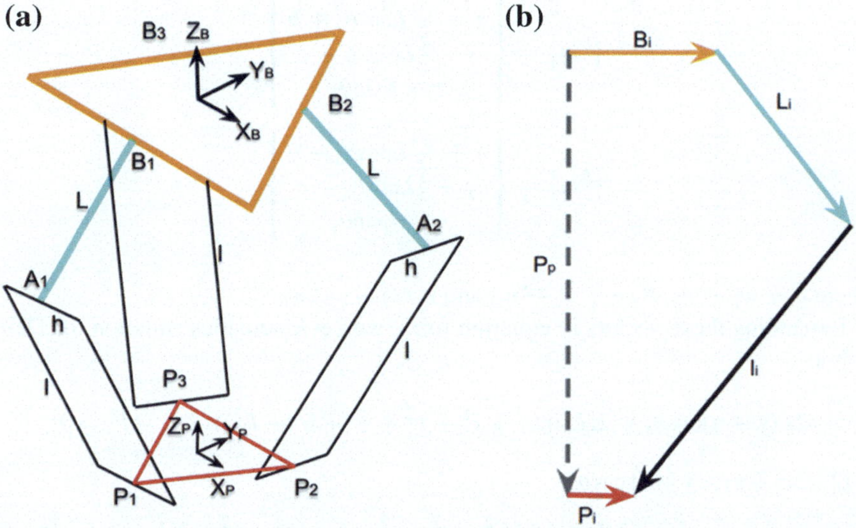

Fig. 2 **a** Delta robot kinematics diagram **b** Vector representation of each links

Besides its complex geometry, mobility of delta robot can easily be proved to be 3 using Chebychev–Grübler–Kutzbach criterion [23] for degree of freedom. Using mobility formula for kinematic linkage:

$$M = 6(N - 1 - J) + \sum\nolimits_{i=0}^{J} f_i$$

where M is the mobility, or number of degrees-of-freedom, N is the total number of links, including ground, J is the number of one-dof joints and f_i is dof of joints. Values of these parameters can be determined from Kinematic Diagram of Delta Robot, where value of N is 11, J is15 and $\sum_{i=0}^{J} f_i$ is 27. On solving the equation by putting these values, the mobility of the delta robot is given out as 3 which is correct.

The kinematics analysis for delta robot can be derived from vector loop closure Eq. [24] of each individual set of links connected to the base and platform shown in Fig. 2b.

$$\{^{B}B_i\} + \{^{B}L_i\} + \{^{B}l_i\} + \{^{B}P_P\} - \{^{P}P_i\} = 0$$

Substituting all the values into the above vector loop closure equation.

$$\{^{B}l_1\} = \begin{Bmatrix} x \\ y + L\cos\theta_1 + a \\ z + L\sin\theta_1 \end{Bmatrix}$$

$$\{^B l_2\} = \begin{Bmatrix} x - \frac{\sqrt{3}}{2} Lcos\theta_2 + b \\ y - \frac{1}{2} Lcos\theta_2 + c \\ z + Lsin\theta_2 \end{Bmatrix}$$

$$\{^B l_3\} = \begin{Bmatrix} x + \frac{\sqrt{3}}{2} Lcos\theta_3 - b \\ y - \frac{1}{2} Lcos\theta_3 + c \\ z + Lsin\theta_3 \end{Bmatrix}$$

where $a = w_B - u_p$, $b = \frac{s_P}{2} - \frac{\sqrt{3}}{2} w_B$ and $c = w_P - \frac{1}{2} w_B$

Expanding these vectors in equation form, we get kinematics equation for Delta Robot:

$$2L(y + a)\cos\theta_1 + 2zL\sin\theta_1 + x^2 + y^2 + z^2 + a^2 + L^2 + 2ya - l^2 = 0$$

$$-L\left(\sqrt{3}(x + b) + y + c\right)\cos\theta_2 \\ +2zL\sin\theta_2 + x^2 + y^2 + z^2 + b^2 + c^2 + L^2 + 2xb + 2ya - l^2 = 0$$

$$L\left(\sqrt{3}(x - b) - y - c\right)\cos\theta_2 \\ +2zL\sin\theta_2 + x^2 + y^2 + z^2 + b^2 + c^2 + L^2 - 2xb + 2ya - l^2 = 0$$

These are the driving equation for kinematics of delta robot, as these equations contains dimensions parameters of the robot (i.e. L, l, a, b, and c), actuators angles (i.e. θ_1, θ_2 and θ_3) and three coordinates of end effector (i.e. x, y, and z). Understanding and effectively solving the kinematics of Delta robots is crucial for their successful operation in various tasks, including pick-and-place, sorting, and packaging. By grasping these equations and algorithms, engineers can accurately control the robot's movements and ensure precise positioning, optimizing its performance in diverse applications.

All the calculations of kinematics and dynamics equations are carried out using dimensions specified in Table 1. Based on the requirements the forward kinematics and inverse kinematics can also be calculated using these equations. Unlike serial robots, in which calculation of forward kinematics are straightforward and inverse kinematics are difficult to analyze, delta robot's inverse kinematics can be easily calculated using above mentioned equation and analysis of forward kinematic is difficult to synthesize. To address the complexity of forward kinematics in Delta robots, a specialized algorithm called the "three sphere intersection algorithm" is employed. This algorithm leverages the geometric relationships between the robot's linkages and spheres centered at the actuator joints to efficiently compute the end-effector position.

To check for our derived the forward and inverse kinematic equation for the delta robot. The tool used for modeling and simulation is MATLAB Simulink 2018b (Academic license). Simulink is a block-based modeling environment used to test model-based design of robots. The model of delta robot was created by adding

Table 1 Dimensional parameters of robot

S. No	Meaning	Value (mm)
1	Base equilateral triangle side (S_B)	173
2	Platform equilateral triangle side (S_P)	52
3	Upper arm length (L)	150
4	Lower arm parallelogram length (l)	350
5	Lower arm parallelogram width (h)	45
6	Distance from {0} to near base side (r_B)	50
7	Distance from {P} to near platform side (r_p)	30

elements of appropriate dimensions (as specified in Table 1), specifying joint types (revolute and U-joints) in between the links, and finally implementing kinematic equation using mathematical function blocks to express forward and inverse kinematics relations. After modeling, simulations are carried out using motion input to actuate joints using trajectory generators blocks (Sine wave and steps).

Figure 3b reveals the intricate workings of the Delta robot's Simulink model, divided into five key sections. Sine wave generators act as digital conductors, composing a circular trajectory for the end effector. This desired path is then fed into the inverse kinematics block, which translates it into precise joint angles for each actuator. These angles animate the robot model (detailed model shown in Fig. 3a), a digital twin mirroring its real-world counterpart. Meanwhile, the forward kinematics block acts as a clairvoyant, predicting the end-effector position based on simulated joint angles and providing valuable feedback for control adjustments. Finally, scopes visualize the data symphony, displaying joint angles, end-effector positions, and other crucial metrics for analysis and optimization. This harmonious interplay of blocks allows engineers to test design iterations, anticipate potential errors, and ultimately refine the Delta robot's performance [25], paving the way for exceptional precision and agility in real-world tasks.

The simulation results are accurate. The calculations for the Delta robot's movement are spot-on. Figure 4b shows the end effector smoothly tracing out the planned circular path in 3D space, just like we intended. Figure 4c reveals the joint angles calculated for each arm, the secret sauce behind the robot's graceful movement. Finally, when we compare the planned movements with the actual movements, everything lines up perfectly with zero errors. This means our kinematic equations are working flawlessly, ensuring the robot will move exactly as we expect it to. This paves the way for a super-precise Delta robot, ready to be used for any operations like pick and place, sorting, and even adhesive dispensing.

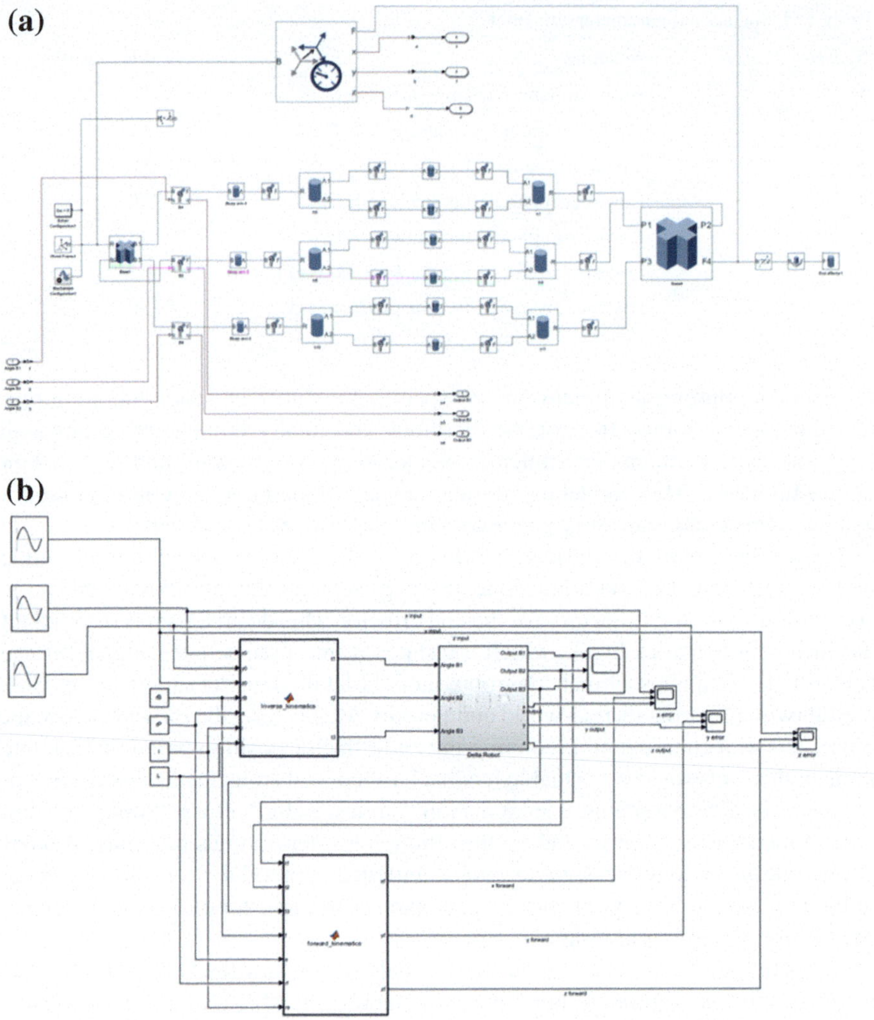

Fig. 3 **a** Modeling of delta robot using blocks of cylinders and various joints type block. **b** Complete Simulink model for delta robot containing inverse kinematics, forward kinematics, and trajectory blocks

4 Fabrication

With the optimal dimensions for the desired workspace obtained using MATLAB simulation, Next step is to create a digital working model of the delta robot which can be easily fabricated using materials and method available to us. Using Autodesk Fusion 360 (Education License) CAD software, we meticulously crafted a precise

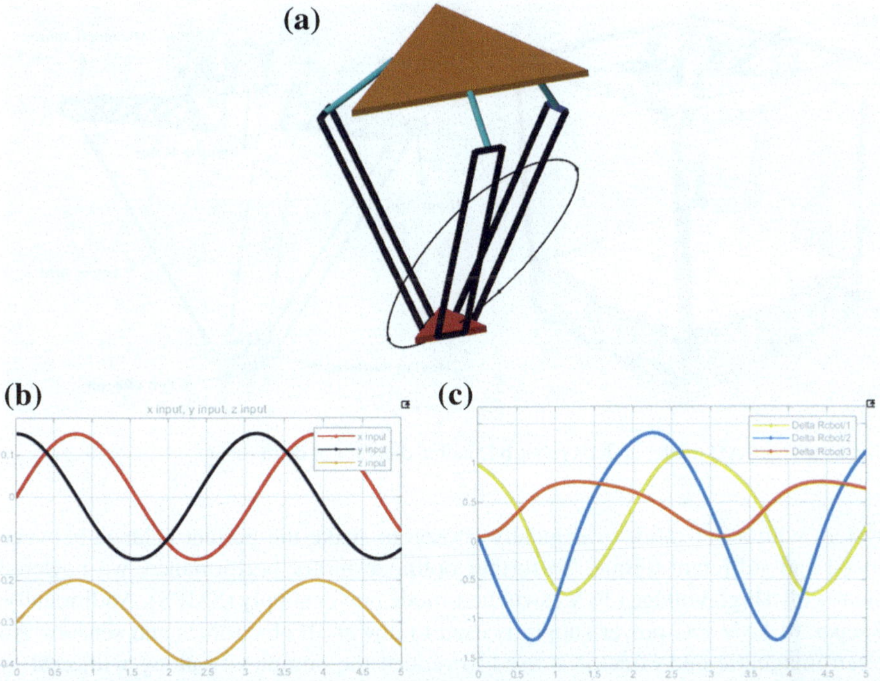

Fig. 4 **a** Image of motion simulation of delta robot Simulink model **b** Trajectory output signal to inverse kinematics function **c** Delta robot's actuated joint angle plot as received from inverse kinematics block

digital model of the Delta robot as shown in Fig. 5a. To fabricate this design, we embraced a diverse array of fabrication techniques.

For the robot's outer frame and base, sheet metal proved the ideal choice. Compared to options like aluminum extrusion profiles, it offered both cost-effectiveness and additional benefits. The resulting structure was not only sturdy and rigid, but also bulky enough to dampen vibrations effectively. However, for the robot's moving parts, minimizing inertia was paramount for achieving high acceleration and deceleration. This led us to utilize Fused Deposition Fabrication (FDM) additive manufacturing for all three upper arms and the moving platform. This technique not only allowed for complex shapes, but also ensured lightweight components crucial for optimal performance.

The six lower arms, tasked with connecting the moving platform to the base, were crafted from sturdy 4 mm carbon fiber rods. To ensure smooth and precise rotation for the upper arms, we opted for compact 8 mm inner diameter needle bearings. These offered a smaller form factor compared to other bearing types while still boasting a high load rating.

Now, to power the robot's movement, we selected three Nema 17 stepper motors, each with a torque rating of 4.2 kg-cm. These were paired with a belt and pulley

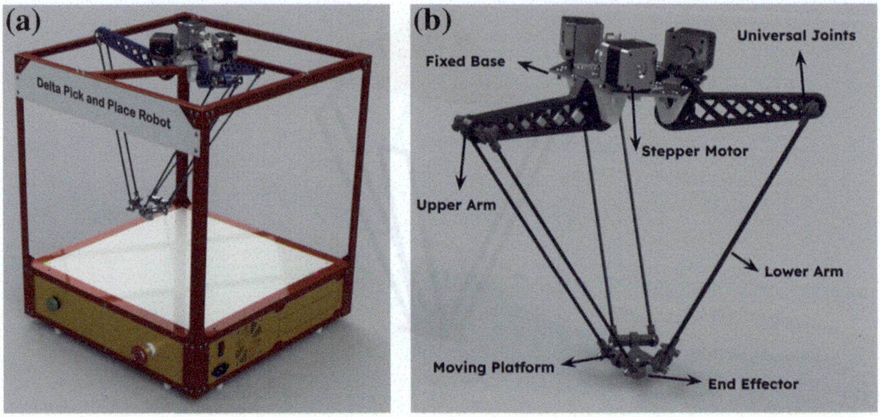

Fig. 5 **a** Full CAD model of delta robot **b** Detailed diagram of delta robot

system with a 3:1 ratio, strategically chosen to strike the perfect balance between speed and sufficient torque. To further optimize motor performance, we powered them with a high voltage (36 V) switched-mode power supply (SMPS). Additionally, a separate 12 V DC power supply is used to power all electronics and sensors. An electromagnetic end-effector is used to reduce the complexity along with cost of overall system.

5 Materials

It is essential to reduce the inertia of moving links of the robot to enable pick and place operation with high speed and fast acceleration and deceleration. Due to this reason various materials are used for different links using Fused Deposition Modeling (FDM) technology. Various thermoplastics polymer used for various links are Polylactic acid (PLA), Acrylonitrile Butadiene Styrene (ABS) and Polyethylene terephthalate glycol (PETG). The greatest advantage of using FDM process for fabrication of links is that components can be allowed to be partially hollow (by using infill parameter) from inside without compromising strength. That's why three upper arms and moving platform are manufactured using FDM process with 20 percent infill. To further reduce the inertia of the lower arms carbon fiber rods are used instead of conventional aluminum tubes. The frame of delta robot is manufactured by mild steel sheet of 1.5 mm thickness using sheet metal fabrication technique including bending which increases the stiffness of frame.

6 Conclusion

The evaluation focused on critical performance metrics, including speed, accuracy, and force capacity. Comparisons were drawn against existing solutions, shedding light on the robot's capabilities.

- **Speed Evaluation**: The measured the time required for the robot to complete a pick and place cycle, transferring a 100 g object from a conveyor belt to a container. The robot achieved an impressive average speed of 2.5 cycles per second, comparable to or surpassing other delta robots documented in the literature [8–11].
- **Accuracy Assessment**: The deviation of the robot's end-effector position from the desired trajectory was meticulously analyzed.
  Figure 6 shows the testing result of fully functional Delta Robot. The black square blocks are intended trajectory and blue square blocks are the actual trajectory resulting from end effector motion. The analysis revealed an maximum error of less than 0.8 millimeters. This demonstrates the robot's ability to move with a high degree of control and repeatability, ensuring reliable task execution.
- **Force Capacity Testing**: The determined the maximum payload the robot could handle without compromising stability or damaging components. The robot demonstrated the ability to lift up to 500 g, meeting the requirements for most pick and place tasks.

Fig. 6 Accuracy test result

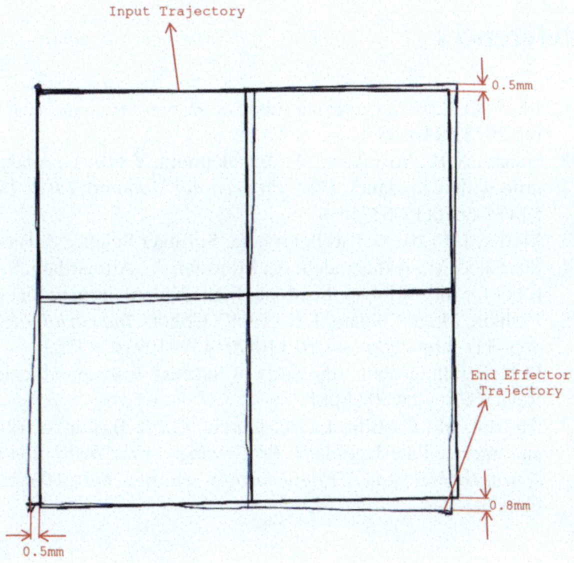

Advantages and Limitations: The design boasts several advantages:

- **Cost-Effectiveness**: By utilizing materials such as sheet metal, FDM, and a belt and pulley system, they minimized fabrication and maintenance costs.
- **Optimized Dimensions**: The fine-tuned link dimensions and installation configurations helps to maximize pick and place rates while minimizing energy consumption.
- **Enhanced Versatility**: An additional degree of rotation for the end-effector enhances the robot's adaptability.

However, certain limitations exist:

- **Workspace Size**: Sacrificing workspace size for speed and agility may restrict applicability in specific scenarios.
- **Complex Kinematics**: Solving forward and inverse kinematics problems required specialized algorithms and intensive computations.

Future Directions:

- **Workspace Expansion**: Adjusting link lengths and base/platform dimensions to increase the workspace.
- **Vision Integration**: Implementing a vision system for object recognition and localization.
- **Advanced Control Algorithms**: Enhancing accuracy and stability through sophisticated control systems.

In summary, the innovative design offers a promising solution for industrial pick and place applications, with room for further enhancements.

References

1. Bates, D. (2013). Cartesian robotics. *Representations, 124*(1), 43–68. https://doi.org/10.1525/rep.2013.124.1.43
2. Hanafusa, H., Yoshikawa, T., & Nakamura, Y. (1981). Analysis and control of articulated robot arms with redundancy. *IFAC Proceedings Volumes., 14*(2), 1927–1932. https://doi.org/10.1016/S1474-6670(17)63754-6
3. Merlet, J.P. (2012). Parallel robots. Springer Science & Business Media.
4. Shariatee, M., Akbarzadeh, A., Mousavi, A., Alimardani, S. Design of an economical SCARA robot for industrial applications. IEEE Xplore. https://doi.org/10.1109/ICRoM.2014.6990957
5. Peshkin, M., & Colgate, J. E. (1999). Cobots. *Industrial robot: An international journal, 26*(5), 335–341. https://doi.org/10.1108/01439919910283722
6. Delta Parallel Robot—the Story of Success. www.parallemic.org. https://www.parallemic.org/Reviews/Review002.html
7. Agustín, M., Castillo, E. M., Garcia, G., & Bashir, A. (2006). Delta robot: Inverse, direct, and intermediate Jacobians. *Proceedings of the Institution of Mechanical Engineers, Part C: Journal of Mechanical Engineering Science., 220*(1), 103–109. https://doi.org/10.1243/095440606x78263

8. Kaewkorn, S., Joochim, C., Keeratiwintakorn, P., Kunapinun A. (2020). Development of pick and place delta robot. *Advances in Intelligent Systems and Computing*. Published online January 1, 2020:475–486. https://doi.org/10.1007/978-3-030-40271-6_47
9. Hadfield, H., Wei, L., Lasenby, J. (2020). The forward and inverse kinematics of a delta robot. *Lecture notes in Computer Science*. Published online January 1, 2020:447–458. https://doi.org/10.1007/978-3-030-61864-3_38
10. Khorasani, A., Gholami, S., Taghirad, H.D. (2015). Optimization of KNTU Delta robot for pick and place application. Published online October 1, 2015. https://doi.org/10.1109/icrom.2015.7367772
11. Makwana, M. A., & Patolia, H. P. (2021). Model-based motion simulation of delta parallel robot. *Journal of Physics: Conference Series., 2115*(1), 012002. https://doi.org/10.1088/1742-6596/2115/1/012002
12. Hoai, P., Cong, V., Thai, Hiep. (51). Design a low-cost delta robot arm for pick and place applications based on computer vision. *FME Transactions 51*(1):99–108. https://doi.org/10.5937/fme2301099p
13. Arai, T., Stoughton, R., Homma, K., Adachi, H., Nakamura, T., Nakashima, K. (1991). Development of a parallel link manipulator. Published online January 1, 1991. https://doi.org/10.1109/icar.1991.240570
14. An approach to modeling closed-loop kinematic chain mechanisms, applied to simulations of the da vinci surgical system. *Acta Polytechnica Hungarica* 2019;16(8). https://doi.org/10.12700/aph.16.8.2019.8.3
15. Azmoun, M., Rouhollahi, A., Masouleh, M. T., Kalhor, A. (2018). Kinematics and control of a 4-DOF delta parallel manipulator. Published online October 1, 2018. https://doi.org/10.1109/icrom.2018.8657618
16. Zhang, N.J., Shi, N.L., Gao, N.R. (2009). None Chaoyang Lian. The mathematical model and direct kinematics solution analysis of Delta parallel robot. Published online January 1, 2009. https://doi.org/10.1109/iccsit.2009.5234909
17. Hadfield, H., Wei, L., Lasenby, J. (2020). The forward and inverse kinematics of a delta robot. *Lecture notes in Computer Science*. Published online January 1, 2020:447–458. https://doi.org/10.1007/978-3-030-61864-3_38
18. Hung, N., Rego, F., Quintas, J., et al. (2022). A review of path following control strategies for autonomous robotic vehicles: Theory, simulations, and experiments. Published online December 7, 2022. https://doi.org/10.1002/rob.22142
19. Kansal, S., & Mukherjee, S. (2022). Vision-based kinematic analysis of the Delta robot for object catching. *Robotica, 40*(6), 2010–2030. https://doi.org/10.1017/S0263574721001491
20. Kakkeri, S., Sunil, S., Ashok Kumar, K., Koti, V., Giriswamy, B.G., Nagaraj, P.B. (2022). Singularity localities in robot work envelope comparative analysis. Published online November 18, 2022. https://doi.org/10.1109/iihc55949.2022.10059938
21. Gallardo-Alvarado, J. (2016). Kinematic analysis of parallel manipulators by algebraic screw theory. Springer International Publishing.
22. Stock, M., & Miller, K. (2003). Optimal kinematic design of spatial parallel manipulators: Application to linear delta robot. *Journal of Mechanical Design, 125*(2), 292–301. https://doi.org/10.1115/1.1563632
23. Gogu, G. (2005). Chebychev–Grübler–Kutzbach's criterion for mobility calculation of multi-loop mechanisms revisited via theory of linear transformations. *European Journal of Mechanics—A/Solids, 24*(3), 427–441. https://doi.org/10.1016/j.euromechsol.2004.12.003
24. Escorcia, M., & Chemori, A. (2022). *Hipolito Aguilar Sierra*. Elsevier.
25. Rachedi, M., Hemici, B., & Bouri, M. (2015). Design of an H∞ controller for the Delta robot: Experimental results. *Advanced Robotics, 29*(18), 1165–1181. https://doi.org/10.1080/01691864.2015.1046924
26. Solomon, I.J., Sevvel, P., Gunasekaran, J. (2020). A review on the various processing parameters in FDM. *Materials Today: Proceedings*. Published online June 13, 2020. https://doi.org/10.1016/j.matpr.2020.05.484

Free Vibration Analysis of Variable Stiffness Laminate Composite Shell Panel Using HSDT and FE Approach

Prasoon Kumar, Ravi Kumar, and Chetan Kumar Hirwani

Abstract In this current research work, the modal analysis of variable stiffness composite laminate (VSCL) shell panels is analyzed. This analysis is facilitated by a numerical model based on higher-order shear deformation theory (HSDT) and finite element (FE) approaches. The VSCL configuration is achieved by incorporating curved fibers as reinforcement. The displacement kinematics of the shell are modelled up to the third order of thickness coordinates, ensuring an accurate representation of the shell's behavior. The physical model of the shell panel is discretized using nine-noded isoperimetric elements, each offering a comprehensive depiction of the panel's geometry and behavior with 81 degrees of freedom per element. The governing differential equation is derived using the energy method, providing a robust foundation for the numerical model. To verify the model's utilization characteristics, element convergence and accuracy tests are conducted initially. Further, the capability of the developed kinematic model is expanded by exploring various parametric examples. These examples include investigating the influences of curvature ratio, shell thickness ratio, orthotropic ratio, and boundary conditions on the modal analysis of laminated composite cylindrical structures. Each parametric study is discussed in detail to provide comprehensive insights into the influence of the VSCL panels under different conditions.

Keywords Variable stiffness (VSCL) · Cylindrical panel · Natural frequency · FEM · HSDT

P. Kumar · R. Kumar · C. K. Hirwani (✉)
Department of Mechanical Engineering, National Institute of Technology Patna, Bihar 800005, India
e-mail: chetank.me@nitp.ac.in

P. Kumar
e-mail: prasoonk.phd22.me@nitp.ac.in

R. Kumar
e-mail: ravik.ph21.me@nitp.ac.in

© The Author(s), under exclusive license to Springer Nature Singapore Pte Ltd. 2024
A. Chaudhary et al. (eds.), *Unraveling New Frontiers and Advances in Bioinformatics*,
https://doi.org/10.1007/978-981-97-7123-3_19

1 Introduction

Composite materials are experiencing a surge in demand across various industries such as marine, modern civil construction, and aerospace. The popularity of composite materials arises from their exceptional specific properties compared to traditional materials. Moreover, their flexibility in design, including the ability to manipulate stacking sequences, materials in the direction of thickness, and fabricate complex shapes, allows them to meet a wide range of application requirements. Traditionally, constant stiffness composite laminates (CSCL) with straight fiber reinforcement were widely used to meet these requirements. Now, modern fabrication techniques allow for the placement of reinforcement in curvilinear paths, resulting in variable stiffness composite laminates (VSCL). These VSCLs exhibit significantly improved properties in comparison to CSCLs. A substantial body of work has already been conducted on the structural dynamic analysis of both CSCL and VSCL composites, with numerous researchers contributing valuable insights. The next section will review and discuss the key findings from these important studies.

The buckling behaviour of laminated composite plates have been studied considering higher-order theory by Ghannadpour et al. [1]. The applied method gave satisfactory results and was found to be reliable. Wang et al. [2] investigated the vibration characteristics of laminated composite plates with constant stiffness using layerwise plate theory combined with first-order shear deformation theory (FSDT). Ghayesh et al. [3]] analyzed the modal response of laminated composite beams using a classical laminate theory (CLT) based numerical model. Safaei et al. [4] explored the vibration behavior of functionally graded composite shells and highlighted the crucial role of boundary spring stiffness in determining natural frequency values. Civalik [5] employed a discrete singular convolution method to study the vibration response of the laminated composite conical shell. The effect on natural frequency with the variation in material properties and geometry was investigated. Asadi et al. [6] investigated the dynamic characteristics of carbon nanotube (CNT) reinforced polymer composite shell structure subjected to axial compression. It was concluded that in comparison to the effect of boundary conditions, the volume fraction and the distribution of CNTs have a greater influence on natural frequencies and buckling loads. M A. Shahmohammadi et al. [7] studied the vibration attributes of laminated composites reinforced with carbon nanotube using the isogeometric finite strip method. Significant improvement in dynamic characteristics of the composite panel was observed with the employment of a three-phase composite. Farokhi et al. [8] studied the effects of tapered ratios and gradient index on functionally graded beams in clamped conditions. Bui et al. [9]studied the effect of different parameters such as fiber orientation, stiffness ratio etc. on the vibration behaviour of the laminated composite plate using CLT. Sarangan et al. [10] highlighted the superiority of zigzag models in predicting the vibration and bending behaviour of flat panel composite laminate.

Among the work in the field of VSCL, Leissa et al. [11]studied composite plates with non-uniformly spaced straight fibers to analyse the vibration responses. Hyer et al. [12] developed a model using the finite element method (FEM) to study the variation in load on a laminated plate with a change in fiber angle. It was observed that the failure load and its location could easily be changed by simply combining plies that have relatively different fiber angle. They also concluded that variable stiffness laminates had a larger failure load compared to traditional straight fiber laminates. Kumar et al. [13, 14] investigated the model analysis of a VSCL plate utilizing the numerical approach using HSDT and FE steps. Rodrigues et al.[13]presented an experimental investigation on the VSCL Plate. Natural frequencies as well as mode shapes are obtained using frequency response functions, compared with the finite element model and found in good agreement with the same. Kumar et al. [16, 17] analysed nanocomposite hydrogel adopting atomistic and experimental technique. Tatting et al. [14] analysed the stress and initial buckling of the curve fiber-reinforced cylindrical shell panel under different loading conditions using a classical lamination theory-based numerical model. Waldhart C et al. [15] investigated the structural responses of the laminate with varying stiffness and found that the buckling load significantly improved in composite with curved fiber than that of straight fiber. Setoodeh et al. [16, 17] also investigated the laminated flat panel in order to achieve the best possible arrangement of fiber in the laminate. Parthasarathy et al. [18] used a series of control points to define a fiber path. After several iterations, it was found that the buckling load in the case of ply that has curved fiber was more than that of the buckling load in ply having linear fiber. Gurdal et al. [19] used linear fiber orientation to define a path. A reference path having fiber orientation varying linearly along the geometric axis and passing through the center of the panel was created. A significant improvement in the performance of the variable stiffness laminates was found in comparison to ply having linear fiber in terms of load-carrying capacity and stiffness. Kumar et al. [17, 24] studied tensile property of nanocomposite hydrogel. Wu et al. [20] analysed the effect of position on the vibration behaviour of variable stiffness panel. Akbarzadeh et al. [21] studied the effect of manufacturing defects such as gaps and overlaps on the natural frequency using HSDT and CLT. Yazdani et al. [22] studied the vibration behaviour of symmetric and unsymmetric VSCL plates through a layerwise formulation. Ribeiro et al. [23] analysed the linear modal behaviour of VSCL plates using p-version finite element considering planes and the rotational shape functions. Honda et al. [24] used the Ritz solution to analyse the fundamental frequency behaviour of composite laminate with varying stiffness. Devarajan et al. [17] applied the isogeometric method to study the effects of vibration on curvilinearly stiffened laminated plates.

The dynamic behavior of constant stiffness composite laminates (CSCL) with uncurved fibers has been well-documented in several existing research. However, variable stiffness composite laminates (VSCL) panel offers a vast design domain and numerous applications. Despite that, there has been comparatively limited study on VSCL panels. Most of the study on VSCLs is related to buckling analysis and static deformation, leaving a gap in the understanding of their vibration behavior. The vibration behavior of VSCL panels has not been extensively studied, according to the author's knowledge. Most existing research on the vibration analysis of VSCL panels relies on either analytical or numerical models based on lower-order theories. To fill this research gap, the present study aims to analyze the vibration behavior of composite reinforced with curve fiber using HSDT and FE approaches. A detailed description of the numerical model can be found in Sect. 2. In Sect. 3, the performance of the developed model is assessed through consistency and comparison tests. Following this, parametric investigations are conducted using various numerical illustrations. Section 4 contains a detailed discussion of all the conclusions drawn from the present investigation.

This approach offers a more comprehensive analysis of the dynamic behavior of VSCL curved panels, contributing valuable insights to the design and optimization of advanced composite materials.

2 Theoretical Formulation

The initial step in creating a variable stiffness composite laminate (VSCL) with curved fibers as reinforcement is to define a fiber path which could be further considered as a reference for other fibers. This reference path serves as the foundation for establishing the fiber orientation across the laminate. By determining the reference path, the orientation of fibers can be strategically manipulated to achieve the desired variation in stiffness and mechanical properties. The following equation defines the fiber path orientation and the same can be seen in Fig. 1.

$$\theta(x) = 2(P_1 - P_0)|x|/a + P_0 \tag{1}$$

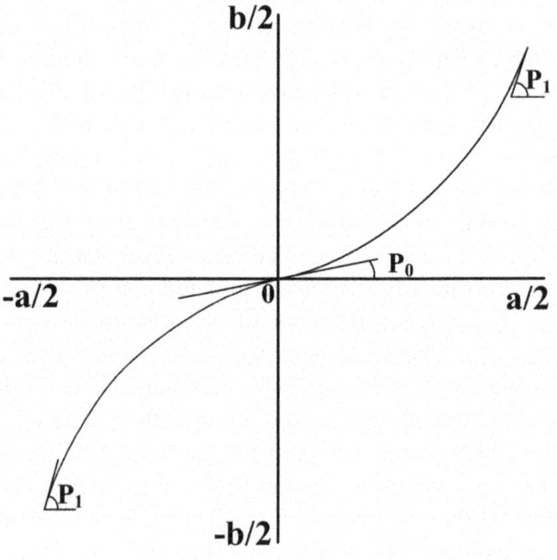

Fig. 1 Reference fiber path in VSCL with curvilinear fiber

here, P_1 indicates the orientation of path on the vertical edges of the shell panel and P_0 is the path orientation at the centre. The length and width of the shell panel are denoted by 'a' and 'b'. After reference path determination, other paths can be traced mainly using two methods. One is the parallel path method and the other is the shifted path method. In the parallel path method, the other fiber path is determined by drawing a path by connecting different points which are at a certain constant distance from the preferred path. In the shifting path method, the other fiber paths are determined by drifting the referred path to a different location. In this analysis, the shifted path method is used.

The variable stiffness laminate creation requires the fiber paths to be curved. There is a limitation to which the fiber can be curved. If the fiber is curved beyond a certain limit, then the kinking effect will develop in the fiber. To restrict the kinking effect in fiber, the curve in the fiber's path should be within the set maximum value. Mathematical expression to explain the bend in the fiber's path is presented [31] as:

$$K = \frac{f''(x)}{\left(1 + (f'(x))^2\right)^{3/2}} \tag{2}$$

In a given lamina, the function $f(x)$ describes the path of the fiber. To determine the curvature of the respective fiber, path equations can be applied in the above function $f(x)$. The reference fiber path is symmetric only about the x-axis (horizontal axis) and not the y-axis (vertical axis), then the curvature of the fiber path needs to be calculated specifically for the portion that lies on the positive side of the panel.

To make the laminate, the curvature of the reference fiber path at each point should be less than the maximum allowed value of 3.28 m^{-1} [15]. In case the curvature of the fiber path exceeds the maximum allowable value, in that case, the laminate cannot be manufactured and there is no need to perform the analysis. The below constraint inequality must be respected in determining the curvature of the reference fiber path in a laminate:

$$K = \frac{2(P_1 - P_0)}{a} \cos\left((P_1 - P_0)\frac{x}{a/2} + P_0\right) < 3.28 \tag{3}$$

2.1 Displacement Kinematics

In the study, the geometry of the cylindrical shell panel is defined using a three-dimensional cartesian coordinate scheme (0, x, y, and z). The x and y axes are situated on the shell's middle plane, with the z axis extending in the positive upward direction as depicted in Fig. 2. To model the displacement field of the variable stiffness composite laminate (VSCL) shell panel, the study adopts Reddy's third-order shear deformation theory (TSDT). The displacement field is described according to the formulation of TSDT, as expressed in Eq. (4) [26].

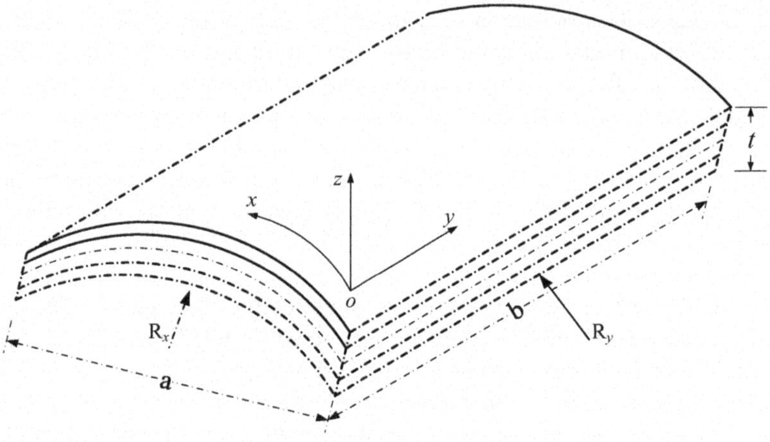

Fig. 2 Coordinates of a laminated cylindrical shell

$$\begin{aligned} u(x, y, z, t) &= u^0(x, y, t) + z\theta_x(x, y, t) + z^2\phi_x(x, y, t) + z^3\varphi_x(x, y, t) \\ v(x, y, z, t) &= v^0(x, y, t) + z\theta_y(x, y, t) + z^2\phi_y(x, y, t) + z^3\varphi_y(x, y, t) \\ w(x, y, z, t) &= w^0(x, y, t) \end{aligned} \quad (4)$$

Here, the u, v and w are displacements along x, y and z directions, respectively. u^0, v^0 and w^0 represents the mid-plane displacements, along x, y and z, respectively. θ_x and θ_y refer to the rotations of the shell about the y-axis and x-axis, respectively. These rotations describe the tilt of the shell perpendicular to the mid-surface, providing a more comprehensive understanding of the shell's deformation under loading. The Taylor expansion series' higher-order terms are represented by the remaining terms in Eq. (4).

2.2 *Stress–Strain Relation*

In a composite laminate, each lamina (or layer) can have different material properties and fiber orientations. When analyzing the stress–strain relationship in the material coordinates of a kth lamina, we usually refer to the local coordinate system of the lamina. For a composite material, the lamina can often be considered orthotropic, meaning its properties vary along three orthogonal axes [19] and is given by:

$$\begin{Bmatrix} \sigma_1 \\ \sigma_2 \\ \tau_{23} \\ \tau_{13} \\ \tau_{12} \end{Bmatrix}^{(k)} = \begin{bmatrix} Q_{11} & Q_{12} & 0 & 0 & 0 \\ Q_{21} & Q_{22} & 0 & 0 & 0 \\ 0 & 0 & Q_{44} & 0 & 0 \\ 0 & 0 & 0 & Q_{55} & 0 \\ 0 & 0 & 0 & 0 & Q_{66} \end{bmatrix}^{(k)} \begin{Bmatrix} \varepsilon_1 \\ \varepsilon_2 \\ \gamma_{23} \\ \gamma_{13} \\ \gamma_{12} \end{Bmatrix}^{(k)} \tag{5}$$

$$\{\sigma\}^{(k)} = \left[Q_{ij}\right]^{(k)} \{\varepsilon\}^{(k)} \tag{6}$$

Here, the matrices $\{\sigma\}^{(k)}$, $\{\varepsilon\}^{(k)}$, $\left[Q_{ij}\right]^{(k)}$ represent the stress, strain and material properties.

2.3 Strain Displacement Equation

In the linear regime, the strain displacement relation can be given [25] by:

$$\{\varepsilon\} = \begin{Bmatrix} \varepsilon_x \\ \varepsilon_y \\ \gamma_{xy} \\ \gamma_{xz} \\ \gamma_{yz} \end{Bmatrix} = \begin{Bmatrix} \frac{\partial u}{\partial x} + \frac{w}{R_x} \\ \frac{\partial v}{\partial y} + \frac{w}{R_y} \\ \frac{\partial u}{\partial y} + \frac{\partial v}{\partial x} + \frac{2w}{R_{xy}} \\ \frac{\partial v}{\partial z} + \frac{\partial w}{\partial y} - \frac{u}{R_x} \\ \frac{\partial u}{\partial z} + \frac{\partial w}{\partial x} + \frac{v}{R_{yy}} \end{Bmatrix} \tag{7}$$

$$\{\varepsilon\} = \{Z\}\{\overline{\varepsilon}\} \tag{8}$$

2.4 Energy Equation

The strain energy of the composite panel with curve fiber is given by:

$$U = \frac{1}{2} \int \varepsilon^T \sigma \, dxdydz \tag{9}$$

The kinetic energy of the VSCL panel will be expressed as follows:

$$T = \frac{1}{2} \int \rho \left\{\dot{d}\right\}^T \left\{\dot{d}\right\} dxdydz \tag{10}$$

Here ρ represents the density of the material, which is a key parameter in analyzing the dynamic behavior of the cylindrical shell and $\{\dot{d}\}$ denotes the time derivative of the displacement vector, also known as the velocity vector. This vector represents the rate of change of displacements over time for each point within the shell structure.

2.5 Finite Element Formulation

The studying VSCL shell panel has been discretized using a nine-noded isoperimetric element, resulting in a thorough representation of the panel for mathematical analysis. The discretization elements consist of nine nodes with 81 degrees of freedom (9 nodes × 9 degrees of freedom per node). These elements provide a higher level of detail and accuracy compared to lower-order elements. Here, the displacement field of the shell can be represented in terms of nodal values and element shape functions (N). This representation is key to performing finite element analysis (FEA) and understanding the deformation of the shell structure. The displacement field is expressed as[33]:

$$\{d_0\} = \sum_{i=1}^{n} [N_i]\{d_{0i}\} \tag{11}$$

here $\{d_0\} = \begin{bmatrix} u_0 & v_0 & w_0 & \theta_x & \theta_y & \phi_x & \phi_y & \varphi_x & \varphi_y \end{bmatrix}^T$ and $\{d_{0i}\} = \left[(u_0)_i \; (v_0)_i \; (w_0)_i \; (\theta_x)_i \; \left(\theta_y\right)_i \; (\phi_x)_i \; \left(\phi_y\right)_i \; (\varphi_x)_i \; \left(\varphi_y\right)_i \right]^T$

In the finite element modelling of a curved structure, the nodal displacement field may be used to describe the reference plane strain vector.

$$\{\overline{\varepsilon}\} = [B]\{d_{0i}\} \tag{12}$$

The general strain–displacement relation matrix, often denoted as $[B]$, is a key component in this transformation.

2.6 Governing Equation

Hamilton's concept is a variational technique for determining the equations of motion for a system. In the context of the free vibration of a composite structure, the governing equation of motion can be derived by applying Hamilton's principle, which states that the integral of the difference between kinetic energy (T) and strain energy (U) over time is stationary (i.e., the variation is zero) for the actual motion of the system and is expressed as:

$$d \int_{t_1}^{t_2} (T - U)dt = 0 \tag{13}$$

Now inserting the parameters of U and T from the Eq. (9) and Eq. (10) obtained as:

$$[m]\{\ddot{d}_0\} + [k]\{d_0\} = 0 \tag{14}$$

here $[m]$ denotes the elemental mass matrix, $[k]$ denotes the elemental stiffness matrix and $\ddot{d}_0, d_0$ denotes the acceleration and displacement respectively.

Additional, $[m]$ and $[k]$ are denoted in Eqs. (15) and (16), respectively.

$$[m] = \int_A [N]^T \rho [N] dA \tag{15}$$

$$[k] = \int_A [B]^T [D][B] dA \tag{16}$$

After deriving the elemental form of the equation of motion for free vibration using Hamilton's principle, the next step is to assemble the global form of the equation. This involves combining the equations for each finite element to create a single, system-wide equation that describes the motion of the entire composite structure.

The elemental form of the equation can be expressed as:

$$[M]\{\ddot{d}\} + [K]\{d\} = 0 \tag{17}$$

here $[M]$ and $[K]$ symbols represent the global mass and stiffness matrix respectively. $\{\ddot{d}\}$ and $\{d\}$ are represent global acceleration and global displacement vectors.

To evaluate the frequency response of the system, the governing equation is converted into an eigenvalue problem:

$$([K] - \lambda^2[M])d = 0 \tag{18}$$

Further, the boundary condition used to solve the equation is:

Simply supported:

$$v_0 = w_0 = \theta_y = \phi_y = \varphi_y = 0 \text{ at } x = 0 \text{ and } a \tag{19}$$

$u_0 = w_0 = \theta_x = \phi_x = \varphi_x = 0$ at $y = 0$ and b.

Clamped:

$$u_0 = v_0 = w_0 = \theta_x = \theta_y = \phi_x = \phi_y = \varphi_x = \varphi_y = 0 \tag{20}$$

at $x = 0$ and a, at $y = 0$ and b

Free:

$$u_0 \neq v_0 \neq w_0 \neq \theta_x \neq \theta_y \neq \phi_x \neq \phi_y \neq \varphi_x \neq \varphi_y \neq 0 \tag{21}$$

at $x = 0$ and a, at $y = 0$ and b

3 Numerical Results and Discussion

The numerical model discussed in the above section is now converted into an algorithm in the MATLAB platform to solve the numerical illustrations and to get the desired modal responses. Furthermore, the performance of the produced numerical model is initially evaluated to confirm the model's reliability and correctness. The model is then used to thoroughly explore the parametric behavior. The analysis focuses on a square ($a = b = 1$ m) cylindrical shell panel with specific lamina sequences [(45°, 0°), (135°, 90°), (135°, 90°), (45°, 0°)]. This configuration serves as the basis for the investigation unless mentioned otherwise.

For all numerical illustrations, the material properties are assumed as follows:

Major elasticity modulus (E_1) = 173 GPa.

Lower elasticity modulus (E_2) = 7.2 GPa.

Shear modulus $G_{12} = G_{13} = G_{23} = 3.76$ GPa.

Poisson ratio (ν_{12}) = 0.29 and density (ρ) = 1540 kg/m^3

These material properties are consistent throughout the analysis to maintain uniformity and facilitate comparison between different parametric variations.

3.1 Convergence Study

A simply-supported four-layered cylindrical shell panel (thickness ratio, $r = 60$, curvature ratio, $R/a = 40$) with lamina sequences [(45°, 0°), (135°, 90°), (135°, 90°), (45°, 0°)] in considered in this example. The fundamental modal characteristics (five modes) are evaluated for different sets of elements and represented in Fig. 3. It has been discovered that as the number of elements rises, the difference in frequency between successive elements diminishes, resulting in responses with good convergence behavior from 81 elements forward. In this regard, a hundred elements are unisized in forthcoming illustrations, for more accurate responses.

3.2 Validation Study

After successfully analyzing the convergence behavior of the current mathematical model, the accuracy behaviour is examined in this illustration. The study examines a simply supported three-layer variable stiffness composite laminate (VSCL) plate with varying lamina sequences and thicknesses. The plate's behavior is examined

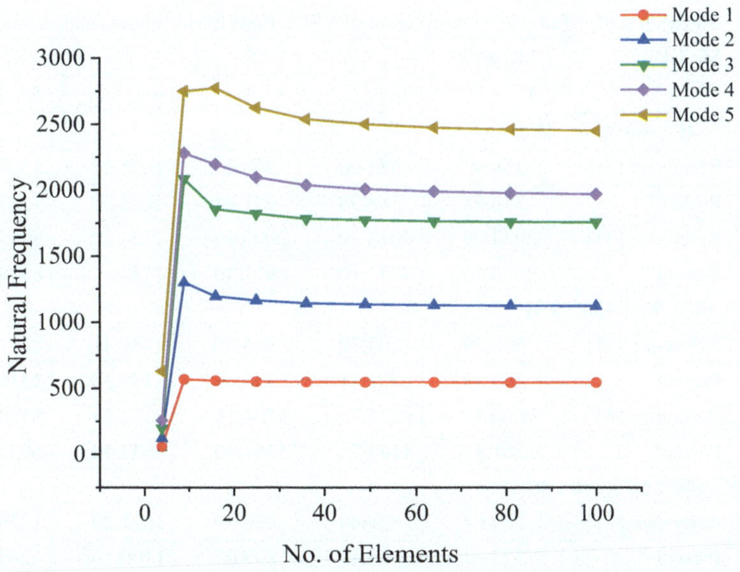

Fig. 3 Convergence behaviour of modal frequency of VSCL cylindrical panel

and solved using a numerical model that combines third-order shear deformation theory (TSDT) and finite element (FE) approaches. Five different mode values are evaluated and a comparison with the reference [31] results is presented in Table 1. All the geometrical and material data are used as provided in the references. The acquired findings from the present numerical model and recorded data are in good agreement, with just slight deviations noted in the upper modes. These slight discrepancies may be attributed to differences in the types of elements used in the current analysis compared to the reference analysis.

3.3 *Numerical Illustration*

3.3.1 Effect on the Natural Frequency with Variation in Curvature Ratio

In this study, the influence of the curvature ratio on the frequency characteristics of a variable stiffness composite laminate (VSCL) panel is examined. The analysis focuses on a simply supported shell panel with a thickness of 0.01 m. The investigation involves evaluating five modal frequencies for different shell curvature ratios (R/a = 10, 30, 50, 70, and 90), and the results are presented in Table 2. The setup for this analysis includes maintaining a constant shell length while linearly increasing the shell radius as the plate curvature ratio increases. This approach allows for a

Table 1 Validation study of first five mode values of VSCL Plate having simply support conditions

h	Model	Modes				
		1	2	3	4	5
[(0°, 45°), (-45°, -60°), (0°, 45°)]						
0.01	Reference [31]	358.48	589.90	960.36	1075.21	1327.88
	Present	356.68	588.98	971.84	1078.30	1331.30
0.1	Reference [31]	2934.69	4688.30	7000.96	7324.22	8471.78
	Present	2836.60	4533.10	6927.30	7182	8318
[(30°, 0°), (45°, 90°), (30°, 0°)]						
0.01	Reference [31]	308.79	503.79	845.50	1131.31	1279.85
	Present	306.39	501.11	856.64	1135.30	1310.30
0.1	Reference [31]	2620.4	4225.74	6704.11	7121.26	8383.48
	Present	2581.8	4145.7	5354.80	6633.40	6633.60
[(90°, 45°), (60°, 30°), (90°, 45°)]						
0.01	Reference [31]	329.68	539.40	886.39	1091.20	1279.90
	Present	327.09	536.49	897.03	1100.10	1283.90
0.1	Reference [31]	2746.66	4402.32	6915.87	7058.72	8254.38
	Present	2663.20	4268.60	6817.70	7026.40	7449

systematic investigation of how different curvature ratios affect the frequency characteristics of the VSCL panel. The analysis shows that as the curvature ratio increases, the natural frequencies of the laminated composite shell panel decrease across all modes. The decrease in natural frequencies suggests that the stiffness of the laminated structure reduces with increasing shell curvature ratios. These results provide valuable insights into the relationship between the curvature ratio and the frequency characteristics of VSCL panels. Understanding this relationship can aid in the design and optimization of laminated composite panels for specific applications.

Table 2 Effect on natural frequency with variation in curvature ratio

R/*a*	Modes				
	1	2	3	4	5
10	577.5	1178.9	1790.6	2057.8	2555.2
30	552.8	1147.2	1787.6	2041.1	2542.7
50	550.8	1144.6	1787.3	2039.7	2541.7
70	550.3	1143.9	1787.3	2039.3	2541.4
90	550	1143.6	1787.3	2039.2	2541.3

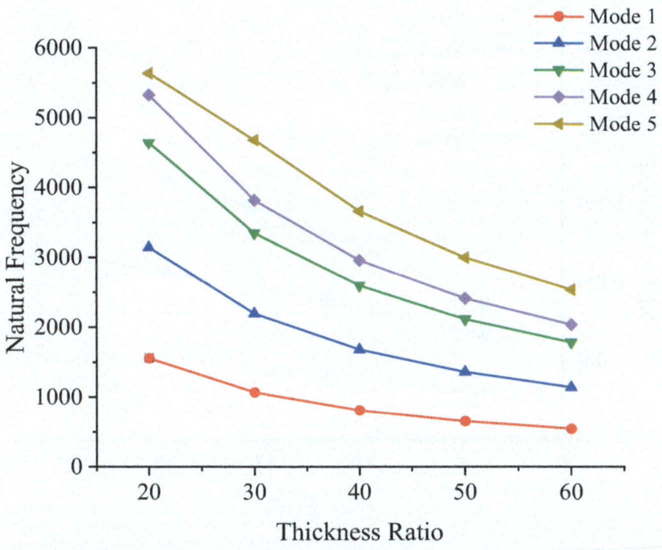

Fig.4 Effect of *a*/*h* ratio on natural frequency

3.3.2 Effect on Natural Frequency with Variation in Thickness Ratio

Further, the current FE model is applied to check the variation in natural frequency with the increase in thickness ratio. To check the said variation, a simply supported panel is considered and the first five modal frequencies are obtained for various plate thickness ratios (20, 30, 40, 50, 60). The obtained responses are plotted in Fig. 4. The analysis of the variable stiffness composite laminate (VSCL) cylindrical shell reveals an important relationship between the natural frequency and the thickness ratio (*a*/*h*). Here's an overview of this relationship and its implications. As the thickness ratio increases, the natural frequency of the VSCL panel generally decreases. This inverse relationship occurs because a larger thickness ratio (*a*/*h*) typically indicates a thinner panel relative to its length. Thinner panels tend to have lower stiffness, which in turn leads to lower natural frequencies. The relation between thickness ratio and natural frequency has been depicted with the help of the plot. It allows for the visualization of how changes in panel thickness affect the panel's vibrational characteristics.

3.3.3 *Effect on the Natural Frequency with Variation in Young's Moduli (E_1/E_2)*

The illustration investigates the effect of the ratio of Young's moduli (E_1/E_2) on the frequency responses of the VSCL panel. This study uses a shell structure with a thickness of 0.01 m and all edges are simply supported. The FE model is applied to analyze the modal frequencies for various ratio sets of Young's moduli (E_1/E_2

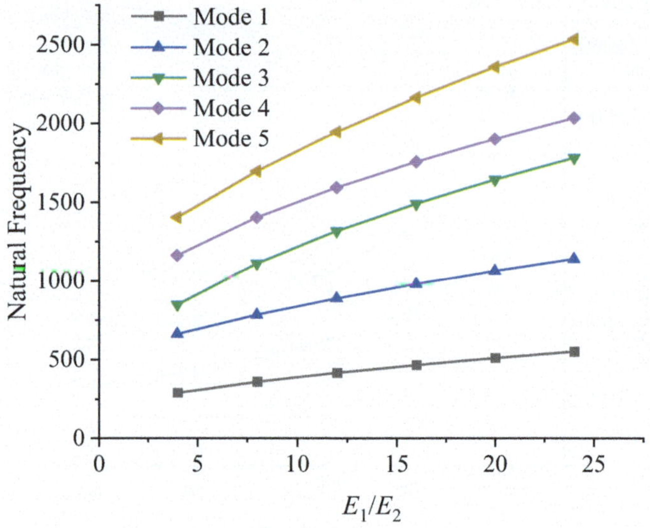

Fig.5 Effect on the natural frequency with the variation in Young's moduli (E_1/E_2)

= 4, 8, 12, 16, 20, and 24), and the results are plotted in Fig. 5. In the study, the minor Young's modulus (E_2) is kept constant while the major Young's modulus (E_1) is gradually increased, leading to different E_1/E_2 ratios. The plot shows that as the E_1/E_2 ratio increases, the natural frequency of the VSCL panel increases across all modes. This is a clear trend that can be observed in the analysis. The increase in natural frequency can be attributed to the increase in the stiffness of the plate as E_1 increases. This increase in stiffness enhances the rigidity of the panel, resulting in higher natural frequencies.

3.3.4 Variation in Natural Frequency with Constraints in Degrees of Freedom of VSCL Shell

The effect of different boundary conditions (BCs) on the natural frequency of the VSCL panel is investigated with the derived numerical model. The analysis focuses on the variation in natural frequency for different combinations of BCs such as CFFF (cantilever-free-free-free), CFCF (clamped-free-clamped-free), SSSS (simply supported on all sides), SFSF (simply supported-free-simply supported-free), SCSC (simply supported-clamped-simply supported-clamped), and CCCC (clamped on all sides) as shown in Fig. 6. The first five modal responses for these various edge conditions are tabulated in Table 3. The natural frequencies vary significantly depending on the boundary conditions applied to the VSCL panel. Different BCs lead to different stiffness and constraints in the system, which in turn affects the frequency responses. The Table 3 shows that the lowest natural frequencies are observed in the case of the CFFF (cantilever-free-free-free) boundary condition, where one side is clamped while the other three are free. This setup provides the least constraints and results in lower stiffness. The highest natural frequencies are observed in the case of the CCCC (clamped–clamped-clamped–clamped) boundary condition, where all sides of the panel are clamped. This condition provides the most constraints and results in higher stiffness. The variation in natural frequencies observed under different BCs indicates that the degree of freedom (DOF) constraints significantly impacts the stiffness of the VSCL panel. Higher constraints, such as in the CCCC case, lead to increased stiffness and higher natural frequencies.

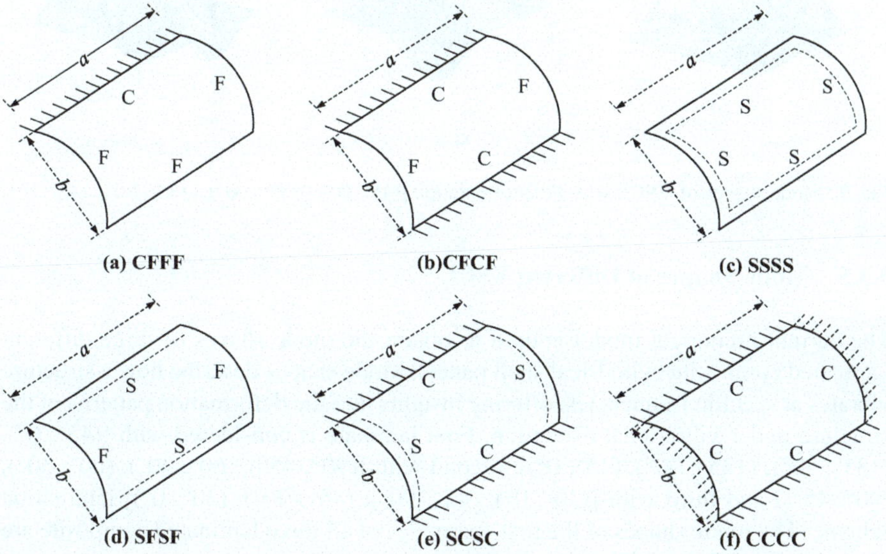

Fig.6 Representation of applied boundary conditions

Table 3 Variation in natural frequency with different boundary conditions

BC	Modes				
	1	2	3	4	5
CFFF	73.1	197	429.6	606.2	886.8
CFCF	292.8	353.1	732.2	1342.7	1592.9
SSSS	551.4	1145.4	1787.4	2040.1	2542
SFSF	567.2	1149.7	1788.1	2041.6	2542.2
SCSC	722.7	1579.3	1915.8	2536.3	3250
CCCC	1162.9	1940.6	2749.6	3165.3	3695.5

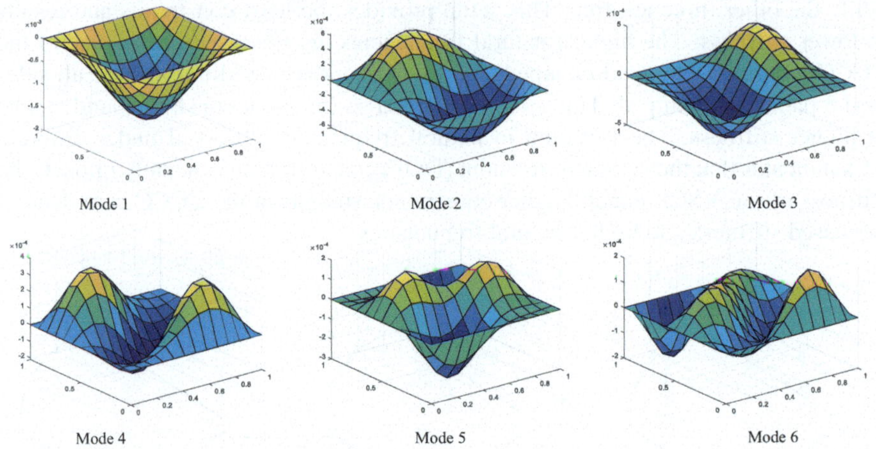

Fig. 7 Mode shapes of VSCL with Fiber orientation [(45°, 0°), (135°, 90°), (135°, 90°), (45°, 0°)]

3.3.5 Mode Shapes of Different VSCL

The current numerical model is used to obtain the mode shapes of three different laminated composite cylindrical shell panels. Mode shapes describe how a structure vibrates at specific frequencies, offering insights into the deformation patterns of the structure under vibrational excitation. First laminate is considered with [(45°, 0°), (135°, 90°), (135°, 90°), (45°, 0°)], second with [(90°, 45°), (60°, 30°), (60°, 30°), (90°, 45°)] and third with [(30°, 0°), (45°, 90°), (45°, 90°), (30°, 0°)] lamination scheme. The mode shapes of the first six modes of all three laminated composite are plotted in Figs. 7, 8, and 9 respectively.

4 Conclusions

The present study derives a two-dimensional (2D) numerical model to investigate the dynamic behaviour of a VSCL cylindrical shell panel. This model is based on HSDT and employs FE approaches for accurate analysis. The use of nine-noded elements for discretization allows for a detailed representation of the cylindrical shell and its dynamic characteristics. The study of the derived numerical model included initial convergence and validation tests to ensure its accuracy and reliability. Following this, the model was used to explore the influence of various geometrical and material parameters on the frequency characteristics of the VSCL cylindrical panel. Several numerical illustrations were conducted, and based on the findings, the following conclusions have been drawn:

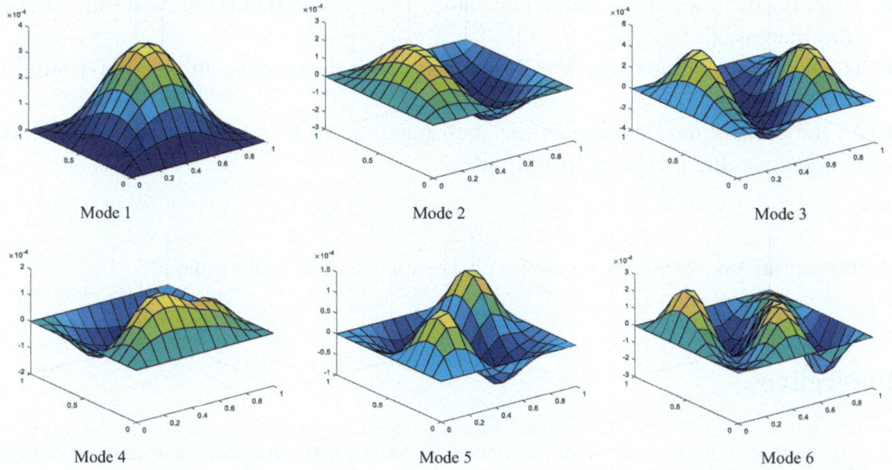

Fig. 8 Mode shapes of VSCL with fiber orientation [(90°, 45°), (60°, 30°), (60°, 30°), (90°, 45°)]

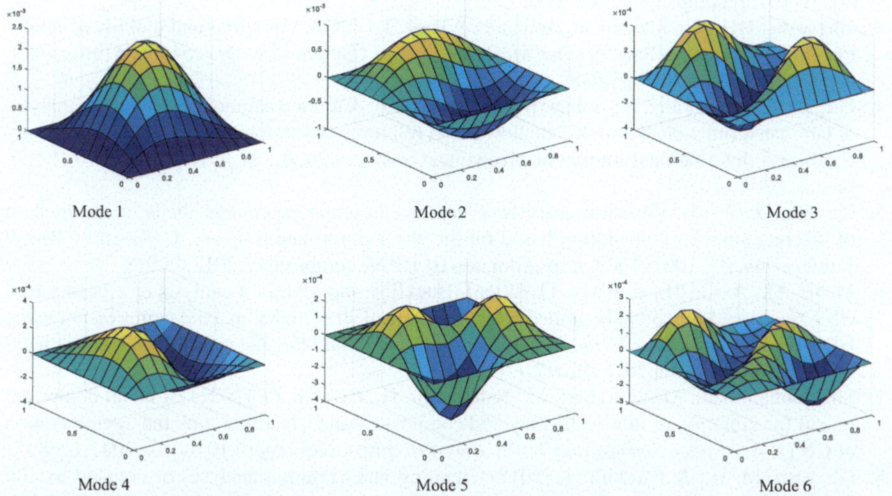

Fig. 9 Mode shapes of VSCL with fiber orientation [(30°, 0°), (45°, 90°), (45°, 90°), (30°, 0°)]

- Element convergence as well as the comparison test of the derived numerical model showcases its capability to effectively analyze the modal frequency characteristics of VSCL cylindrical panel structures. This validation underscores the reliability and accuracy of the model, establishing its suitability for studying the vibrational behavior of VSCL panels and aiding in their design optimization.
- With the increase in the shell's curvature ratio, there is a decrease in the frequency values of the VSCL curved panel.

- There is a decrease in the frequency value of various modes as the shell's thickness ratio increases.
- The frequency value of the VSCL increases with the increase in ratios of Young's modulus.
- As the constraint condition of the shell panel at all four edges is increased, the frequency value increases in all modes.

Conflict of Interest "There are no conflicts of interest to declare by the authors."

References

1. Ovesy, H. R., Ghannadpour, S. A. M., & Sherafat, M. H. (2010). Buckling analysis of laminated composite plates using higher order semi-analytical finite strip method. *Applied Composite Materials, 17*, 69–80. https://doi.org/10.1007/s10443-009-9098-2
2. Wang, S., & Zhang, Y. (2005). Vibration analysis of rectangular composite laminated plates using layerwise B-spline finite strip method. *Composite Structures, 68*, 349–358. https://doi.org/10.1016/j.compstruct.2004.04.001
3. Ghayesh, M. H., Yourdkhani, M., Balar, S., & Reid, T. (2010). Vibrations and stability of axially traveling laminated beams. *Applied Mathematics and Computation, 217*, 545–556. https://doi.org/10.1016/j.amc.2010.05.088
4. Qin, Z., Zhao, S., Pang, X., Safaei, B., Chu, F. (2020). A unified solution for vibration analysis of laminated functionally graded shallow shells reinforced by graphene with general boundary conditions. International Journal of Mechanical Sciences 2020;170. https://doi.org/10.1016/j.ijmecsci.2019.105341.
5. Civalek, Ö. (2013). Vibration analysis of laminated composite conical shells by the method of discrete singular convolution based on the shear deformation theory. *Composites Part B Engineering, 45*, 1001–1009. https://doi.org/10.1016/j.compositesb.2012.05.018
6. Mehri, M., Asadi, H., & Wang, Q. (2016). Buckling and vibration analysis of a pressurized CNT reinforced functionally graded truncated conical shell under an axial compression using HDQ method. *Computer Methods in Applied Mechanics and Engineering, 303*, 75–100. https://doi.org/10.1016/j.cma.2016.01.017
7. Shahmohammadi, M.A., Azhari, M., Salehipour, H., Civalek, Ö. (2021). A novel composite model for vibration of thin-walled layered composite panels incorporating the agglomeration of CNTs. *Aerospace Science and Technology 116*. https://doi.org/10.1016/j.ast.2021.106897.
8. Ghayesh, M. H., & Farokhi, H. (2018). Bending and vibration analyses of coupled axially functionally graded tapered beams. *Nonlinear Dynamics, 91*, 17–28. https://doi.org/10.1007/s11071-017-3783-8
9. Bui, T. Q., Nguyen, M. N., & Zhang, C. (2011). An efficient meshfree method for vibration analysis of laminated composite plates. *Computational Mechanics, 48*, 175–193. https://doi.org/10.1007/s00466-011-0591-8
10. Sarangan, S., & Singh, B. N. (2017). Evaluation of free vibration and bending analysis of laminated composite and sandwich plates using non-polynomial zigzag models: C0 Finite element formulation. *Aerospace Science and Technology, 68*, 496–508. https://doi.org/10.1016/j.ast.2017.06.001
11. Leissa, A.W., Martin, A.F. (1990). Vibration and buckling of rectangular composite plates with variable fiber spacing. vol. 14. 1990.
12. Hyer, M. W., & Charette, R. F. (1991). Use of curvilinear fiber format in composite structure design. *AIAA Journal, 29*, 1011–1015. https://doi.org/10.2514/3.10697

13. Kumar, P., Arya, R., Sharma, N., Hirwani, C. K., & Panda, S. K. (2023). Curved fiber-reinforced laminated composite panel and variable stiffness influence on eigen frequency responses: A higher-order FE approach. *Journal of Vibration Engineering and Technologies, 11*, 2349–2359. https://doi.org/10.1007/S42417-022-00706-6
14. Kumar, R., & Hirwani, C. K. (2022). Investigation on nonlinear bending behavior of sandwich shell panel with cutout using HSDT and FE approach. *Mechanics Based Design of Structures and Machines,* 1–19. https://doi.org/10.1080/15397734.2022.2158868
15. Ribeiro, P., Akhavan, H., Dias Rodrigues, J.F., Akhavan, H., Rodrigues, J.D. (2013). Experimental and finite element modal analysis of variable stiffness composite laminated plates.
16. Kumar, R., Tewari, A., & Parashar, A. (2024). Thermal transport phenomena in PEGDA-based nanocomposite hydrogels using atomistic and experimental techniques. *The Journal of Physical Chemistry B.* https://doi.org/10.1021/acs.jpcb.4c01376
17. Kumar R, Parashar A. (2013). Atomistic simulations of pristine and nanoparticle reinforced hydrogels: A review. *Wiley Interdisciplinary Reviews: Computational Molecular Science* 13. https://doi.org/10.1002/wcms.1655.
18. Tatting, B.F., Gürdal, Z., Hyer, M.W., Johnson, E.R., Librescu, L., Meirovitch, L. (1998). Analysis and design of variable stiffness composite cylinders in engineering mechanics.
19. Waldhart, C., Jones, R.M. (1996). Analysis of tow-placed, variable-stiffness laminates.
20. Setoodeh, S., Abdalla, M. M., & Gürdal, Z. (2006). Design of variable-stiffness laminates using lamination parameters. *Composites Part B Engineering, 37*, 301–309. https://doi.org/10.1016/j.compositesb.2005.12.001
21. Setoodeh, S., Abdalla, M.M., IJsselmuiden, S.T., Gürdal, Z. (2009). Design of variable-stiffness composite panels for maximum buckling load. Composites and Structures 87:109–17. https://doi.org/10.1016/j.compstruct.2008.01.008.
22. Nagendra. S., Kodiyalam, S., Davis, J.E., Parthasarathy, V.N. (1995). Optimization of tow fiber paths for composite design.
23. Guerdal, Z., Olmedo, R. (1992). Composite laminates with spatially varying fiber orientations—"Variable stiffness panel concept." *American Institute of Aeronautics and Astronautics* (AIAA). https://doi.org/10.2514/6.1992-2472.
24. Kumar, R., Chaurasia, A., Tewari, A., Parashar, A. (2014). Atomistic modelling and experimental study of tensile strength of nanocomposite hydrogel. *The International Journal of Mechanical Sciences* 109397. https://doi.org/10.1016/j.ijmecsci.2024.109397.
25. Wu, C.-P., Lee, C.-Y. (2001). Diierential quadrature solution for the free vibration analysis of laminated conical shells with variable stiiness. vol. 43.
26. Akbarzadeh, A. H., Arian Nik, M., & Pasini, D. (2014). The role of shear deformation in laminated plates with curvilinear fiber paths and embedded defects. *Composite Structures, 118*, 217–227. https://doi.org/10.1016/j.compstruct.2014.07.027
27. Yazdani, S., & Ribeiro, P. (2015). A layerwise p-version finite element formulation for free vibration analysis of thick composite laminates with curvilinear fibres. *Composite Structures, 120*, 531–542. https://doi.org/10.1016/j.compstruct.2014.10.030
28. Ribeiro, P. (2012). Non-linear free periodic vibrations of variable stiffness composite laminated plates. *Nonlinear Dynamics, 70*, 1535–1548. https://doi.org/10.1007/s11071-012-0554-4
29. Honda, S., & Narita, Y. (2012). Natural frequencies and vibration modes of laminated composite plates reinforced with arbitrary curvilinear fiber shape paths. *Journal of Sound and Vibration, 331*, 180–191. https://doi.org/10.1016/j.jsv.2011.08.019
30. Devarajan, B. (2021). Free vibration analysis of curvilinearly stiffened composite plates with an arbitrarily shaped cutout using isogeometric analysis.
31. Akhavan, H., & Ribeiro, P. (2011). Natural modes of vibration of variable stiffness composite laminates with curvilinear fibers. *Composite Structures, 93*, 3040–3047. https://doi.org/10.1016/J.COMPSTRUCT.2011.04.027
32. Reddy, J.N. (Junuthula N, Reddy JN (Junuthula N. (2004). Mechanics of laminated composite plates and shells : theory and analysis. CRC Press.
33. Kumar, R., Tiwari, S., & Kumar, H. C. (2023). On transient responses of sandwich plate with cutout using FEM. *Mater Today Proc.* https://doi.org/10.1016/j.matpr.2023.05.409

SPRINGER NATURE

GPSR Compliance

The European Union's (EU) General Product Safety Regulation (GPSR) is a set of rules that requires consumer products to be safe and our obligations to ensure this.

If you have any concerns about our products, you can contact us on ProductSafety@springernature.com

In case Publisher is established outside the EU, the EU authorized representative is:

Springer Nature Customer Service Center GmbH
Europaplatz 3
69115 Heidelberg, Germany

The manufacturer's authorised representative in the EU is Springer Nature Customer Service Centre GmbH, Europaplatz 3, 69115 Heidelberg, Germany. If you have any concerns regarding our products, please contact ProductSafety@springernature.com

Printed and bound by CPI Group (UK) Ltd, Croydon, CR0 4YY
01/10/2026
02225915-0014